FORD DESIGN DEPARTMENT

CONCEPT & SHOWCARS

1932 - 1961

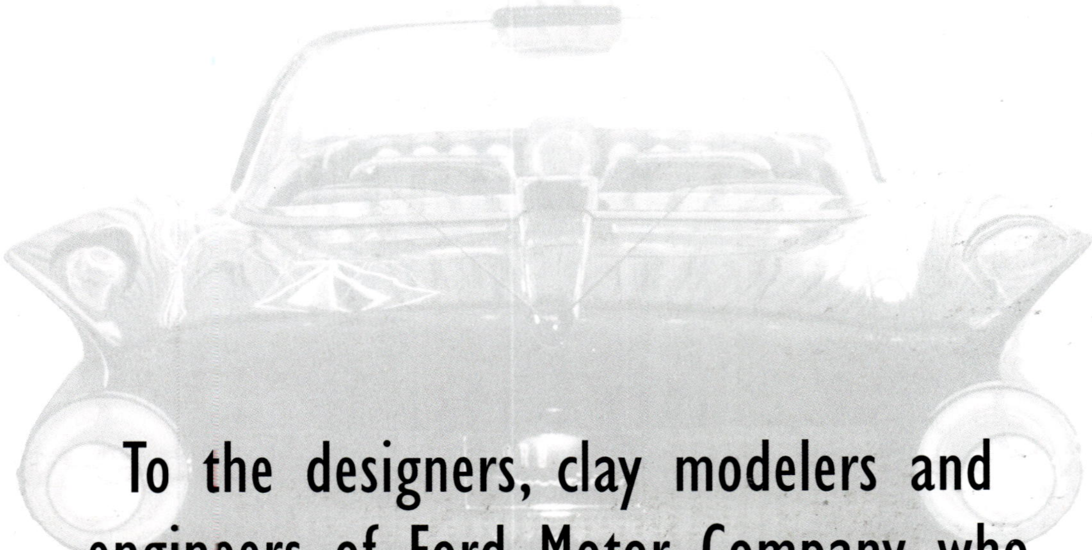

To the designers, clay modelers and engineers of Ford Motor Company who helped create these cars.

Graphics by Geno Edwards and Paul Zegers, Creative Images Printing & Design, Roseburg, OR, USA
Printed in Hong Kong through World Print Ltd
First Edition
ISBN# 0-9672428-0-0
Library of Congress No. 99-094911

Foreword

Design is a lot like music. Whether you're writing a rock tune or a symphony, you start with the same basic eight notes. It's how you arrange them that really makes the difference. Arrange the design cues one way, and you get a two-seat sports car. Arrange them another way, and you get a formal sedan. Turn the same design theme over to the Italians, and you get the equivalent of a grand opera. It's the same today as it was 50 years ago.

Today, aerodynamics is an integral component of design. In the 1940s and '50s, aerodynamics was a new concept we didn't understand all that well. Today, designers universally recognize that cars must be compatible with the air they negotiate to achieve maximum efficiency. Some people believe that if all cars employ the same aerodynamic principles, they will all look alike. As can be seen on the pages of this book, nothing could be farther from the truth. Although all birds are aerodynamic, you can still tell the difference between a canary and an eagle; they all fly, but they have different missions. It's up to the designer to imagine and innovate new ways to arrange the notes, so to speak, to come up with new designs that are modern, aerodynamic, ergonomic and take advantage of new technologies. The designs of the cars featured in this book paint a road map of where we have been and how we got where we are today — and yet the notes are all the same.

Ford currently has seven design studios worldwide. Even though they are all interconnected by an electronic umbrella and use the most modern technological advances, Leonardo da Vinci or Michelangelo would feel right at home in any one of these studios. Designers still use pen and paper and sketch in the traditional way and create clay models by hand, in much the same way it has been done throughout history.

Bob Puffer

Jack Telnack (left) and Frank Hershey (right) in a 1955 Thunderbird at Pebble Beach, 1996.

I started at Ford in 1958. I was encouraged to become a designer by Alex Tremulis and hired by Buzz Grisinger. My first assignment was in Joe Oros' Ford Studio working as a designer in the pre-production area. Many of the designers in this book are my heroes. They are some of the finest artists ever to design cars. You may not find every concept or showcar these artists created to be your type of music, but some of them are likely to become your favorites. That's the difference between concept and production cars. Concept cars are pure art, while production cars -- although art -- are designed to appeal to as many people as possible. ■

Jack Telnack

Table of Contents

6 Continental Division Design Studio

7 Aerodynamics and Alex Tremulis

8 Stylerama

9 Special Vehicles Department

10 The Rise and Fall of George Walker

Introduction

The complete story of Ford's Design Department has not yet been told and, admittedly, this book tells only a part of it. This book is about the concept and showcars designed by or for Ford Motor Co. during the period 1932-61, and some of the designers, clay modelers and engineers who created them. Concept and showcars are not often thought of as works of art, nor are their creators usually given the artistic credit due them. Hopefully, this book will help change that perception.

Ford's concept and showcars were sometimes full-sized; sometimes, they were less than full-sized. They were made from metal, wood, plastic, clay, plaster and fiberglass. Some accurately forecast design features found on future production models; some were pure blue sky. A few of the cars covered in this book were pre-production models or engineering prototypes that were never produced. Except for the Briggs Prototypes and maybe a few others, pre-production prototypes of cars actually produced have generally been excluded. Because Ford produced far too many concept and showcars to include in one book, this book is limited to those cars created through 1961. As of 1998, there have been an estimated 400 Ford Motor Co. concept and showcars. We hope to cover the rest in future works.

Because this book relies on the recall of individuals about what they did 40 to 60 years ago, there may be inaccuracies. We have tried to eliminate as many inaccuracies as possible by relying on multiple recollections. On occasion and, as might be expected, conflicts were encountered in the memories of otherwise reliable and trustworthy informants. As much as possible, we have tried to reconcile those conflicts. Where we could not do so, they are reported. Some concept or showcars may have been overlooked, together with the designers, modelers and engineers who created them. It was unintentional, and we trust that someday they will be rediscovered and their stories told.

Designers often see the production cars they design as compromises intended to appeal to as many prospective buyers as possible. Concept and showcars, however, have traditionally been a welcome opportunity for designers to let their imaginations run wild, uninhibited by the marketplace. This book started out to be only about those cars. As our research progressed, we felt compelled to include the designers and some of the clay modelers and engineers, because we concluded that their stories were as interesting as the cars they created. This book is not a commentary on design. Rather, it is the story of a group of talented artists and the rolling sculpture they created.

We wish to thank the many people without whose help this book could not have been completed, including John Najjar, Gil Spear and Tucker Madawick, all retired designers who are among the most organized of Ford's "retirees." Our sincere thanks also go to Jack Telnack, former vice president and general manager of Ford World Wide Design, Bob Puffer of the Design Center, whose help cannot be underestimated, and Jac Nasser, his boss, who is president and CEO of Ford Motor Co. We are also indebted to Darlene Flaherty, one of Ford's former corporate archivists; Dan Erikson, librarian for Ford Photomedia; the Ford Design Center; the Ford Engineering Department; the Edsel and Eleanor Ford Home; the Henry Ford Museum and Greenfield Village; and many individual designers, clay modelers and engineers for making their archival material, including the pictures in this book, available to us. Of the world's great automobile manufacturers, Ford Motor Co. has diligently preserved its history and readily makes it available to historians as well as to the interested public. Most of all, we are indebted to Edsel Ford, Bob Gregorie and all Ford designers, modelers and engineers, past and present -- for the very special cars they created. ■

Jim and Cheryl Farrell

At age 25, Edsel B. Ford became president of Ford Motor Co. At least in name, he remained in that position until his untimely death, in May 1943, from cancer, undulant fever and overwork. He was largely self-educated, and his life-long passion was art in all its forms. Edsel Ford and his wife, Eleanor, were collectors and patrons of many artists and museums. He was also an accomplished artist, who took lessons all of his life. At a time when others did not yet recognize it as such, Edsel Ford saw the automobile as an artform. Because of a natural modesty, he never claimed to be a designer. In reality, he was a far better designer than most who claimed the title. He knew design theory and history and spoke the technical language used by the most proficient designers of the day. He was Ford's design director in the same sense that Harley Earl was design director at GM. Edsel Ford, always modest about his abilities, preferred to direct and encourage Ford's stylists in the creation of automobile designs rather than physically design them himself.

He was also practical enough to encourage car designs he thought had a realistic chance of being approved by his father and Ford's production people. Henry Ford, a mechanical genius with less interest in design, recognized his son's talent and deferred to him on most design matters. Starting with the 1923 Lincoln and the 1928 Model A Ford, Edsel Ford was responsible for all design at Ford. The big Lincolns of the 1920s and '30s, the Model A Ford, the Lincoln-Zephyr, the early Ford V-8, the Mercury and the V-12 Continentals are all products of Edsel Ford's immense design talent. Although Ford had only one-tenth the number of designers that were employed at GM, the cars designed at Ford during the Edsel Ford years consistently displayed an understated elegance and the sculptured simplicity he insisted on. They have aged well because of him.

Edsel Ford established Ford's Design Department in March 1935. He selected as its head, E.T. "Bob" Gregorie, a designer who was a good sketch artist and particularly adept at interpreting Edsel's design ideas. Although Gregorie is credited with the design of many Ford Motor Co. products, most of them were the result of design sketches completed by him from proposals made or approved by Edsel Ford.

When in town, Edsel Ford visited the Design Department daily. He usually ate lunch in the EEE Building with his father and other Ford engineers and executives. Immediately after lunch, he would walk to the Design Department, which was at the other end of the building, where he spent as much of the afternoon as he could discussing design proposals. Although Edsel Ford always greeted all the designers and clay modelers, he would generally not ask their advice about a particular design or clay model. He usually spoke directly with Gregorie about design proposals, and often Gregorie would sketch those ideas on the spot. He also regularly discussed the clay models under construction with head modeler Dick Beneicke.

Edsel Ford's death at age 49 was the most tragic chapter in the long history of Ford Motor Co. When his son, Henry Ford II, took control of his birthright in 1945, a rebuilding process began. Although the new regime meant new designers and a new way of doing things, Edsel Ford's influence can still be felt by his successors and all designers who work in his shadow. To this day, a life-sized portrait of Edsel Ford dominates the foyer of Ford's Design Center, reminding all who enter of the man who gave birth to Ford's rich design heritage. ■

Ford Photomedia

Edsel B. Ford

1

Edsel Ford and Bob Gregorie, a Design Partnership

1932 Gregorie Roadster

Briggs Prototypes

1934 Special Speedster

1935 Special Sports

Jensen 3 1/2 litre

1938 Ford Landau

1939 Special Speedster

Lincoln 05L

1940 Lincoln-Zephyr Towncar

Light Ford

1932 Gregorie Roadster

The starting point for this book is 1932. That year, a special car was built for Edsel Ford by a Lincoln Division employee named E. T. "Bob" Gregorie. The beginnings of the Ford Design Department can be traced directly to that car and Gregorie, its designer, after whom it was named. The Gregorie Roadster was the first concept car Edsel Ford and Bob Gregorie worked on together and, three years later, it led to Gregorie's appointment as Ford's first design director -- although in those days no one at Ford had titles as such.

Historians correctly point out that the Model T was the first car designed at Ford, but looks had very little to do with the Model T's shape. As with all production cars built before the mid-1920s, mechanics and reliability completely overshadowed most opportunities to make the automobile look pleasing to its potential customers. That all began to change by the mid-1920s. Although Edsel Ford worked closely with several coachbuilders and, even though there were other Ford employees who had previously proposed streamlined Fords, it was the design partnership between Edsel Ford and Bob Gregorie that would make design really happen at Ford.

There were several reasons Edsel Ford picked Gregorie to design and build a special car for him. Gregorie had originally been hired as a Lincoln draftsman in 1931. He was also the first employee of Ford Motor Co. who called himself a designer. Immediately after he started at Ford, Design Department folklore has it that Gregorie placed his drawings of boats on the corner of his desk, where he knew they would be noticed by Edsel Ford. Edsel Ford soon noticed those drawings and decided to test Gregorie's talents. He quickly found that Gregorie had a unique ability to understand his design directions, and that they had similar interests in European cars and fast boats. He also soon came to believe that collaboration with Gregorie might lead to Ford's own design department, and at least one new production model sporty car, which it did. The design department came in 1935. A sporty new Ford product with European flair came in late 1939. It was the Continental. Although none of the three special cars Gregorie built for Edsel Ford look like the Continental, they are its design predecessors.

In 1932, after he returned from Europe, Edsel Ford asked Gregorie to design and supervise the building of a special car for his personal use. Mr. Ford told Gregorie that he wanted a car similar to the sporty-type cars he had seen "on the continent." Gregorie readily agreed, and the skilled personnel at the Ford Airport, where Ford Trimotor airplanes had been built, were made available to assist in building the roadster. The car was built at the old Air Frame Building, but the finish work was completed in the Lincoln plant. During the time the Gregorie Roadster was at the Lincoln plant, it was the only car being worked on, because the plant had been temporarily closed due to a lack of new Lincoln orders.

Even though they had not yet discussed their ultimate goal when the project was started,

Kathryn Buehrig

Elmer Benzin and the Gregorie Roadster after he purchased it from Edsel Ford.

Henry Ford Museum and Greenfield Village.

This is the first special car built for Edsel Ford by Bob Gregorie. It had an aluminum body and 1932 Ford V-8 running gear.

(top) Like many racecars of the day, Edsel Ford's special 1932 Ford had a boattailed rear end.

(bottom) The fenders were made from leftover Trimotor airplane wheel pants. The 1932 Ford grille is veed at bottom, and the hood extends all the way to the windshield.

Edsel Ford and Gregorie both hoped that this first special car might eventually lead to a new production Ford. Given the current state of the economy, both also realized that any new car built at Ford had to use as many available production parts as possible, otherwise it would never gain necessary approval from Henry Ford and his head of production, Charles Sorensen.

The only specific direction Edsel Ford gave Gregorie was to design a car that was "long, low and rakish." The design Gregorie came up with was of a car with a two-passenger body and a boattailed rear section that reflected his boatbuilding background. After he settled on the design, Gregorie started with a standard 1932 Ford chassis and running gear and used as many other '32 Ford parts as possible.

Henry Ford Museum and Greenfield Village.

The front of the Gregorie Roadster came out looking much like the production '32 Ford, except for its vee-shaped grille and windshield. It had chrome-plated bullet-shaped headlights, disk wheelcovers over stock spoke wheels, no runningboards, no top and no exterior door handles on its two suicide doors. The body was fabricated from hand-hammered sheet aluminum, and it had fenders modified from Trimotor airplane wheel pants. Gregorie recalls that a man named Rosie, who worked at the Lincoln plant and was an expert panel beater, hand shaped the fenders on his knee, occasionally using a big leather bag filled with shot. When completed, the car was painted a dark gun-metal gray.

In the summer of 1932, Gregorie drove the completed car from the Lincoln plant back to the Air Frame Building and personally delivered it to Edsel Ford who, according to Gregorie, was "pleasantly amused" by its boattail design. Although Edsel Ford didn't say so, his reaction was probably an indication that the car was not exactly what he had in mind.

Later, after Edsel Ford had received a breakdown for the cost of work done by the Air Frame and Lincoln plant personnel, Gregorie recalls that Mr. Ford remarked how expensive the car had been to build. Although Gregorie estimated the roadster cost approximately $100,000 to build including labor, it probably cost no more than $10,000, which was a large sum in the early 1930s. Edsel Ford drove the car back and forth to work for a while and, when not in use, he parked it in the gardener's shed at his estate. Because Henry Ford did not approve of his son's interest in sporty-type cars, the Gregorie Roadster was seldom parked in the garage at Edsel Ford's estate, where his father or those in his father's employ might see it.

Henry Ford Museum and Greenfield Village.

In late 1933 or early 1934, an updated Ford V-8 engine was installed to give it more power. Gregorie drove the car several times after the new engine was installed, and reports that it had a top speed of about 90 mph.

Unfortunately, the car did not survive. In 1934, at about the time the Model 40 Special Speedster was finished, Edsel Ford sold the Gregorie Roadster to an Elmer Benzin, who drove it for several years and then sold it to a young designer at General Motors, who promptly wrecked it. The car was a total loss, and what was left of it was junked. Many years later, after Mr. Benzin died, his widow, Kathryn, married designer Gordon Buehrig. As she was glancing at one of her new husband's car magazines one day, she saw a picture of the Gregorie Roadster, recognized it and dug out her old pictures of it. ■

On October 21, 1933, the first Briggs Prototype was introduced at the Ford Exhibition of Progress in Detroit, Michigan. A sign overhead described it as "... a design conception of a streamlined automobile that has no reference to current or future Ford production." That description was probably the best way of telling people who saw it that it really *was* being considered for production, which it was. This was also the first proposed Ford vehicle referred to as a concept car. Two years later, the offspring of the Briggs Prototypes went on sale as the all new 1936 Lincoln-Zephyr.

The first Briggs Prototype represented just one of a whole series of streamlined-car designs proposed for production by different automobile manufacturers during the 1930s. The idea of streamlining a car's exterior design started in Europe and was imported to North America by a small group of engineers -- they weren't called stylists or designers in those days — who saw greater opportunities in the New World for their ideas for improving the automobile's configuration and shape.

Paul Jaray, an Austrian engineer, was one of the first to propose application of streamlining to automobile design. By 1920, he had come to the now obvious conclusion that cars not only move along the ground, they also cut through the air.

Briggs Prototypes

Detroit Public Library, National Automotive History Collection

A side view of the wooden Briggs Prototype No. 1 on display at Briggs Manufacturing in July 1933. The Briggs Prototypes, and the success of the 1936 Lincoln-Zephyr they became, influenced the design of all new cars produced at Ford from 1937 until at least 1940.

The 1936 Lincoln-Zephyr was the first commercially successful American unibody car, and it started as the Briggs Prototypes, a series of four concept cars built by Briggs Manufacturing Co., which used the cars to convince Ford Motor Co. it should produce a new medium-priced Lincoln — a streamlined Lincoln that was called the Zephyr.

By the time the Depression caught up with the automobile industry, Briggs Manufacturing had already been a long-time supplier of Ford and Lincoln bodies. The Depression cut Briggs' Ford business almost in half, and its Lincoln business was rapidly grinding to a halt. Edsel Ford was looking for a way to save the Lincoln name, and Briggs was seeking to bolster its business with Ford Motor Co. To compound the problem, Henry Ford and his production boss, Charles Sorensen, believed Briggs was favoring its Chrysler business at Ford's expense.

By the spring of 1932, W. O. Briggs, president of Briggs Manufacturing, had come up with a plan. Lincoln had excess plant capacity, and Briggs knew that Edsel Ford was searching for a way to utilize it. Briggs hired designer John Tjaarda to design a streamlined, medium-priced Lincoln that could be produced by Ford Motor Co. using Briggs-built bodies. Tjaarda saw the opportunity offered him by Briggs as more promising than his so far unsuccessful efforts to produce his own car. (At the same time, work in Briggs'

Henry Ford Museum and Greenfield Village

(above) Briggs Prototype No. 1 at the 1933 Ford Exhibition of Progress in Detroit, Michigan. The sign above the car reads "This car is a design conception of a streamlined automobile that has no reference to current or future Ford production." Because the 1934 Chrysler Airflow, which was referred to by Chrysler as streamlined, did not sell well, Ford chose not to refer to the production Lincoln-Zephyr as streamlined in its advertising.

(below) The taillights, center fin and the porthole backlights on Briggs Prototype No. 1 were eliminated on the production Lincoln-Zephyr.

Detroit Public Library, National Automotive Collection

other design studio was progressing on a redesigned Chrysler Airflow, with Tjaarda's former teacher, Alexander Klemin, as an advisor.).

When he hired Tjaarda, Briggs knew that Edsel Ford was already interested in his streamlined car proposals. To increase his chance of persuading Henry Ford and Ford's production chief, Charles Sorensen, that the new car Tjaarda was designing should be produced, Briggs appointed Howard J. Bonbright, a long time friend of Edsel Ford, to help coordinate the project. Bonbright was placed in charge of a new Briggs department called the Ford Policies and Relations Department.

By early summer 1932, Briggs presented Edsel Ford with three preliminary sketches and a structural model of a streamlined unibody car. The proposals were for a rear-engined sedan, a front-engine sedan and a convertible that could be built either with a front or rear engine. Edsel Ford liked what he saw and asked Briggs to build a non-operating prototype of a rear-engined four-door sedan to use primarily for market research.

The first Briggs Prototype was built from hardwood and was finished and painted to look like a real car. It had a 122-inch wheelbase, was built as a rear-engined sedan, had a small dorsal fin down its back, four opening suicide doors, two porthole windows for the backlight, a radio aerial running down the middle of the roof (often mistaken for a rearview periscope), chrome-framed seats, rear fender skirts and a short, sloping front end.

After the wooden Briggs Prototype was completed, it was displayed jointly by Ford and Briggs at the Detroit Exhibition of Progress. After that, it was shown at many east coast Lincoln and Ford dealerships. Because the wooden prototype proved a popular attraction, starting in June 1934, it was also displayed in the Ford Rotunda during the second year of the Chicago Century of Progress exhibition. Each time the wooden Briggs Prototype was shown, Ford employees mingled with the crowd asking questions about the car's design and features. This was the first recorded use of a market survey to determine how various features of a proposed car were perceived by the public. Eighty percent of those who responded liked the streamlined shape and the chrome-tube-framed seats, but only 50% liked the rear engine.

Even as the first wooden prototype was being finished, Briggs began building a second car. This second Briggs Prototype had a 112-inch wheelbase and was the first operable mechanical prototype of the Lincoln-Zephyr. It was powered by a Ford V-8 engine located in the front and had an otherwise conventional drivetrain. Testing determined that it was underpowered, assuming a final projected weight of 3200 pounds for the production car. (As built, the 1936 Lincoln-Zephyr weighed 3600 pounds.) Ford engineer Frank Johnson was then asked to develop a 12-cylinder engine for use in the production car. The V-12 engine Johnson developed was based on Ford's new flathead V-8. A third Briggs Prototype, this one with a wheelbase of 124 inches, was then built. This second mechanical prototype was the first to have the Lincoln HV-12 engine in it. Both mechanical prototypes were handbuilt at Briggs, and both were used to test roadability, mechanical reliability and the proposed unibody structure. On the second mechanical prototype, because it had the longer HV-12 engine in it, the front axle and wheels were moved 10 inches forward.

This bridge-truss model was used by Tjaarda to explain unibody construction.

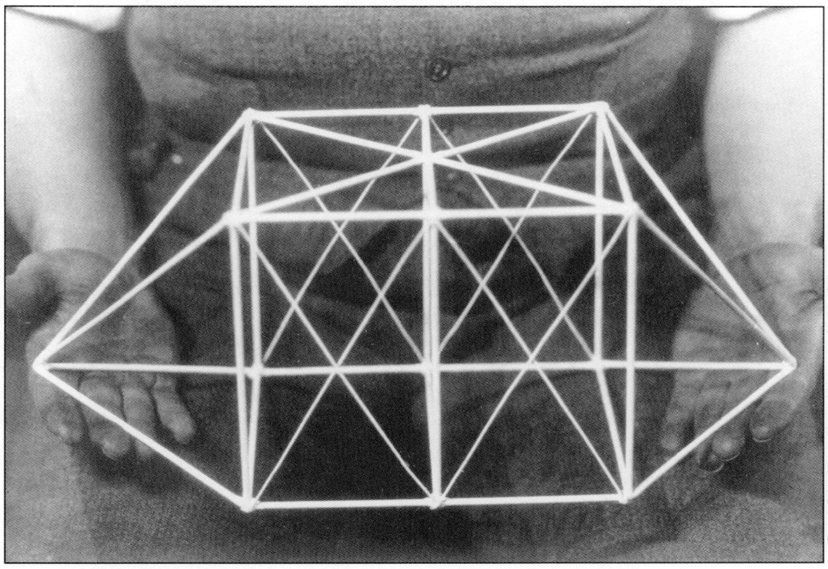

Detroit Public Library, National Automotive History Collection

In late 1933 or early 1934, another mechanical prototype, the fourth and last Briggs Prototype, was built. This one had a rear engine. Edsel Ford felt that the market survey results had been promising enough to at least test Tjaarda's rear-engine proposal, and his father and Charles Sorensen apparently agreed. The fourth prototype closely followed the design of the first wooden mock-up. It had the two-porthole backlight, but it did not have the roof aerial. The fourth prototype weighed 2500 pounds, had a 125-inch wheelbase and a rear-mounted Ford aluminum-block V-8 engine that weighed 300 pounds and, on the test track, reportedly took the car to 110 mph. Adjustable air intakes behind the rear doors provided cooling for the rear-mounted radiator. A semi-automatic transmission with automatic overdrive was also used on the fourth prototype.

The two front-engined mechanical prototypes were never publicly displayed. By the time the fourth Briggs Prototype was publicly shown in October 1935, it was only one month before the production Lincoln-Zephyr was introduced. By that time, its backlight had also been changed from two portholes to a single window.

At the beginning of the project, Edsel Ford told Briggs that Tjaarda's design would likely be built (if at all) as a front-engined car without the sloping front end. There are also reports that the sloping front end would not properly cool the HV-12 front-mounted

Briggs Prototype No. 1 was made from wood and appeared completely finished inside and out.

Detroit Public Library, National Automotive History Collection

Henry Ford Museum and Greenfield Village

Briggs Prototype No. 1 was also displayed at the 1934 Chicago Century of Progress exhibition. By this time, wide-whitewall tires had been added.

engine. Although the second mechanical prototype (third prototype and first car to be equipped with the HV-12 engine) was approved for production, it was conditioned on redesigning the sloped front end. As much as Tjaarda tried, he could not bring himself to change the front end. Each time he came up with a new design, it looked the same and was rejected by Edsel Ford. After various front-end redesigns by Tjaarda, Edsel Ford concluded that Tjaarda was incapable of making the changes he had requested, and that someone else would have to do it.

Ralph Roberts was in charge of a second design studio at Briggs. By mid-1934, the Roberts studio had completed the redesign of the Chrysler Airflow, so all of the designers in both the Roberts and the Tjaarda studios were put to work on a crash program to design a new front end that was acceptable to Edsel Ford. Holden N. "Bob" Koto, a designer in the Roberts studio, designed a 1/4-sized clay model with a pointed front end, a boat-prow hood and front fenders that terminated in the middle of the front suicide doors. Koto's design was approved by Edsel Ford and then incorporated as the front end on a new, full-sized clay model of the entire car. Only after it was completed, did Ford Engineering conclude that the front doors required special hinges that were too expensive for production. In the meantime, Roberts returned to England, where he was doing other Briggs work for Ford. This gave Tjaarda, who was noted for his stubbornness, an excuse to propose his own sloping front-end design again. As a last resort, a frustrated Edsel Ford turned to Bob Gregorie who, at the time, was a Lincoln body draftsman. After telling Briggs he proposed to bring in an outsider, Edsel Ford told Gregorie he wanted him to assist Tjaarda and salvage the design by coming up with a commercially acceptable front end for the car. Gregorie went to Briggs Manufacturing, met with Tjaarda, and explained that he had been asked to redesign the front end of the prototype for production. He also told Tjaarda that it was nothing personal but that, unless something was done, Tjaarda's car was not going to be produced. Gregorie "explained the facts of life" to Tjaarda and says that, after that, "they got along just fine."

Detroit Public Library, National Automotive History Collection

Briggs Prototype No. 4 showing engine placement and the structure of the car.

The same day he arrived at Briggs, Gregorie drew his proposed redesign for the front end of the Briggs Prototype on the back of an old blueprint. Edsel Ford came by later, approved Gregorie's design, and it became the front end of the production Lincoln-Zephyr.

The Way of the Zephyr, Dec-Jan '72

(left) This front-end proposal and other similar designs helped convince Edsel Ford that Tjaarda was incapable of designing an acceptable front end for the car he wished to produce.

(below) Briggs Prototype No. 1 was designed for a rear engine.

Detroit Public Library, National Automotive History Collection

(below) One of the few known photographs of Briggs Prototype No. 4.

The Way of the Zephyr, Dec-Jan '72

(below) The instrument panel of Briggs Prototype No. 1 was symmetrical and had two gloveboxes.

Detroit Public Library, National Automotive History Collection

Detroit Public Library, National Automotive History Collection

Tubular framed seats restricted entry on Briggs Prototype No. 1.

The total time Gregorie spent at Briggs redesigning the front end of the Briggs Prototype was about 10 days. First on paper and then in clay, Gregorie mated his front end and fenders to the Briggs Prototype in a way that made it work perfectly. Gregorie actually salvaged Koto's design by blending the fenders into the overall design in a way that terminated the fenders short of the front doors. That did away with the need for the overly-complicated and expensive hinges. Koto credited Gregorie with resurrecting and saving his design by the use of what he called "contrived" catwalks that were different from the catwalks he had earlier proportioned to the beltline of the car. Gregorie also made other minor changes to Tjaarda's design to make the car look larger and more integrated.

This December 1933 photograph is probably of Briggs Prototype No. 4 under construction at Briggs Manufacturing.

Detroit Public Library, National Automotive History Collection

Tjaarda is the designer most responsible for the Lincoln-Zephyr, but as much as anyone, Koto deserves credit for its pointed grille and boat-prow hood design. Gregorie deserves credit for the design of the front fenders, for refining Koto's grille and hood design and for overall integration of the front-end design. The Koto-Gregorie modifications to the Tjaarda-designed Briggs Prototype resulted in a production-ready 1936 Lincoln-Zephyr, which was introduced in November 1935.

By 1941, Briggs Manufacturing had lost all of its Ford business and, as a result, became even more dependent on its business with Chrysler. In 1952, when W.O. Briggs died, estate taxes forced his family to sell the business to Chrysler. In 1960, the remaining Briggs Prototypes were cut up and destroyed. Even the plans and pictures were destroyed. Except for Ford Motor Co. photographs and those in private hands, there is little record left of the Briggs Prototypes. ■

Dave Cole

Detroit Public Library, National Automotive History Collection

(above) The Briggs display at the 1933 Detroit Exhibition of Progress. A model of Briggs' futuristic train is shown in the background to the right of the Briggs Prototype.

Independent rear suspension on Briggs Prototype No. 4 had "sea legs" which placed shock absorbers and a Ford engine in front of the rear axle. This proposal also used conventional transverse springs with swing axles, similar to what was later used by Volkswagen and Porsche.

Detroit Public Library, National Automotive History Collection

John S. Tjaarda

John Tjaarda was born in the Netherlands. He attended college in England, where he studied aircraft engineering and aerodynamics under Alexander Klemin, an early advocate and teacher of aerodynamics. Tjaarda returned to his homeland after he completed his formal education and, just after World War I, joined the Royal Dutch Air Corps as a pilot and instructor. Later, he was employed by the Fokker Aircraft Co. as an engineer. Tjaarda was greatly influenced by the work of Paul Jaray and, by 1919, he was sketching automobiles that incorporated Jaray's ideas for streamlining, monocoque construction and rear-mounted engines.

In 1923, Tjaarda came to the United States. He first worked customizing new cars for wealthy Southern Californians. In 1926, he moved to Rochester, New York, where he took a job with Locke & Company, a respected coachbuilder. In 1928, he left Locke to take a job at Duesenberg, where he worked with Harry Miller designing racecar bodies and chassis. In 1930, Tjaarda moved to Detroit where he worked in GM's Art and Colour Section under Harley Earl. Earl didn't think Tjaarda's streamlined design proposals were very practical and, in late 1931, Tjaarda left to concentrate his full-time efforts on raising money to produce his own cars. He was not successful.

In 1932, W. O. Briggs of Briggs Manufacturing hired Tjaarda to head one of his studios, and to design a new mid-sized Lincoln that could be built by Briggs. Tjaarda's proposal was shown to the public in 1933-34. After Edsel Ford decided to produce the car, Tjaarda could not bring himself to change the sloped front end as Edsel Ford wanted. Bob Koto and Bob Gregorie then redesigned the front end. With the new front end, the car became the 1936 Lincoln-Zephyr. It, in turn, saved Lincoln and influenced Ford designs for the rest of the decade.

After he redesigned the 1936 Cord body into the Hupp Skylark and the Graham Hollywood, Tjaarda left Briggs in 1941. Following the war, he was unsuccessful in promoting a world car. Dissolutioned with the automobile business, he moved to California, where he opened his own industrial design firm. ■

Holden N. Koto

Bob Koto's father was part owner of a hosiery mill in Beloit, Wisconsin. Koto attended Cornell University, where he studied engineering and architecture. In 1933, after completing three years at Cornell, Koto took a job at Briggs Manufacturing as an automobile designer. Koto updated the body for the 1936 Ford, and he and Bob Gregorie share credit for the front-end redesign of the Briggs Prototype. In 1939, Koto left Briggs to take a job with Hudson.

By 1943, the war had dried up Hudson's car-design work, and Koto was laid off. He and Ted Pietsch, another designer laid off by Hudson, attempted to secure a design contract with Nash by presenting Nash management with their proposals for an inverted bathtub-style car. Although they weren't hired, they later learned that their model became the basis for the design of the post-war Nash.

In 1943, Koto was hired by Raymond Loewy to design Studebakers. In 1946, Koto and Bob Bourke helped designer Richard Caleal design what became George Walker's winning 1949 Ford proposal.

From 1953-55, Koto was assigned to England with other Loewy employees, where they redesigned the Rootes Group cars, including the Sunbeam, Hillman and Humber. Koto left Loewy in August 1955, after Studebaker and Packard merged, and Loewy lost the Studebaker account.

In late 1955, Koto took a job at Ford's Styling Center. His first assignment was to work on the Edsel. In February 1957, he was made head of the Mercury pre-production studio and, in November 1964, he was transferred to the Lincoln-Mercury studio as executive stylist for Lincoln and Comet. Koto was one of the designers of the 1966 Falcon, the 1966 Lincoln and the 1967 Mercury. He retired in April 1968. ■

Ford Design Center

(below) A full-sized clay model of the Briggs Prototype with Bob Koto's front end. This car would have become the Lincoln-Zephyr but for front door hinges, which were too costly to produce.

Henry Ford Museum and Greenfield Village

(below) The 1936 production Lincoln-Zephyr was restyled by Gregorie. A prow-shaped front end, similar to the one originally designed by Bob Koto, was added. The front fenders, front doors and rear slope of the car were also redone by Gregorie to overcome production problems and make the Lincoln-Zephyr appear longer and better proportioned.

Ford Photomedia

Ford Photomedia

Eugene T. "Bob" Gregorie

Bob Gregorie was born in New York City and grew up on Long Island. His father graduated from Clemson University as a mechanical engineer and was in the import business. Although he did not complete high school, Gregorie inherited an appreciation for fine European automobiles and a love of tinkering from his father. From his mother, Gregorie inherited a passion for boating and his artistic ability. As a teenager, Gregorie modified several Model Ts and developed a familiarity with European cars.

He grew up with a desire to design things. In 1927, at age 19, he got his first job as a draftsman at the Elco Boat Co. In 1928, he left Elco to take a job with Cox and Stevens, a larger boatbuilding firm closer to home. There he learned basic boatbuilding skills, methods and shortcuts he was later able to use in designing automobiles. In 1929, Gregorie decided there were better opportunities designing automobiles, so he took a job with the coachbuilding firm of Brewster and Co. Later that same year, after unsuccessfully applying at Franklin and Pierce-Arrow, Gregorie moved to Detroit, where he obtained a job in GM's Art and Colour Section as an apprentice designer. The Depression ended his job with GM after two months, and Gregorie again returned to Cox and Stevens for several months before their work ran out. He then returned home and, in December 1931, while at his family's South Carolina vacation home, he was hired as a body draftsman at Ford's Lincoln plant by Henry Crecelius, Lincoln's chief body engineer.

Gregorie was the first person employed at Ford Motor Co. who called himself a designer. After working as a body draftsman on new Lincolns, Gregorie was asked by Edsel Ford to design and build what became the Gregorie Roadster. His next major assignment was designing a new English small Ford called the Model Y, which was the first new car totally designed at Ford since the Model T. John Crawford, Edsel Ford's production advisor, was impressed with Gregorie's design work and recommended Gregorie's talents to his boss. Gregorie was ambitious, talented, politically astute and articulate. If the design of the Continental was the high point of Gregorie's career, the Model Y and the Model Y-inspired 1933 Ford were the designs that launched it. In 1934, Gregorie designed and built another special car, the Model 40 Special Speedster, for Edsel Ford

In March 1935, Edsel Ford asked Gregorie to form Ford's first design department and become the first head of it. Edsel Ford died in May 1943 and in September 1943, after a disagreement with Joe Galamb, his new boss and Ford's chief body engineer, Gregorie left to open his own industrial and marine design business. In September 1944, at Henry Ford II's request, Gregorie returned to Ford, believing he would be able to work directly with Henry Ford II as he had with his father. After Ernest Breech was hired as executive vice president, Gregorie was required to follow a GM-style chain of command, in which he reported to Breech, whom he did not get along with. Gregorie left Ford for good on December 15, 1946, after his design for the 1949 Ford lost out to the one proposed by independent design consultant George Walker.

After he left Ford, Gregorie moved to Florida, where he occasionally designed boats, but he never again worked as an automobile designer. ■

Henry Ford Museum and Greenfield Village

Like the Gregorie Roadster, the Model 40 Special Speedster was built of aluminum and had fenders made from Ford Trimotor wheel pants. The fenders were bolted to the brake-drum backing plates, and the body was joined to the longitudinal covers over the chassis rails by a row of aircraft rivets.

1934 Model 40 Special Speedster

Throughout the time Bob Gregorie worked for Edsel Ford, both before and after he was appointed head of the Ford Design Department, he designed several special cars to demonstrate concepts that one or the other of them thought might lead to a new production Ford. Although there was more than enough work to keep both of them busy after the Gregorie Roadster was finished, by the summer of 1934, Edsel Ford and Gregorie were again talking about building another special car. Edsel Ford decided that this one should be longer, lower and more streamlined than the first. This second special car Gregorie built for Edsel Ford was started in mid 1934 and was known as the Model 40 Special Speedster. By the time it was finished, Gregorie and Edsel Ford both thought it might be just the car that would lead to a new production-model sporty Ford.

As with the Gregorie Roadster, Gregorie designed the Model 40 Special Speedster himself. He started by sketching several different designs, one of which Edsel Ford approved. After design approval, Gregorie built a 1/25-sized model of it, which he tested in a small windtunnel at the Air Frame Building. After he was satisfied with the design of the body, Gregorie next designed a special chassis to accommodate the lower, longer body.

With the help of Robbie Robinson, superintendent of the Lincoln plant, a stock 1934 Ford frame was lowered 6 inches by cutting off the back end, just ahead of the rear-axle kickup, turning the cut-off section upside down, and then welding it all back together, so that the frame curved under the axle. In the front, 15 pounds of special suspension parts, designed by Gregorie, were substituted for the stock suspension, lowering the front end and moving the front axle forward by about 10 inches.

When he was building the chassis, Gregorie had to constantly keep in mind that what he ultimately came up with had to be fairly simple and had to be made from stock Ford parts. Otherwise, it would never make it past the expected objections from Henry Ford and Charles Sorensen. Gregorie and Robinson spent many hours of trial and error modifying the stock 1934 Ford chassis for use under the Model 40 Special Speedster. After it was done, Gregorie concluded it would probably have been easier if they had built the chassis from scratch. When Gregorie test drove the running chassis, however, he was surprised how roadable it was and how well it handled.

Gregorie and the Air Frame personnel next built the two-passenger body Gregorie had designed from sheet aluminum, with cycle fenders again modified from Trimotor wheel pants, full wheeldisks over spoke wheels and special hubcaps. The front fenders were made so that they turned with the front wheels. The aluminum body was attached to a tubular frame and, like the Gregorie Roadster, the Model 40 Special Speedster had a boattailed rear end. When finished, the car was painted Pearl Essence Gunmetal Dark (gray), had a gray leather interior and a machine-turned instrument panel. With standard 1934 Ford running gear installed, the car weighed 2400 pounds. It had no mufflers on dual exhaust pipes that passed through holes cut in the frame. The passenger compartment and the area behind the bucket seats, where the spare tire and the battery were stored, were fully carpeted.

The Model 40 Special Speedster had several design innovations that were predictive of future production cars. In addition to the unique body construction, the car had a single-piece hood, headlights that were integrated into the body, a completely enclosed radiator with no exposed filler or radiator cap, a starter button on the instrument panel and the absence of runningboards. None of those features were incorporated into production Ford Motor Co. products before 1936.

Henry Ford Museum and Greenfield Village

The front fenders on the Model 40 Special Speedster turned with the wheels.

The Model 40 Special Speedster was delivered to Edsel Ford on September 21, 1934. Mr. Ford, who took title to the car personally, liked the way it handled and was generally pleased with its design. Over the course of the next several months, he and Gregorie spent many of their spare moments discussing the car's design and, for the first time, both felt they had a car that could be built, somewhat modified, as a new, limited-production, sporty Ford.

Although Edsel Ford eventually decided that the Model 40 Special Speedster was too streamlined to be a production car, he was especially pleased with the general design formula Gregorie had come up with — long hood, short deck, little trim, no runningboards and a lower, longer stance. As their discussions evolved, Edsel Ford suggested to Gregorie that if he could combine the styling of the more conservative body from the

Because the car overheated, Edsel Ford asked Gregorie to redesign the front end. This one-tenth sized model Gregorie built incorporated features used when the front end was rebuilt in 1940.

Henry Ford Museum and Greenfield Village

Henry Ford Museum and Greenfield Village

(above) From the rear, the Model 40 Special Speedster looked like a boat.

Gregorie Roadster with the design formula followed on the Model 40 Special Speedster, they might just have the production car they were looking for — and one that would get past the objections expected from Henry Ford and Charles Sorensen. In order that production of the new-model Ford would not interfere with regular production cars built at the Rouge, Edsel Ford decided that whatever body was eventually used on the chassis under the Model 40 Special Speedster, it would have to be built and the car assembled by one of Ford's custom body suppliers. That decision led Gregorie to build a third special car to show to the coachbuilders from whom they planned to solicit bids.

Edsel Ford continued to drive the Model 40 Special Speedster until shortly before his death in 1943. It was licensed for the street and, in good weather, he could be seen driving it around Detroit and Dearborn. The car had a tendency to overheat and, over the years, several attempts were made to solve the problem, all without success. Gregorie concluded that the car overheated because of a design problem — the sheetmetal below the grille closed-off air flow to the bottom of the radiator, thereby trapping hot air.

Just as he had done with the Gregorie Roadster, and to keep the Model 40 Special Speedster from causing a confrontation with his father, Edsel Ford generally stored it out of sight in the unheated gardener's shed at his estate. During a winter freeze in 1939-40, someone forgot to drain the radiator, and the block cracked. At Edsel Ford's direction, a 1940 Mercury flathead V-8, with Edelbrock heads, two Stromberg 97 carburetors, a Thickston intake manifold and a special cam were installed, making the car very fast.

In early 1940, probably at the same time the new engine was installed and, to prevent it from overheating, Gregorie proposed shortening the top grille and adding a separate horizontal grille at the bottom of the radiator. The redesign Gregorie came up with also moved the headlights slightly outboard, converted them to Sealed Beams and incorporated them into either side of the lower grille. A small wooden model of the redesigned Model 40 Special Speedster with the proposed grille changes was built and a photograph of it given to Edsel Ford, so he could see the changes Gregorie proposed. Edsel Ford questioned Gregorie's proposed changes, but when the front end of the car was revamped later that year, the grille area was as Gregorie had redesigned it. At the same time the front-end changes were made, the car's instrument panel was also modernized.

The Model 40 Special Speedster still exists. When Edsel Ford died, the car was one of six included in his estate. In June 1944, the Model 40 Special

Henry Ford Museum and Greenfield Village

(above) The Instrument panel was finished in burnished aluminum and featured Lincoln instruments. Gregorie added a pair of Brookland windscreens, which gave the car a British touch.

Henry Ford Museum and Greenfield Village

The frame rails were completely enclosed by sheetmetal. The canted hood louvers matched the angle of the radiator and the windscreen.

Bob Puffer

The Model 40 Special Speedster at the Amelia Island Concours, May 1999

Michael Lamm

(below and left) The Model 40 Special Speedster in 1958.

Michael Lamm

Speedster was listed in an estate inventory filed with the Macomb County, Michigan, Probate Court as a "Ford Special Speedster #18-1022711, thirty horsepower." The Model 40 Special Speedster was not immediately sold. It was later driven to Miami, Florida and then to Atlanta, Georgia, where it was eventually sold to its second owner for $1000.

In 1947, the second owner shipped the car to Los Angeles, California, where it was placed in storage. In the May 1948 issue of *Road and Track* magazine, the second owner advertised the car for sale for $2500, but it did not sell. In 1952, the Model 40 Special Speedster surfaced again in Hollywood, California, where it was photographed with actress Lynn Bari for a magazine called *Auto Sport Review*. After that, it was back into storage for several more years. In 1957, it was removed from storage and driven back to Georgia. It next surfaced on the Garrard Import used car lot near Pensacola, Florida, in January 1958. By that time, it had been registered in Florida as a 1940 model "Ford custom-built speedster."

That same year, Earl Pallasch, a sailor home on leave, saw the car, fell in love with it, bought it for $603 and drove it home to Sebring, Florida. By the time Pallasch bought the Model 40 Special Speedster, it had been repainted red and the upholstery changed to red leather. Pallasch drove the car for several years, and claims he could peg the 120 mph speedometer.

In the late 1960s, Pallasch disassembled the engine to rebuild it. Before he could finish the job, the Navy sent him to Vietnam for an extended tour of duty. Pallasch placed the Model 40 Special Speedster in storage with the engine still disassembled. When he returned home, several years later, the engine had seized. In 1999, the car was sold to Bill Wright of Jacksonville, Florida. He showed it at the 1999 Amelia Island Concours. It currently has 21,000 miles on it. Wright says he will restore the car to the condition it was in when Edsel Ford owned it. ■

Bob Gregorie and the Model 40 Special Speedster at the Ford Airport test track.

E.T. Gregorie

1935 Special Sports

It wasn't the body design of the Model 40 Special Speedster that Edsel Ford and Gregorie were most interested in — it was the chassis under it, and that led Gregorie to build a third special car for Edsel Ford.

Because of suggestions made by Ford's chief engineer, Larry Sheldrick, Gregorie decided that substantial testing of the longer, lower chassis developed for the Model 40 Special Speedster was needed, just to make sure he hadn't overlooked anything and, that when in production, it would hold up. That required building the 1935 Special Sports. By the time it was built, both Gregorie and Edsel Ford felt they were well on their way to gaining needed approval from Henry Ford and Charles Sorensen to build their proposed continental-style Ford sporty car. They planned to send the modified chassis to an outside manufacturer who would build a special body and assemble the car. When completed, the finished cars were to be sold through the Ford dealer network

Henry Ford Museum and Greenfield Village

Gregorie and Walter Kruke drove the Special Sports, which had no heater or sidecurtains, to New York City in the dead of winter to try to convince John Inskip, who had recently purchased Brewster and Co. from Rolls-Royce of America, to build a sporty car using this car's chassis and running gear.

Gregorie built the chassis for the 1935 Special Sports from a 1935 Ford production chassis. Fortunately, the '35 Ford's chassis had already been modified in such a way that it did not require cutting the frame in two and welding it back together. Edsel Ford and Gregorie were both delighted with the changes already made in the stock '35 Ford chassis, and felt it made needed modifications much less complicated and less expensive. The same front suspension parts Gregorie fabricated for use on the Model 40 Special Speedster could still be used to lower and extend the wheelbase of the '35 Ford chassis.

To work out the bugs on the modified chassis, Gregorie decided to test the car on the road as well as on the Ford test track. To do that, he had to build a body for the car so it could legally be driven on the street.

The body Gregorie built for the 1935 Special Sports was makeshift and was what is commonly referred to as a mule. It was made during the fall and winter of 1934-35 by the Air Frame personnel for the primary purpose of making a street-legal test car. The cowl and doors were from a 1934 Ford roadster, and the grille, front bumper and he adlights came from a production '35 Ford. A special hood was fabricated, and the front fenders were again made from Trimotor wheel pants. The rear fenders were stock '34 Ford. Although the car looked very finished, it was improvised without much attention to its actual design. The craftsmanship, however, was first rate. Gregorie finished the '35 Special Sports off with a set of front driving lights from a Stutz he had once owned He also put polished full wheeldisks on the right side. The left side of the car had stock '35 Ford wire wheels. A make-shift top was installed, seats, upholstery and a tonneau cover were added, but the car had no windows or sidecurtains. Both the chassis and body were built in right-hand-drive configuration for no other reason than because Gregorie liked right-hand-drive cars. (He had recently returned from a trip to England.) The parts used to make the right-hand-drive chassis came from the Windsor, Ontario Ford of Canada plant.

Gregorie tested the car for several thousand miles, until both he and Edsel Ford were convinced the chassis was ready for production. He and Edsel Ford then contacted several coachbuilders in the Detroit area to try to interest them in building bodies for their sporty continental-style Ford but found no takers. The excuses given ranged from too small to too large a volume. The only custom coachbuilder who expressed any interest was Gregorie's former employer, John Inskip at Brewster and Co. in New York. Even though Brewster was several states away, the more they thought about it, Gregorie and Edsel Ford came to believe it was really a pretty good choice, since Brewster had already built a line of custom Ford towncars. Inskip asked to see the modified chassis, and Gregorie decided to take it to New York himself. In February 1935, he recruited Walter Kruke, another designer at Ford, and they drove the car to New York during a snowstorm. The car had no heater and no sidecurtains, but Gregorie says they took plenty of applejack to keep them warm. When they showed the car to Inskip, he agreed to build the bodies and assemble the cars, but he wanted Ford to build a factory for him to do it in. That was out of the question and, when Gregorie returned to Dearborn, a disappointed Edsel Ford called it quits -- in this country. In early March, he sent a letter to Ford of England inquiring whether any of the leading Ford dealers there were interested.

E.T. Gregorie

Gregorie's modifications to the front suspension of a production 1935 Ford chassis lowered it six inches and moved the front axle forward by ten inches.

After Gregorie returned to Dearborn, Edsel Ford gave him the 1935 Special Sports as a gift. Gregorie drove it for several years, usually only in good weather. He then sold it to a friend on Long Island for $500, and it dropped out of sight. In 1954, automotive writer Richard Kelly rediscovered the car on the Lynn Mayfield used car lot in Burbank, California. The 1935 Special Sports had been repainted in a two-tone color scheme, and the body had been mildly customized by adding a LaSalle grille and a Carson padded top. The story Kelly got from Mayfield was that the car had been purchased in 1952 by a Burt Felz, another used car dealer, for $250. Felz then sold what he thought was a customized 1934 Ford Victoria to a cab driver who, in turn, sold it to Mayfield. Kelly didn't buy the car and only later learned of its pedigree. Although he has been actively searching for it since 1954, it has yet to resurface. ■

Although completely finished, the makeshift body used on the 1935 Special Sports was pieced together by Gregorie to test the chassis. Gregorie referred to the body as makeshift but the workmanship was first rate.

E.T. Gregorie

Richard Kelly

(above) In 1954, the 1935 Special Sports was spotted in a Southern California used car lot. It had been modified but was recognizable. Automotive writer Richard Kelly took this photograph and has been looking for the car ever since.

(below) This aerial photograph of the EEE Building and surroundings was taken in the mid 1930s. (1) is the location of the Design Department; (2) is the Henry Ford Museum; (3) is Greenfield Village under construction; (4) is the location of the Air Frame Building; (5) is the Ford Airport and; (6) is the Airport Terminal Building. After World War II, the Design Department was moved to the other end of the EEE Building.

Henry Ford Museum and Greenfield Village

Given the devastating economic effect of the Depression on the coachbuilders he had previously worked with, Edsel Ford was probably not surprised when no domestic firm was willing to build his new Ford sporty car. In March 1935, which was within weeks, if not days, of Gregorie's return from his unsuccessful visit with John Inskip at Brewster, Edsel Ford wrote to Sir Percival Perry, head of Ford of England, sending him drawings and photographs of the Special Sports, and inquiring whether he or one of his leading dealers would be interested in building the car. Perry begged off claiming a lack of a market in England for a new sporty Ford. He also told Edsel Ford that none of the dealers he talked to were interested either.

Perry didn't let it drop there, however, and sent the photos and drawings to an acquaintance named John Moore-Brabazon. Brabazon was a famous racecar driver and World War I pilot, who would later become a member of Parliament, a government minister and, eventually, Lord Brabazon. When Brabazon got a look at the photos of the Special Sports, he wrote Perry a letter expressing an interest in building such a car. Perry sent the letter on to Edsel Ford. When Edsel Ford got Brabazon's letter, he asked Gregorie to build another Special Sports chassis and ship it, with a full set of plans, to England so Brabazon could see the chassis firsthand. After Brabazon got a look at the chassis, he wrote directly to Edsel Ford telling him how impressed he was, and that he wanted to produce a car using that chassis.

Jensen 3 1/2 Litre

Jensen Motors used Ford running gear, Gregorie's special chassis and some modified Ford sheetmetal to make the Jensen 3 1/2 Litre.

E.T. Gregorie

By January 1936, Perry, acting on behalf of Edsel Ford and Brabazon, put together the final details of an agreement giving Brabazon the exclusive right to build cars using the Special Sports chassis. Brabazon then set out to find a coachbuilder. He eventually reached an agreement with Jensen Motors Limited, a small coachbuilding firm owned by Richard and Allan Jensen, who agreed to pay Brabazon a royalty for all cars they built using the Special Sports chassis.

One reason for Edsel Ford's trip to England in July 1936 was to meet Brabazon and the Jensens, and to get a firsthand look at two prototype Jensen 3 1/2 Litre cars they had built using the Special Sports chassis. Edsel Ford was impressed with what he saw and was very encouraging to Brabazon and the Jensens.

E.T. Gregorie

A double firewall was used in the Jensen 3 1/2 Litre. The battery was in the compartment behind the Ford V-8 engine.

The Jensen brothers were relatively new to the custom-car business. As teenagers, Richard and Allan Jensen thought they could build better looking custom bodies than most British manufacturers were then offering. In short order, they acquired W. S. Smith

E.T. Gregorie

Pricelist for the Jensen 3 1/2 Litre body styles and for the chassis only.

and Sons, an old-line coachbuilding firm, renamed it Jensen Motors Limited and began building sports car bodies. The special bodies the Jensens built were installed on other manufacturers' production chassis but generally used the same running gear that came with the chassis. In 1934, the Jensens began building special-bodied cars using Ford chassis and V-8 engines. When Brabazon offered them the use of Ford's Special Sports chassis on a royalty basis, it was exactly what they were looking for.

From 1936-39, Jensen Motors produced and sold several models of what they called their 3 1/2 Litre car line. The car used Gregorie's modified chassis, a Ford V-8 engine and a typically English sporty-type body designed and built by the Jensens. The Ford V-8 engine used in the Jensen 3 1/2 Litre was modified to produce 120 hp by adding shaved aluminum heads, special ignition, and two S.U. carburetors on an aluminum intake manifold.

Although Edsel Ford initially hoped to incorporate the Special Sports chassis under a new production-model Ford for the domestic market, there was no real consideration given to selling the English-built Jensen 3 1/2 Litre cars through U.S. Ford dealers. (Pursuant to a special order from Californian Percy T. Morgan in 1936, Jensen Motors built three Jensen Fords using modified 1936 Ford sheetmetal, the 120-horsepower Jensen version of the Ford V-8 engine and Gregorie's special chassis, and they exported them to the U.S.) The outbreak of World War II in 1939 curtailed all British automobile production. Although Jensen Motors resumed production after the war, because English cars were taxed by the size of their engines, the Jensen 3 1/2 Litre was no longer offered. Fewer than 100 Jensen 3 1/2 Litre cars were built. For various reasons, it was not the success Edsel Ford, the Jensen brothers or Lord Brabazon hoped it would be. ■

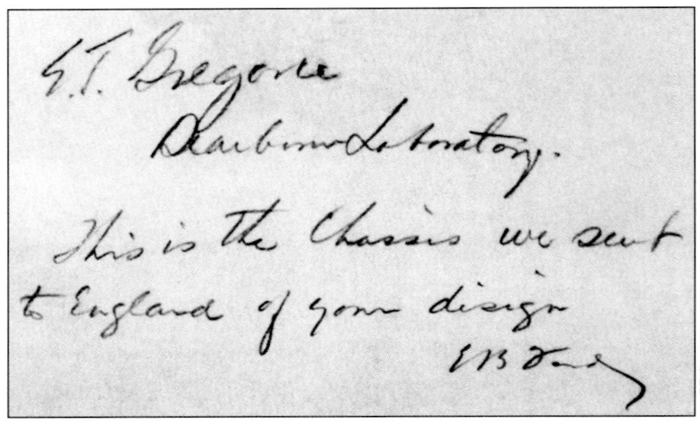

E.T. Gregorie

(right) Edsel Ford wrote this note and sent it to Bob Gregorie with a picture of the special chassis as it was modified at the Jensen factory for use in the Jensen 3 1/2 Litre.

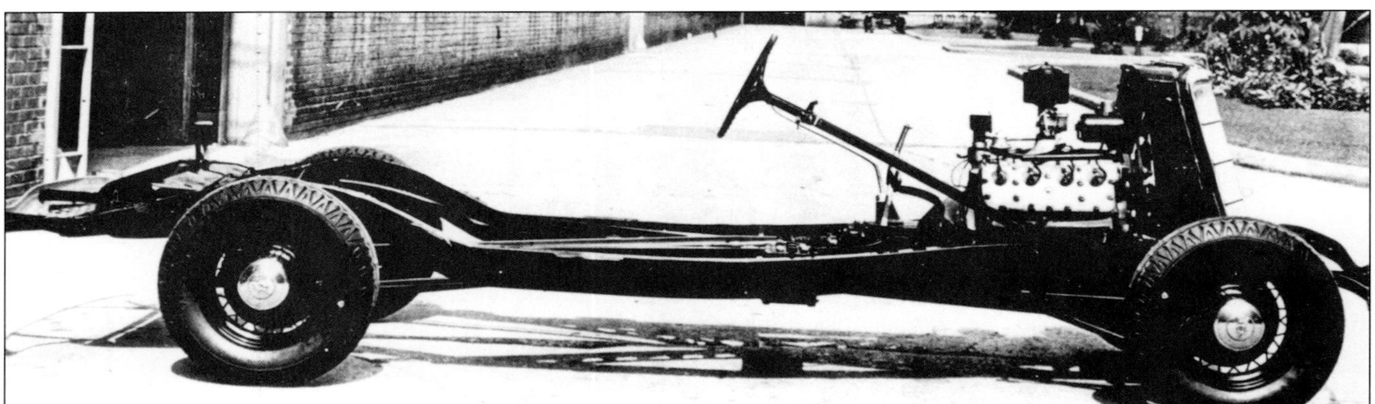

(right) The running chassis as it was when sent to England after Gregorie modified it.

E.T. Gregorie

(left) Gregorie's special chassis as it was modified at the Jensen factory.

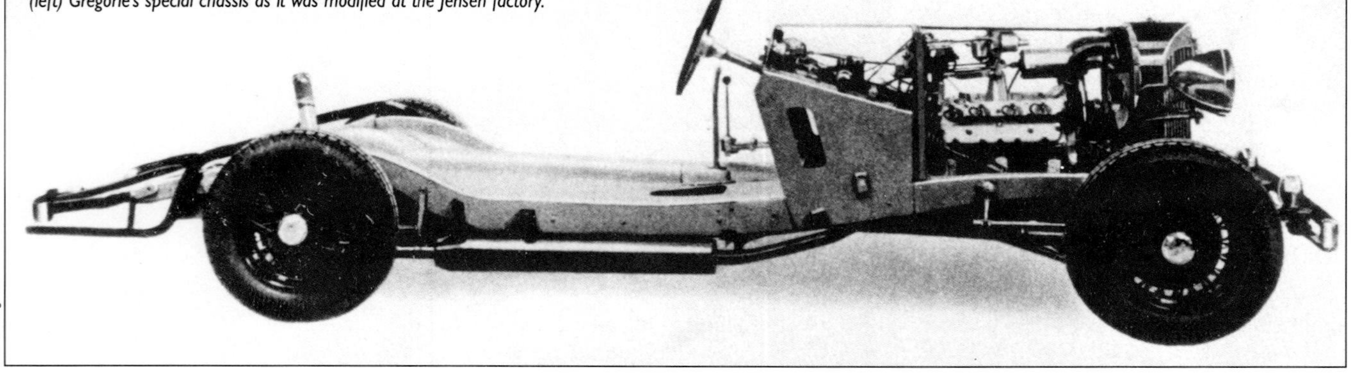

E.T. Gregorie

Tucker Madawick

Special nose trim and Lincoln-Zephyr wheelcovers were used to dress up this partially finished 1938 Ford Landau. There are no known photos of the finished car.

1938 Ford Landau

Edsel Ford encouraged his chief designer, Bob Gregorie, to continually experiment by making cars with new body styles and by customizing production cars. Edsel Ford had a soft spot for special or custom-built Fords and was always interested in new ways to make them a little fancier or a little more sporty.

In late 1937, Gregorie built another special car. This one was a customized 1938 Ford Deluxe two-door sedan. The car had blanked-out rear-quarter windows and was fitted with a leather landau top. According to Gregorie, the car also had a special paint job and was upholstered in pigskin.

Gregorie first prepared a sketch of the car. He then showed it to Edsel Ford, and Mr. Ford gave him the go ahead to build it. Both Ford and Gregorie liked the looks of the design, and it was built so they could determine how the car looked in person. When the sketch was first shown to Edsel Ford, Gregorie remembers him commenting, "We have to do those things, don't we?" As Gregorie remembers it, when Edsel Ford authorized building the car, all he asked was to see it when it was finished. When the car was finished, Mr. Ford drove it home and, the next day, he told Gregorie it rode well and was especially quiet with all the padding on the roof. Ford designer Tucker Madawick recalls that the car was later used by Mrs. Edsel Ford for her personal transportation. Gregorie estimated the cost of the 1938 Ford Landau at $50,000 which, according to Madawick, is probably too high, given the value of the dollar in 1938.

What happened to the 1938 Ford Landau is unknown. ■

John R. Hay

Ford Design Center

John Hay grew up in East Monroe, Ohio. As a teenager, he worked in the family butcher shop and built detailed car models from coffee cans and solder. The model cars Hay built also had working brakes, suspension, lighting and steering. A local Ford dealer sent Hay's model of a 1937 Ford phaeton to Ford Motor Co. and, as a result, Hay received an invitation from the Henry Ford Trade School to join 300 students at a summer work/study program. In May 1937, Hay started in the foundry and then in the drill and machine departments at the Rouge plant. A friend showed Hay's model to Bob Gregorie, who asked him to come to work at the Design Department. By October 1937, Hay was working in the Design Department shops as a wood and metal worker.

As World War II approached, Hay was assigned to the Air Frame Building to work on Ford's amphibious Jeep. He left Ford in 1941 to work for Jam Handy making models used in training films and, in June 1942, he enlisted in the Aviation Engineers and spent the war building airfields across North Africa and Italy. Hay returned to Ford in late 1945 and, except for a year with an independent engineering firm, he worked at the Design Department until retirement in March 1974. At retirement, Hay was head of Ford's metal and paint shops.

Hay built or supervised the building of the metal and plastic parts used on almost all Ford concept cars, including the grilles, bumpers, mirrors, moldings, windshield frames and taillights. He made the metal ornamentation and some of the metal body panels for the number 1 and 2 Continentals and the first model of the Continental hood ornament. He also built the special trim used on the 1938 Ford Landau, and later made the tooling used to make special upholstery items, including the headliner used in the X-100. ■

Tucker Madawick

At right is Continental No. 2 after it was sold to Bob Gregorie. Gregorie repainted the car ivory and accessorized it. The 1935 Ford phaeton at left belonged to Tucker Madawick, a staff designer at Ford from 1939-42.

1939 Special Speedster (Continental)

The original Continental, at first referred to simply as a Special Speedster, was designed in Ford Motor Co.'s three-year-old Design Department during the winter of 1938-39, to demonstrate to Edsel Ford the skills his 30-year-old head of design, E. T. "Bob" Gregorie and his design staff had acquired. The car was also built to provide a unique European-looking sporty car for the boss' upcoming vacation.

Soon after Lincoln Motor Co. was purchased by Ford in 1922, Edsel Ford set out to make Lincoln a styling leader. Several years later, he also took on responsibility for the design of Fords. The Model A was the first new Ford to be styled from the ground up. The design was done by Murray, one of Ford's outside suppliers. Four years later, an all-new 1932 Ford was designed, again from the ground up, but this time by Briggs Manufacturing, another supplier of car bodies to Ford. The new V-8 engine available in '32 Fords was Henry Ford's doing, but the pleasing looks of the Model A and the '32 Ford reflected Edsel Ford's excellent taste. By 1933, the exterior appearance of the Ford automobile was again changed for the third time in as many years, something previously unthinkable.

Before the Depression, Ford employed a modest number of body draftsmen to design specific body parts, mostly for Lincolns. If major design changes or completely new car designs were needed, they were done by designers employed by one of the independent body suppliers Ford used. This proved to be a less than satisfactory arrangement to Edsel Ford. The problem was further aggravated by the Depression, which drove many of the independents out of business at the very time frequent design changes were becoming commonplace. To make matters worse, Edsel Ford usually found that if he liked a particular part of one supplier's design proposal and another part from someone else's, he couldn't pick and choose. By relying on outside suppliers, he didn't have the flexibility he wanted. It's little wonder then that Edsel Ford began to lay the groundwork for his own in-house design department at Ford.

Henry Ford Museum and Greenfield Village

Henry Ford Museum and Greenfield Village

(above and right) According to Gregorie, the original one-tenth sized clay of the Continental built by Bud Adams, under his supervision, was never made into a plaster model, and it was eventually destroyed. These photographs, taken at the Design Department in 1952, are of a one-quarter sized plaster model made by Adams several months later.

Before 1931, what little in-house design work there was at Ford was done by Henry Crecelius, Lincoln's chief body engineer, a staff of no more than three other body draftsmen and a few assistants. In 1931, 23-year-old Bob Gregorie was hired as one of those body draftsmen and assigned to the Engineering Department. Gregorie was a high-school drop out with only modest prior experience, mostly designing boats. Within four years of his employment, Gregorie was appointed the first head of Ford's Design Department and, during his 15 year career at Ford, 11 as head of the Design Department, he was responsible for the design of some of the most beautiful cars ever produced by any automobile manufacturer — including the Continental.

In 1928, Henry Ford's favorite test pilot, Harry Brooks, was killed, and four years later, the 14-A replacement airplane for the Ford Trimotor proved fatally flawed. Henry Ford's response to these two setbacks was to discontinue production of all new Ford-built aircraft. From then on, the airport facilities were kept open, but with only a skeleton crew, to sell parts and provide service for existing Ford airplanes. Because they could fabricate any part needed, the remaining aircraft personnel were some of Ford's most skilled employees. Beginning in 1932, Edsel Ford made the aircraft plant and its experienced personnel available to Gregorie for the experimental work he was asked to do for him.

George E. "Bud" Adams

Bud Adams was born and raised in Detroit. He was the son of a Ford tool and diemaker. Adams entered the Henry Ford Trade School at age 13, after his father's death, and became a straight A. student. In 1937, he was one of the original five graduates of the Trade School hired as Design Department apprentices. (As an apprentice, Adams was paid 65 cents an hour.) While at the Design Department, Adams built the one-tenth sized clay model of the first Continental. He is also responsible for the design of the 1941 Ford front end.

John Najjar

In 1938, Adams received a full-tuition scholarship to the Lawrence Institute of Technology to study engineering. He attended school at night and graduated with honors in 1942 as a mechanical engineer. After two years active duty in the Navy and two years doing military design work at Briggs Manufacturing, Adams returned to Ford in 1946, where he became one of Ford's leading chassis and suspension engineers.

As a Ford engineer, Adams developed the chassis and suspension for several concept cars, including the X-100. He also helped develop Ford's power-brake, power-steering and air-suspension systems. For three years in the 1970s, Adams was assigned to Ford of Europe as chief transmission and chassis engineer. He retired in 1981. ■

Before it was shipped to Edsel Ford in Florida, Continental No. 1 was repainted Eagle Gray instead of the original ochre-yellow.

Henry Ford Museum and Greenfield Village

It might not have been as a reward for salvaging the Lincoln-Zephyr, but it did convince Edsel Ford that Gregorie was the right person to head Ford's own in-house design department. In March 1935, Edsel Ford called Gregorie from Florida, where he was vacationing, and offered him the job. Edsel Ford told Gregorie he had already talked to chief engineer Larry Sheldrick, and that a studio was being prepared for him in the EEE Building. Gregorie didn't hesitate, telling Edsel Ford he would be "overjoyed" to accept the job.

Jesse Haines

No. 2 Continental about 1967.

It took Gregorie some time to establish an organization, gather personnel and design his first production cars. Gregorie also had to build all of his own equipment, including the large blackboards and the styling bridges used for modeling cars. (A bridge is a device used to obtain accurate measurements from the exterior surface of a clay model so it can be duplicated for production.) When Gregorie started the Design Department, he had two staff designers, John Walters and Walter Kruke. By the time the Continental was under way, he had hired two more designers, Willys "Bill" Wagner and Bruno Kolt. (During the time the Continental was designed, designers were paid $250-300 a month, which were good wages at the time.) Body engineer/layout man Edward A. Martin had also been with Gregorie since before the Design Department was started. In 1937-38, Martin Regitko, another body engineer, and several more apprentice designers were added to Gregorie's small staff. Gregorie's shop foreman was James Lynch, who had previously worked at the Ford aircraft factory and his head clay modeler was Richard Beneicke. Gregorie also supplemented his small staff by hiring the most promising apprentices from the Trade School. The 1937 Ford and the 1938 Lincoln-Zephyr were the first cars entirely designed at Gregorie's new Design Department. By the time the Continental was designed three years later, the Design Department was also working on facelifts for the 1939 Ford and the 1939 Zephyr, the all-new Mercury and on designs for a new truck, a new tractor and a new bus.

Jesse Haines

No. 2 Continental after Jesse Haines had the bodywork completed.

Surprisingly, it was Gregorie and not Edsel Ford, who took the lead in proposing new models. Gregorie and Ed Martin, who functioned as Gregorie's assistant manager, prepared elaborate charts comparing the cars and prices offered by GM, Chrysler and Ford. Those charts emphasized the gap in the mid-priced market Gregorie thought Ford was ignoring. Typically, when Gregorie suggested a new product, Edsel Ford responded with a noncommittal "we'll see," and that would be it for awhile. Gregorie instinctively felt that Edsel Ford was also looking for acceptable proposals for new cars to make to his father and Charles Sorensen.

In 1937, at the conclusion of one of his trips to Europe, Edsel Ford shipped back to the Design Department a small European car he liked the looks of. Although it was not meant to have the effect it did, when the car arrived, it hurt Gregorie's feelings. The end result was that Gregorie started to think seriously again about building another special, European-style car for Edsel Ford.

Edsel Ford had always assumed that any new production vehicle would be built as a Ford or a Ford derivative. Although they were all Fords, the special cars Gregorie designed for Edsel Ford were made at the Lincoln plant, because Henry Ford, Charles Sorensen, Ford's chief of production, Pete Martin, production manager at the Rouge plant, where Fords were built, and Joe Galamb, Ford's chief body engineer, all objected to building anything new at the Rouge plant. They thought it would interrupt their routine, high-volume production schedule.

In the fall of 1938, Gregorie concluded that the odds of producing a new model would be better if it were a Lincoln rather than a Ford. By this time, the Mercury was almost finished, and it already shared much with the Ford. The Lincoln occupied much of Gregorie's time anyway, because he was working on a replacement for the Lincoln K. Gregorie also instinctively knew that Lincoln needed a sales boost. More importantly than is generally recognized, for 1938, the Zephyr platform had been re-engineered and strengthened so convertible Zephyrs could be produced for the first time. That turned out to be a primary reason a Lincoln-Zephyr-based Continental was considered.

Gregorie's rationale in proposing the new Lincoln-Zephyr convertible platform as a basis for a new production car had several other advantages: 1) There was room at the Lincoln plant where there were underutilized, skilled workmen; 2) the new Mercury would fill the lower medium-price range when it went into production; and 3) there was more money to be made from a new Lincoln than a Ford or Mercury.

According to Ed Martin, one afternoon in mid-October 1938, after he and Gregorie returned from lunch, Gregorie asked him to get out the 1/10-scale blueprints (package drawings) of the new Lincoln-Zephyr coupe convertible. Martin then sat down beside Gregorie at Gregorie's desk. Gregorie took a piece of clear vellum, put it over the blueprint and began to draw the outline of a lower, longer-looking, sporty convertible he thought could be built by sectioning and chopping the current Lincoln-Zephyr convertible. In a 1951 letter to Ford designer John Najjar, Martin recalled that he had previously seen other perspective sketches of such a car on Gregorie's desk, indicating to him that Gregorie had been working on the project for some time.

The same day Gregorie did the drawing, he showed it to Edsel Ford, who gave the go-ahead to build it as a special Zephyr convertible for his personal use. Gregorie then had Ed Martin make a full-sized, sideview line drawing of the car. When it was done, the drawing was glued to Masonite and cut out. The full-sized profile of the car was then shown to Edsel Ford, who was by this time becoming very enthusiastic about it.

Henry Ford Museum and Greenfield Village

Continental No. 1 after it was repainted, and before it was shipped to Edsel Ford, who was vacationing in Florida.

According to George E. "Bud" Adams, one of the Trade School boys assigned to work with head clay modeler Richard Beneicke, Gregorie then moved a wooden modeling table over next to his desk, which was in one corner of the studio, and partitioned off the area with several large blackboards. There, Adams spent the next week building a 1/10-sized clay model of the Continental. The model was about 20 inches long. Gregorie would occasionally direct Adams by suggesting he smooth out a line here, or make an edge less sharp there. While the 1/10-sized clay was under way, Edsel Ford came in often to look at its progress. Adam's recollection was that, at that time, Gregorie and Edsel Ford discussed plans to build three cars, one for Edsel Ford, one for his oldest son Henry Ford II and one for his middle son, Benson Ford. Although there are persistent rumors that a third 1939 Continental prototype was built, Gregorie confirms that only two were actually made, and that none were built for Edsel Ford's sons, who already had their own cars. The 1/10-sized clay model was completed towards the end of October 1938 and was painted Eagle Gray Metallic, Edsel Ford's color of preference.

After Edsel Ford saw the finished 1/10-sized clay, he asked Gregorie if it could be built as a finished car in time for his planned March vacation at Hobe Sound, Florida. After checking with Robbie Robinson, superintendent of the Lincoln plant, Gregorie told him it could be done, even though time was short. Gregorie, Ed Martin and body engineer

John Najjar

Placid "Bennie" Barbera

Bennie Barbera began his career at Ford in 1932 as a student at the Henry Ford Trade School. His first job was repairing goggles. He was later promoted to repairing valves and then became an operator of a lathe and other machinery. As with other Trade School boys, part of each day was spent in school and part learning a trade.

After graduating in 1936, Barbera was assigned to work in the Soy Bean Laboratory in Greenfield Village, where he had almost daily contact with Henry Ford. One year later, he returned to the Henry Ford Trade School and, in February 1938, he became one of five students chosen by Edsel Ford to apprentice in the Design Department.

Barbera was assigned to designer Bill Wagner as a detailer/draftsman. He assisted Wagner in determining dimensions for bumpers, taillights and other trim items that were then made into soft metal prototypes to be used on full-sized models being designed. In 1939, Wagner and Barbera began modeling bumpers, taillights and other trim from clay. As a result, Barbera became a clay modeler. He assisted Wagner in building Ford's first styling bridges used for measuring clay models. Barbera also assisted in clay modeling the first production Continental.

From 1941 until 1945, Barbera was an infantry soldier stationed in New Guinea. After the war, he returned to the Design Department and worked on Gregorie's 1949 Ford proposal.

In May 1955, George Walker appointed Barbera a modeling manager, a position he held until 1974. Barbera celebrated his 61st year as a full-time employee at the Design Center in February 1999. ■

Martin Regitko then discussed how to build the car using as many existing Lincoln-Zephyr body panels as possible. They also decided to build it on the Lincoln-Zephyr sedan-convertible platform. Regitko was assigned the task of making a full-sized paper draft from offset measurements obtained by placing the 1/10-sized clay model under a 1/10-scale styling bridge.

The No. 1 Continental was built from 1939 Lincoln-Zephyr sheetmetal.

Henry Ford Museum and Greenfield Village

In late October and early November, Regitko laid out the internal structure for the body, including the door pillars and the sections around the windows and developed all the exterior body surfaces. He also determined the design of a special A-pillar front windshield frame, which was the only part of the body structure that had to be built from scratch. Because there was some concern whether the convertible top would collapse into the space available for the top well, apprentice John Najjar was given the task of building a full-sized top buck. Najjar made the top bows from cardboard and confirmed that the top collapsed sufficiently into the well and could be covered by a top boot. Lincoln body engineer Henry Crecelius and Ford body engineer Joe Galamb were also asked to review the 1/10-sized clay model and Regitko's paper draft, and both confirmed that the car could be built on a Lincoln-Zephyr platform as proposed.

The paper draft and measurements prepared by Regitko, and the 1/10-clay model Adams built were sent to the Lincoln factory in mid-November 1938 and were used to build the first Continental. Construction began in mid-November and, by mid December, the body itself was under way. While the first Continental was being built, Gregorie discovered that, in his haste, he had overlooked the spare tire — there was no room for it in the trunk. After discussing the problem, Edsel Ford and Gregorie decided to indent the back panel of the car and attach an exposed spare tire. Regitko later made sketches of an alternate rear end that allowed for an enclosed spare tire, but that proposal was rejected by Edsel Ford, who thought that the exposed spare tire looked "more continental." Because they were in a hurry, Edsel Ford instructed Gregorie to put no excess trim on the car, because fabricating it would take too much time.

Henry Ford Museum and Greenfield Village

The No. 1 Continental had an old style exposed spare tire mounted on the back of the trunk.

The usual process for building a new-model car included making multiple renderings, one or more small clay models, a full-sized clay model, detailed engineering studies and drawings, and only then were one or more prototypes built for testing. The normal process for building and testing a car from scratch usually took three years and sometimes longer. As Gregorie soon found out, the shortcuts he took did have an adverse effect on the roadworthiness of the first Continental.

Henry Ford Museum and Greenfield Village

No. 2 Continental had a higher trunk, four inches less stretch in the front fender and hood area and much less lead in the body.

The first Continental was built on a 1939 Lincoln-Zephyr sedan-convertible chassis that carried serial number H74750. The finished car was 210 inches long, 62 inches high (top up) and weighed about 5000 pounds. The body was made lower by removing the 3-inch subfloor of the Lincoln-Zephyr. It was built with a reworked firewall box, four-inch sections were removed from the doors and the front fenders, and the hood was lengthened by 12 inches. A new rear section was hand-hammered over a hardwood buck, and the runningboards were completely removed. Top bows from a 1939 Lincoln-Zephyr coupe-convertible, with some modifications, were used. On the inside, a stock '39 Lincoln-Zephyr instrument panel was used, with the lower center section removed, and the clock was relocated to the glovebox door. A radio was installed, but the speaker grille and aerial were not. The ball and spear hood ornament appeared for the first time on this car. Otherwise, there was very little trim on the car -- even the chrome strip between the grilles was left off. The car was finished on February 23, 1939, and returned to the Design Department, where everyone took an immediate dislike to the color Gregorie had chosen -- an ochre-yellow.

During the drive from the Lincoln plant to the Design Department, which was only about ten miles, the lead used in the cowl area began to crack out. The car had been engineered and built in such a hurry, that it had several hundred pounds of lead in it. Gregorie immediately took the car out on the test track, which was about a block away, where it proved to be "real shaky." Gregorie suggested welding I-beam channels in an X fashion, just ahead of the cowl, to give it better support and hopefully stop the shaking. This remedy was only partially successful. After the X-member was welded in place on the engine side of the cowl, the paint was in need of repair, so Gregorie heeded the suggestions of his designers and had the car repainted Eagle Gray metallic and reupholstered in gray leather. On March 3, 1939, the first Continental was shipped to Edsel Ford in Florida, probably on a factory-owned truck.

Henry Ford Museum and Greenfield Village

The interior of No. 2 Continental used a modified 1939 Lincoln-Zephyr instrument panel, and a prototype column shift mechanism.

A few weeks later, Gregorie received a call from Edsel Ford in Florida, who reported that the car leaked badly around the mid-section and was a "bucket of lead," but it was drawing rave reviews from everyone who saw it. He told Gregorie he thought he could have sold "a thousand of them," and that he wanted Gregorie to contact Robbie Robinson to see if several hundred 1940 Continentals could be put into limited production. This was the first time Gregorie says that he and Edsel Ford talked about producing the car. To work out the mechanical details of production, Edsel Ford suggested that Gregorie start construction of a second car as a mechanical prototype to be used for testing, among other things, the column shift, which was introduced on Ford passenger cars for 1940. By the time he returned to Dearborn, Edsel Ford had decided to include the Continental in the regular Lincoln-Zephyr production lineup for 1940.

Henry Ford Museum and Greenfield Village

The front of No. 2 Continental is very similar to front of No. 1 Continental.

William Clay Ford, the 14-year-old youngest son of Edsel and Eleanor Ford, remembers driving the first Continental in Florida during his 1939 spring vacation. He drove it 15 to 20 times during the several weeks he was in Florida and remembers the car as very attractive but not very good mechanically — especially the brakes. The first Continental was not seen again at the Design Department after it was returned from Florida. (After Edsel Ford's vacation was over, the car was probably returned to the Lincoln factory and destroyed in late 1939.)

Soon after Edsel Ford's telephone call to Gregorie, John Najjar recalls that Lincoln engineers came to the Design Department to work out how they could best produce the Continental from existing Lincoln-Zephyr sheetmetal, and to determine what non-production parts they needed to build from scratch. Both at the Design Department and at the Lincoln factory, Ed Martin spent long hours with Robbie Robinson and Lincoln engineers Walter Street and C. Bartholomew planning for production of the Continental.

Much more care was taken in building the second Continental, which used nowhere near as much lead. The second Continental was started in mid-March 1939, and it was completed by June 5, 1939. At the time the second Continental was built, several design changes were made, but it was also built on a 1939 Lincoln-Zephyr sedan-convertible chassis, number H82410. The trunk on the second Continental was higher and more squared-off, the front fenders and hood were 4 inches shorter, and the car was equipped with a column shift. It was painted black and had tan leather and whipcord upholstery. From June through October 1939, it was used primarily for testing.

Henry Ford Museum and Greenfield Village

The rear end of No. 2 Continental featured a higher trunk area.

During the time the second Continental was being built at the Lincoln plant, a full-sized clay model, based on the design for the 1940 Lincoln-Zephyr, was under way at the Design Department, so that accurate measurements could be taken from it for production. The full-sized clay model was built using a production '40 Lincoln-Zephyr chassis and sheetmetal as a base. It had stock wheels and full suspension. The hood and fenders were sectioned and the gaps filled over with clay. The trunk and rear end were formed from wood slats covered with sculpted clay. Because the buck started as a production car, the inside could be worked on at the same time as the outside. Apprentice designer Bob Thomas worked on the instrument panel of the full-sized clay.

Other designers who worked on the design of the full-sized clay were Walter Kruke, who did the interior trim, John Walters, whose assignment was the instrument panel, and Bill Wagner, who was in charge of bumpers, lights and exterior hardware. John Najjar, under Wagner's direction, worked on the spare-tire deck and the rear-fender area. Martin Regitko did most of the body surfaces and the body engineering, and Ed Martin did the drafting and planning. The full-sized clay model was finished in mid-July 1939, a month after the completed second Continental had been returned to the Design Department.

E.T. Gregorie

Gregorie's home on Gross Ile. The No. 2 Continental, repainted ivory, is in the middle. On the left is Gregorie's 1940 Mercury four-door convertible and, on the right, is his 1939 Ford station wagon.

When the full-sized clay model was finished, offset measurements were taken from it, and the engineering drawings were sent to the Lincoln factory, where a make-shift production line was set up to manufacture the 1940 Continental.

Two pre-production 1940 Lincoln-Zephyr Continental cabriolet bodies were built in September 1939 and, on October 3, 1939, Edsel Ford drove the first production 1940 Lincoln-Zephyr Continental cabriolet off the assembly line. Gregorie later drove the second one off the assembly line. The 1940 Continental cabriolet introduced to the press at the Ford Rotunda in Dearborn on October 2, 1939, was supposedly a pre-production model. It was introduced to the public on October 6, 1939.

Until just before production, the car was called the Special Speedster. The name Continental was originally used to refer to all of the special, European-style cars built for Edsel Ford. That Continental became the name for the car is a fortuitous accident that probably happened because no one could come up with a more fitting name.

The second Continental still exists. In late September 1939, even before testing was completed, it was sold to Gregorie. Gregorie drove the car to the New York Auto Show, where the 1940 Ford product line, including the '40 Continental, was displayed at the Astor Hotel, beginning October 15, 1939. Gregorie's 1939 Continental was not displayed. Gregorie kept the number two Continental for about two years, during which time he repainted it an ivory color, added rear fender gravel shields, front mud flaps and a metal covering over the otherwise exposed spare tire.

The second Continental also turned out to be structurally weak in the cowl area. On a trip from Buffalo back to Dearborn, the car got the cowl shakes so bad, Gregorie stopped at a lumberyard along the way and bought several two-by-fours, which he wedged under the instrument panel to help strengthen the cowl. Later, metal X-braces were welded in to replace the two-by-fours. The shifting rods on the second Continental had a habit of locking up, disabling the car when it happened. It apparently happened often enough that Gregorie had a long metal rod made, so he could kneel down, reach under the car and hit the shifting rods on the side of the transmission until they unlocked. On one occasion, Gregorie took the car to a special affair at the Cadillac Hotel in Detroit, where a valet parked the car in the parking lot. Gregorie was called out of his meeting because it was stuck in the middle of the parking lot with the shifting rods locked up, blocking traffic. (Gregorie on his knees in dress clothes trying to unlock the shifting mechanism on his Continental must have been one of those Kodak moments.) Gregorie also tells of driving the car on a dark road in Pennsylvania at night and, realizing that he

John Najjar

Martin Regitko

Martin Regitko was born in Czechoslovakia. Before emigrating to the United States, he lived in Italy, where he helped design and build racecars.

From 1914 until 1938, Regitko was a supervisor at the Willoughby Company, a custom body builder located first in Rome and then Utica, New York. In 1938, when Willoughby closed its doors, Regitko took a job at the Ford Design Department as a body engineer, where he supervised surface layout and development work.

Regitko was proficient at preparing full-sized body drafts, and did the layout on many concept cars from the original Continental to the Futura. Before he retired in the mid-1960s, Regitko was manager of all Styling Center draftsmen. ■

Willys P. Wagner

Bill Wagner was an architect, who began styling cars for Murray Corp. of America, one of Ford's body suppliers. In May 1935, Wagner was hired by Ford to conduct a styling display at the Ford Pavilion at the San Diego International Exhibition. In November 1935, after the display closed, Wagner was assigned to Bob Gregorie's Design Department as a designer. Wagner designed Ford, Lincoln and, later, Mercury bumpers, lights, door handles and external hardware. The first production car Wagner worked on was the 1937 Ford, which was also the first production car designed entirely by Ford's new Design Department. Wagner convinced Gregorie to incorporate the headlights into the fenders of the '37 Ford.

During World War II, Wagner supervised a group of Ford draftsmen, who checked B-24 drawings after they were completed .

Wagner preferred to design trucks and not cars. In 1947, he was appointed head of the truck and international studio. Wagner was involved in the design of most Ford trucks from 1949 to 1965, several models of the Canadian Meteor and Monarch, the 1949 Ford tractor, the briefly produced Ford Metropolitan bus, the 1957 Ranchero pickup and the 1961 Econoline van. Wagner's design innovations include the wide pickup box and the hood that covered the tops of the fenders on pickups.

In his off-time, Wagner became one of the inventors of dry-roasted peanuts.

When he retired in 1963, at the age of 65, Wagner had the greatest seniority at Ford's Design Center. He was succeeded as chief stylist in the truck and international studio by John Najjar, the Ford designer with the second-longest longevity. ■

John Najjar

was on the wrong road, he tried to turn around, only to have the shifting mechanism lock on him while he was sideways in the middle of the road. It's little wonder that Gregorie kept the second Continental for only two years before selling it to a used car dealer in Detroit for $800.

The car showed up in Los Angeles, California in 1943, where it was purchased by an airplane manufacturer, who used it as a VIP car. (During the war, many convertibles, including Continental cabriolets, were bought on the east coast and driven to California, where they were sold for premium prices.) After the war, the car was sold to Jack Romas, a fireman who, in his off-time, was building a new home in Woodland Hills. Mr. Romas used the car as a pickup to haul building materials. After Romas completed his new house, he put the car up for sale. Ralph Ball's brother, who lived in Tarzana, California at the time, heard about the car and told his brother about it. In 1951, Ball purchased the second Continental from Mr. Romas for $125. Ball drove the car in its rather deteriorated condition until December 1956, when a motorist hit the car on the driver's side putting Ball in the hospital and disabling the car. Beverly Hills attorney, Bob Gottlieb, at that time a columnist on classic cars for *Motor Trend* magazine, persuaded the insurance company that the car had historical value. Ball was paid repair costs, rather than the $50 salvage value initially offered by the insurance company. Rather than repair the car, Ball placed it in storage, where it remained until April 1963, when it was purchased by Jesse Haines.

Haines dismantled the car and later had extensive metal work done to the body at a shop in Haileysville, Pennsylvania, but he did not complete the car's restoration. In 1969, Haines sold the second Continental to Bob Anderson of Houtzdale, Pennsylvania. Anderson currently has the car in storage and plans to restore it. ■

Henry Ford Museum and Greenfield Village

Henry Ford Museum and Greenfield Village

(left and above) A full-sized clay model of the proposed 1940 Lincoln-Zephyr Continental.

Walter Kruke, John H. Walters and E. B. Kolt

Walter Kruke, John H. Walters and Bruno Kolt were the first designers hired by Edsel Ford to work with Bob Gregorie when Ford's new Design Department was formed in March 1935. From 1935 to 1947, the design process required that each designer be responsible for the design of the same parts on all Ford Motor Co. cars and trucks.

Walter Kruke came from a long-time Dearborn family and was a graduate of the Yale School of Fine Arts. Before he became a designer, he was a band leader in competition with Rudy Vallee. After his band broke up, Kruke took a job on one of Ford's Great Lakes freighters. That ultimately led to a job in the Ford aircraft factory, where he designed interiors for the Ford Trimotor airplane. When the Design Department was formed in 1935, Kruke was recruited by Edsel Ford. Kruke designed interior trim, door panels and chose the colors and upholstery for Ford products. He was still a designer at Ford when he died from undulant fever in the late 1940s or early 1950s.

John Walters was a long-time industrial designer. In 1935, Edsel Ford invited him to work in Ford's new Design Department. During his tenure at the Design Department, Walters designed steering wheels and instrument panels. Walters felt it necessary to circulate in order to become an experienced industrial designer. He left Ford in 1941 to start his own industrial design business in Ann Arbor and later in Detroit. As an independent industrial designer, Walters did work for Whirlpool and other appliance manufacturers. He also employed several Ford designers, who moonlighted designing appliances for him. Later in his career and, even after he retired, Walters worked for Bill Schmidt and Associates.

Bruno Kolt was trained as an architect in Germany. After World War I, he came to the United States. His first job was in Hollywood, California, where he worked as a designer of movie sets. He didn't like the movie industry and left to work in Detroit designing cars. His primary assignment at Ford's Design Department was sheetmetal surface development and grilles. Many young designers and apprentices at Ford considered Kolt a mentor and father figure. Kolt's skills as a designer were generally well respected by other Ford designers. When George Snyder became general manager of the Ford Design Department in August 1948, Kolt and Ed Martin were let go. Kolt took a job with Budd Manufacturing but died soon thereafter. ■

Henry Ford Museum and Greenfield Village

Bruno Kolt at the Design Department sometime in 1939.

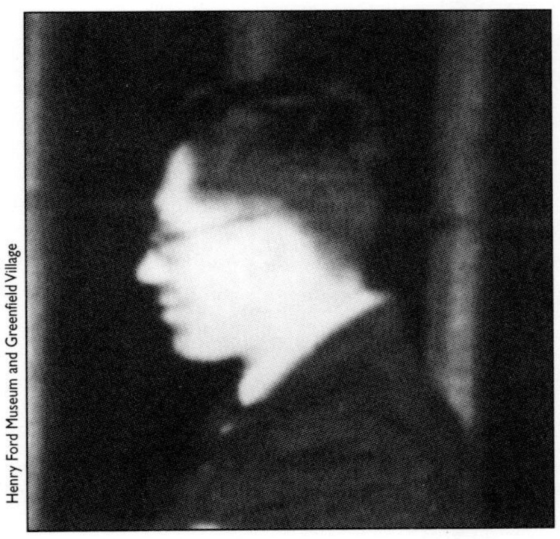

Henry Ford Museum and Greenfield Village

John Walters

Henry Ford Museum and Greenfield Village

Walter Kruke

Henry Ford Museum and Greenfield Village

An informal design review of an early production Mercury. Bruno Kolt is second from the left with his hands on his hips, Frank Francis is in front of the pillar, John Walters is second from the right, and Walter Kruke is on the right.

Lincoln 05L

Once the economy started to recover from the Depression, automobile manufacturers and the custom coachbuilders who had been able to hold on, expected the market for custom-built luxury cars to return. In the meantime, Cadillac, Packard, and Lincoln, the strongest of the companies that produced luxury cars, hedged their bets by bringing out mid-sized models. The Zephyr, introduced in November 1935, saved Lincoln, but Edsel Ford still hoped the big Lincoln would make a comeback.

Believing the market for big, prestigious cars would rebound, Cadillac introduced a new V-16 model in 1938, and Packard followed suit with a new V-12. After seeing what Cadillac and Packard were doing, Edsel Ford felt it absolutely necessary to replace the Lincoln K to compete with Cadillac and Packard.

In 1935, the Lincoln K got new sheetmetal, but it was still riding on the same basic chassis and using an engine and running gear based on the same car purchased from Henry Leland in 1922. By 1938, Edsel Ford decided it was time to start designing a replacement for the Lincoln K. Not only would the body be totally redesigned, he wanted the new Lincoln to have an all-new V-12, a new chassis and new suspension. As far as Edsel Ford was concerned, this was to be the grandest Lincoln ever. He also decided that the design of the new Lincoln, designated the 05L, would be a total departure from the current Ford, Mercury and the Lincoln-Zephyr automobiles, all of which shared a strong family resemblance.

Design work on the 05L started in September 1938. It continued over a two-year period but was sandwiched in between other projects. During most of the time the 05L was being designed, a 1938 Cadillac 60 Special was available in the studio as a benchmark against which to measure progress on the 05L's design. Although the 05L was constantly compared with the design of the Cadillac 60 Special, and it was meant to compete with the V-16 Cadillac, neither Cadillac was the primary design influence for the 05L. Tucker Madawick, one of the designers who worked on the 05L, while acknowledging that the 05L "looked like a German command car," maintains that Edsel Ford and Gregorie wanted to leap-frog Cadillac by patterning the 05L after the big Mercedes. The Continental was apparently not the only car influenced by the European cars Edsel Ford admired so much.

There were several problems never overcome during the two years the 05L was being designed. The first problem was that Gregorie was inexperienced in designing large cars. The full-sized clay model of the 05L had a 142-inch wheelbase. Most previous Ford cars that big had been designed and built by the many coachbuilders who specialized in cars of that size. This was the largest car Gregorie or any of the other designers at Ford had ever tried to design, and they found it tough sledding.

The 05L was planned as a replacement for the venerable but outdated Lincoln K. The rear of the full-sized clay had strong Buick and Cadillac overtones.

Henry Ford Museum and Greenfield Village

Henry Ford Museum and Greenfield Village

Martin Regitko and Ed Martin, Gregorie's most experienced design engineers, were assigned to work on the 05L. Bruno Kolt designed the body surfaces, and Gregorie worked closely with him. Dick Beneicke, Ford's modeling manager, supervised the clay modeling of the car. Madawick remembers that he and designer Duncan McRae spent days under Gregorie and Beneicke's supervision modeling the long sides of the car. No matter how hard or long they worked on it, Gregorie was never happy with the look of the long expanse between the A and C-pillars.

Except for the grille, front end and the sides, the design of the 05L was complete. Very little development work was done on an engine or suspension for the 05L.

Because Gregorie was never satisfied with the design of the 05L, it was all too easy to put it aside to work on something else — and there were plenty of other projects under way. Madawick recalls that each time Edsel Ford returned from his frequent trips, Gregorie could see his disappointment at the lack of progress on the 05L, and they would be pulled off other projects to start working on the 05L again, only to go through the same frustrating cycle several weeks later.

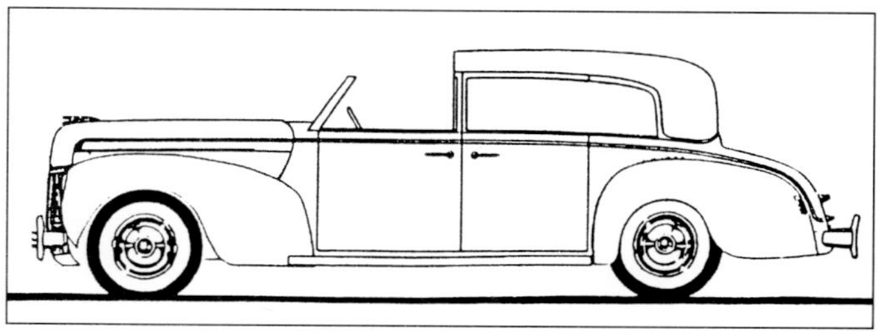

This is one of many proposals for a custom-built body based on the Lincoln 05L.

As war with Germany approached, both Edsel Ford and Gregorie became concerned about the strong teutonic, military look of the 05L, and a conscious effort was made to tone it down. As can be seen from the design changes made in the 05L's front end and grille, alterations were made, but a final design was never settled before work was stopped on the car in late 1940. The end came due to the impending war and the mountain of war work Ford was being asked to do.

When the 05L was started, it was tentatively scheduled for production in the 1940 model year. The "0" was for 1940, "5" represented the fifth production engine to be offered by Ford Motor Co. and "L" stood for Lincoln. The design work done on the 05L was the cheapest part of its development, but little was apparently done on the more costly chassis, suspension or the new V-12 engine intended for the car.

Edsel Ford wanted the V-12 in the 05L to be more technologically advanced than Cadillac's V-16. At the time the Lincoln 05L was being designed, Ford Engineering was in fact working on a new V-12, but it wasn't for the Lincoln 05L. The flathead, V-8-based V-12 developed for the 1936 Lincoln-Zephyr was engineered by veteran Ford engineer Frank Johnson, whose duties were cut back shortly after the Lincoln-Zephyr went into production. Cornelius Van Ranst, who had earlier in his career developed the Cord L-29's front-wheel-drive system and had more recently engineered Packard's

Changes to the grille and front end of the 05L kept designers and clay modelers busy for months.

Henry Ford Museum and Greenfield Village

Tucker Madawick

Tucker P. Madawick

Tucker Madawick was born and raised in New York City and attended Pratt Institute. He majored in automotive design. In 1938, after spending three years at Pratt, the Ford Design Department recruited him while he was visiting the New York World's Fair. He started that year as a staff designer and spent most of his time rendering and clay modeling. Madawick arrived at Ford after the Continental was completed but worked on the 1941 and 1942 Fords, Mercurys and Lincolns, the proposed Lincoln 05L and the clay model of the Brunn Lincoln-Zephyr Towncar.

At the start of World War II, Madawick was assigned to the Willow Run B-24 inspection department. Later, he worked as a production coordinator on the B-24, B-32 and B-36 bombers for Consolidated Vultee (Convair). After the war, Madawick worked for Convair as production coordinator on the B-36. In late 1946, when he left to resume his automotive design career. Bob Gregorie was in the process of leaving Ford by that time and, because it was apparent Ford's Design Department was going through many changes, Madawick took a job with industrial designers Lippencott and Margolius and was assigned to the team that designed and clay modeled an alternate Tucker proposal.

In 1947, Madawick was hired by Raymond Loewy to manage his London office, where he did design work on the Austin and Rootes Group automobiles. In 1950, Madawick returned to the United States and worked on many of Loewy's accounts, including Studebaker.

In 1959, he joined RCA as vice president of design for their Consumer Electronics Division. He retired in 1981.

Madawick served as president of both major professional industrial design associations, the Industrial Designers Institute and the Industrial Designers Society of America. ■

V-12, was recruited by Ford. Van Ranst was assigned to work on a new 1650 CID, water-cooled V-12 aircraft engine that, when built, produced over 900 horsepower. The new V-12 Van Ranst developed was almost four times as large as the V-12 that powered the Lincoln K, but it weighed only 500 pounds more than that engine. Although the V-12 aircraft engine developed by Van Ranst was never produced, it was later modified to become an 1100 CID water-cooled V-8 tank engine, which was used during World War II to power Sherman tanks.

Van Ranst's V-12 aircraft engine and the derivative V-8 tank engine were of modular design, had dual overhead cams, four valves per cylinder and made extensive use of aluminum. According to Ford production chief Charles Sorensen, the tank engine required re-engineering only 90 of the 550 tools used to produce the V-12 aircraft engine. It's not too much of a stretch to imagine that except for World War II and the lack of a market, the V-12 aircraft engine Van Ranst developed could have become the basis for a substantially smaller, mechanically superior V-12 engine used to power the proposed Lincoln 05L.

Edsel Ford also wanted a modern suspension system for the Lincoln 05L, but his father dug his heels in at any suggestion that the suspension on any Ford product be different than the buggy springs he favored. Nevertheless, Edsel Ford quietly authorized development work on a modern torsion-bar suspension system for the proposed Lincoln 05L. That work was also stopped by the war.

Edsel Ford originally planned to offer the Lincoln 05L as a seven-passenger sedan, a limousine and a convertible sedan on the 142-inch wheelbase model, and a five-passenger sedan, a smaller towncar, a coupe and a sporty sedan on a 136-inch wheelbase model. Depending on the wheelbase, the Lincoln 05L was designed to be 222 or 228 inches in overall length and about 70 inches high.

In September 1939, John Najjar made several 1/10-sized engineering drawings of the Lincoln 05L. They were detailed except for the lack of mechanical components, which were still unknown. The next month, designer Ross Cousins made five rendering of proposed models of the Lincoln 05L. After that, planning for the Lincoln 05L came to a stop. As time passed, the full-sized clay model of the Lincoln 05L cracked, and it was eventually destroyed. World War II and Edsel Ford's death, and not the lack of an engine or suspension for the car, ended plans to produce the Lincoln 05L.

The only feature of the proposed Lincoln 05L used on a later Ford product was the first grille design, which became the grille used on the 1942-46 Ford truck line.

In 1937, only 990 Lincoln Model Ks were produced. By 1939, the last year of production, only 224 models were sold, and some of those were sold as 1940 models. Even if an acceptable design, modern suspension and a new V-12 had been available, the Lincoln 05L was probably doomed because of changed market conditions. Although World War II finally ended the Depression, it also changed American society so much that the market for the 05L disappeared. Most of those who could afford the Lincoln 05L no longer felt comfortable in that type of car. In addition, the more moderately-priced cars were becoming as reliable, comfortable, and as powerful as top of the line, luxury models had traditionally been. ■

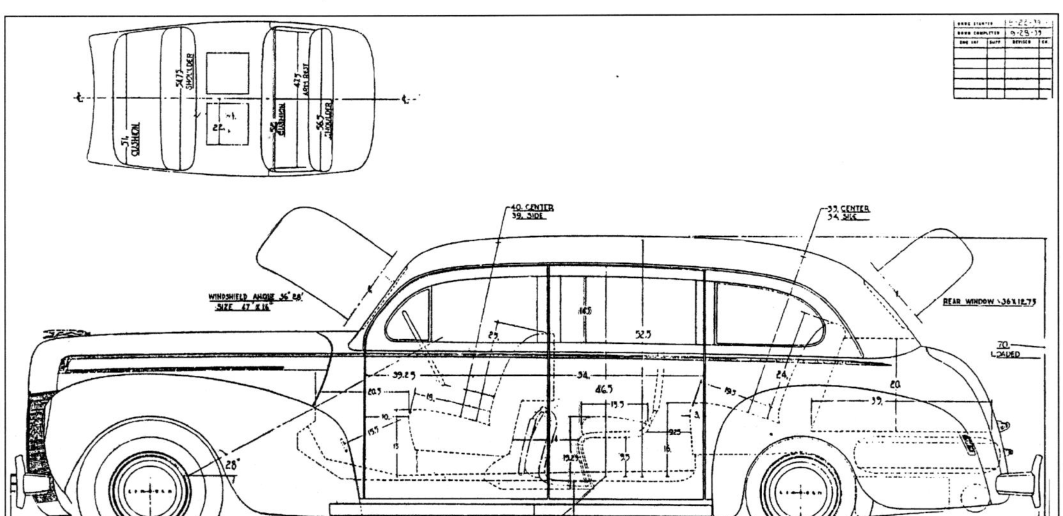
John Najjar

This one-tenth sized engineering drawing of the 05L was prepared by John Najjar in September 1939. The drawing included no mechanical components.

Henry Ford Museum and Greenfield Village

(above) The Cadillac 60 Special in the background influenced the design of many automobiles, including the 05L.

Henry Ford Museum and Greenfield Village

(above) The grille design on the Lincoln 05L was redone at least three times before the project was cancelled.

Henry Ford Museum and Greenfield Village

The final grille/front-end design for the Lincoln 05L featured more rounded shapes, Sealed Beam headlights and a grille similar to the one eventually used on the 1940 Ford Deluxe.

Richard Quinn

Duncan McRae

Duncan McRae was born and raised in Lansing, Michigan. He was the son of a Michigan state senator. He loved cars and was a naturally talented artist. McRae spent two years studying engineering at the University of Southern California and graduated from Michigan State University in 1940. He was hired at Ford's Design Department in 1940 and worked as a modeler and designer on production cars as well as the 05L, busses and tractors. In 1942, McRae left Ford to work as an illustrator and layout man at Chrysler's design department. From 1943-45, he was a B-29 flight engineer in the Army Air Corps. After the war, McRae returned to Chrysler as a designer in the Plymouth studio but left in 1946 to work as a designer at Kaiser-Frazer. While at Kaiser-Frazer, he assisted Dutch Darrin with the design of the 1951 Kaiser.

In 1949, McRae was rehired by Ford and spent the next three years as a designer in the Ford and Mercury studios. In 1952, he left Ford to work for Reynolds Metals as sales manager in their automotive department. In 1953, he returned to Kaiser-Frazer and, when they got out of the automobile business in this country, McRae became a consultant to Studebaker-Packard with Bill Schmidt. (Schmidt designed Packards and McRae designed Studebakers.) McRae stayed at Studebaker-Packard until 1959 and designed the Lark and the 1958 Packard Hawk. In 1960-61, McRae worked for Curtiss-Wright designing and developing the Air-Car, a hovercraft-type vehicle.

For three years, beginning in 1961, McRae worked as an independent industrial designer. In 1964, he returned to the Ford Design Center as a designer in the international studio. Later that year, he was sent to Ford of England as executive designer. While in England, he redesigned the Cortina, Escort, Anglia and Zephyr. In 1967-68, McRae headed the Ford design studio in Germany and was responsible for redesigning the Taunus. McRae returned to Dearborn for a year and was then assigned to Ford of Australia as executive designer. In 1973, he returned to Dearborn and retired in 1974. ■

1940 Lincoln-Zephyr Towncar

Edsel Ford had a strong attachment to the elegant limousines and towncars that had traditionally been custom built in limited numbers by the best of the contemporary coachbuilders. He also felt an obligation to the coachhouses he had worked with since Ford Motor Co. acquired Lincoln in 1922. He was disheartened when the effects of the Depression began to drive many of them out of business. After 1939, the Lincoln Ks, many of which had traditionally been built with custom bodies, were discontinued and, although a factory-built Lincoln-Zephyr Custom model was available in 1941-42, the coachbuilt era was also pretty much finished at Ford Motor Co. and everywhere else.

Illustrator Ross Cousins was hired as a Ford designer in 1938 to prepare expert renderings of design proposals for presentation to Edsel Ford. One of the renderings Cousins prepared was of a towncar based on a 1940 Lincoln-Zephyr. Edsel Ford saw the rendering, liked it and soon contracted with custom coachbuilder Hermann A. Brunn to build towncars and limousines based on the 1940-42 Lincoln-Zephyr. Ten of the 14 towncars and limousines were built as 1940 models, three as 1941s, and one was built as a 1942. Some reports indicate that as few as 12 cars were actually built.

Because there was concern about mating what was basically an old-fashioned carriage body to the streamlined Lincoln-Zephyr, Bob Gregorie personally supervised the design work on the Brunn-made Lincoln-Zephyr towncars. After Cousins' rendering, a 1/10-sized clay model of the proposed car was built by Tucker Madawick under Gregorie's supervision. It was approved by Edsel Ford and then sent to the Brunn factory with the design drawings.

(below and bottom) Edsel Ford's Lincoln-Zephyr Towncar has been restored to the way it was when it was delivered to him.

Ed Spagnolo

To turn the Lincoln-Zephyrs into towncars and limousines, the Brunn shop started with a Lincoln-Zephyr four-door sedan but discarded the body from the cowl back. The body was replaced with a custom version made from a hardwood framework and skinned with sheet aluminum pounded out over a wooden buck. All four doors were custom fabricated from hardwood and aluminum. A bronze casting was made for the windshield frame, which required that the otherwise stock Lincoln-Zephyr instrument panel be shortened approximately one-inch at each end.

The top of the passenger compartment was made from fabric over a chicken wire and wood framework, just like 1936 and prior year closed Fords. The hood, front sheetmetal and all four fenders were stock and apparently required little or no modification. The bustle, or trunk area, where the exposed spare tire sits, became an

Ed Spagnolo

Henry Ford Museum and Greenfield Village

This rendering by designer Ross Cousins encouraged Edsel Ford to build 14 Lincoln-Zephyr towncars and limousines.

aluminum casting. The running gear and the 125-inch chassis remained stock. Probably because the cars were primarily for in-town use, they did not come with the Columbia rear-axle option.

Edsel Ford owned both a 1940 and later a 1941 Lincoln-Zephyr Brunn Towncar, and his wife Eleanor had a 1940 Lincoln-Zephyr Brunn Limousine for her personal use. (A limousine has a closed chauffeur's compartment; the chauffeur's compartment in a towncar can be opened.) Mrs. Ford is believed to have kept her 1940 Lincoln-Zephyr Limousine until she replaced it with a 1952 Lincoln towncar. Several of the Brunn-built Lincoln-Zephyr Towncars and Limousines were updated before or after sale to look like subsequent model-year Lincolns. Mrs. Henry Ford's Brunn Towncar was retrimmed to look like a 1946-48 Lincoln.

John and Marjorae Murphy of Miami, Florida, are the current owners of Edsel Ford's 1940 Lincoln-Zephyr Brunn Towncar. As it was when Edsel Ford took delivery, the car is painted black and silver. The chauffeur's compartment is upholstered in maroon leather, and the passenger compartment is done in a gold-tan broadcloth with dark-brown rugs. Heat is supplied to the passenger compartment from a hot water heater via a heat register. A buzzer and separate intercom are used to communicate with the chauffeur.

Fourteen towncars and limousines were not enough to forestall the inevitable. Hermann A. Brunn died in 1941, and his son, Hermann C. Brunn, chose to close down and liquidate the business, or what was left of it. In 1944, after two years working as a tool designer, Hermann C. Brunn, like so many of the other skilled employees from the closed-down coachbuilders, took a job in the Ford Design Department. ■

Hermann C. Brunn

Gale Halderman

Founded in 1908, Brunn and Company of Buffalo, New York, was one of the premiere coachbuilders of the classic era. Brunn was probably best known for the Lincoln towncars and limousines built at Edsel Ford's direction during the 1920s and '30s. The Brunn name was more closely associated with Lincoln than any other vehicle, and it is the only coachbuilder's name to appear on Lincolns produced every year until 1941, when Brunn closed its doors, a victim of changing times.

Hermann C. Brunn, son of the founder, joined his father's company after graduating from high school in 1927. He apprenticed in all aspects of Brunn's business. Later, his father sent him to France, where he studied auto design at Kellner and Co. From 1934 until the senior Brunn died in 1941, father and son were partners in Brunn and Company.

After his father's death, the younger Brunn liquidated the business and, from 1942-44, he worked as a tool designer for Houde Engineering in Buffalo, New York.

In 1944, the younger Brunn took a job at the Ford Design Department as a body layout draftsman. In 1945, he became an interior and trim designer. Until his retirement in 1970, Brunn selected materials and was responsible for many of the interior designs used on production Lincolns.

Before he joined Ford, Brunn supervised the production of the Brunn-built 1940 Lincoln-Zephyr Towncars and Limousines. At Ford, he was one of the designers of the interior appointments on the presidential Lincoln limousines, the X-100, Mrs. Ford's 1952 X-400 Lincoln towncar and on the 1969 Lincoln Continental Mark III. ■

Tucker Madawick

This one-tenth sized clay model of the 1940 Brunn Lincoln-Zephyr Towncar was made by designer Tucker Madawick.

Light Ford

As the so called low-priced cars increased in size, they also gradually increased in price. That left an even lower-priced market waiting for some car maker to exploit. Henry Ford recognized that opportunity and, in the late 1930s, began planning for a new smaller Ford. Thus, at 75 years of age, Henry Ford became the champion of the small Ford at the same time his son was championing new mid-sized models.

The 1941 Plastic Car, developed at Henry Ford's direction, was preceded by other small-car studies he ordered in the late 1930s. Although a plastic car was not considered practical by everyone at Ford Motor Co., Henry Ford was still the boss, and if he thought a small Ford made of plastic represented the direction his company should go, that's the way it went. Unfortunately, World War II intervened, and most new-car development was put on the back burner.

After Edsel Ford's death in May 1943, Henry Ford, then 81 and in failing health, reclaimed the presidency of Ford Motor Co., while the head of Ford security, an ex-boxer named Harry Bennett, assumed overall control. Bennett, sometimes acting on his own and sometimes for Henry Ford, forced out or fired many of the men who made Ford Motor Co. what it was. Among those forced out was Larry Sheldrick, Ford's chief engineer, who left in mid 1943. One of the unexpected consequences of Sheldrick's departure was that long-time Ford body engineer Joe Galamb was placed in charge of Bob Gregorie and his Design Department. Gregorie and Galamb didn't get along and, in September 1943, Gregorie left to open his own independent design business.

Ford Industrial Archives

Left to right, John Wharam, chief automobile engineer, Dale Roader, chief truck engineer, Hudson McCarroll, executive engineer, Val Tallberg, secretary and co-ordinating engineer, Clyde Paton, consulting engineer and Bob Gregorie head of Design Department with a one-tenth sized model of Light Ford.

Galamb immediately appointed Thomas L. Hibbard to replace Gregorie. Hibbard was working at the Rouge plant when Galamb reassigned him to the Design Department. After Henry Ford II was elected a company vice president in April 1944, and it looked as though he was going to soon force Bennett out, Gregorie was persuaded to return as head of the Design Department, with Tom Hibbard as his assistant.

Amid all this confusion, in 1945, a well respected 55-year-old chemical and metallurgical engineer and long-time Ford employee named R. Hudson McCarroll was appointed Ford's temporary chief engineer by Henry Ford II. McCarroll had been an employee and a trusted lieutenant of Henry Ford since 1915. There was no question of McCarroll's loyalty, and he was well liked. Henry Ford II thought him a good choice for Ford's chief engineer, even though there weren't many left to choose from. (When McCarroll was a child, his father had been Henry and Clara Ford's minister. Edsel Ford and McCarroll attended the same school, although McCarroll was a few years older.)

McCarroll's pet project was the Light Ford, which was started before the war and had never ended as far as Engineering was concerned. In 1942, plans were quietly dusted off and work on the Light Ford got going again. Hibbard actually began design work on a small postwar Ford sometime in 1943 and built several clay models of it in fastback form. By mid-1944, a fastback prototype on a 98-inch wheelbase had been built. There was nothing revolutionary about the design, because it had already been decided that the Light Ford should follow the looks of the rest of the postwar Ford lineup. Since the 1949 Mercury had originally been designed as the new postwar full-sized Ford, the Light Ford was very similar in its exterior design, except that the grille was, as Gregorie put it, "improvised."

In late 1944, Bill Schmidt was transferred with 16 other people who had design skills from the Willow Run bomber plant directly to Gregorie's Design Department. War work was winding down, and preparations were getting underway to eventually resume automobile production. Schmidt had never designed cars before, but he was placed in a makeshift advanced studio in the tower of the Terminal Building at the Ford Airport, where he started designing future Fords. Schmidt's workspace was in the cupola at the top of the tower, reached by climbing a spiral staircase. Under Gregorie's supervision, Schmidt's group reworked prewar designs, and then made one-tenth sized clay models of 13 proposed new-car models, including the Light Ford. After they were approved by Gregorie, some of the models were made into full-sized clays. The first complete car that Schmidt designed was a revised Light Ford, which he also helped make into a full-sized clay. During the time Schmidt was in Gregorie's advanced-design area, Henry II, Benson and Bill Ford each spent two weeks working in the Design Department with Schmidt, who showed them how the place operated, and what designers did.

In September 1944, *Ford Field* magazine, in a report confirmed by Henry Ford II, said that, as soon the war was over, Ford intended to produce a modern version of the Model A, which would sell for around $500. Although the report did not say so, the Light Ford was not to have a plastic body or a monocoque construction as Henry Ford had originally wanted, but the price was close to what he had proposed before the war.

Henry Ford Museum and Greenfield Village

An advertising illustration prepared for the Light Ford program by Ross Cousins and Bud Kaufman.

After the war, Clyde R. Paton, a former Packard engineer, was hired to take charge of turning the prototype into a production vehicle. After considering several new powerplants, including four and five-cylinder, water and air-cooled, rear and front-mounted aluminum and iron engines and everything in between, he ultimately decided to use the small V-8 "60" engine that had been discontinued in Ford cars in 1941. The suspension on the Light Ford was also to follow the general design of the rest of the planned postwar Ford product line with independent front suspension, Hotchkiss driveline, and longitudinal rear leaf springs.

In the meantime, General Motors was working on its own small Chevrolet on the assumption that, after the war, there would be an economic downturn, and thus a big market for smaller, less expensive cars. In May 1945, General Motors announced it would build a small car after the war was over -- and the race was on.

Before the resumption of general automobile manufacturing in late 1945, it was assumed that new cars with new body styling would be an absolute necessity. Nobody figured that the car-starved public would buy anything manufacturers produced. Prewar models were dusted off and, with little change, they sold as fast as they could be manufactured. During this time, Ford learned that the small Chevrolet was planned with a wheelbase of 108 inches, so the Light Ford's wheelbase was increased, first to 100 inches and finally up to 106 inches. Each of these increases in wheelbase resulted in design revisions, which were done at Gregorie's Design Department.

In 1946, this photo of an operable prototype of the Light Ford was taken in Greenfield Village.

Detroit Public Library, National Automotive Collection.

In April 1946, a new Light Car Division was formed to build the Light Ford, but no firm date for production was decided on. In the meantime, two things happened that sealed the fate of the Light Ford in the United States. J.R. "Jack" Davis, a confidant of Henry Ford II and Ford Motor Co.'s sales manager, concluded there was no need for a Light Ford, if at all, until at least 1949. He figured out he could sell every car Ford was producing, and it made a lot of sense when he recommended against

Richard Beneicke

Dick Beneicke was trained in Germany as a sculptor and decorative plasterer. He came to the United States in 1923 and worked as an ornamental plasterer in the Detroit area. Because the Depression dried up construction jobs, in about 1933, Beneicke took a job as a clay and plaster modeler at Briggs Manufacturing, one of Ford's largest body suppliers. At Briggs, Beneicke worked with and came to know both Henry and Edsel Ford.

When Ford Motor Co. established its own Design Department in 1935 and with Briggs' agreement, Edsel Ford hired Beneicke as Ford's first clay modeling manager. While at Briggs, Beneicke was the lead modeler on the Briggs prototypes and, at Ford, he was the lead modeler on many of the special cars, including the first Continental. Beneicke was one of the few modelers with whom Edsel Ford regularly discussed the design of models being worked on at the Ford Design Department.

Beneicke continued as modeling manager at Ford during World War II and worked full-time as a modeler after the war. In 1946, Charlie Stobar wcs recruited from GM to take over Beneicke's workload as modeling manager. Beneicke suffered a stroke and died in 1952. He worked at the Design Department until the day before he died. ■

John Najjar

spending the money to build a new car that was untried in the market. In the summer of 1946, the Engineering Planning Committee, Hudson McCarroll's group, which had originally been set up to study the proposed Light Ford, concluded that the Light Ford would cost only about 17% less to build than the full-sized Ford. At that point, with no production or manufacturing experience, McCarroll didn't know what to do, so he tried to please everyone and wound up pleasing no one. Even though the Light Ford could not be changed to make it less expensive to build, because it had approximately the same number of parts, McCarroll was not inclined to drop the car he had championed for so long, and there was still a lot of support in the Engineering Department for the Light Ford.

In July 1946, the Light Ford was still very much alive, when Ernest R. Breech was hired as executive vice president and as a mentor to teach Henry Ford II all about the automobile business. Breech was also hired to give the Company a new direction, and he soon determined that McCarroll was not suited for the job of Ford's chief engineer.

When Breech started at Ford, the Light Ford, with its 106-inch wheelbase and the regular Ford, with a 114-inch wheelbase, were both being tooled for production. Soon after that, Breech hired Harold T. Youngren from GM and, by August 1946, McCarroll had been replaced by Youngren as Ford's chief engineer. (McCarroll returned to chemical and metallurgical engineering and died of a brain hemorrhage in March 1948.) Youngren told Breech that, in his opinion, both the Light Ford and the 114-inch wheelbase Ford were too costly to build for what they could be sold for. Youngren then took Breech to look at the full-sized clays models of both cars and, like Youngren, Breech came away convinced that neither would make money.

Breech then suggested a design contest between Gregorie and George Walker, an outside design consultant and his friend, to come up with a new design for a smaller 112-inch wheelbase 1949 Ford. (Originally Ford's new models were scheduled for production for the 1948 model year.) Breech wanted to get rid of Gregorie and, not too surprisingly, Walker's proposal won. Gregorie, who by then had decided he didn't like Breech or the GM-style committee system Breech brought with him, resigned. The end result was that Walker's proposal for the 1949 Ford went into production and was marketed starting in June 1948.

In September 1946, the Purchasing Department was ordered to stop buying parts for the Light Ford. That effectively killed it.

An early engineering prototype of the Light Ford.

Henry Ford Museum and Greenfield Village

Ford Industrial Archives

Harold T. Youngren, executive engineer, George Walker, styling consultant and Thomas L. Hibbard, head of the Design Department are shown with a model of the Light Ford. This photograph was taken December 18, 1946, three days after Bob Gregorie resigned.

Hudson McCarroll with a one-tenth sized model of the Light Ford.

Ford Industrial Archives

Thomas L. Hibbard

Tom Hibbard began his design career at 16 in Cleveland, Ohio, working for Leon Rubay designing special bodies for Packard and other luxury cars. He then worked for C.P. Kimball in Chicago. After serving in the Army in France during World War I, he returned to New York and took a job designing cars at Brewster, where he met Ray Dietrich. In 1920, Hibbard and Dietrich left Brewster to establish LeBaron, their own coachbuilding business. Hibbard soon left for Paris, France intending to open a LeBaron office there. Instead, he opened a separate business in Paris with Dutch Darrin designing and building custom cars. By 1931, Hibbard and Darrin were out of business, victims of the Depression.

Hibbard returned to Detroit and, in 1932, took a job at GM's Art and Colour Section as head of the Cadillac Studio. In 1934, Hibbard left GM to form his own design-consulting firm. He returned to GM in 1939 and stayed for two years.

Starting in 1941, Hibbard worked on war projects at Ford's Rouge plant under Joe Galamb, Ford's chief body engineer. In September 1943, after Edsel Ford died, Bob Gregorie resigned as head of the Design Department because of a disagreement with Galamb. Galamb named Hibbard as Gregorie's replacement. In April 1944, Henry Ford II asked Gregorie to return. Galamb was retired, and Gregorie was given his old job back at a substantial increase in salary. Upon Gregorie's return, Hibbard became his assistant.

Gregorie resigned again in mid-December 1946. He recommended Hibbard as his replacement and, on January 1, 1947, Hibbard was reappointed head of Ford's Design Department. In July 1947, GM body engineer John Oswald was hired as Ford's chief body engineer. Because Oswald's job was to supervise the Design Department, Hibbard became his assistant. Several months later, Hibbard was fired, when former G.M. designer George Snyder was appointed chief designer. Between 1947 and 1966, when he retired, Hibbard worked for Kelsey–Hayes. While at Kelsey-Hayes, one of the last projects he worked on was the accessory rally wheels for the 1965-66 Mustang. ■

Surprisingly, the Light Ford went on to become a 1949 model Ford also, but it was not sold in the United States. Before the war, Ford built cars in France under the Matford name. In May 1945, the war was finally over in Europe, and the French were anxious to get back to automobile production. The president of Ford of France, a banker named Maurice Dolfus, came to Dearborn in the summer of 1945 seeking help in restarting automobile production in his war-torn country. Dolfus saw and immediately fell in love with the Light Ford. Breech solved the Light Ford problem by selling it, tooling and all, to Dolfus, who produced it as the all-new French postwar model Ford. Dolfus' engineers changed the tooling to metric measurements, substituted a French electrical system, and had the '49 Light Ford ready for the October 1948 Paris Auto Show.

Henry Ford Museum and Greenfield Village

The production Vedette borrowed its greenhouse from the 1939 Mercury.

Dolfus renamed the Light Ford the Vedette after a small, fast boat used by the French Navy. Unfortunately, the Vedette was only moderately successful. The V-8 60-horsepower engine, although too small for American cars, was not small enough or economical enough for the French, where postwar fuel prices were very high. The Vedette never sold as well as Dolfus or Ford of France hoped and, by 1952, Ford of France was in financial trouble. To stave off receivership and, as a condition of guaranteed loans, Ford sent "Whiz Kid" Jack Reith to Ford of France as manager. Reith made changes to the then-dated design of the old Vedette and, by 1954, he brought out an all-new Vedette, which went into production in 1955. Reith not only improved the product, he made Ford of France profitable again, but Ford Motor Co., tired of the labor and management problems peculiar to the French automobile business, wanted out. So, in November 1954, Reith negotiated the sale of Ford's French automobile business to Simca in exchange for stock. Reith returned to Dearborn a hero and, four years later, Ford sold its Simca stock to Chrysler at a substantial profit.

After he returned to Dearborn, Jack Reith started spending his spare time at the Design Department, helped design the Mercury XM-Turnpike Cruiser, later became general manager of Mercury Division and proposed building a new product ultimately named the Edsel. ■

Clay modeler Clyde Trombley with a full-sized clay model of the Light Ford.

Henry Ford Museum and Greenfield Village

Henry Ford Museum and Greenfield Village

(above) This photograph was taken on June 6, 1945, when Ford's dealer network got their first look at Ford's proposed 1948 new-car lineup. On the right, from front to back, are the Ford and the Mercury proposals. In the middle are two models of the Light Ford and on the left are the Lincoln and Cosmopolitan. Also planned for Ford's 1948 new-car lineup were a Continental and a new Lincoln Custom. They were not ready at the time of this showing.

(below) The grille on this Light Ford proposal was designed by Bill Schmidt and is similar to the production Vedette.

Henry Ford Museum and Greenfield Village

2

Henry Ford's Last Hurrah

1937 Ford Coupe Utility

1941 Plastic Car

1937 Ford Coupe Utility

Starting in 1932, Ford produced a pickup that looked like a car. Unfortunately, it was not produced or sold in North America but only in Australia as the Ford Utility Pickup or "ute." Folklore has it that an Australian farmer's wife wrote Henry Ford asking him to build a pickup her family could use during the week and also be proud to drive to church on Sunday. That seemed perfectly logical to Henry Ford and, although utes were produced in Australia, he found no market for such a vehicle in North America. Except for a few Studebaker and Hudson utility pickups produced in the United States in the late 1930s, no other U.S. automobile manufacturer, including Ford, marketed anything similar to the ute until the 1957 Ford Ranchero.

In 1937, Ford offered an optional pickup box that could be slipped into the trunk cavity after the trunk lid on a Ford coupe had been removed. The slide-in box, although an inexpensive option, did not sell well, and it was quickly discontinued.

Ford tried to find a market for a utility pickup in North America. On February 26, 1937, Ford's sales department sent a letter to Ford dealers asking them for their input on a proposed coupe utility pickup model Ford wanted to add to their 1937 passenger-car lineup. The proposed new model was a five-window coupe with a built-in pickup box that featured both a tailgate and a lid that closed. The rear compartment was 36 1/4 inches wide, 55 inches long and 18 1/2 inches deep. The deck lid was as wide as the rear compartment and 37 3/4 inches long. The tailgate was 13 1/4 inches high. Because the rear part of the coupe utility generally followed the looks of the rest of the car, the back of the box was curved where the deck lid and tailgate met. The deck lid was also counterbalanced so that it would stay open in any position. The rear compartment was lined with wood and had steel skid strips on the floor. Like the 1957 Ranchero of some twenty years later, the spare tire in the '37 Ford ute was located behind the seat inside the passenger compartment.

Although a prototype was built as a Deluxe model, the letter to the dealers announced that, if built, the coupe utility would be available only as a Standard model with either a 60 or an 85 hp engine. The car was assigned model number 74-778-A, and it was estimated that it would cost about $100 more than a Ford Standard coupe.

Dealers were asked to respond to a questionnaire with an estimated number of sales they expected over the next year. They were also asked to report whether they expected any adverse effects on the sale of Ford coupe models, and whether they thought the ute would increase Ford retail sales in general. Lastly, dealers were asked if they thought it was a good idea to offer a coupe utility as part of Ford's new-car lineup. There is no information as to what response the sales department received, but it probably wasn't favorable, because the Ford Coupe Utility was not offered for sale in 1937 or heard of again until the 1957 Ranchero.

According to Bob Gregorie, head of Ford's Design Department, at the time the prototype model 74-778-A Coupe Utility was designed and built at Briggs Manufacturing. Gregorie felt its "see-saw" design was badly proportioned and that it looked "just terrible." Although Gregorie disliked the design, the reason it was not produced was because there was just no market for it. It would take another 20 years for a North American market to develop for a Ford ute. ■

Bob Gregorie did not think much of the design of the proposed 1937 Ford Coupe Utility. He felt it was badly proportioned.

Henry Ford Museum and Greenfield Village

Henry Ford Museum and Greenfield Village

(left) Through 1949, all Ford cars were required to have trunks in which upright milk cans could be carried with the deck lid closed.

Henry Ford Museum and Greenfield Village

(above) Henry Ford was interested in making the farmer's life easier. He thought a fancy pickup was a great idea. It was in Australia but not in North America until 1957.

Henry Ford Museum and Greenfield Village

(above) With the lid closed, owners of the proposed 1937 Ford Coupe Utility could carry their cargo without concern for the weather.

Henry Ford Museum and Greenfield Village

(right) Whether it was a 50 gallon barrel, bushel baskets or milk cans, the proposed 1937 Ford Coupe Utility was designed to appeal primarily to a rural market.

1941 Plastic Car

If Henry Ford had a second love, it was probably farming. His desire to help the farmer came to the forefront in the mid-1920s after he made several trips to Europe. With each trip he returned home more impressed with the agricultural prosperity of the European countries he visited, and he soon made up his mind to do something to encourage that same kind of prosperity at home.

In 1928, he set up an experimental chemical laboratory in Greenfield Village and put a young man named Robert Boyer in charge of it. Henry Ford had known the Boyer family for years. While Robert Boyer was growing up, his father worked as a bookkeeper at Ford's Highland Park plant and later at the Henry Ford Hospital. Just as young Boyer started high school, Ford transferred his father to Massachusetts as the manager of the historic Wayside Inn he had recently purchased. During Ford's winter visits to the Wayside Inn, he and young Boyer ice skated together on the mill pond near the Inn and became friends. When Boyer finished high school, he returned to Dearborn and, at Mr. Ford's urging, enrolled in the Henry Ford Trade School, where he studied mathematics and mechanics. A year later, Ford put him in charge of his new experimental chemical laboratory at Greenfield Village. In short order, Boyer became a very proficient chemist, and soon had 15 other chemists working for him, all looking for new ways to use farm products. Mr. Ford's specific directive to Boyer was to come up with new ways to use farm products in the automobile industry.`

In 1931, after investigating almost every kind of plant capable of being raised as a farm crop, Boyer began concentrating his efforts on the soy bean. Soon, soy bean oil was being used to make the paint that went on Fords and Lincolns and, before long, soy bean meal was also being used to make plastic steering wheels, coil cases and dash knobs.

With the help of Ford's resident chemical expert, Hudson McCarroll, Boyer also developed soy bean-based substitutes for leather, wool and mohair, although he was never able to perfect any of them to the point where they were durable enough for use in automobiles. By 1937, Henry Ford was even wearing suits made from soy fiber material Boyer had manufactured. That same year, Ford began construction of a factory to produce his new soy bean fabric, which he expected to eventually use in his automobiles. Henry Ford's ultimate objective, however, was to develop a plastic car he could mass produce from soy beans.

Henry Ford's last major car related project was not the 1932 Ford V-8 engine, as is often reported but, rather, it was a plastic car made from soy beans, and it combined his desire to help the farmer and his car making genius. Henry Ford wanted to build a plas-

At the behest of an aging Henry Ford (right), Robert Boyer (left) became an industry advocate for the use of plastics in automobile body construction. Plastic panels weighed only 25% of their steel counterparts. Boyer foresaw the day when plastic panels would result in both weight and cost savings, something only now being fully realized.

Henry Ford Museum and Greenfield Village

John Najjar

Ed Martin and the Design Department staff developed the chassis and tubular body profiles for Henry Ford's Plastic Car. He is holding a one-fourth sized model of the tubular body framework.

Edward A. Martin

Ed Martin was born in the Philippines. He was almost entirely self-taught. In his late teens, Martin shipped out as a junior radio operator on a tramp steamer sailing the Pacific Ocean between Asian ports and San Fransisco. His visits to the west coast resulted in a job at Ford's Long Beach assembly plant, where his drafting and planning abilities were soon noticed. He was eventually transferred to Dearborn, where he worked with Bob Gregorie as an engineering draftsman and planner even before the Ford Design Department was started in March 1935.

After Ford's Design Department was organized, Martin became responsible for teaching apprentice designers and the Trade School boys how to design cars. Martin taught many young designers how to design cars the "Ford way" and is remembered as an effective teacher, a mentor and a friend to many of them.

Martin also produced detailed charts and graphs comparing Ford products with the competition. He constantly urged that Ford build new models to compete with GM and Chrysler's mid-priced models. Martin served as Gregorie's unofficial assistant manager. Henry Ford frowned on titles and organizational charts, so the one Martin prepared, which was one of the few available to Ford engineers and designers, was kept hidden in his desk.

After Gregorie left in December 1946, Martin continued in his position as unofficial assistant manager under Tom Hibbard, Gregorie's replacement. When John Oswald was hired, Hibbard again became assistant manager. After George Snyder took over as head of the Design Department, he made a conscious effort to get rid of as many Ford designers as he could. Ed Martin was one of those Snyder dismissed. ■

Design apprentice John Najjar was assigned to assist Martin. He is working on a quarter-sized model of the chassis for Henry Ford's Plastic Car.

John Najjar

tic car he could sell for less than $400. Many people believe he would have done just that if the war and his advancing age had not prevented it.

The first car-related products Boyer made from soy beans were tractor seats and, by 1938, the process had been developed sufficiently to make experimental glovebox doors from soy bean fiber compressed under heat and pressure in molds he designed. Within several months, Boyer was also successfully making large sheets of soy bean plastic, strong enough to support Henry Ford jumping up and down on them. Encouraged by the results and eager to get started on an experimental car made from soy bean fiber, Ford reassigned Lowell E. Overly from one of his other manufacturing plants to work with Boyer. Overly was a tool designer who made the molds for the soy bean plastic instrument panel knobs used on Fords and Lincolns.

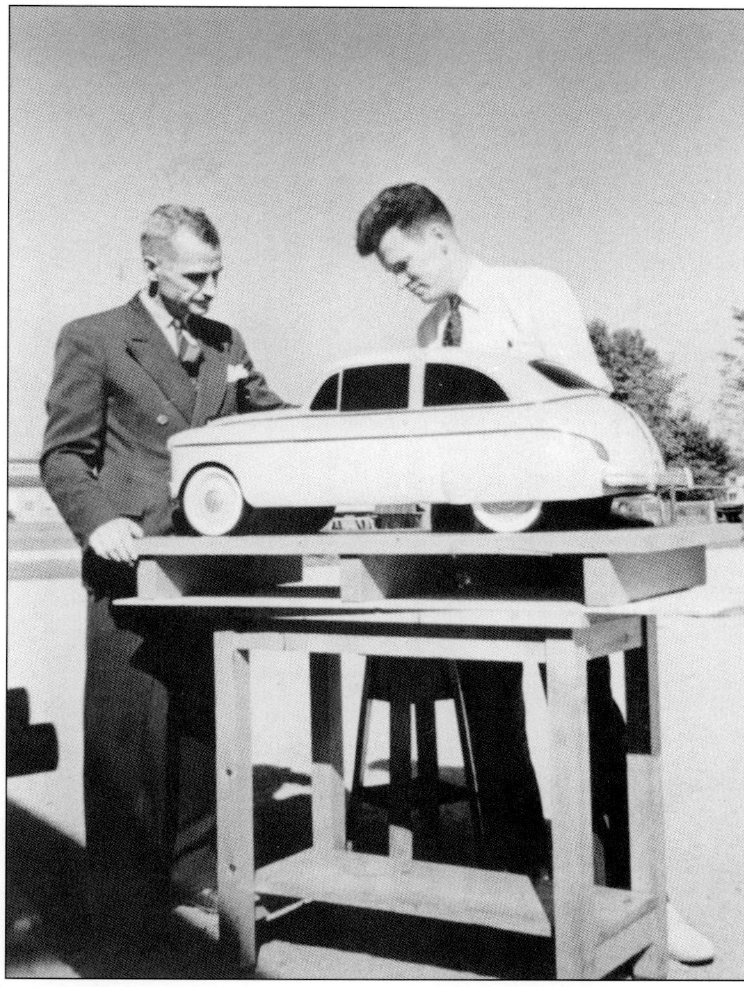

Henry Ford Museum and Greenfield Village

Lowell Overly and a young chemist looking at a clay model of the Plastic Car.

The first usable automobile body panel Boyer and Overly made was a deck lid Henry Ford had installed on his personal 1941 Ford in late 1940. To demonstrate the strength of the plastic deck lid, Henry Ford held a news conference during which he hit the plastic deck lid with an axe. Unfortunately, the blade split the plastic and actually penetrated the deck lid. For the next press conference glass fiber was mixed into the batch of soy bean plastic used to make the replacement deck lid for Mr. Ford's car. For good measure, a rubber boot was also put over the blade end of the axe Mr. Ford used in the next demonstration. When he swung at the replacement deck lid, the axe bounced off, rebounded out of Mr. Ford's hands and landed about 15 feet away. As a result of that demonstration, Mr. Ford gave Boyer and Overly orders to immediately start making a small Ford car with a plastic body.

Because Overly had experience in designing plastic molds, Mr. Ford assigned him to work with Ford's Design Department to come up with an acceptable design for a plastic car. Bob Gregorie, who was in charge of the Design Department, assigned the project to Ed Martin, one of the design engineers working for him. Martin was instructed to do a complete layout for the proposed Plastic Car. John Najjar was assigned to work with Martin on the project. It was Gregorie's intention to do the entire project in his Design Department. That included building a monocoque frame for the car, which Gregorie determined was necessary to meet the 2000 pound weight limitation imposed by Mr. Ford.

Before starting on a full-sized car, Gregorie built a quarter-sized clay model he designed himself. Martin, with the help of Najjar, designed a chassis that consisted of a singular tubular beam running down the middle of the car with front and rear radius rods and springs attached. The car Gregorie designed was meant to have a rear engine and his own unique suspension system.

Henry Ford Museum and Greenfield Village

Body engineering personnel who built the frame demonstrate the strength of its double-side rail construction.

The body framework was made separately. It was also made from a tubular framework that was to be attached to the chassis. The soy-bean plastic panels from which the body surfaces were made were designed to be attached to the tubular body framework. The tubular body framework was designed to form an X, starting at the rocker panels, proceeding up the A and C-pillars and crossing at the top of the car before continuing down the other side. Gregorie had a quarter-sized model of the tubular body framework fabricated in the Ford metal shops. When it was finished, he demonstrated its strength by supporting the framework at either end, off the ground on blocks of wood and then jumping up and down on it.

Gregorie and Overly both had strong personalities, and they were soon at odds over Gregorie's proposed styling of the Plastic Car. Overly thought it was his job to determine the car's styling, and he thought Gregorie's design proposal was impractical because of the reverse curves required by the separate fenders Gregorie proposed. Gregorie thought Overly's design proposal was "grotesque." Overly's design proposal was of a more streamlined, slab-sided car with integral fenders. As work progressed, Overly thought Gregorie was ignoring his expertise and he complained to Boyer — and to Mr. Ford. Gregorie finally told Boyer "either I do it all or take it away." Mr. Ford finally had enough of the complaints he was getting from Gregorie, Boyer and Overly and stopped the project. Without any explanation to Gregorie, Boyer or Overly, Mr. Ford ordered all the models, drawings and parts gathered up, and he had them shipped to one of his warehouses, where they were placed in storage. Mr. Ford was tired of the bickering!

Three months later, Mr. Ford came to see Boyer at the Soy Bean Laboratory, and asked him if he thought Overly was capable of finishing the design and building the Plastic Car himself. While Mr. Ford waited downstairs, Boyer went upstairs to Overly's office to ask him. Overly said he'd give it a try, so Boyer went back downstairs and told Mr. Ford that Overly thought he could do the job. Mr. Ford gave orders to get started immediately, and everything he had placed in storage was then delivered to the Soy Bean Laboratory. Overly's office was moved downstairs to make it easier for Mr. Ford to follow his progress. According to Overly, Mr. Ford didn't pay much attention to what was on paper but, once it was being built and he could see and touch it, he really understood it, and he was

The Plastic Car in front of the Soy Bean Laboratory at Greenfield Village.

Henry Ford Museum and Greenfield Village

Henry Ford Museum and Greenfield Village

Lowell Overly at the wheel of the Plastic Car.

Henry Ford Museum and Greenfield Village

Robert Boyer is holding a quarter-sized model of the car's tubular body framework next to a quarter-sized clay model of the Plastic Car.

there all the time. During the actual building process, Overly recalled occasions when Mr. Ford came by two or three times a day to check on his progress.

According to Overly, Mr. Ford was a strong proponent of the tubular body framework, the unique chassis and the streamlined body design Overly proposed. According to Overly, Mr. Ford ordered him to eliminate all drag from the design of the Plastic Car and to make the window glass flush with the body panels. Henry Ford repeatedly emphasized to Overly that he wanted the car to be as streamlined as possible. He also told Overly that, in his opinion, cars were getting far too big, and that he was a believer in small cars like his proposed Plastic Car. Mr. Ford did change Gregorie's rear-engine layout to a conventional front-engine/rear-wheel-drive configuration and required the use of standard Ford V-8 "60" running gear.

Overly first built a 1/4-sized clay model of the car he proposed to build, and Mr. Ford approved the design. Lines were scribed on the quarter-sized model to help determine where the plastic panels would be joined. Once that was done, the surface of the model was smoothed out, and it was taken over to the Design Department shops to be painted. While there, Overly and Gregorie talked about Overly's new design. Gregorie made some suggestions and, as a result, Overly did make a few changes. After the clay model was painted, Overly used it to work out the actual design of the body panels. In the meantime, the Design Department constructed a full-sized clay model based on Overly's design. When it was finished, the full-sized clay was used to make a plaster mold which, in turn, was used to make dies for the body panels.

When the full-sized Plastic Car was built, the greenhouse had to be made slightly taller than on the quarter-sized model, which gave the finished car a different silhouette.

Henry Ford Museum and Greenfield Village

Because of the war in Europe, tool-quality steel was in short supply, so Overly made the final molds out of concrete. Plastic resin was then poured in the concrete molds under pressure at 240 degrees, which tended to cause the concrete to break up. Originally, Mr. Ford's plan was to impregnate the plastic resin with color. That plan had to be abandoned because the uneven pressure that had to be used on the fragile concrete molds made the color in the plastic look uneven.

After the plastic panels hardened, they were removed from the concrete molds, but much additional hand finishing was required. Once finished, each separate panel was attached to the tubular body framework

This was Gregorie's clay model of the proposed Plastic Car. Overly thought the separate fenders proposed by Gregorie were too complex, given the molding restrictions.

with metal clamps and screws, and the joints were sealed with more resin compound. After the body was finished, the car was painted. Acrylic was used for the windows, which were set in rubber seals recessed about 1/8-inch away from the exterior body surfaces. Only the driver's window could be opened. The upholstery, including the seat padding, was made from soy bean fiber. When finished, the Plastic Car weighed in at 2300 pounds.

Ford Motor Co. issued a press release, dated August 14, 1941, claiming that Henry Ford's Plastic Car was going to revolutionize the automobile industry, and that Mr. Ford believed that future plastic cars could be made from any fibrous material. Whether that was true or not, many years later a dispute arose between Boyer and Overly as to just what the soy bean car had really been made from. Boyer claimed that there was very little soy bean fiber in the "phenol formaldehyde resin" used to make the plastic panels. Overly disagreed. He said the material used to form the panels was a "mixture of soy bean fiber in a phenolic resin with formaldehyde used for impregnation." Whatever it was made from, everyone who got near the Plastic Car agrees that it had the distinct odor of formaldehyde — it smelled like a mortuary!

Lowell Overly working on a clay model of his own design for the Plastic Car.

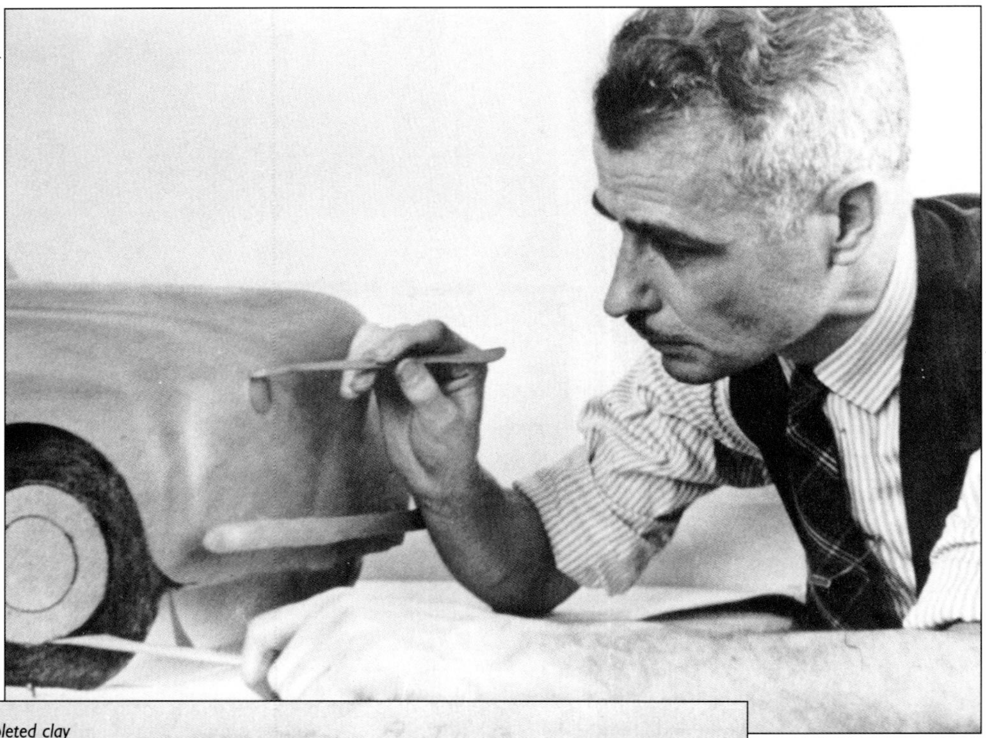

The Plastic Car was introduced to the public August 13, 1941, at the annual Dearborn Days celebration, and it was displayed later that fall at the Michigan State Fair. The car was shown both times in the open air. In its August 24th issue, the *New York Times* predicted that Ford would market a Plastic Car in 1943, and that a price of $400 was probable. Although publicity was generally favorable, the coming war ended any further development work on Henry Ford's Plastic Car.

After the Michigan State Fair, Boyer drove the Plastic Car for awhile, and then it was put in storage in the basement of the Design Department. The panels soon shrank and it smelled.

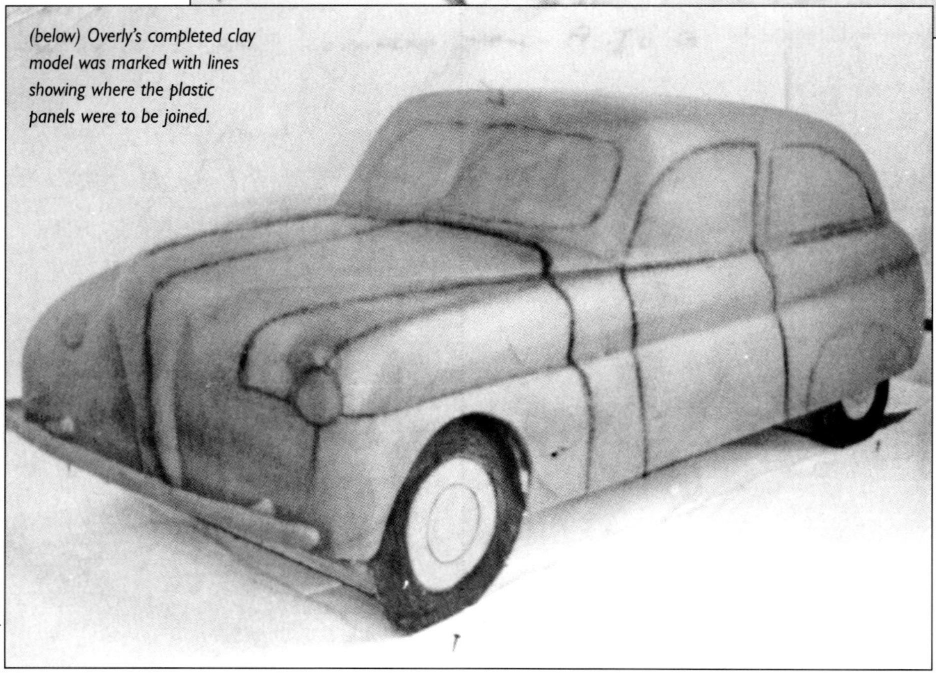

(below) Overly's completed clay model was marked with lines showing where the plastic panels were to be joined.

The car was thought destroyed sometime during the war, so the steel from it could be used for the war effort, but John Najjar recalls seeing the deteriorated Plastic Car in the basement of the old Design Department building as late as 1971. The building has since been remodeled, but the Plastic Car is no longer anywhere to be found.

In 1941, Henry Ford was 78 years old and had suffered several small strokes. After the war, he was not capable of continuing any automotive development work. He died in 1947. Very few people now associate Henry Ford with newfangled ideas like streamlining or tubular monocoque construction. His Plastic Car tells a far different story and, with good reason, some people believe that if Henry Ford had started working on his Plastic Car ten years earlier, we might now be driving plastic Fords, Lincolns and Mercurys. ■

Ross Cousins

Ross Cousins' father was a car illustrator from Toronto, Canada. In 1926, the family moved to Michigan, so his dad could take a job illustrating Packards. Cousins studied art at Cass Technical High School. After graduation, he took a job at a printing and graphics firm, where he prepared illustrations for car brochures and the Buick magazine.

In 1938, Richard Beneicke, head clay modeler at Ford, told Cousins that Gregorie was looking for an illustrator. Cousins applied and was soon hired. Gregorie was a good sketch artist but not the best illustrator. He wanted someone to render proposals for presentations to Edsel Ford. Cousins was also the first designer at Ford who was proficient in airbrush technique.

Cousins' rendering of a Lincoln-Zephyr Towncar became the inspiration for the real thing. He also did renderings of production and special cars designed at the Design Department, including the Continental, the 05L, the Light Ford and a series of proposed wooden-bodied cars that led to the Estate Wagon and the Ford Sportsman. In addition to illustrations and renderings, Cousins designed ornamentation and trim for production Fords, Mercurys and Lincolns.

After Gregorie left in September 1943, Cousins took a job at Briggs Manufacturing. When Gregorie was rehired in May 1944, Cousins returned to the Design Department at Gregorie's request.

Believing that change was imminent, Cousins left Ford for good in late 1946, shortly before Gregorie did. He joined a Detroit graphics firm that prepared illustrations for car brochures and magazine ads. Cousins prepared many Cadillac illustrations through the 1950s. His illustrations won many awards and appeared in magazines and advertisements of the day. ■

Ross Cousins

(below) The running gear installed in the Plastic Car's chassis with the body framework attached. Plastic panels were still to be attached to the body's framework.

Henry Ford Museum and Greenfield Village

3

Henry Ford II's Personal Concept Cars

Estate Wagon

Sedanca de Ville

Farina Mercury

Cisitalia Mercury

Indianapolis Coupe

Estate Wagon

During the fall of 1944, on one of his get-acquainted visits to the EEE Building, Henry Ford II spotted an unused 1931 Ford Model A frame. It was near Bob Gregorie's department, so he asked Gregorie what it was doing there. Gregorie explained that Mr. Ford's grandfather had ordered it left there, so that every time Ford engineers tried to overly complicate the chassis or suspension of one of the new cars they were working on, the Model A frame would be there to remind them that their boss preferred simplicity.

At the time of Henry Ford II's visit, the war was not yet over, and no new cars could legally be built. After he found the frame, Henry Ford II asked Gregorie if it could be used to build a custom car for use at his summer home in Southampton, New York. Mr. Ford told Gregorie he understood the difficulties in building such a car given the shortage of materials but, if it could be done, he wanted a beach wagon to take his children back and forth to the beach. Gregorie said he thought he could come up with something from the leftover parts in storage that Mr. Ford could use and enjoy.

Gregorie called the custom car he built for Henry Ford II an Estate Wagon. Metal to build the body was impossible to come by, but there was no wartime restriction on the use of wood, and Gregorie knew a ready source for lots of wood. The body of the Estate Wagon was built from the same wood stock and panels Ford Motor Co. was manufacturing at its Iron Mountain plant and using to build gliders for the war effort.

Henry Ford Museum and Greenfield Village

Designer Doris Dickason with the Estate Wagon before its delivery to Henry Ford II.

Henry Ford II's Estate Wagon was designed by Gregorie and built directly from sketches he prepared. No clay models were made. The car was built at the old Air Frame Building using Air Frame personnel, mahogany stock and hardwood panels ordered by Gregorie from Iron Mountain. Gregorie saw the Estate Wagon as good practice for several of the new models he wanted to produce after the war. Chrysler marketed a wooden-bodied sedan in 1941-42 and, by the time the Estate Wagon was built, the Design Department was proposing to add wooden-bodied station wagons, sedans and convertibles to Ford's postwar new-car lineup.

Building the Estate Wagon was also dependent on whatever leftover parts Ford had in storage. The fenders, mechanical brakes, pedals, radiator shell and the steering column were all leftover Model A parts. The shocks were from a 1932 Ford, and a '32 Ford Model B four-cylinder engine was used. The steering wheel, wheels and Sealed Beam headlights (inside Model A housings) were stock 1940 Ford Standard parts. Bumpers were modified 1941 Ford Super Deluxe, and the instruments were a mix of 1938 and '39 Ford Standard parts. A Model A cowl was reworked and the custom-made hood, formed from a single piece of metal, extended from the radiator shell all the way to the base of the windshield. The floor pan, top bows, instrument panel and windshield frame were custom made. The

From the front, the Estate Wagon looks like a custom-built Model A.

Henry Ford Museum and Greenfield Village

hardware used came from leftover Ford station wagon parts from several past years. The convertible top was made from khaki, the only material then readily available. The door sidecurtains were made from Plexiglas and the rear-quarter sidecurtains from flexible poly-vinyl-chloride (PVC). The seats were upholstered in maroon leather in a similar pattern to the later production 1946 Ford Sportsman. The back curtain rolled up and could be fastened to an inside top bow with snaps. The spare tire was bolted to the outside of the tailgate and covered by a metal tirecover. The metal parts on the Estate Wagon were painted maroon, and the car was accessorized with a driver's-side spotlight and a toolbox welded to the engine side of the firewall.

Henry Ford Museum and Greenfield Village

After the Estate Wagon was completed, it was discovered that if anyone sat on the open tailgate, it caused the doors to open. The problem was corrected by adding a second set of frame rails.

The tailgate, with the spare tire attached, had no counterweights to help open and close it. After the car was completed, the Air Frame personnel discovered that when anyone sat on the lowered tailgate, the car doors sprang open. The problem was traced to the relatively weak Model A frame, and the solution Gregorie came up with was to put two additional Model A frame rails under the car.

The Estate Wagon was finished in January 1945, but it was registered as a 1941 Ford probably because of wartime restrictions on the manufacture of new cars. When it was completed, Gregorie personally delivered it to Henry Ford II. Gregorie also told Henry Ford II that he would like to buy the car from him when he was done with it, and Mr. Ford said "sure."

The following summer, Henry Ford II used the Estate Wagon for several months at his Southampton summer home. By the fall of 1945, new cars were again rolling off the Ford assembly lines, so Henry Ford II gave the Estate Wagon to Gregorie. When the car was delivered to Gregorie, he was at the Design Department hosting Ford's new executive vice president Ernest Breech and a group of Ford executives. According to Gregorie, the meeting was not going all that well but, when the Estate Wagon was delivered, Breech's attention turned to it and, after he had carefully looked it over, he told Gregorie he had to have it. Gregorie told him he had already spoken for it several months before, and it had been delivered to the Design Department so he could take it home that evening. Gregorie says he got great pleasure out of telling Breech he couldn't have the car.

That Henry Ford II disposed of the Estate Wagon after only one season's use should not be taken as an indication that he didn't like the car. Even though he and Gregorie both had to admit that the Estate Wagon rode like a truck, Henry Ford II liked the concept well enough to authorize the addition of a new wooden-bodied convertible, designated the

Model 71 Sportsman, to the Ford lineup for 1946. By October 1945, the first production Sportsman was built without benefit of blueprints at Ford's Iron Mountain plant, by attaching finished wooden panels to a regular convertible body with the doors, rear quarters and trunk-area sheet metal removed and a hand-formed steel skeleton attached. The first Sportsman was displayed in Dearborn, then it was heavily accessorized and delivered to actress Ella Raines in Hollywood, California on Christmas Day 1945.

The Estate Wagon still exists. After Gregorie left Ford in December 1946, he moved to Florida and took the Estate Wagon with him. The Florida summers were hard on the wood and paint and, in 1949-50, Gregorie tried to trade the Estate Wagon in on a new Ford station wagon. He showed the dealer an old registration listing Henry Ford II as the previous owner and told the dealer that the Estate Wagon was a one-off hand-built car that originally cost $50,000. The dealer wouldn't give Gregorie what Gregorie thought the car was worth, so he kept the Estate Wagon and had it repainted and revarnished. Sometime after it was refurbished, Gregorie sold the Estate Wagon to Carl Griner of Ormond Beach, Florida. Not much is known about Mr. Griner, except that he was a retired Packard executive, a friend of Gregorie's, and he didn't use the Estate Wagon very much.

During the late 1940s and early 1950s, Ken Coulter, a florist from Maryville, Tennessee, began collecting Model As. In 1951, he heard about an unusual Model A convertible owned by a former Ford designer living in Florida. Coulter tracked down Gregorie and tried to acquire the one open Model A Ford he didn't have. How, when, and under what circumstances Gregorie reacquired the Estate Wagon is unknown, but Gregorie told Coulter he didn't want to sell it and, for over a year, Coulter couldn't get Gregorie to change his mind. On one of his many calls to Gregorie, Coulter mentioned that he had an all-original, 9000-mile Model A roadster. Gregorie agreed to trade Coulter the Estate Wagon for his Model A roadster and some cash. After some additional negotiation, a price was agreed on, and Coulter loaded his Model A roadster on a trailer and left for Florida. When he got there, Coulter and Gregorie exchanged titles, and Coulter gave Gregorie $3000. He then left for home with the Estate Wagon on his trailer. After the sale was completed and while Coulter was loading the Estate Wagon, Gregorie told him that 22 parts suppliers had contributed to the building of the Estate Wagon. He also gave Coulter pictures and literature to go with the car.

When Coulter bought the Estate Wagon, it had less than 8000 miles on it and was in good original condition. During the time he owned it, Coulter put less than 300 additional miles on it. He also added to the pictures and literature Gregorie had given him and wrote a detailed history of the car based on his additional research. Unfortunately, in the early 1980s, one of his employees accidently burned all of Coulter's documentation on the Estate Wagon.

In 1986, Coulter sold the Estate Wagon to Roy Frost, an Alabama plastics manufacturer. Frost bought the Estate Wagon the same way Coulter bought it from Gregorie -- by persistence. Frost and Coulter agreed to a price, and the Estate Wagon went to Alabama where it is today. It is still in good original condition and has not been restored. ■

Robert H. Doehler

John Najjar

Bob Doehler was born in Milwaukee, Wisconsin. As a young boy, his family moved to Iron Mountain, Michigan, when his father was hired to manage Henry Ford's power plants in northern Michigan. After graduating from high school, Doehler's father arranged an interview at Ford's tool shop. When the supervisor saw Doehler's drawings, he referred him to Bob Gregorie, who interviewed and hired Doehler as a clay modeler. During World War II, Doehler was transferred to the tool shop and earned his journeyman card as a tool maker. Shortly after the war, Doehler returned to the Design Department, where he worked as a layout man under Martin Regitko. One of the first jobs Doehler worked on after he returned to the Design Department was Henry Ford II's Estate Wagon.

By the early 1950s, Doehler had became a designer. In 1953, he was offered and accepted a job at Studebaker as a designer. After the merger with Packard, he remained at South Bend, until Studebaker stopped producing cars in the United States in late 1963. While at Studebaker-Packard, Doehler helped design the Studebaker pickup and the Lark. After he retired, he opened a body shop in Milwaukee specializing in restoration work. He also did some free lance clay modeling. In the early 1980s, he sold his shop to an employee and retired. ■

The instruments used on the Estate Wagon were 1938 and '39 Ford Standard.

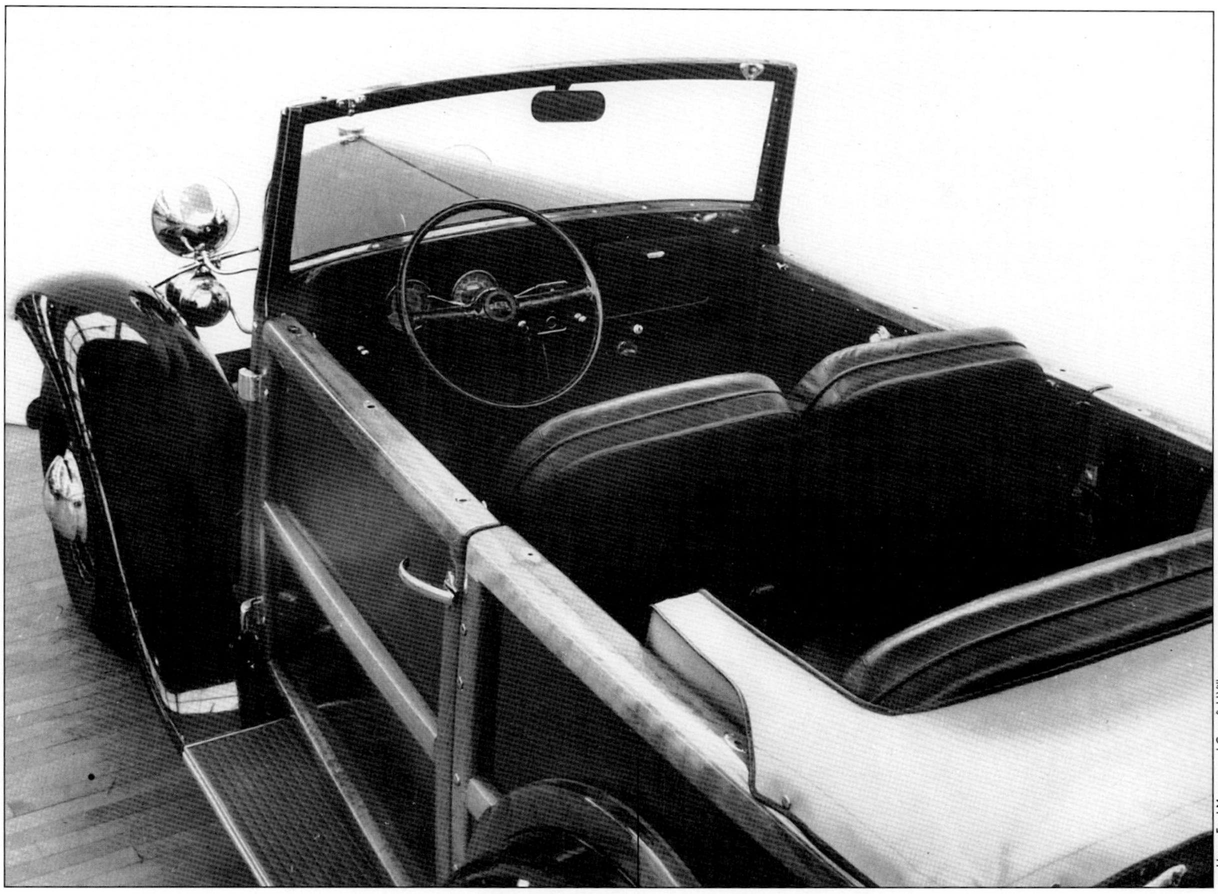

Henry Ford Museum and Greenfield Village

The First Women Designers At Ford

Until recently, design and just about every other job in the automobile business was a man's world. Joe Oros' wife, the former Betty Thatcher, was a prewar designer at Hudson and is reputed to have been the first woman designer hired by a major automobile manufacturer. With no regrets, Betty married Joe and retired in 1939 in deference to Joe's career and to raise a family, which eventually included five children. Other women designers did follow Betty Thatcher — but slowly.

Ford's first women designers were hired towards the end of World War II just as automobile design work was starting up again. Several of them were fine arts graduates from the University of Michigan. They were hired by Bob Gregorie because men were in the service or in war work. The first woman designer hired at Ford was Leota Carroll, who was soon joined by Letha Allan, Elre Campbell, Beth O'Rourke, Doris Dickason and Florence Henderson.

Leota Carroll, a talented illustrator, was assigned to work with Ross Cousins. She was very photogenic and was often used as a model in Design Department photographs. Beth O'Rourke and Doris Dickason were assigned to John Najjar and designed steering wheels and instrument panels. Florence Henderson, whose father worked in the Ford aeronautics department, designed ornamentation. Elre Campbell and Letha Allan were assigned to work with Hermann Brunn and Ted Hobbs in color and trim.

At the time they were hired, Ford's wartime women designers were all single. As the war ended, men returning to the job market were given job preference. By 1948, all of Ford's women designers had married or departed for other reasons. Over the next 20 years, Ford employed several women as part-time designers or as outside design consultants, but it was not until 1970 that there was another full-time woman designer at Ford. Today, several of Ford's best designers are woman including the manager of one of the major design studios. ■

Henry Ford Museum and Greenfield Village

Leota Carroll

John Najjar

Letha Allan

John Najjar

Beth O'Rourke

John Najjar

Elre Campbell

Tucker Madawick

Florence Henderson

Henry Ford Museum and Greenfield Village

Leota Carroll and Steve Galla demonstrating a folding seat proposal developed by Tom Hibbard.

John Najjar

Doris Dickason

Dave Cole

Henry Ford II and his Sedanca de Ville.

Sedanca de Ville

Planning was under way for Ford's postwar new-car line up even before the war was over, and it gathered steam with victory in Europe in May 1945. All-new body styles and a new Light Ford were planned for 1948. A new Continental was originally planned for the postwar new-car lineup, based on Lincoln Cosmopolitan sheetmetal, but there were design problems that could not be overcome. A Continental based on the sheetmetal of the Lincoln Cosmopolitan didn't look right. Gregorie's designers couldn't come up with an acceptable design, and there just wasn't enough volume to justify building a Continental with its own unique sheetmetal. The final blow came when executive vice president Ernest R. Breech pushed through a plan for a postwar Ford different from the one Gregorie had already designed. Because of Breech's plan, all available development money and manpower were directed towards a crash program to develop and produce the new Ford. One of the casualties of that decision was the new Continental. Introduction of Ford's postwar new-car line was postponed until the 1949 model year and, by the time the new Ford's design was approved in early 1947, it was fairly certain there would be no new Continentals after the 1948 model year -- at least for awhile.

A year after the Estate Wagon was built, Henry Ford II commissioned another special car for his own use. This one was a modified Continental and, this time, he quietly bypassed his own Design Department and asked Coachcraft, a custom car builder in Southern California, to build it. Movie star Gary Cooper had earlier introduced Henry Ford II to Rudy Stoessel, Burt Chalmers and Paul Erdos, the principals in Coachcraft. By the time he ordered his custom Continental from Coachcraft, Henry Ford II had already seen Gregorie's design for the proposed postwar Continental and was less than pleased with it. Whether that influenced him to select Coachcraft to build his custom Continental is unknown.

In early 1946, two 118-inch Mercury convertible chassis were shipped to Coachcraft with a request to build two open cars on them. The Mercury chassis shipped to California had experimental 337 CID flathead V-8s in them with standard Lincoln three-speed transmissions without overdrive. The Lincoln 337 CID V-8 engine was based on a Ford truck engine and was later used in 1949 production Lincolns. Coachcraft was instructed to build two custom-bodied cars that were smaller than, but looked like updated 1946 production Continentals. For unknown reasons, only one car was built, and the second chassis was returned to Ford. The car Coachcraft built for Henry Ford II was called a Lincoln Cabriolet Sedanca de Ville by Coachcraft.

The necessary design work for the Sedanca de Ville was done by Bob Walker, a local Southern California designer, who prepared several sketches of the car Coachcraft proposed to build. After the design was approved, the standard body parts needed were shipped from Ford Motor Co. to Coachcraft including standard 1946 Lincoln Continental

Dave Cole

(above) Left to right, Burt Chalmers, Paul Edros, Rudy Stoessel and other Coachcraft employees who built the Sedanca de Ville.

(Right) The deck lid and chassis are the only Mercury parts used to build the Sedanca de Ville.

(below) The top of the instrument panel had padding that covered the opening for the radio, so the radio controls had to be moved to the steering wheel.

Dave Cole

body, doors, hood, grille, bumpers and front fenders. Rear fenders and fender skirts from a 1941 Lincoln Continental were also sent, probably because they were a better match for the sloped rear deck of the car. The only Mercury part used on the car, besides the chassis, was a much-modified deck lid. In addition to furnishing all of the parts, Ford Motor Co. paid Coachcraft $20,000 to build the Sedanca de Ville.

In February 1947, the Sedanca de Ville was completed and shipped to Henry Ford II in Dearborn. (Gregorie resigned December 15, 1946.) It was titled as a 1948 Lincoln. The VIN number assigned to the car was 99A-741103, which was an early 1946 Mercury serial number probably taken from the car's frame number. The car was painted medium-metallic green and had a cream-colored or light-tan leather interior done in a custom pattern. As built, the Sedanca de Ville had a removable hardtop, with separate sections over each front passenger area that could be removed independently from the top. As originally built, there was no separate convertible top. The front fenders on the Sedanca de Ville were swept back into the doors, taillights from a 1940 Chevrolet were used, and the back license plate was indented into the deck lid. The front windshield was custom-built and mildly raked back towards the driver and veed in the middle. Because padding was used on the top of the standard Continental instrument panel, and the place where the radio controls were usually located was covered over, a new radio control head was installed on the steering column.

Dave Cole

In the early 1950s, Henry Ford II had a curved-glass one-piece windshield and a power convertible top installed on the Sedanca de Ville. The work was done at the old Lincoln plant. At the time, the removable hardtop was discarded.

Henry Ford II kept the Sedanca de Ville until 1956, when he sold it to Lottie Collier of the *Collier* magazine family. In 1959, she sold it to Kurt Brinker of Mt. Clemens, Michigan. By the time Brinker bought the Sedanca de Ville, it had been repainted white and reupholstered in two-tone blue leather. A lumberyard owner from Bad Axe, Michigan purchased the car

Dave Cole

(Right) Movie star Gary Cooper with the Sedanca de Ville in front of Coachcraft.

(below) The Sedanca de Ville in 1998 after it was restored .

Fred Hunter

from Brinker in 1967 and repainted it blue. In July 1988, the car was sold to Alonzo and Victor Favot of Livonia, Michigan. By the time the Favots acquired the car, it was in need of complete restoration, which they undertook. In 1995, the restoration completed, the car was sold to Fred Hunter of Fort Lauderdale, Florida and is now a part of his collection.

It is unlikely Ford Motor Co. intended to build any new product that looked like the Sedanca de Ville, which was most probably a special one-off car built for the boss and nothing more. ■

Ford began building Model As in France in 1928 and, in 1934, Ford and Mathis joined forces to build the Matford. In 1946, Ford's joint venture with Mathis was dissolved, and Ford formed Ford Societe Anonyme Francaise, or Ford SAF for short. Ford of England owned 60% of Ford SAF. Maurice Dolfus, head of Ford SAF, convinced Dearborn to let him build and market the Vedette which was introduced in France in October 1948 but, for many reasons, the Vedette didn't sell well.

When he first assumed command of Ford Motor Co., one of Henry Ford II's priorities was to increase sales of the mid-sized Mercury. He knew that Ford was losing out to the competition's mid-sized offerings and, in 1948, he commissioned a formal study to come up with a way to remedy that problem. That same year, Henry Ford II and Ernie Breech visited Europe. Although their trip was for the purpose of encouraging a revitalization of Ford's war-torn subsidiaries in Germany, France, Italy and elsewhere, they also visited several of the Italian carrozzerias, including Stabilimenti Farina, which was one of Italy's most respected. Either before or after that visit, Ford and Breech tried to hire a talented Italian car designer named Battista Farina. He turned them down but, the next year, Ford sent a Mercury chassis to Stabilimenti Farina and commissioned the building of the Farina Mercury for Mr. Ford's personal use. When it was finished, it was delivered to Dearborn to be enjoyed by Mr. Ford and tested by Ford engineers. Cost probably prevented Ford from seriously considering it for production.

Farina Mercury

Gary Wright

Originally built as a personal car for Henry Ford II, the Farina Mercury was briefly considered for production and sale through Lincoln-Mercury dealers in the United States. The car was eventually produced in France as the Comete.

In 1949, Dolfus retired and was succeeded by Francois Lehideax as managing director of Ford SAF. Shortly after the Vedette was introduced, Simca brought out a sporty luxury model that Lehideax thought Ford SAF had to match. Lehideax, to say the least, was not a team player. Without telling anyone, he authorized production of a new sporty-luxury car based on the design of Henry Ford II's Farina Mercury. The car was built by Facel Metallon, a French custom body builder, and it was added to the French SAF new-car lineup in the spring of 1951 as the Comete. The Comete used a Vedette chassis, the Ford V-8 "60" engine, and it sold for over $5000 — which made it the most expensive French production car made during the 1951 model year.

Published reports claim that the Comete was introduced to the French automotive press even before Henry Ford II learned of it. Those same reports claim that, although Henry Ford II was disappointed that the Comete was not designed in Dearborn, he was none the less impressed with it. Henry Ford II was apparently so impressed that he replaced Lehideax the next year with Jack Reith. Reith was never accepted at Ford SAF but, by 1954, he had arranged the sale of Ford SAF to Simca and returned to Dearborn a hero to champion the Mercury Turnpike Cruiser — and the Edsel.

An article in the May 1953 issue of *Motor Trend* magazine claimed that, in early 1953, Ford made plans to import the Comete, put a bigger V-8 engine in it and sell it as a Ford sporty car, while they were deciding whether to build their own sports car. Plans were apparently canceled, because the larger 239 CID flathead engine adversely affected weight distribution. How serious those plans were is unknown, but the car was displayed in the showroom at the Design Department in Dearborn. ■

Battista Pininfarina

Battista Pininfarina was the youngest of 11 boys. Although he loved to draw cars as a child, he was expelled from school twice. His first job was as a mechanic in the Fiat plant where, at age 17, he designed his first car, the Fiat Torpedo. In 1920, he spent his savings for a trip to the United States, where he visited the Ford plant. After he returned to Italy, he raced cars and became a designer for his brother, Giovanni Farina. In 1930, when his brother's business closed, Pininfarina borrowed money and opened his own carrozzeria in Turin. In 1958, he built a new factory and, in 1959, he turned the company over to his son and son-in-law.

As a child, he was given the nickname "Pinin" which means "little" or "the kid." In 1961, his family surname of Farina was legally changed by presidential decree to Pininfarina, a name he had used professionally for many years.

Ferrari

During his long career, Pininfarina was hired as a consultant by most major American car companies and held design contracts with many European manufacturers, including Alfa Romeo, Lancia and Ferarri.

In addition to designing the Farina Mercury and the Cisitalia 202, which was the basis for the Cisitalia Mercury, Pininfarina designed custom as well as production automobiles, including a 1952 Nash proposal, the 1959 Cadillac Eldorado Brougham and most Ferraris built after 1953. His design innovations were regularly adopted by American and European car makers. ■

Henry Ford Museum and Greenfield Village

(above) The Farina Mercury arrived at Ford in late 1950, after it was shown at the Turin Auto Show.

Henry Ford Museum and Greenfield Village

(above) The Farina Mercury had a V-8 engine and was capable of speeds of up to 130 mph.

(below) This car unwittingly became the prototype for the French Comete.

Henry Ford Museum and Greenfield Village

Cisitalia Mercury

Cisitalia was started in Turin, Italy, in 1939, by wealthy industrialist Piero Dusio for the purpose of manufacturing machine tools and garage equipment. World War II devastated Italy and its manufacturing base. Dusio owned a car-racing team before the war, so he turned to building racecars again, once the war was over. His first racecar, the single seat D-46, was designed and built in 1945-46 under the direction of engineer/designer Giovanni Savonnuzi. Although Dusio spent too much money building and promoting his racecars, in the late 1940s, he decided to market a two-passenger sports car designated the Cisitalia 202. It was also engineered by Savonnuzi, but it was designed, and the first one was built by Pininfarina.

In 1948, the first Cisitalias 202s were imported to the United States by car enthusiast Max Hoffman and sold for $6800 through his New York City dealership. Although the price was high, Hoffman was able to sell all the Cisitalias he could get his hands on. Henry Ford II admired the design of the Cisitalia 202 and purchased both a coupe and a convertible from Hoffman for his personal use.

By 1949, Cisitalia was developing another racecar, this one for the Grand Prix circuit. When money ran short, Piero Dusio moved to Argentina leaving Cisitalia and its problems to his son, Carlo. Carlo Dusio decided that the best way to improve Cisitalia's financial fortunes was to approach Henry Ford II, who he knew to be a Cisitalia enthusiast, with a joint-marketing proposal. He discussed the matter with Savonnuzi and decided to make a proposal directly to Henry Ford II. The proposal Dusio came up with was to build and market Italian-made hybrid cars in the United States using Ford engines, transmissions and suspensions with Cisitalia bodies and chassis Savonnuzi agreed to design. Dusio's original projections were that they could produce 500-1000 Cisitalia Fords each year at a substantial profit.

In 1950, while the X-100 was being designed, Henry Ford II unexpectedly showed up at the Design Department driving one of his Cisitalia sports cars. Mr. Ford explained to the designers present that the Cisitalia was what he thought a well-designed car should look like, and then he pointedly asked why Ford's designers couldn't design cars that looked like the Cisitalia. After Mr. Ford left, John Najjar was assigned the task of writing a report to answer Mr. Ford's question. Najjar's report explained that the American public generally wanted bigger cars than the Cisitalia, and that our laws required automobiles with bumpers, lights and other features at uniform heights, something Italian car makers didn't have to worry about. Henry Ford II never reconciled himself to the explanation he received from his Design Department, but his preferences did have an indirect influence on the design of the X-100.

Singer Perry Como with the Vignale-built Cisitalia Ford and Mercury in 1952, after they were delivered to Peter Sullivan II.

Peter Sullivan III

Ghia

Ghia

(top and above) The second prototype Cisitalia Mercury, code named the 808 XF, was built in 1951.

As did his father before him, Henry Ford II had a strong appreciation for European cars, and he especially liked cars from the Italian school of design. Henry Ford II spent substantial periods of time in Europe both for business and pleasure. In April 1950, Carlo Dusio and Savonnuzi read in a local paper that Henry Ford II was in Italy. Dusio immediately telephoned Mr. Ford's secretary, James Cumming, and arranged for an appointment. When they met several weeks later, Dusio made his joint venture proposal, and Henry Ford II eagerly agreed to it with little discussion and no hesitation. They agreed that Dusio would construct four or five prototypes and ship them to Dearborn for testing. Dusio and Ford also agreed that their joint venture would produce a Ford and a Mercury model coupe and convertible, and that they would be imported and sold through Ford and Lincoln-Mercury dealers in the United States. The Cisitalia Ford was to be powered by a Ford six-cylinder engine, and the Cisitalia Mercury was to use the flathead V-8. Ford Motor Co. provided development money and sent running gear to Cisitalia for use in the cars they were to build.

By August 1950, Dusio was on the way to the United States with the first Cisitalia Mercury prototype. After going over the car carefully and doing a complete cost breakdown, Ford's sales department told Mr. Ford that the price the Cisitalia Mercury would have to sell for was too high. The engineers who inspected the Cisitalia Mercury prototype also determined that there were problems with the car that had to be remedied before the Cisitalia Mercury was anywhere near ready for production.

To reduce the proposed selling price, Ford engineers proposed using Cisitalia bodies on a modified Ford chassis and assembling the cars in Dearborn rather than Italy. At first, that proposal wasn't acceptable to Savonnuzi, but he finally agreed to design two prototypes, a coupe and a convertible, using modified Ford chassis. The prototypes Savonnuzi designed using the modified Ford chassis were not as low as the cars he had originally designed but, when they were done, even Savonnuzi reluctantly agreed that the use of a Ford chassis was an acceptable compromise.

R. Peter Sullivan II, Henry Ford II's brother in law and a Long Island, New York Ford dealer, was placed in charge of the project. Even though the new prototypes sent to Dearborn proved unacceptable, Henry Ford II apparently remained hopeful production and cost problems could be overcome. Prototypes of the newly-redesigned Cisitalia Ford and Mercury coupe and convertible models, both built by Vignale, were delivered to Sullivan in 1952. The Cisitalia Mercury produced an estimated 200 horsepower and had a top speed of 130 miles an hour.

All total, at least seven prototypes of the Cisitalia Ford/Mercurys were built by Ghia, Vignale and Pininfarina. Even though there were minor differences in the designs of the Cisitalia Ford and the Cisitalia Mercury, in general, there were three separate designs submitted by Cisitalia over the years the project was considered, not all of which were the product of Savonnuzi or Pininfarina. The body of the prototype built at Ghia was designed by Mario Boano. In June 1953, Vignale produced the last coupe and convertible prototypes which were sent to the United States. Ford engineers again found problems with the prototypes, and Ford's sales department never felt the cars would sell in sufficient numbers, given the high production costs which were never reduced to an acceptable level.

The interior of the 808 XF had standard Ford gauges and steering wheel. The car was also equipped with a Fordomatic transmission.

Ghia

Peter Sullivan III

The 1952 prototype Cisitalia Mercury at Garden City, New Jersey. Left to right, are Ronnie Gartuier, Perry Como and Bill Martin. Gartuier and Martin were employees at Peter Sullivan's Long Island, New York Ford dealership.

In November 1953, an article appeared in a U.S. trade magazine reporting that Ford had agreed to market a Cisitalia sports car constructed on a modified Ford chassis and using a Ford six-cylinder engine. The car was described as having an aluminum two-passenger sports car body built by Cisitalia at their Turin, Italy factory. The article also reported that Cisitalia planned to take a shortened Ford running chassis, shipped to them from Dearborn, beef up the suspension, install dual carburetors, put a Cisitalia body on it and ship it back to Dearborn, where Ford planned to sell it starting with the 1954 model year for about $3200. Unfortunately, by the time this overly-optimistic report was published, the program had already been canceled, and Ford's own sporty car, the Thunderbird, was planned for introduction in the fall of 1954.

Sullivan advertised the Cisitalia coupe for sale in the April 1953 issue of *Road and Track* magazine for $7500, and eventually sold it. For many years after the program was cancelled, Sullivan drove the Cisitalia Mercury convertible as his personal car.

By 1953, Cisitalia had even more serious financial problems. When Ford decided to build the Thunderbird, it meant the end of the joint venture with Cisitalia and probably the end of Cisitalia. Cisitalia's financial difficulties, the projected sales price of the Cisitalia Mercury -- undoubtedly higher than the $3200 reported in the news article -- and the differences in production and quality standards doomed the Cisitalia Mercury project. By the time the production and cost problems were close to resolution, assuming they ever were, Ford decided to pursue its own sporty car. Given the limited market for an American sporty car, Ford decided there was probably not a market for two sports cars from Ford. Cisitalia made several more efforts to market a successful car, but market conditions had changed, and Carlo Dusio was forced to permanently close down Cisitalia in 1965. ■

The Cisitalia Mercury 808 XF during construction. Small metal panels were hand formed over a hardwood framework, and then welded together into the finished car.

Ghia

Indianapolis Coupe

Until a disagreement ended his partnership with Luigi Segre, Mario Boano was one of the owners of Ghia, a long-established carrozzeria, or coachbuilder, in Turin, Italy. After Mario Boano and his son, Gian Paolo, left Ghia in 1953, they formed their own carrozzeria, which they operated until 1956, when it merged with Fiat. At Fiat, Gian Paolo Boano served as the first director of their design department, and his father served as a design consultant.

Henry Ford II met and became friends with Mario Boano during one of his many trips to Italy in the early 1950s. He also saw Mario Boano on Boano's yearly trips to Dearborn. After the Boanos left Ghia and established their own custom body building firm in 1953, Mario Boano solicited business from both Ford and Chrysler, and his firm designed at least one Chrysler concept car as well as the Lincoln Indianapolis Coupe.

At Henry Ford II's direction, Lincoln-Mercury Division shipped a 1954 Lincoln chassis equipped with an experimental Continental Mark II engine and transmission to the Boanos' Turin shop in mid-1954. The Indianapolis Coupe was built on that chassis. Ford agreed to let the Boanos show the Indianapolis Coupe on the European show circuit, including the Turin Auto Show, before it was sent to the United States, but it was not favorably received. Critics thought the front end of the Indianapolis Coupe was too bulbous and the car too gimmicky.

These touched up photographs of the Indianapolis Coupe are from the cover of the November 1955 issue of Auto Age *magazine.*

When the car arrived in the United States in late 1955, the only publicity it generated was an article in the November 1955 issue of *Auto Age*. There was also a single picture of the Indianapolis Coupe at the Turin Auto Show in the August 1955 issue of *Road and Track*. The *Auto Age* article indicated that the Indianapolis Coupe was built as a personal car for Henry Ford II, although it hinted that the car represented "a pilot design of a radical new Lincoln style." What little publicity there was about the Indianapolis Coupe in the United States was also generally unfavorable. Henry Ford II kept the Indianapolis Coupe for only a short while and, in 1956, he sold it to a friend, movie star Errol Flynn.

The Indianapolis Coupe passed to Flynn's niece when he died in 1958, and she later sold it to The Vintage Auto Store in Nyack, New York. The next owner bought the car with plans to restore it but, after getting the first bill from the restoration shop he had chosen to do the work, he quickly sold it to friends, Chuck and Rita Hannah of Hawthorne, New Jersey, who have owned the Indianapolis Coupe since 1972. They believe they are the fifth or sixth owners. When the Hannahs bought the Indianapolis Coupe, it ran well but had been damaged by a fire that scorched the instrument panel and driver's seat. The backlight was also broken out and the windshield was cracked. (The backlight was apparently broken at the time of the fire to gain access to the inside of the car.) The Hannahs, who owned a restoration shop, were slowly restoring the car, but restoration had not yet been completed when Chuck Hannah died in 1997.

After they bought the car, the Hannahs contacted Ford Motor Co. in an effort to confirm the car's history. Ford executives expressed some interest in repurchasing the Indianapolis Coupe, but told the Hannahs that most of the records on the car had been destroyed when the Ford Rotunda burned in 1962.

Photograph from August 1955 Road & Track.

No Ford designer can now be located who remembers the car or acknowledges the possibility that its design was reviewed or approved by the Ford Design Department. Given the Boanos' design skills and the car's general similarity to other cars they designed, it is probable the car was designed entirely by them. The Indianapolis Coupe was apparently never shown at any auto show in the United States, and it had no influence on any production Ford product.

The Indianapolis Coupe has a 123-inch wheelbase, the same as the production 1954 Lincoln. The body is all-steel and the car is fully functional. At the openings at the back of the front fenders are fake exhaust pipes. The openings at the front of the back fenders are also fake. Contemporary news articles explained that if the car was put into production, the front fender openings could be used to exhaust engine heat, and the rear fender openings could be used to cool the brakes. The spare tire is located in a drawer below the trunk. The center of the back bumper, between the taillight pods, pulls out with the drawer for access to the spare tire.

Ghia

Mario Felice Boano

Mario Boano began his working career as an assistant to a bronze-sculptor who was a friend of his father's. He also studied drawing, bas-relief and sculpture. When he completed his schooling, Boano took a job with Farina & Co. as an assistant body designer. He was promoted to designer and then made supervisor of all draftsmen. Later, he was appointed chief designer for Farina. In 1930, the Farina brothers broke up, and Boano joined Pininfarina as its chief designer. In 1934, Boano opened his own business building wooden automobile bodies for Pininfarina, Ghia, Bertone and others.

When Ciacinto Ghia, the company's founder, died in 1943, his family invited Boano to take over the business. Boano rebuilt the war-devastated business and turned it into one of the premier Italian carrozzerias. During the time he ran Ghia, Mario Boano designed cars for Fiat, Lancia, Alfa Romeo, Delage, Talbot Lago, Delahaye, Ferrari and others. Starting in 1950, Ghia designed and built a number of Chrysler show cars. (Ghia was purchased by Ford in 1971.)

In 1949, designer Luigi Segre bought an interest in Ghia. He, Boano and Boano's son, Gian Paolo, who was also a designer, worked well together until 1953, when a disagreement led to the Boanos' departure. Thereafter, the Boanos formed Carrozzeria Boano, also located in Turin, Italy. The Indianapolis Coupe was designed by Mario and Gian Paulo Boano and built at Carrozzeria Boano. In 1956, Carrozzeria Boano was bought out and merged into Centro Stile Fiat, where Mario Boano served as a consultant, and his son Gian Paolo Boano served as design director. ■

(above) Road and Track described the Lincoln Indianapolis Coupe as an "unfortunate jazzy design needing little comment and less notice."

(left) The Indianapolis Coupe in 1972.

(below) The Indianapolis Coupe in 1996.

Chuck Hannah

Chuck Hannah

The car has a only a front seat with a one-piece seat bottom and separate seat backs that fold forward to give access to a small luggage area behind the seats. When built, the Indianapolis Coupe was painted bright orange and upholstered in orange and white pleated leather. Like the Futura, the instrument panel on the Indianapolis Coupe can be closed-off by a single metal panel that, when unlocked, slides down and out of sight. The car has a metric speedometer, and the odometer shows 12,000 kilometers on it. (about 7200 miles)

The design of the Indianapolis Coupe was not one of Henry Ford II's favorites, and it was not indicative of the direction of any future Ford Motor Co. designs. ■

4

Gil Spear's Advanced Design Studio

Muroc

Volcano

Chesterfield

Madagascar

Syrtis

Cougar

Ventura

Cavalier

Bolero

Marlin

Lancer

Cutlass

X-Turbo

Gran Turismo

Henry Ford Museum and Greenfield Village

Concept cars on display at a 1952 Advanced Studio open house. The Muroc is followed by the Cutlass, Bolero, Marlin and the Cavalier.

After the appointment of Harold Youngren as Ford's chief engineer in mid-1946, personnel at the Engineering Department increased six-fold in as many years. The Design Department was still a part of Engineering and, with Youngren's appointment, the number of designers also began to increase just as fast as Ford's offer of more money could recruit them from GM, Kaiser-Frazer and elsewhere.

Tom Hibbard was appointed Ford's design director on January 1, 1947. Although he was experienced and well respected, most Design Department employees thought he was a nice guy, who was in over his head. More importantly, Ford's new management soon concluded that Hibbard was past his prime and not the aggressive or effective leader they needed to catch GM. Body engineer John Oswald was placed in charge of the Design Department, and he soon replaced Hibbard with George Snyder, another recent GM alumni, in an effort to totally remake the Design Department from top to bottom.

Designer Eric Ramstrum had been in charge of Gregorie's small advanced studio but, by the time Oswald and Snyder took charge, Gregorie's advanced studio had already gone by the wayside as did Ramstrum. Snyder instituted a new studio system, which included walled-off, separate studios where designers worked behind locked doors on Fords or Mercurys or Lincolns, but not on all three as had been past practice.

Next in line for change was creation of a new advanced studio. Although some of the old guard in Ford's Engineering Department didn't like the GM-style changes that were taking place, George Snyder recommended re-establishing an advanced studio. Youngren and Oswald thought Snyder's recommendation made a lot of sense, so Snyder's assistant manager, former GM designer Gene Bordinat, asked A.G. "Gil" Spear to step down as head of the Ford Studio and take on the responsibility for the advanced studio.

From Spear's standpoint, the timing was excellent. He had just finished the design of the production 1950-51 Fords, and his assistant, former GM designer Frank Bianchi, was there to take over as head of the Ford Studio. In his new position as head of the Ford Advanced Studio, Spear's job was: 1) to try to forecast what Ford vehicles would look like in five to ten years; 2) to develop ideas the production studios could incorporate in their current designs and; 3) to train Ford's new designers to design cars "the Ford way." Spear soon made the Advanced Studio a place where the heads of the production studios could observe the progress made by the newly-hired designers and, when they were ready, invite them to join the production studios.

Because they felt that 3/8-sized models showed better detail, Spear and Snyder decided that future models should be built in 3/8-scale, rather than the quarter-scale previously in use at the Design Department.

Muroc

A. G. Spear Jr.

Gil Spear grew up in New York where both of his parents were commercial artists. Spear started as the first industrial design student at Pratt Institute in 1934 but had to drop out in 1935 because of a lack of funding. He took a job as a cartoonist and, in 1937, he was hired at GM as a designer in Bill Mitchell's Cadillac studio. When Spear was laid off at GM in 1938, he took a job with Norman Bel Geddes designing theatrical sets. When Bel Geddes was hired to design the GM exhibit at the 1939 New York World's Fair, Spear assisted in building an elaborate city of the future ride. After the World's Fair was over, Spear took a job in Chrysler's design studio and was responsible for the design of the 1942 Chrysler grille. He left Chrysler in June 1942.

During World War II, Spear worked for Briggs Manufacturing as an aircraft salvage engineer. Towards the end of the war, he also started his own industrial design business.

In April 1947, Spear decided to build a new house for his family. To qualify for a construction loan, the bank told Spear he needed a regular pay check. Answering an ad in the newspaper, Spear ap-plied at Ford's Design Department, where he intended to work only long enough to make it look good to the loan officer.

Spear's first assignment at Ford was to complete the 1949 Ford clay model that Oros and Engel had started, so engineering drawings could be finalized from it. When Oswald instituted the studio system in 1947, Spear was appointed first head of the Ford Design Studio and he was in charge of updating the 1950 and '51 Fords.

In May 1948, Gene Bordinat asked Spear to organize Ford's Advanced Studio as a means of getting fresh design talent into Ford and providing new ideas for future Ford products. In late1954, Spear became head of the Special Projects Studio, where he supervised the design of the D-523 and the D-524. In mid-1955, Spear left the Styling Center to become the head of the Engineering Department's separate design studio. (Special Development Studio)

While at the Special Development Studio, Spear supervised completion of the D-523 and D-524 and development of the Bimini, the Midshipman and the Personal Safety Vehicles. In February 1958, Spear returned to the Styling Center as an executive designer in the Ford pre-production studio. He later became an executive designer for international operations and in the Lincoln-Mercury studio.

From 1967 to 1971, Spear was chief designer at Ford of England. He retired in 1974. Spear continues to work with patent attorneys perfecting new products. ■

The Muroc was designed and built in late 1948. It was the first 3/8-sized concept car specifically designed by Spear for review by his still-skeptical bosses in the Engineering Department. Spear knew it had to be "hot," different and an attention getter, but that it also had to be within the package set by Engineering. The hood could be lower than package, because no one knew how low engines or radiators would be in the future, but Engineering wouldn't let the roof or windshield be as raked as Spear wanted. As the clay model was being built, Spear reluctantly asked clay modeler Larry Wilson to raise the roof. From the perspective of 50 years later, Spear still thinks the roof and windshield look stuffy.

The finished Muroc was built from plaster, and the top was made from hammered metal. The car was named after the dry lake where California hot rodders raced. As with many of Spear's later concept cars, the Muroc's features were purposely exaggerated for effect. The exterior had blisters over each wheel and reverse C-pillars that were very dramatic. It also had front bucket seats that rotated 90 degrees outward when the doors were opened, and a steering wheel that could be rotated from one side to the other. The speedometer incorporated a light-projected gunsight idea used on World War II fighter planes, that could not be used on the production 1949 Ford because it was too costly and, as Spear puts it, because there was a "fear of the unknown." Outside suppliers helped develop an illuminated disk that projected the speed on the windshield. To make sure the feature could be seen on the 3/8-sized Muroc, Spear put numerals all the way across the windshield. The idea was that the numerals showing the car's speed would move across the windshield and be illuminated in front of the driver as the car's speed changed.

The Muroc was shown several times at the Design Department and at the Ford Rotunda. For several years it was also on display in the Advanced Studio. No one knows what happened to it, but Spear believes that after it became outdated or chipped, it was probably destroyed or taken home by someone. ■

Ford Photomedia

(above) Gil Spear with the Muroc.

(below) The speedometer was at the base of the windshield. The Muroc also had a pivoting steering wheel and a rolling map box.

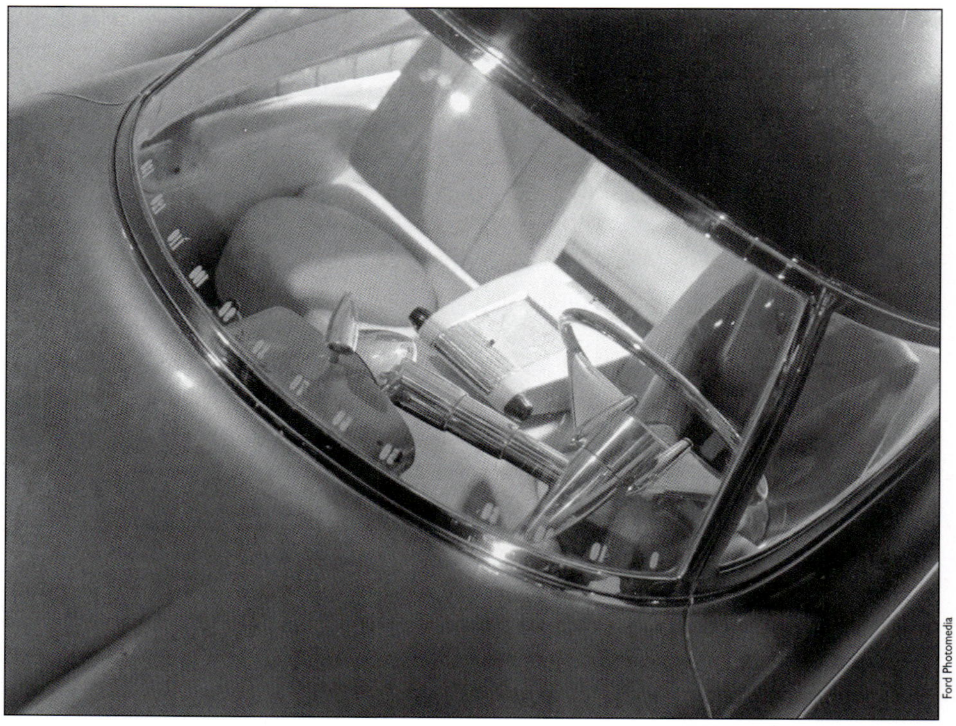

Ford Photomedia

Gil Spear

(above) The body of the Muroc was as low to the ground as Spear could make it.

(below) Fender blisters over the wheels was a design theme Spear favored because it allowed the car's mass to be lower to the ground.

Gil Spear

(below) Harold Youngren, Ford's chief engineer, with the Muroc at a 1952 Advanced Studio open house.

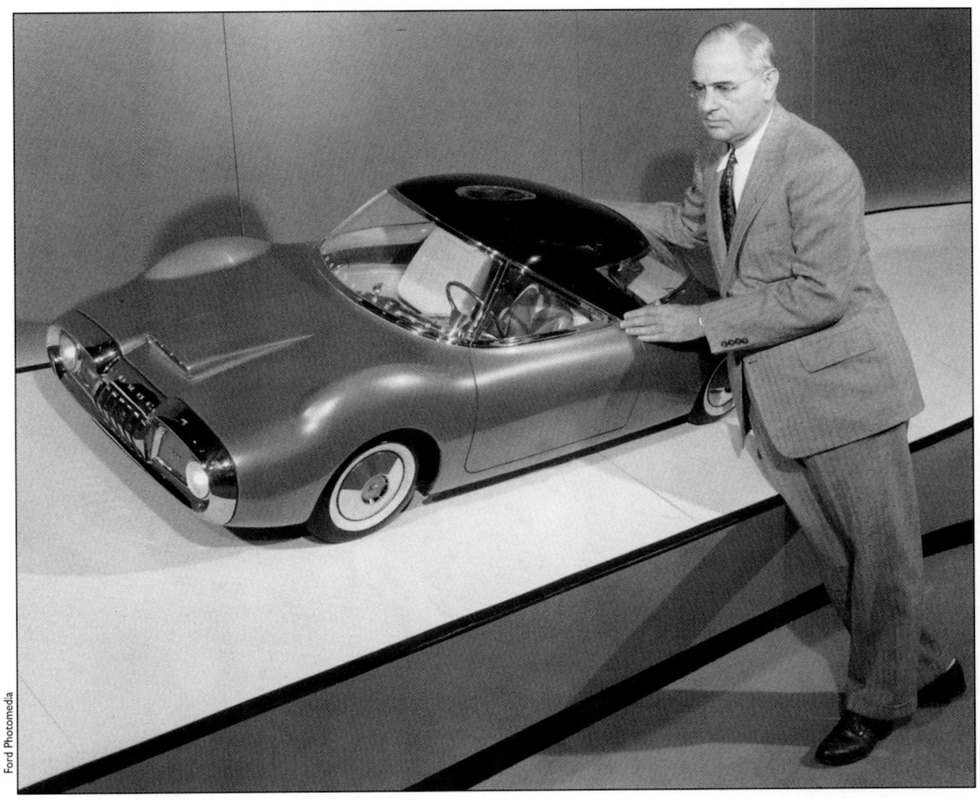

Ford Photomedia

Harold T. Youngren

Harold Youngren started his automotive career in 1910 as a draftsman at Allis-Chalmers Manufacturing Co. In 1929, he was hired as an engineer at GM's Oldsmobile Division and, in 1933, he was appointed the division's chief engineer. In 1944, he left Oldsmobile Division to become chief of engineering development at Borg-Warner. When Ernest R. Breech left Borg-Warner to become Ford's executive vice president, he and Henry Ford II asked Youngren to become Ford's chief engineer. Youngren accepted the offer because he felt his job at Borg-Warner did not allow him enough time to be an engineer.

Youngren started at Ford August 1, 1946. Even before he began, he reviewed the Ford Motor Co. cars planned for the 1948 model year and determined they were generally too big and heavy. At Breech's request, Youngren drew up specifications for a new postwar Ford, which he based on the measurements of the 1947 Studebaker. It was eventually produced as the 1949 Ford.

Youngren was also responsible for the redesign of the flathead V-8 used in 1949-52 Ford products. He also initiated and supervised development of Ford's first overhead valve V-8 which was introduced in 1952. Before he retired in 1952, Youngren increased the size of Ford's Engineering Department from 300 to 1,800 engineers and started construction of many of Ford's new engineering buildings. ■

Ford Photomedia

The Advanced Studio in1952. This photograph is from the book Ford at Fifty. *On the left, and partially hidden by the Muroc, is clay modeler Larry Wilson. Behind Wilson is the Bolero. To the right of the pillar is the Chesterfield. At the right, the man standing between the seated designers is Jim Huggins. To the right of the pillar is clay modeler Andy Lesko. According to Spear, the three-eighths sized model in the middle of the photograph was used by Bill Schmidt as inspiration for the XL-500.*

Ford Rocketeers

In the early 1950s, Ford studio designers built an elaborate three-foot high rocket out of cardboard, put a fake fuse in it and lit it next to the desks of managers John Najjar and Duncan McRae. After the rocket fizzled, and everyone had a good laugh, Najjar and McRae thought it would be funny if they did the same thing to their boss, Bob Maguire. When they returned from lunch, Najjar and McRae took the rocket into Maguire's office, propped it up on a chair, lit the fuse and ran out closing the door behind them. What they didn't know was that while they were at lunch, Lyle Briggs and Al Mueller put two cherry bombs and a lot of red paper steamers inside the rocket. As Najjar and McRae were waiting in the hall, they heard a huge explosion. Shaken, they ran back into Maguire's office to find him still sitting at his desk, but his office was full of thick blue smoke, with red paper and cardboard particles raining down from all over the room. All Maguire said was that he presumed they were going to clean up the mess. In the meantime, Charlie Waterhouse, head of the Design Department, came running from the other end of the building to find out what the explosion had been. Maguire told him that it must have been a tire that blew out. ■

Otto M. "Pete" Renz

Pete Renz

Pete Renz grew up less than a mile from Ford's Twin Cities Assembly Plant. As a four year-old, he remembers looking out the window at Model A chassis being driven down the street from the assembly plant. During World War II, Renz served in the U.S. Navy. Because he loved to draw cars, after he was discharged from the Navy, he wrote letters to the Ford and GM design departments. Based on the recommendations he received in response, he applied at Pratt Institute but was unable to start classes until 1947.

When he graduated in1950, he and classmates Bill Wheeler and John Weisbrod interviewed at the Ford Design Department and were hired. His first assignment was in Gil Spear's Advanced Studio, where he did design work on the Cavalier. He also did renderings of the Muroc and designed the stylized M used on the wheelcovers on one side of the Muroc. In 1951, Renz was transferred to the Mercury exterior and later to the Mercury interior studios.

In 1955, Don Stumpf moved to American Motors and persuaded Renz and Jim Pappas to join him. Renz stayed at American Motors until August 1964. After he left American Motors, he worked in several Detroit-area job shops. He was rehired at the Ford Design Department in 1966 and was later promoted to supervisor in the interior trim department. Renz retired in August 1981. ■

The Volcano in the Advanced Design Studio in 1953. In the foreground, from left to right, are clay modeler LaMarr Ferguson, clay modeler Larry Wilson, designer Bill Balla, clay modeler Cecil Perkins and clay modeler Irving Stollman.

Volcano

The Volcano does not look like a revolutionary design proposal until you realize it was designed in 1949. When he designed it, Ken Spencer had just graduated from Pratt Institute and was one of several recently-hired designers assigned to Gil Spear's Advanced Studio. The Volcano was the first car designed by Spencer after he started at Ford. Although most cars designed by new designers never got farther than 1/8-sized clay models, Spencer's Volcano was eventually made into a 3/8-sized plaster model. Although other designers were designing cars that had individual features that showed up on production cars built several years later, the Volcano accurately forecast the long, low, streamlined look that eventually became the 1956 Lincoln. The Volcano also forecast the two-tone paint schemes and the compound curves of the wrap-around windshield that predominated on the mid-fifties cars.

The Volcano started as a design exercise for Spencer. He was asked to render many different models of the car, but it is likely the Volcano was intended to be a two-door hardtop from the very beginning. The purpose of rendering different versions of a car was to take into account the fact that production automobiles had to look good not only as two-door hardtops but also as four-doors, and to exhaust the design to the point where the designer and his managers were certain the design selected was the best it could be.

Because Spencer was new, clay modeler Larry Wilson was assigned to help him complete the design and the presentation that went with it. As Spencer remembers it, Larry Wilson was not only a clay modeler, he sometimes acted in the capacity of a designer, especially in the Advanced Studio.

The Volcano was used in studio photographs starting in 1950, when it was completed, until sometime after mid-1953, when the new design building was occupied. Spencer does not know what happened to the Volcano after that, but believes it was destroyed when it became too chipped to photograph. ■

C. Kenneth Spencer Jr.

Ken Spencer grew up in New York and New Jersey, and always wanted to be an automobile designer. After serving in the Army during World War II and fighting in many European ground battles, including the Battle of the Bulge, Spencer attended Pratt Institute on the G.I. Bill majoring in industrial design.

He graduated in 1949 and took a job as a designer at GM. After he was hired, but before he could report for work, he was told he could not work at GM because he was the nephew of Bob Maguire, a designer at Ford. When Maguire found out his nephew had applied at GM and, before he knew the job offer had been revoked, he went to Gene Bordinat, and they both offered Spencer a job at Ford.

Spencer started his design career in Gil Spear's Advanced Studio in July 1949. He was soon transferred to the Ford studio working under his uncle. While in the Ford studio, he worked on the 1953-55 Ford production cars. In late 1953, he was assigned to the Advanced Studio again, while he was waiting to transfer into the Lincoln-Mercury studio.

Spencer designed the Volcano in 1949. He also worked on the Futura and later supervised construction of the 3/8-sized model of the Turnpike Cruiser sent to Ghia, so a fully-operable car could be built.

In November 1955, he was appointed executive designer in the Edsel studio. He was responsible for the exterior design of the Edsel. While in the Edsel studio, he and Romeyn Hammond designed the Shrike concept car for the Stylerama program.

Later, Spencer was assigned to the interior studio under David Ash. He and Jim Sipple also designed the Cougar II (Bordinat Cobra), the LaContessa, and the Mini-Max concept cars.

Throughout his career, Spencer worked in all studios and managed several of them, including a tour as chief designer in the truck studio. He retired in 1982. ■

Ford Design Center

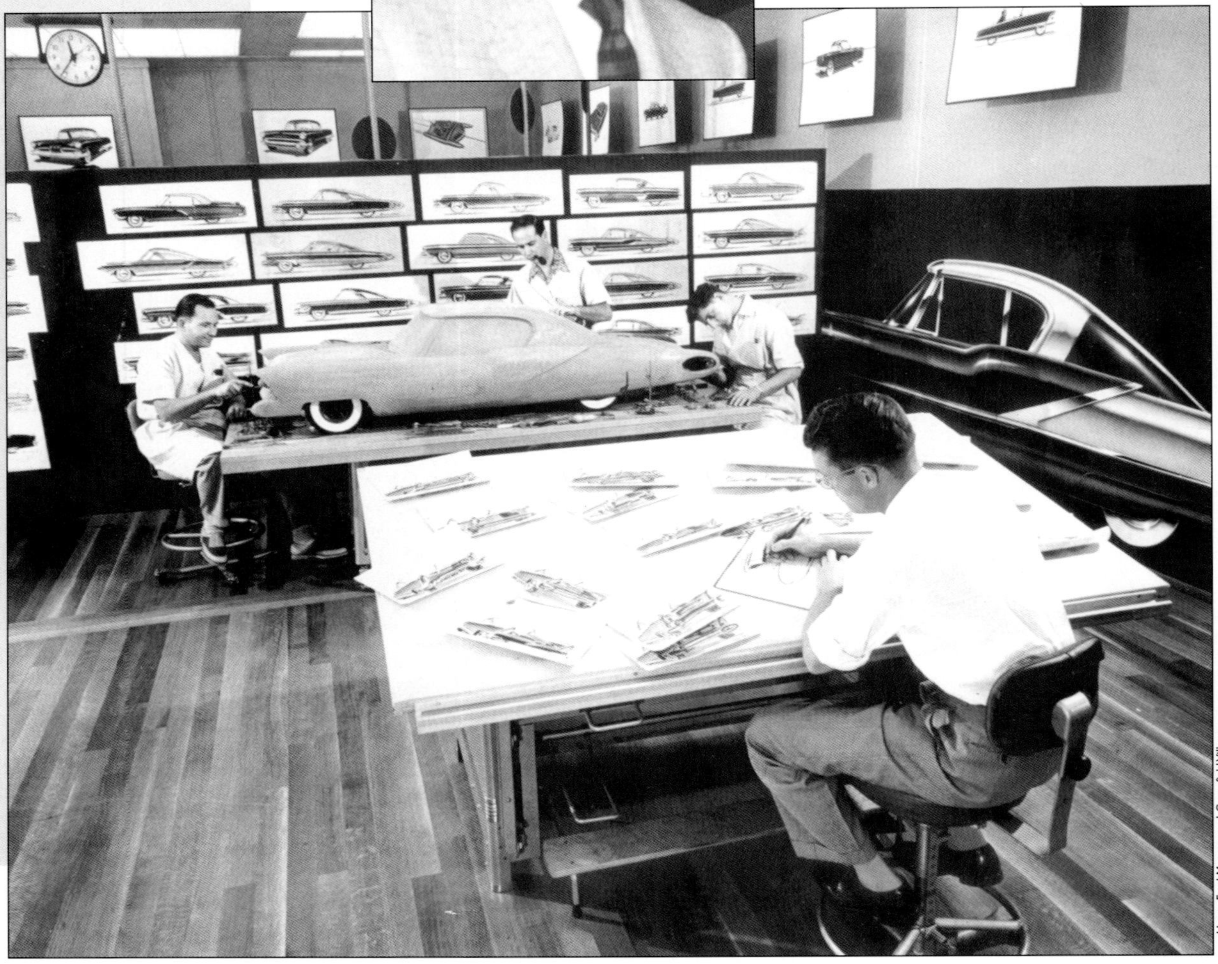

Henry Ford Museum and Greenfield Village

Ken Spencer is at his drafting table. While designing the Volcano, Spencer prepared the alternate renderings shown behind the clay modelers. Left to right are clay modelers LaMarr Ferguson, Larry Wilson and Bob Genslak.

The front fender lines of the Volcano forecast the 1956 Lincoln.

Chrissie Tremulis

The postwar new-car line up as proposed by Bob Gregorie was the same one proposed by Edsel Ford before he died. It included a Light Ford, the Ford, the Mercury, two Lincoln models, a Continental and a formal Lincoln Custom. Gregorie's designers were ultimately unsuccessful in designing a Continental using the sheetmetal of the proposed Lincoln Cosmopolitan. To compound the problem, newly appointed Ford executive vice president Ernest R. Breech soon became Gregorie's nemesis at Ford. He ultimately cancelled the Light Ford, the Lincoln Custom, the Continental and the new Ford, as Gregorie had designed them, because he didn't believe they would make a profit. Although Gregorie usually disagreed, whatever the argument was about, Breech prevailed. Gregorie has always maintained that the 1949 Mercury, which he originally designed as the new postwar Ford, was his all-time best design.

Gil Spear

The Chesterfield was originally designed in 1949 as a proposed Continental. By the time this model of it was made, the grille had been changed and it became a convertible.

Chesterfield

Probably the most obvious explanation for the absence of the Continental, when the 1949 models were unveiled in mid-1948, was that there had been just too much to do, given the resources needed to design and build Breech's new Ford in the short time available. However, that wasn't the end of the Continental -- even temporarily.

Between 1945 and 1952, the Design Department came up with many Continental proposals. If nothing else, they indicate that it was always hoped that next year there would be a new Continental, even if it had to be postponed from year to year. It is not an exaggeration to say that every designer who worked at Ford between 1945 and 1952 came up with his or her own Continental proposal. Ford management even found it necessary to caution those in charge at the Design Department not to squander time or money on new Continental designs and, yet, renderings and clay models of Continentals proliferated during those years. Many of them were done by designers on their own time. The Continental was definitely far from dead.

The Chesterfield was designed by Gil Spear in 1949. It was originally designed as one of four 3/8-sized clay models being considered for the proposed 1951 Continental. In 1952, the name of the model was changed to the Chesterfield, and it was shown in an open house held in the Advanced Studio for families of Design Department personnel, Ford executives and the press. According to Spear, the Chesterfield was named after the English Lord Chesterfield who was noted for his elegance and suaveness. The design of the Chesterfield was changed at least once. It started as a coupe but was later made into a convertible. The grille and front bumper were probably changed at the same time. What happened to the Chesterfield is unknown. ■

Gil Spear

(above) The vent windows on the Chesterfield were designed as a continuation of the windshield.

Gil Spear

(above) Taillights were on the top of the Chesterfield's fenders.

Fred Hoadley

(above) The Chesterfield started as a Continental two-door sedan.

(below) Fred Hoadley is clay modeling the car that eventually became the Chesterfield.

Fred Hoadley

Madagascar

As World War II was winding down in Germany, Robert M. "Bob" Thomas got leave and, in March 1945, returned to the United States aboard the Queen Mary. When his leave was over, he reported to Fort Sheridan, Illinois, and was discharged. He applied for a job at GM, was hired and assigned to George Snyder's Oldsmobile Studio.

Like Snyder, Thomas was a blackboard designer who relied less on artistic illustration and more on drawing full-sized layouts. Those layouts were like a blueprint in that they showed all the sections and highlights of the car being designed.

When Snyder left GM and then got a job at Ford in 1947, Thomas followed. Thomas' first assignment at Ford was in a special advanced studio working with Snyder on a proposed 1949 Continental. After that project was finished, Thomas remained in that special advanced studio, where he and several other designers made sketches of advanced cars for Snyder. When Gil Spear was appointed head of the reorganized Advanced Studio, Thomas was assigned to help him train Ford's new designers and also to come up with design innovations for future Ford products.

Although Spear thought of Thomas as his assistant manager, Thomas never held that title. Thomas enjoyed working with the new designers and on the advanced cars being built in the studio, but his teaching duties left very little time to work on advanced cars of his own design — except the Madagascar. According to Thomas, Spear thought the design looked good enough to build as a plaster model and, in 1948-49, the Madagascar was built as a 3/8-sized model. The Madagascar was a four-door Ford that featured a Plexiglas roof, a spinner front end, wrap-around bumpers, crown-type C-pillars, a horizontal roof aerial and a sculpted boattail deck lid that came to a point at the rear. The car was painted a medium-blue, and the bumpers and brightwork were covered with paint or foil. Although Thomas acknowledges that his use of the crown over the roof was not the first time that feature had been used, the Madagascar probably influenced the crown on the XL-500 and the 1955-56 Ford Crown Victoria.

According to Thomas, the spinner he designed on the front of the Madagascar was much too heavy and expensive for a production car. He also recalls that naming the car the Madagascar had very little to do with the island off the coast of Africa of the same name. Thomas picked the name because he liked its inherent play on words.

In 1949, when Thomas left Ford for a job at the Nash design department, the Madagascar was still on display in the Advanced Studio. He doesn't know what happened to it after that, but he did see it again in the 1953 book *Ford at Fifty*. That photograph is reprinted at the top of page 80; the Madagascar is partially shown in the lower right hand corner. ■

The Madagascar featured a large spinner on the front end. The engine air intake was below the bumper.

Bob Thomas

Fred Beamish

Bob Thomas

(above) The Madagascar featured a crown over the roof that probably inspired the XL500 and the '55-56 production Crown Victoria.

Robert McGuffey Thomas

Because Bob Thomas was a descendant of the author of the McGuffey Readers, as a teenager, he was hired by Henry Ford as a guide at Greenfield Village. Thomas had training in mechanical drawing and art, so he was soon made an assistant to Irving Bacon, Henry Ford's personal artist. In September 1937, Bacon arranged for Thomas to become an apprentice at the Design Department, where he worked on the interior of the original Continental. In August 1939, Thomas left Ford to take a job in the Hudson design studio.

After serving in the Army Corps of Engineers during World War II, Thomas took a job in GM's Styling Section working in the Oldsmobile studio as George Snyder's assistant. In July 1947, Thomas followed Snyder to Ford but left in October 1950 to work as a designer at Nash.

In October 1952, Thomas received a call from Gordon Buehrig who was the chief body engineer on the Continental Mark II project. Thomas became John Reinhart's executive designer on the Mark II design team. Thomas and Charlie Phaneuf share responsibility for the design of the Continental star.

After the design of the Mark II was completed, design work at Continental Division was slow, so Thomas was transferred to the Design Department in February 1955. Immediately after George Walker's reorganization, Thomas transferred back to Continental Division and become head of their advanced studio, where he worked on the design of the Mark III Berline. In March 1956, just before the demise of Continental Division, Thomas was assigned to the Styling Center as chief design analyst. In November 1961, he was appointed a design manager. In 1965, he became a design executive at Ford of England. He returned to Dearborn in 1968 and became a design executive in the Ford and Lincoln-Mercury studios before he retired in 1974.

Thomas was on the team that designed the 1961 Lincoln. He also worked on the Levacar, the LaGalaxie, the Avventura, the Stiletto and the Aurora concept cars and also the Pinto.

Thomas was an accomplished musician, whose band played at many Styling Center functions. ■

Bob Thomas

(above) The Madagascar also featured a roof-mounted aerial and an unusual headlight configuration.

(below) The wrap-around rear bumper and veed deck lid were considered advanced features when the Madagascar was designed.

Bob Thomas

Syrtis

In late summer 1948, just about the time Gil Spear was moving from the Ford studio into the Advanced Studio, he started hearing rumors about a new Buick model called the Riviera hardtop convertible. He thought the idea was interesting, but felt the name selected for it — hardtop convertible — was false advertising, because the top did not actually retract.

On the morning of August 21, 1948, Spear was on his way to work when he stopped at a red light, right next to a prototype of the 1949 Buick Riviera hardtop convertible. Spear looked the car over carefully and, feeling a little mischievous, asked the driver what the car he was driving was called. When the driver responded that it was a new Buick hardtop convertible, Spear challenged him to put the top down. As the driver of the Buick sat there with a puzzled look on his face, the light changed to green, and Spear drove off chuckling to himself.

That incident started Spear thinking about how he could build a true hardtop convertible. By the time he arrived at the Design Department, he had come up with an idea for a car with a steel roof that actually slid down on top of the trunk. When he got to his desk, Spear quickly sketched his idea using two pieces of tracing paper, one for the body of the car and the other for the top. It took Spear some erasing and redrawing to get the proportions right, but the first retractable hardtop Spear came up with had a long rear deck and a top that slid back and down over the trunk lid. The backlight followed the contour of the back end of the car and covered that area, so the license plate could be seen through the backlight.

Gil Spear

A one-eighths sized model was built to test the retractable-top concept. Newly hired designer Barney Harb is holding a picture of one of Gil Spear's renderings of the Syrtis. The name Syrtis was first used on this rendering which was done after the one-eighth sized model shown on Harb's drafting table.

Part of Spear's job as head of the Advanced Studio was to anticipate future design trends. It was his belief that the next generation of automobiles would adopt the fighter plane look, which meant shorter hoods and longer decks. He thought this trend could easily accommodate a real hardtop convertible without any break in the roof. Although Spear was aware of earlier single seat retractable hardtops, including the Chrysler Thunderbolt, he was trying to come up with one that included a back seat as well as a front seat.

The three-eighths sized Syrtis on display at the Design Department.

Spear next made a rendering on black paper and colored it with an airbrush. He completed the rendering on October 1, 1948, and it was witnessed by designers Bob Thomas and Roland Taylor. Spear entered his finished rendering in the Advanced Studio's design book as "SX (styling experimental) 109" and then hung it on the wall in his studio.

The rendering of Spear's hardtop convertible stayed on the wall in the Advanced Studio for several months before he found time to work on the idea again. Spear's next retractable proposal was in the form of an animated cardboard cut-out built by designer George Krispinsky. The purpose of the cardboard cut-out was to show non-designers, including Ford executives, how the concept worked, and how easily the concept could be incorporated on future production cars. It was finished in November or December 1950 and hung on the studio wall. The animated cardboard model had a top that could be cranked up and down by a bent paperclip, and it soon became the center of attention. Many Ford executives, including Lewis Crusoe and Benson Ford, were fascinated with the cardboard model and often stopped to work the paperclip crank.

When Spear designed his hardtop convertible, he felt there were three basic concepts he had to incorporate in the design to give it the best chance of making it to production. Whatever else, the design had to be uncomplicated. The mechanism Spear came up with needed only three moving parts; the roof, the backlight and the deck lid, and it needed one electric motor and two limit switches; one to stop it when it was fully retracted, and the second to stop it when it was fully raised. Spear also designed his retractable hardtop convertible with ample luggage room, whether the top was up or retracted. Spear even designed a miniature extra-large-sized golf bag to match the one Henry Ford II used. The

Spear's original rendering of a retractable hardtop was completed October 1, 1948, and witnessed by designers Bob Thomas and Roland Taylor.

third requirement Spear felt was necessary for the success of his design was a top that retracted inside a six-foot high garage. Only Earl MacPherson and Henry Ford II asked Spear about retracting the top inside a garage.

It wasn't long before Gene Bordinat, then assistant manager of the Design Department, asked Spear to follow up on his retractable idea, but he suggested Spear redesign the top so that it collapsed into the trunk and not on top of it. Bordinat felt a top that retracted on top of the trunk was too restrictive to incorporate in future designs. At first, Spear resisted Bordinat's suggestion that the top retract into the trunk and insisted that it was much simpler if the top retracted on top of the trunk. After they talked about it over several weeks, Spear became convinced that Bordinat's position was correct, especially if they ever hoped to build a production retractable.

Because Spear thought the Ford Division was most likely to build a retractable model, he drew this 1953 Ford retractable to show management.

Gil Spear

After talking with Bordinat, Spear began to design a working retractable with a roof that slid into the trunk. He also asked Jim Huggins, a body draftsman, who was being trained as a designer, to try to come up with an alternate system to the one Spear had proposed. Huggins' retractable proposal was called the Saturn but, before he could build a model of it, he was transferred to a production studio.

In early 1951, Spear finished the design of a 1/8-sized model of the retractable hardtop. It used two S-shaped tracks upon which the top slid into the trunk. Using technology he picked up during the war, Spear also made the trunk so it could be open from either end. The 1/8-sized car was modeled by Larry Wilson and Mitchell Rukat. By the spring of 1951, Wilson and Rukat had finished the clay model and had started to cast a 1/8-sized plaster model. At the same time, Howard Fifield of the Design Department metal shop began making the metal trim and a roof for the 1/8-sized model. By squeezing it in between the other projects he was assigned, Fifield was able to complete and assemble the finished 1/8-sized model by Christmas 1951. While the 1/8-sized model was being built, Spear began making a loose-leaf notebook to help explain the design and advantages of the retractable -- and every alternate method of technology he thought Ford executives could come up with.

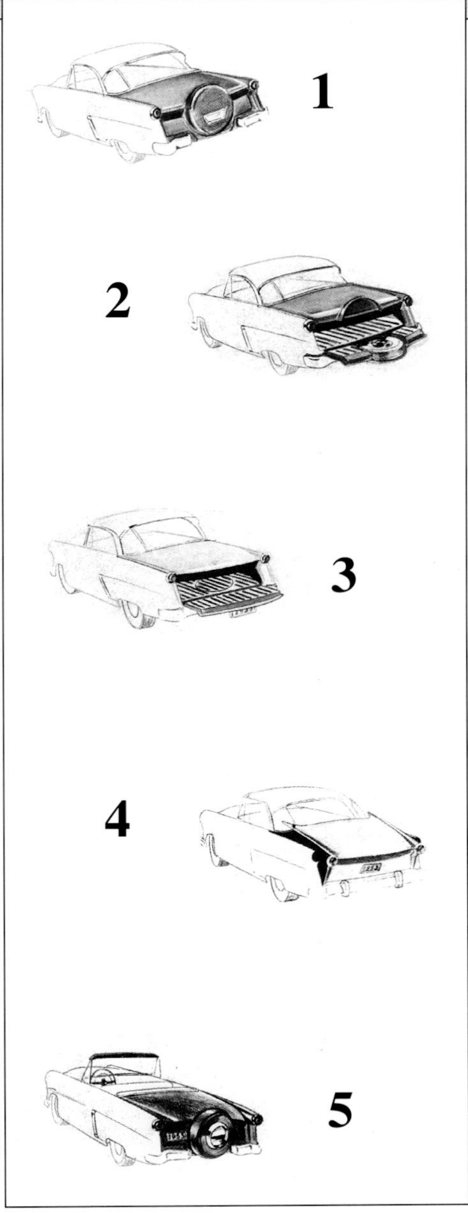

Gil Spear

(1) The deck lid opens at the front for retraction of the roof, which retracts under the deck. (2) Luggage can be loaded through the tailgate, which can also support the spare tire. (3) The tailgate can also be used for loading. The deck lid opens at the front, and the spare tire can be located under the floor. (4) The deck lid opens from the front or rear. (5) The roof-down position is over a fixed rear deck, and the tailgate can be used for loading.

When Spear designed the 1/8-sized model, the Design Department was still a part of Engineering. Earl MacPherson was Ford's assistant chief engineer and, because of his position, he was also second in command at the Design Department. On one of the occasions MacPherson was visiting the Design Department, Bordinat showed him photographs of completed designs, including Spear's retractable. Bordinat also showed MacPherson the notebook Spear had prepared. MacPherson glanced at the picture of the retractable, but became engrossed in Spear's notebook and its explanations and alternatives. After he finished studying Spear's notebook, he asked if he could see the model. After MacPherson saw it, he immediately had a production number assigned to it and allocated money to build a 3/8-sized model with a fully-working roof mechanism. MacPherson also asked Spear to start preparing drawings for a patent application. In August 1951, Spear and Ben Kroll, another designer assigned to the Advanced Studio, began preparing patent drawings which, when completed, were sent to Ford's patent office.

After MacPherson gave the go-ahead, Spear began designing what he thought a 3/8-sized model of the retractable should look like. Although the patent drawings had to be similar to a production Ford product, Spear decided on a different design for the 3/8-sized model. He used the dimensions of the 1952 Ford but, to encourage as much attention as possible, the front fenders and doors were designed to look like what Spear thought future rocketships might look like, with streamlined shapes forming around the headlights and following along the sides of the car to the rear bumper and ending in small skegs or fins at the bottom of each back fender. A vendor suggested hiding the taillights behind see-through chrome plating (probably vacuum metalizing), but the process was not developed in time to use, so the taillights were exposed. The rendering Spear prepared showed the body of the retractable painted white with mother-of-pearl in the paint and with a dark metallic-blue top.

It wasn't until Spear was given the okay to build the retractable proposal as a 3/8-sized model that it was named the Syrtis, which was one of the first names he recorded in the Advanced Studio name book. He came across the name while looking at a map of the moon but, several years later, he discovered there was also an area near the Sahara Desert with the same name. When he chose Syrtis as the name for his retractable hard-

Ford Photomedia

The 1957 Ford Hide-Away-Hardtop was introduced at the New York Auto Show, December 8, 1956, to a standing room only audience. The first production model was delivered to President Eisenhower at the White House on April 14, 1957.

top model, he was fairly certain it was different enough, that it would never be considered for a production car. He picked the name Syrtis because it was unusual enough to encourage people to ask what it meant and where it came from.

Spear named the working retractable roof feature of the finished Syrtis the roof-o-matic. The top was one piece with a separate backlight that could be left in place, or lowered into the luggage area, whether the top was up or down. With the top retracted, the backlight could also be rotated forward to become a tonneau windshield. Spear designed the retractable-top system so that when the top was up, there was no loss of back seat room and, when retracted, there was still enough room in the trunk to carry a few good-sized suitcases. Because there was not as much luggage room as a normal hardtop, the spare tire was designed to fit in the floor of the trunk or in a continental kit.

There were other design features on the Syrtis that Spear thought were novel. From the grille to the windshield, the Syrtis was devoid of chrome trim, and the hood swept up at the back to meet the curved windshield. The front bumper wrapped around and halfway down the sides of the car. At the front, the bumper was integrated into the grille and designed to control air flow into the radiator.

The headlights Spear used on the Syrtis were also unique. They were incorporated into the leading edges of the front fenders in slits with refracting micro lenses behind the slits. The headlights themselves were concealed inside the fenders and, when on, the light beams were focused to cross through the micro lenses and then onto the road. Dual exhausts, called jet tubes, exited from flared housings at the bottom ends of the back bumpers. The four large exhaust ports under the doors were painted red inside the exhaust pipes, but were for decoration only. Above the exhaust exits, at the top of the rear fenders, were similar taillight/turn signal light housings.

Spear prepared a booklet explaining the alternatives and advantages of his all-metal retractable hardtop. The booklet included this cartoon explaining how an automobile manufacturer on Mars might view such a proposal. (Spear was a cartoonist early in his career.)

Gil Spear

In the passenger compartment, to the right of the driver, was a central pylon extending out from the instrument panel with pushbutton controls for shifting the car's automatic transmission. A very thin steel-core steering wheel was used that had only one spoke for support. The spoke was at the bottom during normal straight-ahead driving and permitted a full view of the instruments which were recessed into the dash.

Authorization number EJN-S-015 was assigned to the 3/8-sized Syrtis project on September 18, 1951. Based on that authorization, a special 3/8-sized hollow-clay modeling armature was designed by Ben Kroll and made in the Design Department wood shop. Most of the actual layout work and the mechanical parts of the model were worked out by Jim Huggins. Larry Wilson did the clay modeling of the 3/8-sized Syrtis. The windshield, backlight and windows were first cast in plaster and then molded from Plexiglas by studio engineer George Martin. The top was again pounded out from a thin copper sheet by Howard Fifield in the metal shop. Fifield also made the other metal parts used on the 3/8-sized model. When the clay model was completed, the shop cast the finished body in plaster. Spear then designed a small battery powered hand control that operated an electric motor used to open and close the deck lid and run the top up and down.

Henry Ford Museum and Greenfield Village

(above) The idea of a steel top convertible was not new. Although designers at Ford and other companies had previously designed such a car, Spear was the first to design a workable six-passenger model. This rendering was made in October 1946 by Ford designer Eric Ramstrum.

Although plans were drawn up at the same time the Syrtis was built for a 3/8-sized 1953 Ford with a retractable top, Spear hoped the Syrtis would be made into a full-sized showcar — and then hopefully incorporated on future Fords, Mercurys and Lincolns. (The proposed 3/8-sized '53 Ford with a retractable roof was not built.)

To forestall management concerns that Spear might jump ship to work for a competitor, it was decided to patent four different variations of the retractable and name both Spear and Huggins on the patents. Between November 1952 and February 1953, Spear and Huggins filed patent applications for four different versions of the retracting roof. Because both Spear and Huggins felt that the retractable method used on Huggins' Saturn retractable duplicated one of the proposals already developed by Spear, Huggins' proposal was not made a part of the patent application, something Ford Motor Co. later came to regret. Patents were ultimately granted to Spear and Huggins in 1956 and, as they had agreed when the patent applications were filed, they were assigned to Ford.

Gil Spear

(above) The name Syrtis was placed below the windshield in block letters. The headlights were high-intensity beams that refracted through small openings at the front of the fenders.

(below) The deck lid on the Syrtis opened from both the front and back.

Even before the 3/8-sized model of the Syrtis was finished, Bill and Benson Ford began showing increased interest in the project. When it was finished in June 1952, they arranged for brother Henry Ford II to see it. In the meantime, two things happened that caused Ford to make the 3/8-sized Syrtis and its roof-o-matic retracting-roof system public. Jim Huggins left Ford to take a job as a designer at Chrysler and, right after that, the book *Ford at Fifty* came out with pictures of the 1/8-sized model of Spear's retractable. Photographer Philippe Halsman, who had been given unlimited access to the Ford Design Department, took detailed photographs of the 1/8-sized model. Whoever approved the use of Halsman's photographs didn't realize their significance, and neither Bordinat nor Spear realized they were going to be published, until they saw the book.

Gil Spear

In March 1953, Ford issued a press release and one photograph of the Syrtis. Two months later, the 3/8-sized Syrtis was publicly displayed at the 50th Anniversary opening of

Gil Spear

(above) With the top up, the Syrtis looked like a standard two-door hardtop.

Gil Spear

(above) With the press of a button, the trunk opened to swallow the top, turning the Syrtis into a true convertible.

(below) With the top down and the backlight retracted, the Syrtis looked like a standard convertible model.

Gil Spear

Gil Spear

(above) The backlight could be moved forward to give the Syrtis two windshields.

Gil Spear

(above) Even with the top retracted, there was still some room in the trunk for luggage.

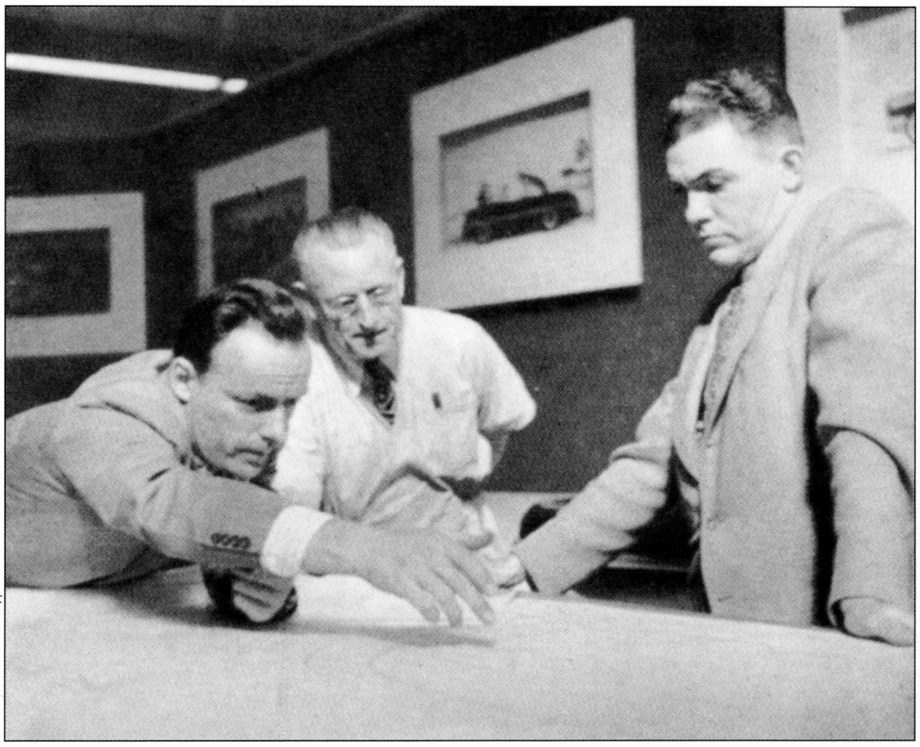

From the book *Ford at Fifty*

Gil Spear (left) and Jim Huggins (right) discuss the design of the Syrtis' rear-deck lid with metalworker Howard Fifield (center). Over Fifield's shoulder is Spear's animated cardboard cut-out of the retractable hardtop.

James A. Huggins

Jim Huggins was a Detroit native. In 1935, he started work at Ford's Rouge plant in the tool drawing department. In late July 1941, he was transferred to the bomber plant to work on the B-24. While at Willow Run, Huggins became proficient in making full-sized layouts. In1945, after the war was over, he was assigned to the Design Department as a layout man.

While working in Gil Spear's Advanced Studio in the early 1950s, Huggins was a layout man and a designer. Huggins assisted Spear in the design and building of the retractable hardtop. As a result of their work on the Syrtis, Huggins and Spear patented several different types of retractable mechanisms. One method that Huggins developed for his own retractable concept car was called the Saturn. The Saturn was not built. In 1953, Huggins was transferred to the Ford studio. Later that same year, he was recruited by Chrysler as a designer in their advanced studio, where he worked on Chrysler's proposed retractable hardtop. Huggins later worked as a designer in Chrysler's station wagon and interior studios and retired from Chrysler in approximately 1980. ■

Ford's new Research and Engineering Center. The press release said that the roof-o-matic was "designed with the thought in mind that its unique roof mechanism might be applied to any of the company's passenger car lines." In May 1954, another press release was issued, this time accompanied by four photographs of the Syrtis. The second press release said that Ford was actually developing a car with a retractable roof, but it ended by stating that "the hardtop convertible design is still in the research and development stage." It did not say that, since January 1954, a full-sized prototype of a Continental Mark II Retractable had been on the road with a hardtop that retracted into the trunk with the pull of a knob.

After Special Projects Division (later Continental Division) was formed to build the new Continental Mark II, Harley Copp, chief engineer on the Mark II project, convinced Bill Ford to produce a Continental Mark II Retractable. After Finance Committee approval was given in May 1953, all of Spear's retractable models, including the 3/8-sized Syrtis, were delivered to Building 19 at the old Lincoln plant, where the Continental Mark II Retractable was built.

In early 1953, Benson Ford also proposed building the Mercury XM-800 showcar with a retractable roof. Spear prepared a proposal, and a rendering was made of what the XM-800 would look like with a retractable roof. The designers of the XM-800 preferred the roof design they had proposed and weren't in favor of moving the windshield back the seven inches Spear's proposal showed was necessary to accommodate a retractable roof. In addition, it was decided that the top-of-the-line Continental Mark II should get first crack at producing a full-sized retractable, especially since they had agreed to spend the money to develop it.

Over the next several years, at Society of Automotive Engineers meetings, it generally became known that Chrysler was also developing a retractable hardtop. Spear was later told, but has been unable to confirm, that Chrysler's prototype retractable hardtop was built in Italy and was aboard the Andrea Doria when it sank in the Atlantic Ocean in 1956.

Several years later, after he had taken a job as chief designer in MacPherson's Special Vehicles Department, Spear got a call from Howard Fifield at the Styling Center. Fifield, who was almost in tears, told Spear he thought he should know that the Syrtis was being busted up out on the loading docks by someone with a sledge hammer. Spear rushed over to the Styling Center in time to see the results of someone's handiwork. He was never able to determine who was responsible for the destruction of the Syrtis. ■

George Krispinsky

George M . Krispinsky

George Krispinsky grew up in Youngstown Ohio. After service in the Navy during World War II, he attended the Cleveland Institute of Arts on the G.I. Bill. After graduation in June 1950, he was hired as a Ford designer by Gene Bordinat and was assigned to the Advanced Studio. While in the Advanced Studio, he helped design and then built the cardboard cut-out of the retractable hardtop used to help demonstrate to Henry Ford II and other executives at Ford the way a full-sized retractable would work.

After about nine months in the Advanced Studio, Krispinsky was transferred to the Mercury exterior studio, where he designed the grille proposal used for the 1954 Mercury.

Krispinsky left Ford in 1953 to work for Chrysler. In 1955, Bill Schmidt recruited him as a manager in the Packard exterior studio. After Studebaker-Packard downsized to South Bend in 1957, Krispinsky returned to Chrysler. In 1963, he was asked by Schmidt to join his private industrial design firm as head of automobile design. In the mid-1960s, Dick Teague, who Krispinsky had met while they were both working at Chrysler, asked him to come to American Motors, where he became head of the Jeep exterior studio. While at American Motors, Krispinsky's studio designed the Jeep Cherokee and the Jeep Grand Cherokee. Krispinsky retired in 1987. ■

From the book *Ford at Fifty*

From left to right are designer Jim Huggins, body engineer Roy Hawver, Gene Bordinat, assistant manager of the Design Department, designer Gil Spear and Design Department manager Charlie Waterhouse. In the middle is the one-eighths sized model of the retractable hardtop.

Cougar

The name Cougar has been used many times on Ford concept cars and proposed production cars. The first time the name Cougar was used was on a 1/8-sized half model designed by Gil Spear in 1948-49. In 1954, the name was again considered for the production Thunderbird. The name was also used on another one of Spear's concept cars in 1962. Later the Mustang was almost named the Cougar, and it was finally used on the 1967 production Mercury Cougar. Designer Ben Kroll's son Robert suggested the name Cougar for one of the cars in his dad's studio, when he was 10 years old and was told by his father that he named the first Cougar.

Just before the first Cougar was designed, Gil Spear visited a yacht club in upstate New York. On the walls of the yacht club, Spear observed several half-hull boat models. Because he knew he was going to be short of space in his small Advanced Studio, the half-hull idea prompted him to build only half a Cougar.

A problem arose with half models Spear couldn't find a way around. When viewed up against a mirror, the illusion of the whole car was created but, because a mirror reflects from the backside of the glass, there was always a gap twice the thickness of the glass, and it made the car look like there was a gap right down the middle of whatever was being reflected. The gap is most noticeable in the center of the windshield on the Cougar. Spear considered substituting polished stainless steel for the mirror, but there was no money in the budget for the sheets of stainless steel needed for the many models he anticipated building in the Advanced Studio.

Even though Spear knew mirrors were not the final answer, he decided to build the half-model Cougar to see what it looked like. Larry Wilson clay modeled the car, and then cast the half-model Cougar in plaster.

The half-model Cougar was not even the original Cougar. The original Cougar started as a smaller three-dimensional model Spear made from spare parts taken from his junk drawer and, when finished, it hung on the studio wall. Only after he got the idea of making half models was the Cougar built as an 1/8-sized half model.

As with many of the concept cars he designed, Spear intentionally made the Cougar a "bit wild," as he puts it, in an effort to attract the attention of his Design Department bosses, who he hoped would approve building it as a full-sized car. A full-sized Cougar was never authorized, but Spear's idea of building half models and placing them against polished stainless steel is still in use at the Design Center. ■

The Cougar was a half model placed against a mirror. The split in the windshield is twice the thickness of the mirror.

Henry Ford Museum and Greenfield Village

Lawrence R. Wilson

Larry Wilson grew up on his parents' Indiana farm. He left school after the 11th grade. He was a self-taught artist and, at age 17, he made a small foundry in his father's barn, where he melted tin cans to pour into molds of cars of his own design.

In 1940, Wilson, then 21 years old, took two of his metal models to Dearborn and, without an appointment, dropped by to ask Henry Ford for a job. Just as Ford's secretary was telling Wilson he could not see him without an appointment, Ford walked in and asked Wilson what he had in the two shoe boxes he was carrying. Inside the boxes were two model cars Wilson had made from melted tin cans. They were well detailed and had working doors, hoods and trunks. After he looked them over carefully, Ford asked Wilson to demonstrate their features, which Wilson did. Ford arranged for Wilson to immediately start work as a clay modeler at the Design Department.

During World War II, Wilson was an infantry soldier. After his entire platoon was killed in the Battle of the Bulge, Wilson, who was badly injured, was hidden for several weeks in a partisan's water-filled cellar. After he was rescued by advancing troops, he spent 18 months in a Veterans Hospital recovering from injuries. After working at Studebaker for several months, Wilson returned to the Ford Design Department, where he was assigned to work with Gil Spear in the Advanced Studio. Spear encouraged Wilson to become a designer. Wilson helped design and build the Muroc, the Cougar and many of the other 3/8-sized concept cars built in the Advanced Studio. He designed the Ventura and clay modeled the X-2000 and many other full-sized concept cars. He also worked on the production 1957 Ford, the 1957 Ford retractable and the 1965 Mustang.

Wilson was plagued by his war injuries throughout his postwar career and retired in 1981. ■

Ford Photomedia

Clay modeler Larry Wilson is measuring the width of the gap in the center of the windshield.

After the Design Department moved across the street to the new building in May 1953, Spear was given permission to convert an unused room on the second floor into a classroom. That room was also used as a showroom for three-eighths sized concept cars, some of which are shown here. From right to left are the Cutlass, Chesterfield, Muroc and two unknown cars.

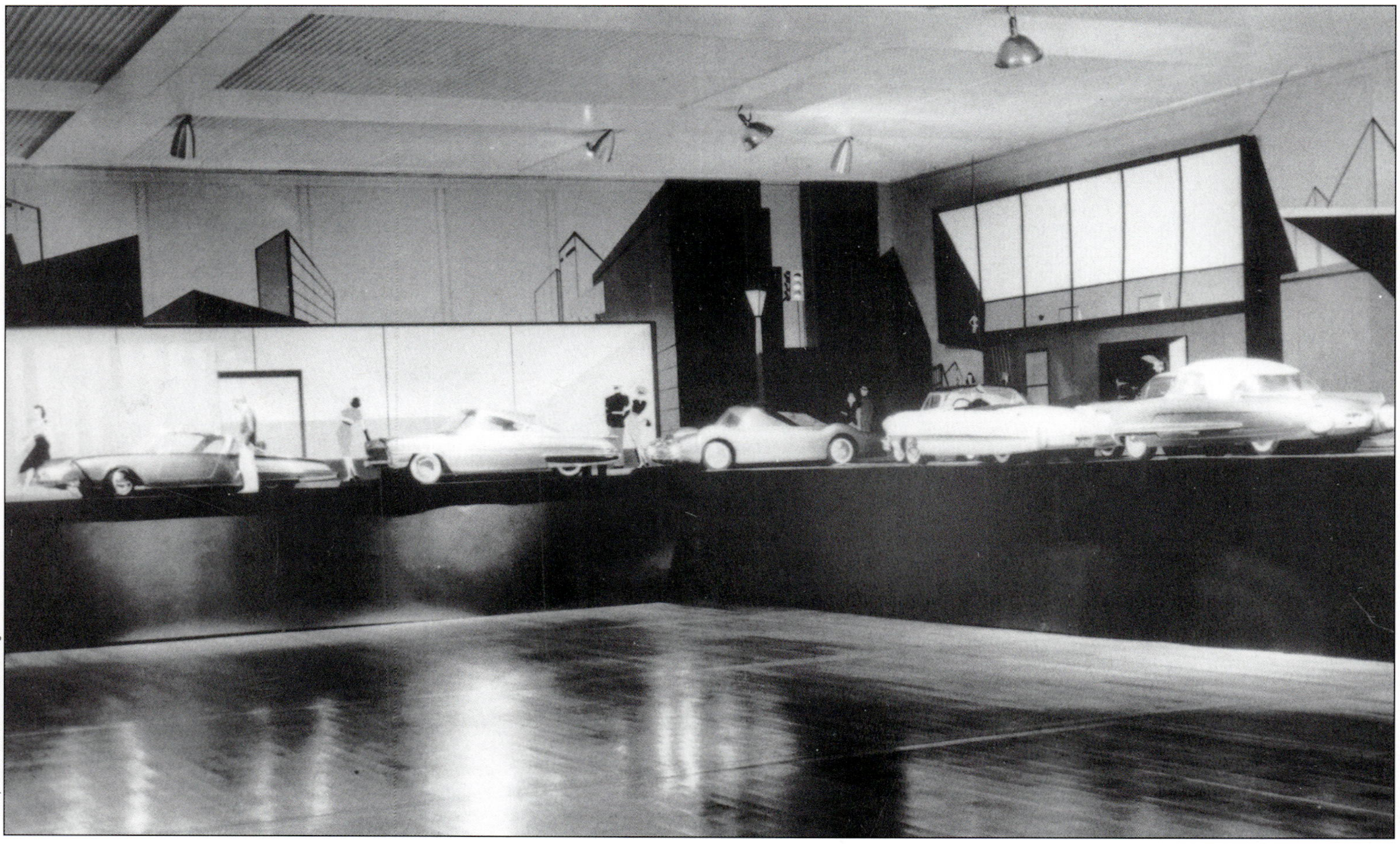

Henry Ford Museum and Greenfield Village

Jim Powers

The Ventura had a bit part in the 1954 movie Woman's World.

Ventura

Not all clay modelers have the skills needed to become designers, and not all designers can sculpt in three dimensions, which is what the clay modeler does. Automobile design depends on both, because very few designers can design on paper and sculpt in clay and, as a general rule, clay modelers are not the best illustrators.

Most clay modelers do not want to become designers but, over the years, several of them have tried to bridge the gap and become designers. If a clay modeler had the talent and the desire, management often agreed to let him train as a designer. In practice, however, it seldom worked. Good clay modelers were in great demand and could actually make more money than designers when overtime was taken into account. Designers were on salary, as opposed to clay modelers who were usually paid by the hour. Throughout the 1950s, clay modelers at Ford regularly worked 12 hours or more a day during the week, 6 hours or more on Saturday and sometimes part or all day on Sunday. Clay modelers routinely report working 68-80 hours a week. They were just too busy to become designers.

When designers graduated from art school, or if they had extraordinary talent but did not attend an art school, Ford hired them and usually placed them in an Advanced Studio, where they would spend 6 to 12 months learning to design cars the Ford way. Even after Ford began its clay modeling school in 1950, skilled clay modelers remained in short supply. Many designers also acknowledge that skilled clay modelers were always in greater demand than designers, and that a skilled clay modeler could save a design, when building the clay model.

Larry Wilson was one of the clay modelers who had both design and clay modeling talent but never found the time to become a designer. During the time he was a clay modeler in Gil Spear's Advanced Studio, Spear encouraged his design skills. One of the 3/8-sized cars Wilson designed while in Spear's studio was the Ventura, which he completed in 1953 as a proposed Lincoln sporty car.

Between the time the 1/8 and 3/8-sized clay models of the Ventura were built, there were several design changes including changes to the greenhouse and elimination of the dorsal fins on top of the rear fenders. The Ventura was designed with a bar headlight across the front, dagmars at the ends of the front fenders, flip-up Plexiglas-roof panels over the doors and stationary bumper extensions on the front wheelcovers.

The Ventura made a cameo appearance in the movie *Woman's World* and was used in promotional still photographs for the movie. It was also displayed in the design department classroom. What happened to the Ventura is unknown. ■

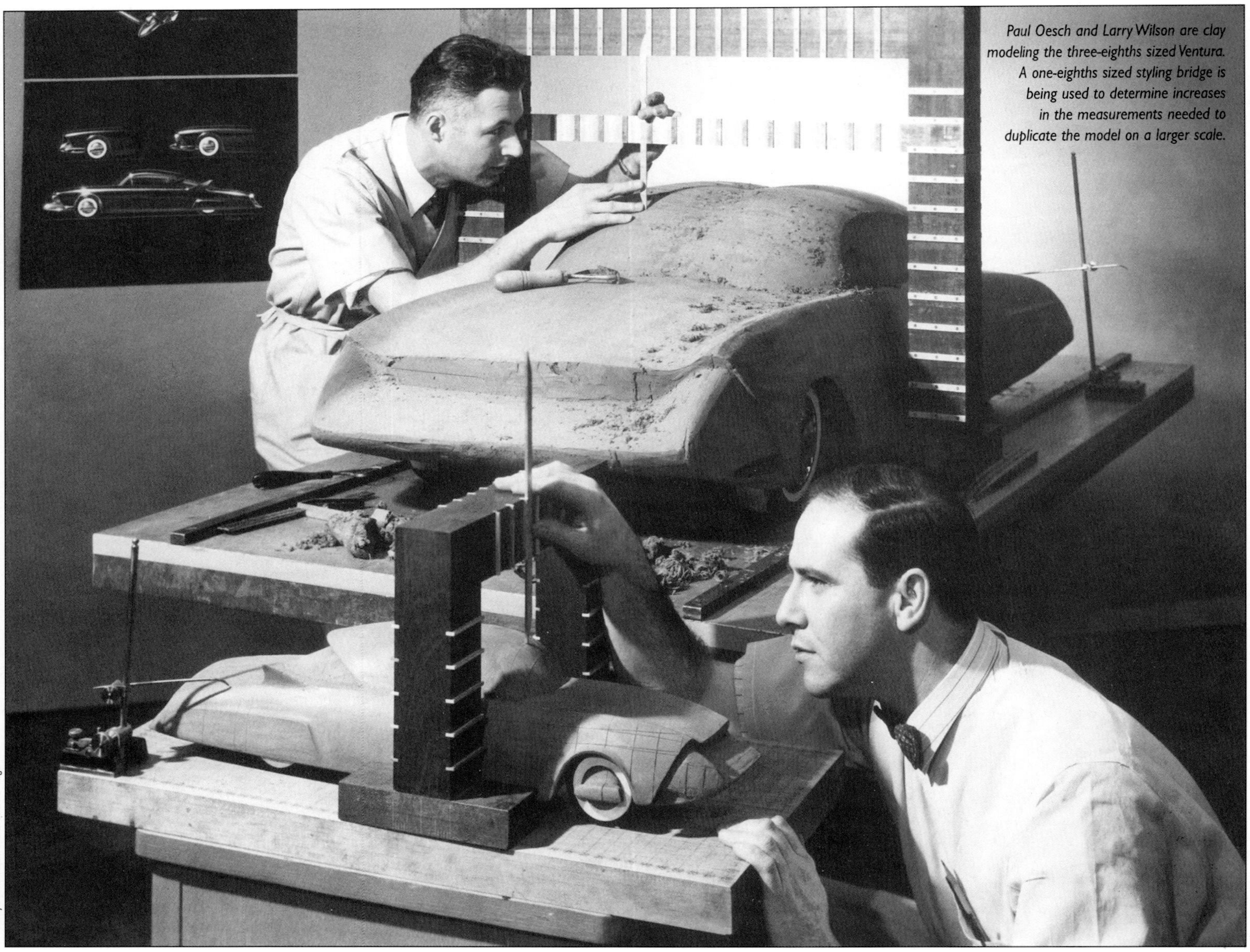

Paul Oesch and Larry Wilson are clay modeling the three-eighths sized Ventura. A one-eighths sized styling bridge is being used to determine increases in the measurements needed to duplicate the model on a larger scale.

Henry Ford Museum and Greenfield Village

Clay modelers Warren Gaymer and John Kruscil with an armature, before application of any clay. The modelers are adjusting a styling bridge. Although this armature was not used for the Ventura, it shows the process and the equipment used. Bill Wheeler is at the right of the photo at his drafting table.

Gil Treweek

Cavalier

Gil Spear and Roy Brown met in the late 1930s, when they were both beginning designers in GM's Cadillac studio. One day, a full-sized clay model of the proposed 1941 LaSalle was wheeled into the Cadillac studio, and several designers, including Brown and Spear, were assigned the task of turning it into a smaller version of the Cadillac. The LaSalle was canceled after the 1940 model year, and that's the reason the proposed 1941 model was moved into their studio. Neither Brown nor Spear ever forgot the proposed '41 LaSalle's long, narrow, vertical grille. Spear and Brown later went their separate ways, but they remained close friends.

Fast forward to 1948, when Gordon Buehrig and several investors commissioned Derham Manufacturing of Rosemont, Pennsylvania, to build a one-off sports car of Buehrig's design. The car was named the Tasco. Buehrig and several other investors proposed to produce it as a new American sports car. In an effort to find financing, the car was shown to automobile manufacturers and others in the Detroit and Dearborn area. While the Tasco was in Dearborn, several Ford designers, including Spear, got a look at it. The tall thin grille on the Tasco reminded Spear of the clay model of the 1941 LaSalle, and he immediately returned to the Advanced Studio, where he designed a proposed Continental incorporating the same narrow, vertical grille used on the LaSalle and the Tasco.

Spear called his updated Continental the Cavalier. The LaSalle and the Cavalier were probably named for the same person, Rene Robert Cavalier Sieur de LaSalle. The original rendering Spear made of the Cavalier was one of his favorites. It showed a woman who had just parallel parked next to a curb opening one of the side-opening deck lids on the Cavalier. To Spear, the side-opening trunk solved one of the major drawbacks of the original Continental, which had a trunk that wasn't easily accessible. After he finished the rendering, it was twice stolen, probably by someone who had a wall just made for it. The first time it was taken, Spear complained to security, and it was returned. Shortly after the complaint was withdrawn, the rendering disappeared for good.

The Cavalier was cast as a 3/8-sized plaster model and had a metal roof. Even though it had Plexiglas windows, it had no interior. The Cavalier was painted dark green and had an off-white convertible top. After it was painted, Spear cut narrow strips of scotch tape and put them on the driver's door to imitate wicker. He then spray painted the scotch tape with a clearcoat to seal it.

When the Cavalier was completed, it was displayed on one of the counters in the Advanced Studio next to a model of the postwar Continental for comparison. It was also displayed during a 1952 Advanced Studio open house held for employees, their families and Ford executives. The Cavalier, although not identified, was also in two of the photographs in the 1953 book *Ford At Fifty*.

When designing the Edsel, designer Roy Brown borrowed the Cavalier.

Ford Photomedia

Henry Ford Museum and Greenfield Village

The Cavalier was a Continental proposal designed by Gil Spear in 1948. The deck lid on the Cavalier was two-part, so it could be opened from curbside. The integrated rear bumper, taillights and bezels influenced the 1952-53 production Mercurys.

In 1953, Roy Brown became a designer at Ford and was soon selected as chief designer of the car that was later produced as the Edsel. His studio was initially called the Special Products Studio. Once it was established, Brown began looking for inspiration for the new car he was designing. Brown asked Spear if he could borrow the Cavalier, Spear said "sure," and the Cavalier was moved to Brown's studio. The 1941 LaSalle and the Cavalier, with their vertical grilles, were the inspiration for the front end Brown designed for the Edsel. ■

The Marlin and The Cavalier at a 1952 Advanced Studio open house.

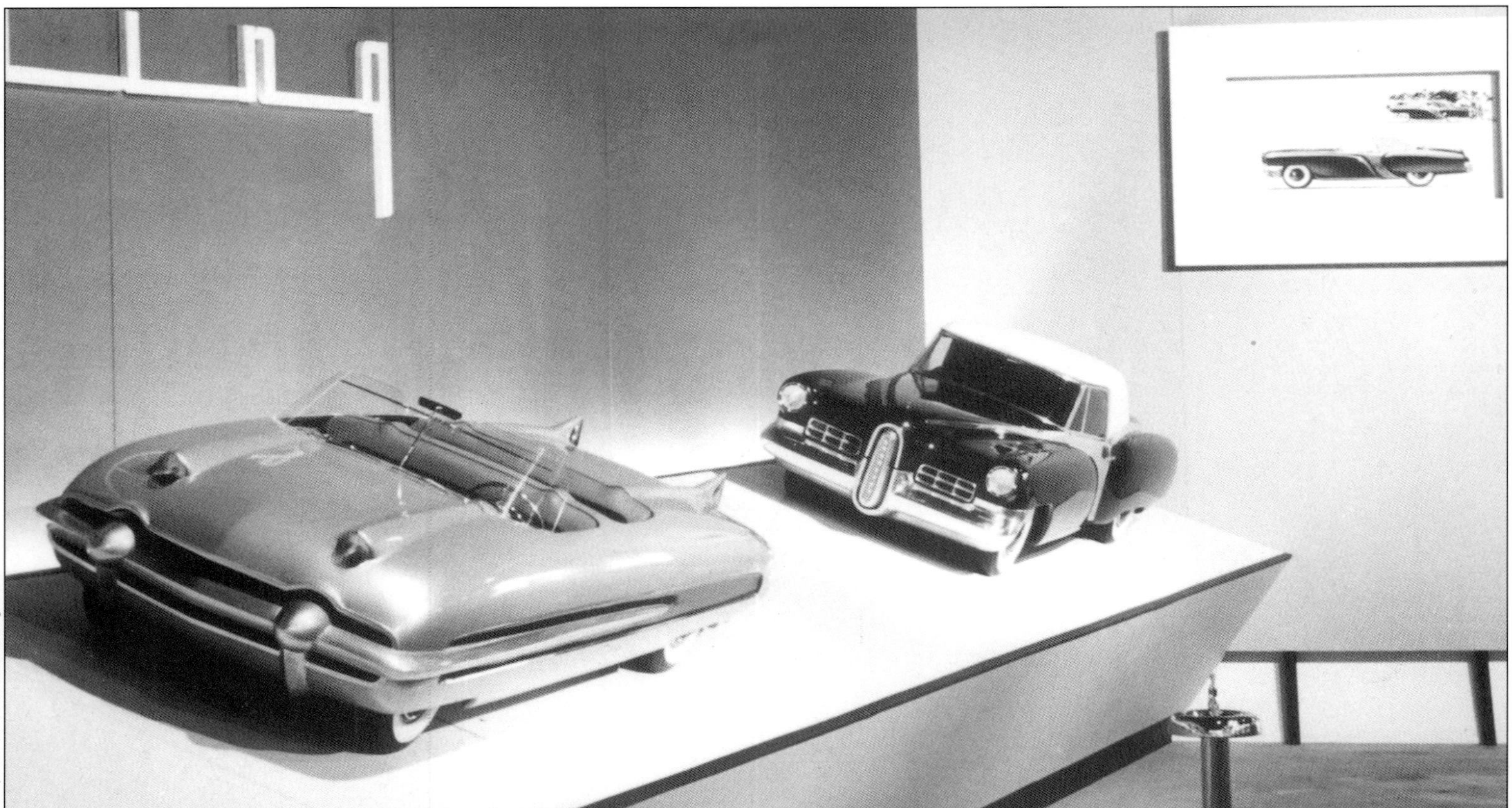

Henry Ford Museum and Greenfield Village

Gil Spear had a small but experienced staff, who worked with him in the Advanced Studio and helped teach the new designers how to design the Ford way. To some degree, Spear was able to hand pick his staff, and he generally felt that the designers and clay modelers assigned to the Advanced Studio had excellent technical and people skills. His small staff usually included one other experienced designer, a seasoned design engineer and at least one senior clay modeler. Throughout the more than six years Spear ran the Advanced Studio, he had three assistant managers.

Designer Bob Thomas was Spear's first assistant manager. When Thomas was transferred to a production studio, he was replaced by Jim Huggins, who had started at Ford's Rouge plant in 1935 drawing tools. Huggins was assigned to the Design Department as a layout man and eventually became a designer. In 1952, Huggins was transferred to a production studio shortly after Alex Tremulis was hired. With Huggins' transfer, Tremulis became Spear's assistant. Tremulis took over most of the teaching duties and proved to be very effective at it.

In mid-May 1953, Spear moved the Advanced Studio into the new design building across the street, even before it was ready for general occupancy. While touring the building, Spear found unused space on the second floor above the office area, and he got permission to set up a classroom there, where Alex Tremulis taught new designers what they hadn't learned in school. In mid-1955, when Spear became head of the Special Vehicle Department, Tremulis was appointed head of the Advanced Studio.

The Bolero was designed by Jim Huggins in 1951, after Huggins became Spear's assistant. Huggins liked the bathtub look and generally designed his cars in that style. Spear made it a practice not to interfere with designs proposed by the other experienced designers in his Studio. The greenhouse on the Bolero is similar to the Saturn retractable proposal Huggins prepared, when he and Spear were working together on the Syrtis.

When it was completed, the Bolero was one of the 3/8-sized models chosen for display at an open house the Advanced Studio held in 1952. The banked track the cars were displayed on during the open house was Spear's idea, and he helped build it. In the photograph on page 102, Huggins is explaining technical details about the Bolero, which is on the banked track. ■

Bolero

The Bolero was designed by Jim Huggins in 1951.

J.R. Samsen

Ford Photomedia

(above) Jim Huggins explaining the features of his Bolero at a 1952 Advanced Studio open house.

(below) Designer Chuck Gale with the Bolero on his second day on the job.

J.R. Samsen

J.R. Samsen

The predominant design features on the Marlin were the fairings running from the headlights to the taillights.

Marlin

During the time Gil Spear headed the Advanced Studio, he named the 3/8-sized concept cars built in his studio rather than assign them numbers as had been past practice at Ford. Spear kept a book so designers could record names they thought could be later used on the cars designed in the Advanced and production studios. That book became the basis for a similar book that is still in use at the Design Center today as a name bank for future production and concept cars.

Marlin was one of the names recorded in that book. It was used as the name of a 3/8-sized model Spear designed in 1950-51. The Marlin's design incorporated fairings or humps starting at the headlights, running along the hood, and then forming raised areas over the speedometer and clock at the top of the instrument panel. The fairings also aligned with the steering wheel and continued through the passenger compartment to form small headrests on the front and rear-seat backs. On the deck lid, the fairings ended in small Cadillac-style fins that housed the taillights. The two spinners on the front of the Marlin were designed as bumper guards. They also lined up with the headlights and fairings.

The taillights on the Marlin lit up when the brake pedal was depressed and, as with most of the 3/8-sized concept cars, the front and rear lights were illuminated by a small battery hidden on the underside of the car. The Marlin was built as an open car so the fairings would appear more pronounced, and so the finished interior of the car could be easier displayed. John Weisbrod, a recently-hired designer assigned to the Advanced Studio, assisted Spear with the design of the Marlin's interior. Because Spear wanted the Marlin to be noticed, it was painted a pearlescent pink/orange. The chrome parts were painted with highly-polished metallic paint.

The finished model of the Marlin was cast in plaster from molds taken from the original 3/8-sized clay model. The wheels were made from painted wood turned on a lathe in the Design Department shops. The windshield was made from sheet Plexiglas heat-formed to shape over a plaster form.

The Marlin was shown at the same 1952 Advanced Studio open house where the Bolero, Cutlass, Muroc, Chesterfield and the Cavalier were displayed. When design trends changed in the mid 1950s, designers were given some of the outdated models. Others were disposed of when they were damaged. The Marlin's fate is unknown. ■

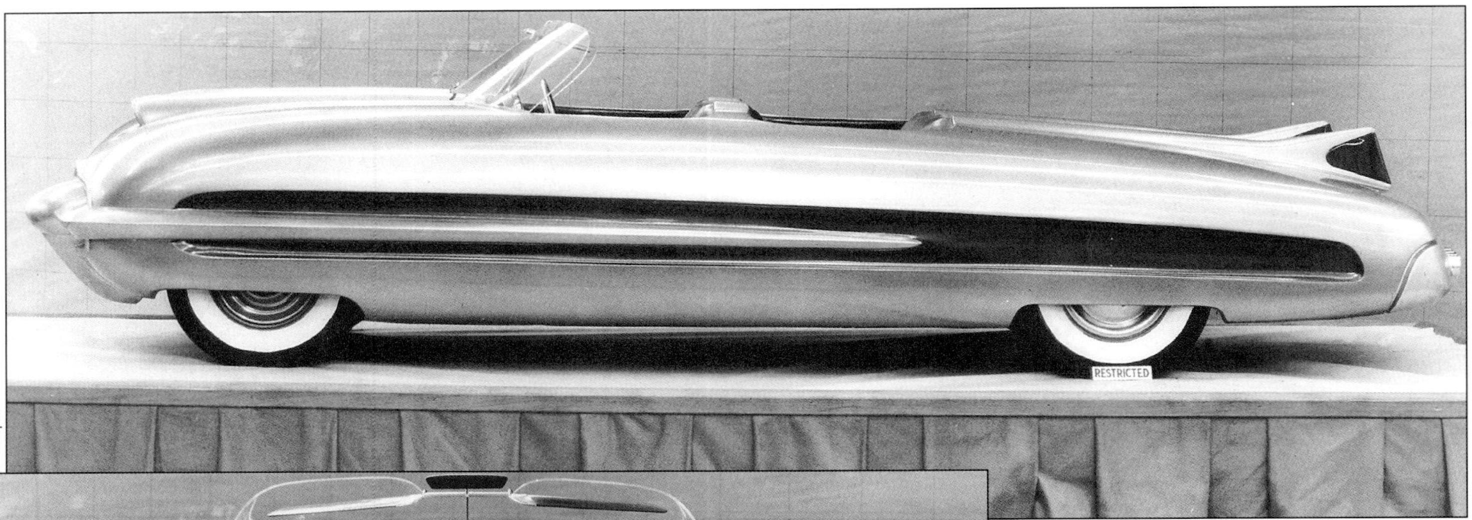

Gil Spear

(above) The Marlin's design was later changed by shortening the wrap-around front bumper and adding a two-toned cove on the sides.

Gil Spear

(above) For display, the headlights and taillights of the Marlin could be illuminated by a battery hidden beneath the car.

(below) The Marlin was built as a convertible so the interior could be seen.

Gil Spear

(Below) John Weisbrod with the Marlin at a 1952 Advanced Studio open house.

Ford Photomedia

John Weisbrod

John Weisbrod grew up in Fenton, Iowa. As a child, he loved to draw cars. Because his uncle was the local Ford dealer, Weisbrod was always partial to Fords. During World War II, Weisbrod served in the Navy. After he was discharged, he attended Pratt Institute on the G.I. Bill and studied to become a car designer. When he graduated in 1950, he was recruited by Gene Bordinat and hired at the Ford Design Department with classmates Bill Wheeler and Pete Renz.

Weisbrod's first assignment was in Gil Spear's Advanced Studio, where he worked on several of Spear's 3/8-sized concept cars. Under Spear's direction, he helped design the interior of the Marlin.

After approximately six months in the Advanced Studio, Weisbrod was assigned to the Ford body development studio, where he worked with Gordon Buehrig on the 1952 Ford steel-bodied station wagon and the 1951 Ford Victoria. He was later assigned to the Ford exterior and then the Ford interior studio. For several years, before his unexpected death in 1962, he was a designer in the Lincoln exterior studio. ■

Ford Photomedia

(above) *Clay modeler John Ebejer is shown preparing the Lancer for a 1952 Advanced Studio open house.*

Ford Photomedia

(above) *Gil Spear's 1951 rendering of the Lancelot.*

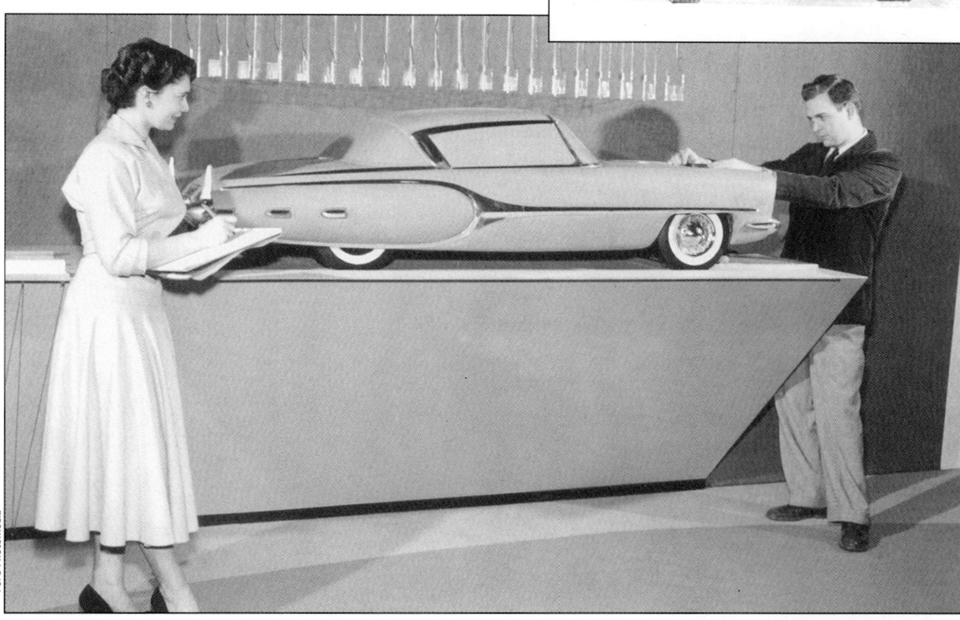

Ford Photomedia

(left) *An unknown model and John Weisbrod preparing the Lancer for a 1952 Advanced Studio open house.*

Lancer

The Lancer was designed by Gil Spear and built as a 3/8-sized plaster model in 1950-51. At the time Spear made the first rendering of the car, it was called the Lancelot. By the time it was made as a 3/8-sized clay model and then cast in plaster, the name had been changed to the Lancer. Spear thought the name change was necessary because of the lack of space on the car's hood for the additional two letters necessary to spell Lancelot. The name was later removed from the hood and everywhere else.

There are several differences between the rendering of the Lancelot and the finished 3/8-sized model. Many times, changes in design were made at the time clay models were built. Another common practice with Ford concept cars, whether 3/8-sized or full-sized, was to develop different designs on each side of a completed model.

The small mannequin standing at the rear of the Lancer in the photograph at the top of this page was named Oscar. He was shown next to many of the 3/8-sized concept cars to give the cars perspective. Similar mannequins were also used in the production studios with the 3/8-sized models they made. Oscar was originally developed in Spear's Advanced Studio, and a uniform-sized Oscar was one of Spear's suggestions. ■

Gil Spear

(above) *The design of the Lancer had similarities to the 1951 Packard.*

(below) *The Lancer was later cast in plaster and painted.*

Gil Spear

(above) *The Lancer had flow-through ventilation.*

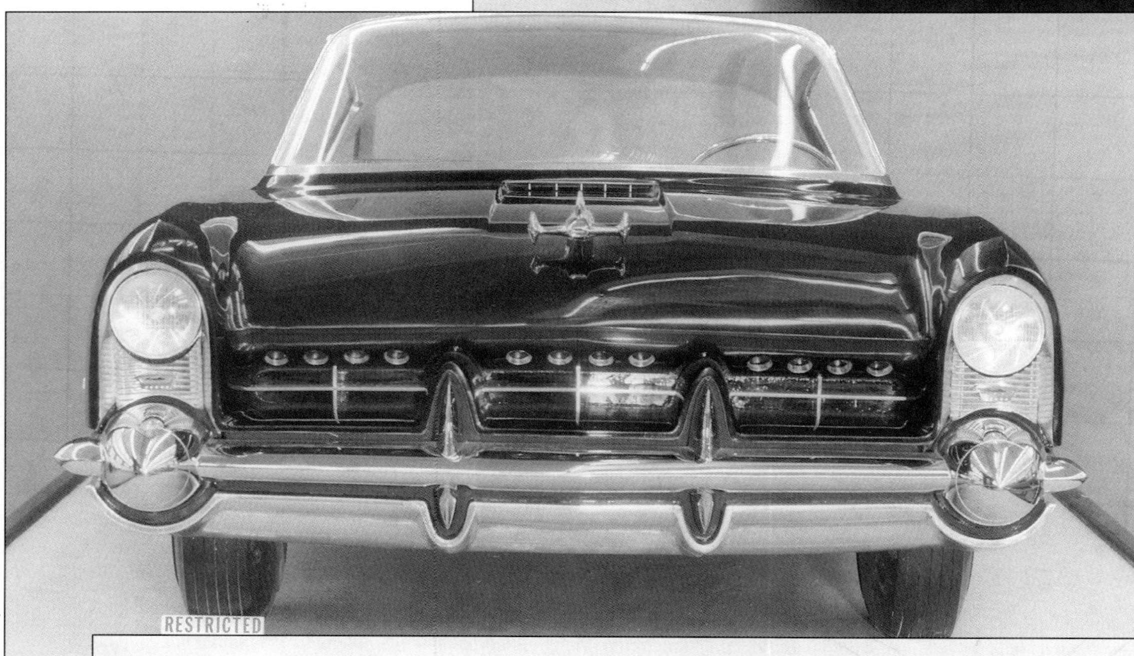

Gill Spear

(below) *As was common practice, the driver's side of the Lancer had different features from the passenger side.*

Gil Spear

Cutlass

The Cutlass was one of the first 3/8-sized concept cars designed by Gil Spear, after he was made head of the Advanced Studio in 1948. Designer Ben Kroll assisted Spear with the design of the Cutlass. A 3/8-sized clay model and, later, a 3/8-sized plaster model with a metal top were made. The fully-skirted front wheels of the Cutlass were inset enough, so they could be turned from side to side at full-lock position without hitting the skirts. If the Cutlass had been built as a full-sized car, Spear expected it would have had removable fender skirts both front and rear.

By today's standards, Spear is surprised at the conservative slant of the Cutlass' windshield. At the time the Cutlass was designed, windshield slant was part of the package dictated by Engineering and, as Spear recalls it, Engineering maintained that making the windshield more slanted distorted vision. Spear disagreed but, nevertheless, had to stay within the package Engineering set.

The hood ornament on the Cutlass is similar to the standard one used on the 1954 Ford. The taillights and rear-end design of the Cutlass are similar to those used on the X-100, because Joe Oros selected the Cutlass as the basis for the design of the X-100. The taillights not only became a focal point on the X-100, they also became one of the predominant features on production Fords of the 1950s and 1960s. While the X-100 was being designed, the Cutlass was moved into the Lincoln studio, and it remained there until the X-100 was completed. After that, Spear lost track of the Cutlass, and believes it was probably destroyed.

Designers constantly tried to incorporate in their designs what they referred to as focal points. A focal point, if it had good recognition qualities, could also be used in succeeding years, slightly altered, to give a car line continuity. The round taillights were not the only focal point on the Cutlass, but they became the best known and longest lasting. The spinner grille, a focal point that first appeared on the 1949 Ford, also appears on the Cutlass. The widow's peak at the top of the Cutlass' windshield was intended as a focal point, but it never appeared on any production Ford, Mercury or Lincoln. ■

The Cutlass was the primary inspiration for the X-100.

Henry Ford Museum and Greenfield Village

Benjamin Kroll

Ben Kroll graduated from Wayne State University in 1935 with a degree in Art Education and, for the next five years taught art in the Detroit public schools. When World War II began, Kroll was hired at Ford's Willow Run plant, where he designed anti-aircraft guns and radar equipment for the B-24 bomber. In late 1945, after the war was over, Kroll was transferred to the Design Department as a draftsman. Because he had been on a leave of absence from the Detroit public schools during the war, he resigned his teaching position on Gregorie's promise he would become a designer.

Because of his rendering, drafting and design skills and because of his teaching background, Kroll was assigned to the Advanced Studio, when it was expanded in 1948. During the time he was in the Advanced Studio, he did design and layout work on most of Gil Spear's 3/8-sized concept cars including the Syrtis and the Cutlass.

Kroll stayed at Ford until 1955, when he left with four other designers to join Bill Schmidt at Studebaker-Packard's design department. When Studebaker-Packard downsized their design department and moved to South Bend, Indiana., Kroll left to open his own industrial design business specializing in fiberglass construction.

In 1959, Kroll was hired by GM's Fisher Body Division as a graphics illustrator and was responsible for many of the factory brochures prepared there.

He retired in 1973. ■

Ford Photomedia

(above) Ben Kroll pointing out the name "Cutlass" on the car's backlight.

Henry Ford Museum and Greenfield Village

(above) The hood ornament on the Cutlass was used in slightly-altered form on the 1954 production Ford.

(below) When the Cutlass was designed, Gil Spear wanted the windshield to have more of a rake, but he was required to stay within the package set by Engineering.

Gil Spear

(right) The X-Turbo was designed to be powered by a gas-turbine engine.

(left) The X-Turbo was designed and clay modeled by Fred Hoadley who, at the time, was attempting to become a designer.

X-Turbo

Early on in his career at Ford, clay modeler Fred Hoadley was promised an opportunity to become a designer but, each time he tried, his clay modeling skills were needed on some project that just couldn't wait. In 1954, Hoadley got another chance, when Charlie Waterhouse, then general manager of the Design Department, transferred him to Alex Tremulis' Advanced Studio to begin his career as a designer. Gil Spear had just left the Advanced Studio for the Special Projects Studio, but Tremulis had not yet been formally appointed head of the Advanced Studio.

The first car Hoadley designed in the Advanced Studio was the X-Turbo. Like most models done in the Advanced Studio, Hoadley's X-Turbo included plans for the chassis and running gear. (He still has those plans.) The X-Turbo was first built as a 1/8-sized clay model, and then it was cast as a 1/8-sized plaster model. It was never made into a 3/8-sized model.

The X-Turbo was designed for a gas-turbine engine with a large air intake in the front and a smaller exhaust outlet in the back just below the trunk. As with many concept cars designed at Ford, the X-Turbo was heavily influenced by design features on GM's LeSabre concept car, which was first introduced in 1951. Hoadley designed two variations of the X-Turbo, one with and one without the tubes on the roof. The X-Turbo was modeled with the roof tubes, which had lights in both ends.

The plaster model of the X-Turbo was built with a full interior and Plexiglas windows. Hoadley shaped the windows over separate plaster molds made from his model. He felt that by using windows and a full interior he could avoid the heavy look of some of the other models without see-through windows.

The 1/8-sized plaster X-Turbo model was saved by Hoadley and is now displayed on top of a bookshelf in his home. ■

Fred Hoadley still has the original X-Turbo.

The new Design Department building was completed in May 1953. It is on the other side of Oakwood Blvd., across from the EEE Building where the old Design Department was located. (1) is the office area; (2) is the Styling Rotunda; (3) is the area where the design studios were located; (4) is the building's entry; (5) was the enclosed courtyard. The house near (6) sat on a knoll and, from the second floor, there was a clear field of vision into the Design Department's enclosed courtyard. A competitor rented the house and, when Ford discovered their spies were using it for a vantage point, Ford bought the house and immediately tore it down.

Henry Ford Museum and Greenfield Village

Gran Turismo

The Gran Turismo was the second concept car Fred Hoadley designed and built while in the Advanced Studio trying to become a designer. Hoadley was among the minority of those at the Ford Design Department who believed that car designs should try to avoid fads like fins and the rocketship look. His cars, especially when compared with those being designed by Alex Tremulis, were very conservative.

The Gran Turismo was eventually built as a 3/8-sized clay model. It was designed to carry four passengers and used standard 1952 Ford dimensions, but it was substantially lower and had headlights that were concealed behind the grille. The right side of the model was never completed, and the car was not painted. There is no significance to the Indiana license plates, which were used on several concept cars.

While Hoadley was clay modeling the 3/8-sized Gran Turismo, another reorganization took place at the Design Department, and Charlie Waterhouse asked Hoadley to help overcome a huge backlog in Gil Spear's Special Products Studio. Unfortunately, those types of emergencies occurred frequently and, even though his pay was regularly increased as an inducement to continue clay modeling, Hoadley never became a designer.

After Hoadley left the Advanced Studio and returned to clay modeling, Alex Tremulis was asked by Henry Ford II to design an aerodynamic car that would stay on the ground at high speeds. (At Daytona Beach, a Thunderbird had become airborne and crashed.) To save time, Tremulis and clay modeler Larry Wilson turned the Gran Turismo into the Mexico. ■

Fred Hoadley

(above) The Gran Turismo was a very conservative design done at a time when fins were coming into vogue.

Fred Hoadley

(above) The trim and bumpers were chrome plated with foil. Wheels on this and other three-eighths sized models were made from wood.

Fred Hoadley

(above) The design of the Gran Turismo was based on the 1952 Ford package measurements.

Fred Hoadley

(left) Because Hoadley was needed as a clay modeler in one of the studios, the passenger side of the Gran Turismo was never finished.

Frederick E. Hoadley

Fred Hoadley learned the basics of design and clay modeling in high school art classes and, by his senior year, he designed and clay modeled his first model car. After he graduated in 1939, he moved from Findlay, Ohio to Detroit intending to apply at GM's Styling Section. A friend suggested he apply at Ford. Because of his knowledge of surface layout, in April 1941, he was hired as an apprentice in the Ford Design Department, where he was assigned to work with Bud Adams as a clay modeler. After Adams left the Design Department, Hoadley took over Adams' clay modeling responsibilities.

During World War II, Hoadley worked on the B-24 bomber and, after serving in the army, he returned to Detroit intending to enroll in design school. In November 1946, he was persuaded to return to Ford with promises that he would be trained as a designer. A month after he returned, John Najjar badly needed a clay modeler in the Mercury studio. In January 1947, Hoadley was transferred to Bill Wagner's truck and international studio.

Gil Treweek

Increases in his salary, and promises that he would eventually be trained as a designer encouraged Hoadley to continue clay modeling. In 1954, Charlie Waterhouse transferred him to the Advanced Studio to study design. While in the Advanced Studio, Hoadley designed the 1/8-sized X-Turbo concept car and had almost completed the 3/8-sized Gran Turismo concept car, when another reorganization returned Hoadley to clay modeling, this time in the special products and international studios working with Gil Spear and Wes Dahlberg. From January 1959 to August 1967, Hoadley was head clay modeler at Ford's German design center.

After he returned to Dearborn, Hoadley revamped the modeling process at Ford, as he had done in Germany. In 1969, he was named Ford's modeling manager, a position he held until retirement. Hoadley is responsible for making Ford's clay modeling process much more precise, and he is credited with designing new equipment including adjustable armatures that can be reused.

Hoadley retired in 1978. He has recently written a book on design and clay modeling. ■

Steve Tremulis

(above) Models of the 1953 Lincoln and Mercury were shown with the Gran Turismo to emphasize the Gran Turismo's lower profile.

(below) Design Department general manager Charlie Waterhouse (left) and assistant manager Gene Bordinat (right) inspect the Rotunda portion of the new Design Department building in May 1953, just before it was occupied.

L.A. Waterhouse

5

Renaissance

L.A. Waterhouse

Charlie Waterhouse (right) and Frank Leonard (left), an outside supplier of exterior bright moldings, at the introduction of the 1955 Ford.

Charles L. Waterhouse

Charlie Waterhouse was a Massachusetts native who began his automotive career at Judkins, a coachbuilder. Waterhouse received his post-high school education at the Wentworth Institute in Boston, the Andrew F. Johnson School for Body Engineering and Designers and, in 1927, he graduated from the Rochester Art Institute. His family also operated the Waterhouse coachbuilding firm from 1928 until 1933. In his early years, Waterhouse also worked for the Merrimac, Walker and Cunningham Body Companies.

Waterhouse was hired by Ford Motor Co. in 1928 as a Lincoln body draftsman. He left Ford in 1933 to accept a position at Fleetwood and at GM's Art and Colour Section. There, he headed the Product Study Department, an experimental body group. In 1942, Waterhouse transferred to the War Aircraft Department Section of GM, where he was in charge of 200 men engineering new fighter planes. After the war, Waterhouse was transferred to the GM Products Study Group.

In September 1948, Waterhouse returned to Ford Motor Co. as Lincoln-Mercury body engineer, a position he held until November 1949, when he was appointed manager of Ford's Design Department. He remained in that position until December 1953, when he was named manager of the Special Projects Studio.

In May 1955, when Design was separated from Engineering, and George Walker was appointed a Ford vice president and general manager of the Styling Center, Waterhouse remained as an employee of the Engineering Department and as manager of the Special Vehicles Department. In October 1956, Waterhouse was appointed chief body engineer for the Mercury Parklane and, in March 1957, he was named manager of Ford International's Foreign Products Department. He remained in that position, which involved substantial European travel, until his retirement in December 1963. ■

A 1997 rendering of George Snyder's proposed 1949 Continental by George Barbaz.

1949 Continental

Edsel Ford's unexpected death in May 1943 was a tragic loss for Ford Motor Co. and, especially for those who had worked closely with him, including Bob Gregorie. After Edsel Ford's death, Gregorie's Design Department was placed under the supervision of Joe Galamb, Ford's long-time chief body engineer. Gregorie and Galamb didn't get along and, in September 1943, Gregorie resigned in frustration and started his own business in Detroit as an industrial designer and naval architect. When Gregorie resigned, Galamb asked Tom Hibbard, who was doing war work at the Rouge plant, to take over as head of the Design Department.

In August 1943, the government arranged for the release of Edsel Ford's oldest son, Henry Ford II, from the Navy, in the hope he would gain control of Ford Motor Co. and bring some sense of order to it. Gaining control wasn't easy but, in April 1944, Henry Ford II was elected a Ford vice president. One of the first things he did after his election was to call Gregorie, who he persuaded to return by offering an increase in his salary and promising that he would work closely with him on the design of future Ford products. What really convinced Gregorie of Henry Ford II's sincerity was his willingness to retire Galamb immediately, which he did.

The month following Gregorie's return, 28 year old Henry Ford II was elected Ford Motor Co. president. Gregorie felt he was on the way to establishing a one-on-one relationship with Henry Ford II, which would continue Ford design the same way it had been with his father. By June 1946, however, Henry Ford II realized he was in over his head and hired GM executive Ernest R. Breech as Ford Motor Co. executive vice president to teach him how to run the Company.

Breech thought Gregorie's proposed 1948 Ford was too big, and he felt it wouldn't make a profit. He convinced Henry Ford II to delay the introduction of all new cars for a year, made Gregorie's Ford the new Mercury and ordered a design face-off between Gregorie and independent design consultant George Walker, who were each to design a proposed new Ford. Walker's proposal won, which did not surprise Gregorie given the long-time friendship between Breech and Walker. Breech then effectively placed himself between Gregorie and Henry Ford II. At the same time this all happened, Gregorie's salary was doubled, which he took as a clear indication that Henry Ford II wanted him to stay.

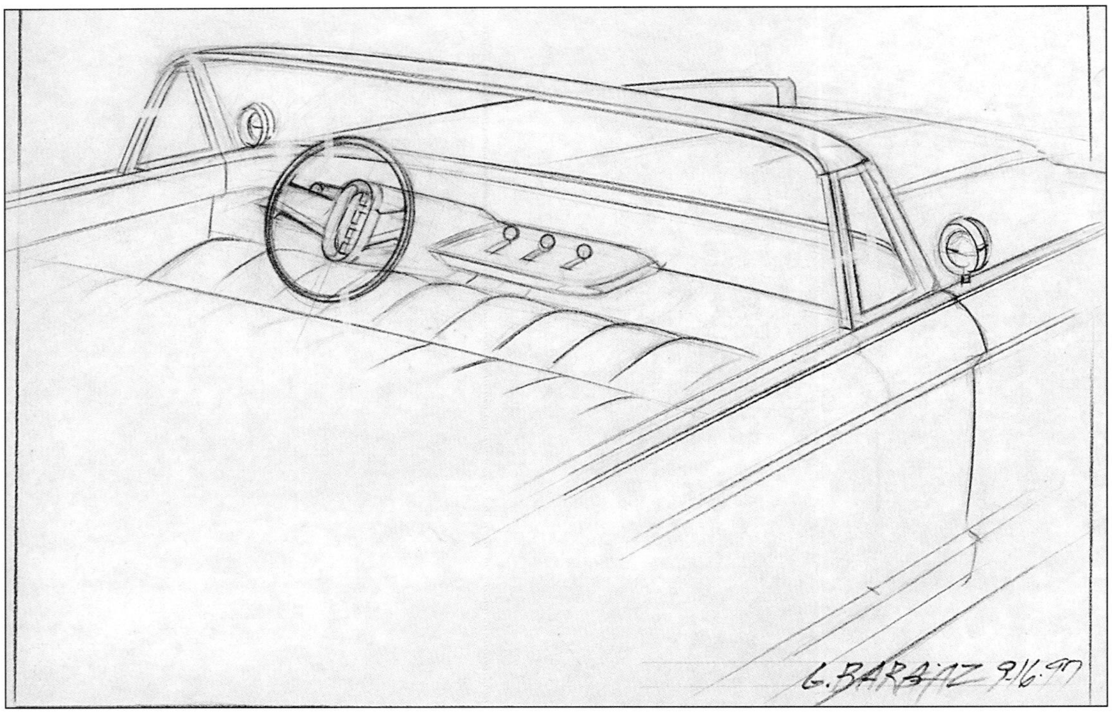

A 1997 interior rendering of the 1949 Continental by George Barbaz.

While Edsel Ford was alive, he and Gregorie made design decisions on the spot. Breech's committee system, and the lack of continued direct access to Henry Ford II, just didn't fit the way Gregorie thought things ought to be done. He approached Henry Ford II on one of his increasingly rare visits to the Design Department and told him he didn't think he fit in at the new Ford Motor Co. According to Bennie Barbera, who overheard much of that conversation, Gregorie told Mr. Ford that he thought he was going to work directly with Mr. Ford in the same manner he had worked with his father. Mr. Ford told Gregorie that was not possible, and that he would have to work through Breech. Gregorie said he didn't think he could do that, but that he didn't want to be an obstacle in the way of changes Breech was proposing, so he thought it best if he and Henry Ford II part company as friends. Unable to talk Gregorie out of leaving, Henry Ford II gave him six months to change his mind and return as head of the Design Department, no questions asked. On December 15, 1946, Gregorie resigned. He recommended and Henry Ford II appointed Tom Hibbard, who had become Gregorie's assistant after he returned, as his successor. Gregorie soon moved to Florida and into a semi-retirement designing boats.

By the time Hibbard took over as head of the Design Department, the full-sized clay model of the 1949 Ford was almost finished. Even though Walker thought he had been promised a long-term consultancy at Ford, Breech didn't agree and, when Walker's designers finished the full-sized 1949 clay model, Walker and his employees left. Hiring Walker as a consultant had created a lot of hard feelings and confusion, and it convinced Breech that it was preferable for Ford to build up its own Design Department rather than rely on outside consultants. Breech then proceeded to reorganize Ford's Design Department along the lines of GM's Styling Section, and Ford feverishly began recruiting new designers. Ads were placed in newspapers and trade publications, and designers were actively recruited from other automobile manufacturers with the promise of more money. Six months later, Breech realized Ford lacked seasoned design and administrative talent, so he asked Walker, his two key employees, Joe Oros and Elwood Engel, and several other members of Walker's staff to return as long-term design consultants.

Immediately after he came to Ford, Breech hired former Oldsmobile chief engineer Harold Youngren to head Ford's Engineering Department. Breech and Youngren personally recruited as many GM engineers and designers as Ford Motor Co.'s money could buy. One of those recruited was Oldsmobile's chief body engineer, John Oswald, who came to Ford in July 1947 as chief body engineer. When Oswald was hired, Breech placed him in charge of the Design Department and gave him directions to clean out the dead wood and to build a new department that was second to none. Hibbard didn't like it, but he then became Oswald's assistant.

At the same time Breech and Youngren began recruiting new engineers and designers, Henry Lauve, who had been head of GM's Buick design studio, was promoted to a supervisory position over GM's other studio heads. George Snyder, who was head of the Oldsmobile studio, felt that Lauve's promotion was his demotion and, early in 1947, he quit. In August 1947, Oswald hired Snyder and assigned him to Ford's small advanced design group, supposedly as just another new designer.

George Snyder

George Snyder began his design career working for the Brewster coachbuilder firm in New York. He was at Brewster eight years, before he moved to GM in 1931. At GM, he was a designer in the Chevrolet, Pontiac and advanced studios before he became head of the Oldsmobile studio. Bob Thomas, among others, credits Snyder with teaching him the real art of car design. Snyder also did much of the design work on GM's 1938 Y-Job concept car.

In early 1947, Snyder left GM because he felt he had been unfairly passed over for promotion. In July 1947, former Oldsmobile Division body engineer John Oswald was hired by Ford's new chief engineer, Harold Youngren, as administrator of the Ford Design Department, which was still under Engineering. In August 1947, Oswald hired Snyder, who had been unable to find another job as a designer after leaving GM. His first assignment was in Ford's advanced design area.

In August 1948, Snyder was appointed Ford's director of styling. Hibbard was let go because he was not in agreement with the direction the Ford Design Department was taking, and how Snyder was treating the Ford designers and clay modelers, who had been there before Snyder came.

Ford Industrial Archives

Snyder was hired to revitalize and modernize the Design Department. He created the studio system and partitioned the studios. Overall, the changes he made were seen by most designers as overdue. The methods he used were not appreciated and were divisive.

One of the first things Snyder did was terminate Ed Martin and Bruno Kolt. Before Snyder could follow through with plans to terminate other long-time employees at the Design Department, including all of Ford's clay modelers, Henry Ford II stepped in and told him he could not terminate any other Ford designers or clay modelers.

Because of continuing problems and his continued harassment of the long-time Ford employees, Snyder was terminated. Ernie Breech and Snyder were both Masons, and Breech felt sorry for him so, with Breech's concurrence and on Walker's recommendation, Oswald rehired Snyder as chief designer at Ford of England. After he arrived in England, his problems grew worse causing the dismissal of both Snyder and Oswald in 1949. ■

Shortly before he left GM, Snyder designed an advanced car he and his executive stylist Bob Thomas built as a 3/8-sized model. When Snyder left GM, he took the drawings of that 3/8-sized model with him, but either didn't care or didn't know that Harley Earl was aware he had them. When Snyder started at Ford, the 1949 Continental had not yet been completely eliminated from the new-car lineup, and it didn't take Snyder long to conclude that Ford's designers lacked the talent to design a new Continental — something he thought he had already done.

The day Snyder started at Ford, layout man Bob Todd was asked to set up another one of the Hamilton desks used in the Design Department for another new designer who had just been hired. At first, the other employees at Ford's Design Department were told that George Snyder was just one of the many new designers being hired. Snyder's desk was set up near Todd's and, during the first week he was there, Todd happened to see one of the designs he was doing. Because it was in 3/8-scale, Todd told him that at Ford they designed in quarter-scale not 3/8-scale. Snyder told Todd that they *used* to do things that way and kept right on working. Todd didn't understand the significance of Snyder's remark until some time later, when Snyder became his boss -- and then they were all working in 3/8-scale.

After he had been at Ford for several months, Snyder started to design his proposed 1949 Continental. Oswald gave the go-ahead, although it is not certain how much he knew about the background of Snyder's design. The Continental Snyder designed just happened to be the same car he had designed and built as a 3/8-sized model before he left GM. The only difference was that, at Ford, it was built as a full-sized model and stretched to the 125-inch wheelbase of the Continental. In November and December 1947, Bob Thomas, who had also recently returned to Ford as a designer, began making a full-sized side-view drawing of Snyder's proposed Continental on one of the blackboards in Bill Wagner's truck studio and, by January 1948, a full-sized clay model of Snyder's Continental was under way in a separate advanced studio set up in an out of the way area in the EEE Building.

In addition to Thomas, Snyder selected several other recent hirees from GM to help build his Continental, including Gene Bordinat, who was in charge of exterior design. George Barbaz, a designer who was fresh out of school, was assigned to assist Bordinat on exterior design. Bob Maguire, who had recently been head of one of the GM interior studios, was placed in charge of the interior soft trim, and former GM designer Don Beyreis was asked to design the instrument panel.

A special armature was built so the designers could clay model a full interior on the same armature used for the exterior. Charlie Stobar, recently hired away from GM to be Ford's modeling manager, was in charge of the clay modeling. According to Thomas, the design was modified as they went along to fit the Continental package. Thomas does not remember the original package dimensions of the 3/8-sized GM design, but recalls that it was somewhat smaller. According to the designers and modelers who worked on Snyder's Continental, it included the following special design features:

- Wheelwell openings that were squared-off and had a slight flair to them. The rear-wheelwell openings slightly covered the wheels, although there was no provision for fender skirts.
- So that a full interior could be clay modeled and then displayed, the car was built with a removable black-fabric convertible top over top bows. A removable top, however, was not proposed for production.
- When completed, the car was painted fire-engine red, inside and out.
- Snyder's Continental was also completely detailed and had full instrumentation including plastic knobs, metal dials and chrome bezels. The most unique features on the instrument panel were two arms, each about eight inches long, extending from the steering column about six inches below the steering wheel. On each arm were controls, including the turn signals. In the center hub of the steering wheel were pushbutton shift controls for an automatic transmission.

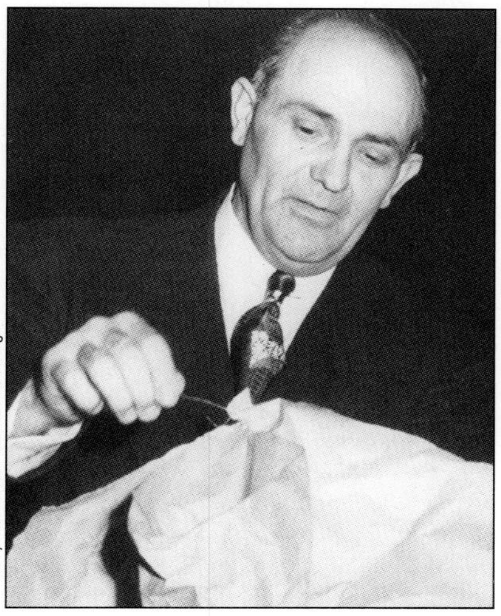

John Oswald

John Oswald started his automobile career at Auburn in 1927 as chief body engineer. He engineered much of the front end of the Cord L-29. Oswald also worked as a body engineer on the Duesenberg Model J. Later in his career, he became chief body engineer at GM's Oldsmobile Division.

In July 1947, Oswald was hired away from GM by Harold Youngren, Ford's new chief engineer. He became Ford's chief body engineer. One of his jobs as chief body engineer was as the administrator of Ford's Design Department. When Oswald became head of the Design Department, Tom Hibbard was demoted and became his assistant..

In August 1947, Oswald hired George Snyder as a Ford designer. When Snyder was hired, Oswald promised him he would become Ford's director of styling. In August 1948, Oswald appointed Snyder director of styling and terminated Hibbard. Oswald encouraged Snyder to establish separate design studios for Ford, Lincoln and Mercury and to expand and reorganize the advanced studio, which Snyder oversaw.

Oswald left Ford in 1949 largely because of his support for Snyder, who was terminated at the same time. ■

This rendering of the proposed 1949 Continental was prepared by Bob Thomas in 1984. Both Thomas and Barbaz did their renderings from memory, which probably accounts for the differences.

Bob Thomas, 1984

A rear view of George Snyder's proposed 1949 Continental rendered by designer Bob Thomas in 1984.

- The door handles, although they were not operable, were hidden in the top character line of the indented scoops on the sides of the car.
- The car was equipped with a car phone. The aerial on the front fender was for the radio, and a second aerial on the top of the car was for the phone.
- Next to the front vent window, there were spotlights on either side of the car. They resembled spherical chrome balls and were cut-off in the front so lightbulbs could be inserted.
- The car had a curved, wraparound windshield made from Plexiglas that was shaped over a plaster mold cast from the clay model. Just above the windshield header, in a style reminiscent of some early Brunn customs, there was a Plexiglas see-through skylight about six inches wide running the whole length of the header.
- The car had a high beltline that was capped with a three-quarter to one-inch wide chrome strip running the full length of the car.
- Indented and scalloped into each side of the car was a scoop-like vent that terminated just in front of each back wheel in small chrome grilles.
- In oval-shaped chrome pods at the end of each front fender were stacked dual headlights. In each pod, surrounding the dual headlights, was textured metal with a Lincoln crest between the headlights.
- In the oval-shaped chrome pods at the end of the rear fenders were three taillights. The top one was red, the middle one was amber, and the bottom one was green. When the car was moving forward under power, the green light was illuminated. When the car was coasting or slowing down, the amber-colored light was on, and when the brakes were activated, the red light went on.
- Snyder himself designed deep-dish wheelcovers that were built in the Design Department shops from chrome-plated spun aluminum. The wheelcovers were designed as part of the wheel hubs and were attached to the car by approximately 16 separate lugnuts located where the tire and the wheelcover met.

Henry Ford Museum and Greenfield Village

A 1948 Lincoln-Mercury Division advertising proposal for a proposed 1950 Continental.

When the full-sized clay model was finished, it was shown to Ford executives, who were genuinely impressed with it, but felt that it was too futuristic to consider it as the new Continental. According to Thomas, the model looked so real that, during one of the showings, Ernie Breech grabbed the door handle and tried to open the door. According to Barbaz, the car was "really sharp." Although Barbaz was very proud of it, he also was fairly certain it would never be built, because it was too advanced for the public's taste.

After Snyder's proposed Continental was finished, it was also shown to several visiting dignitaries, including Prince Michael of Romania. On the day Prince Michael was to visit the Design Department, FBI agents were there early to check out security. Just before the Prince arrived, Barbaz ran down the hall to the rest room. When returning to the studio, he pushed the door open just as Henry Ford II was doing the same thing from the other side. At first, Barbaz thought he had broken Mr. Ford's wrist, but he apologized and, luckily, Mr. Ford was not permanently injured and laughed-off the incident. Prince Michael arrived right after that. When he was shown the Continental, no one thought to tell him it was a clay model. Thinking the car real, he shook it causing the front 2 1/2-feet of clay overhang to fall off on his feet. Barbaz remembers the FBI almost going nuts when that happened. Most designers there at the time enjoyed a good laugh — but only after the Prince, Henry Ford II and the FBI had departed.

This three-eighths sized clay model of a proposed 1950 Continental was an evolutionary design based on George Snyder's proposed '49 Continental. By the time this model was made, some effort had been made to productionize the design.

Henry Ford Museum and Greenfield Village

Because Snyder's Continental had to be completely repainted every time the clay was reworked, even minor changes were difficult and time consuming. (When di-noc became available several years later, it became much easier and faster to make changes to finished clay models.) When the front overhang of the car fell off onto Prince Michael's feet, it may also have become an excuse to change the design of the car. In any case, Oswald sug-

L.A. Waterhouse

Eugene Bordinat

Gene Bordinat grew up in Pleasant Ridge, a suburb of Detroit, where his father was a plant engineer at Willys-Overland and later at Chrysler. Bordinat lived in the same neighborhood and went to high school with George Walker's son and daughter. Bordinat would catch a ride to school with Walker's children in the back seat of Walker's Marmon. On the way to school, Walker let Bordinat go through the designs he had worked on the night before.

In 1936, after two years at the University of Michigan studying Art, Bordinat joined 33 other design students at GM's Art and Colour Section. At the end of his first year, Bordinat was hired as a designer in GM's advanced studio, making $150 per month. His first job was working on the proposed 1941 LaSalle. Later, he transferred to the Chevrolet studio, where he worked on the 1942 Chevrolet. When the war started, Bordinat worked at GM setting up the production line for the M42A tank. During the war, Bordinat served two years in the Army Air Corps.

After the war, Bordinat returned to GM to find that his friends John Oswald and Harold Youngren were leaving to work for Ford. In August 1947, Bordinat had a chance meeting with George Snyder, who had also been hired at Ford's Design Department. Bordinat was anxious to break into management, and Snyder encouraged him to accept a job at Ford, where the salaries and opportunities were greater. In late August 1947, Bordinat accepted a job at Ford's Design Department working as Tom Hibbard's assistant.

After Snyder was appointed director of styling, Bordinat became his assistant and was assigned responsibility for reorganizing the advanced studio, a job he asked Gil Spear to spearhead. When Charles Waterhouse was appointed administrative head of Ford design in late 1949, Bordinat became his assistant manager. In 1954, Bordinat became director of the Lincoln-Mercury studio. In mid-1955, after George Walker became general manager of the Design Department, the Lincoln and Mercury studios were split, and Bordinat became head of the Mercury studio. The studios were later combined again and Bordinat later became head of the Lincoln-Mercury studio.

In 1961, Bordinat was appointed Walker's successor and remained a Ford vice president and general manager of the Design Center until his retirement in late 1980. ■

George Barbaz

George Barbaz grew up in Dearborn and began sketching cars as a child. During his senior year in high school, Barbaz joined the U.S. Army and served in the infantry during World War II. He was wounded twice at Anzio. In 1946, Barbaz returned to Dearborn and enrolled at Meinzinger's Art Academy.

After class, Barbaz worked at a gas station, where Jake Aldrich, a Ford designer, was a regular customer. Aldrich recommended that Barbaz apply for a job at the Ford Design Department when he completed school. In January 1948, Barbaz interviewed with George Snyder, who told him there were no design jobs available. Before leaving, Barbaz asked Snyder if he would at least critique his portfolio. Snyder agreed and, after he looked at it, Snyder asked if he could show the portfolio to other designers. A few minutes later, Snyder returned and hired Barbaz. A week before he finished school, Barbaz started as a designer at Ford. His first assignment was in Gil Spear's Advanced Studio.

Barbaz' next assignment after he left the Advanced Studio, was working on Snyder's proposed '49 Continental. He also worked on the design of the 1949 Presidential Lincoln, the XL-500, the Quicksilver and many other concept cars. He became manager in the Edsel pre-production studio, and, later, he was a manager in the truck studio. He also helped design many of the Canadian Meteors and Monarchs.

Throughout his career, Barbaz was assigned to various exterior studios and was often loaned to the advanced and pre-production studios to work on concept cars. He is an accomplished airbrush artist, who was often asked to do airbrush renderings. He was one of the designers regularly assigned to work at the auto shows the Design Department participated in. Barbaz also did caricatures and cartoons for various functions at the Styling Center.

He retired in 1981. ■

Ford Design Center

gested that the taillight and headlight ovals on each side of the car be squared-off, like a "nun's habit," as he put it. (What Oswald was apparently referring to were the headpieces some orders of nuns wear that form a trapezoid-type opening around their heads.) Some of the other rounded features of the car were also squared-off. After Oswald's suggested design changes were completed, the car was repainted green. Barbaz says that the squared-off look wasn't as bad as it sounds, but the changes did not improve the design.

In his book, *Confessions Of An Automobile Stylist,* Thomas claims that Harley Earl knew that Snyder had taken the plans for the 3/8-sized model with him when he left GM, and that Earl let it be known that if Ford used the design on a production car, GM would sue. Joe Oros recalls that, at the time, he was told by George Walker that someone from GM's design department had been able to describe the car to him without seeing it, which, in part corroborates Thomas' recollection. Thomas also attributes the lack of a photographic record of Snyder's Continental to GM's threat of a lawsuit. (In fact, no photographs can now be located.) Barbaz also recalls hearing something about a threat of a lawsuit connected with the car. Years later, Barbaz made a rendering of Snyder's Continental and gave it to Gene Bordinat, who was then vice president of design. To Barbaz's surprise, Bordinat was very uneasy when given the rendering and pretended not to remember or recognize the car.

Wes Dahlberg, David Ash and Fred Hoadley remember seeing the car at the Design Department after it was finished but, after Snyder and Oswald left, the car disappeared. No one recalls who gave the order to destroy it, but everyone who remembers it is certain it was quickly destroyed after Snyder's departure. Even though the car and all photographs of it were apparently destroyed, Snyder's Continental influenced the design of subsequent Ford production and concept cars for years to come:

- Dave Ash credits Snyder's Continental for the side scoops and vents he designed into the sides and rear fenders of the first Mustang;
- the windshield inspired the shape of those on the X-100 and the Thunderbird;
- the rear integrated taillight housings, rear bumpers and the instrument panel inspired those used on the 1952-54 Mercury;
- the front bumper and grille were similar to those used on the X-100;
- the pushbutton-shift quadrant in the center of the steering wheel was used on the XL-500 and the 1958 Edsel;
- the hidden door handles were similar to the ones used on the 1961 Thunderbird;
- The high, chrome-capped beltline appeared on the Quicksilver, the 1960 Ford, the 1961 Thunderbird and the 1961 Continental.

Snyder was a talented designer. He helped design GM's 1938 Y-Job concept car. It was his performance as an administrator that created problems. After he came to Ford, he tried to pattern his behavior, dress and demeanor after Harley Earl, his former boss at GM. Unfortunately, his style and mannerisms were interpreted by most Ford designers as "just plain meanness." He fired or threatened to fire many of the original Ford designers and all the clay modelers, and he harassed and demoted many of those Henry Ford II prevented him from firing. Mr. Ford told him not to, but Snyder continued to harass the designers and clay modelers left over from Gregorie's Design Department. Clay modeler Werner Framke complained to Oswald, who he had known since their days at Fisher Body. Because he felt his job was also on the line, Oswald told Mr. Ford, who sent word that Snyder was fired. After he learned he had been fired, a visibly-shaken Snyder hurried over to Mr. Ford's office and, according to Framke, got down on his knees to beg for his job back. Mr. Ford was tired of the constant bickering and would not relent -- Snyder was gone.

A serious alcohol problem probably caused much of the meanness that led to Snyder's termination in mid-1949. Although there was a collective sigh of relief when he left the Ford Design Department, with Breech's encouragement and on Walker's recommendation, Snyder was rehired a few months later by Oswald as chief designer at Ford of England. Snyder's problems continued and, soon after he moved to England, he was again terminated. Shortly after that, Oswald was also terminated, primarily because of his support for Snyder. ■

Eleanor Ford became the matriarch of Ford Motor Co. at the same time her sons began to establish themselves as the third generation to run the Company. At the time of her husband's death, in 1943, Mrs. Ford's daily transportation was usually a 1940 Lincoln-Zephyr Brunn Towncar. In 1946, that car was updated to look like a new Lincoln.

By mid-1952, Mrs. Ford decided to replace her 12-year-old towncar and, in the process, she became actively involved in the design of the new car. She began by drawing up requirements for a tall towncar, so she wouldn't have to stoop to enter. She also required that her new car be very dignified, plain and without excess chrome -- and that it be black. Val Tallberg, a co-ordinating engineer at Ford, dealt directly with Mrs. Ford and conveyed her instructions to Bill Schmidt, chief designer in the Lincoln studio. Schmidt, in turn, assigned the job of actually designing the car, according to Mrs. Ford's requirements, to stylist Don DeLaRossa, who prepared several proposals, one of which was approved by Mrs. Ford. Mrs. Ford was involved in all of the details of the car, and even specified that the hood ornament should be from the 1950 Lincoln, because she liked it better than the one that was on the 1952 Lincoln. The car was assigned a code name of X-400 and, because of Mrs. Ford's interest and involvement, it became a major design project.

1952 Lincoln Towncar (X-400)

Henry Ford Museum and Greenfield Village

Henry Ford Museum and Greenfield Village

(top) Designers who worked on Mrs. Ford's towncar tried to persuade her to modernize and streamline its design. Mrs. Ford resisted such suggestions in favor of her more practical and classic preferences.

(above) Although the body on Mrs. Ford's towncar looked shorter than the production 1952 Lincoln, it was actually longer because of the continental kit.

At about the time the X-400 was being designed, John Reinhart was hired as chief designer for the Continental Mark II project. Before Reinhart's design staff was completed and a separate Special Projects design studio was established, Reinhart was recruited to work on the design of Mrs. Ford's new towncar.

Schmidt, DeLaRossa and Reinhart were not pleased with the design specifications Mrs. Ford insisted on, but they all took the train to Grand Rapids, Michigan to explain the design to Raymond Dietrich. The well-respected Dietrich had restarted his custom-body-building business in 1950 and was one of the few remaining custom-body builders still in business. He had worked with Schmidt several years earlier on the Presidential Lincolns and, more importantly, he was a former neighbor and a long-time friend of Edsel and Eleanor Ford. Soon after their visit, arrangements were finalized for Dietrich to build Mrs. Ford's new towncar. The standard Lincoln body panels used were shipped to Dietrich, and he began construction of Mrs. Ford's towncar. Almost immediately after construction began, DeLaRossa was reassigned to another studio, and Schmidt asked Reinhart to take over responsibility for completion of the car to Mrs. Ford's satisfaction.

Reinhart made several suggestions he thought would help improve the looks of Mrs. Ford's towncar, including a wide chrome band along the bottom of the car. The suggestions Reinhart made were conveyed to Mrs. Ford by Tallberg only to be politely rejected. Reinhart was able to incorporate a flat Continental-style windshield on the car, which added about six inches to the hood and helped balance the car's looks. The biggest problem was the high roofline. Although it was for easy entrance by the arthritic Mrs. Ford, who was more concerned with comfort than she was with modern design principles, it still made the car look short and squat. The X-400 was built with a Continental-style spare tire and trunk, old-style runningboards, and it weighed a hefty 5590 pounds. It was also the first Ford car intended to be regularly driven that was built with factory-installed air conditioning.

Dietrich built Mrs. Ford's towncar body the same way coachhouses had built special-bodied cars in the 1930s, with a hardwood framework and metal body panels attached directly to the wood. Production doors, trunk lid, hood, fenders and front grille were the only stock Lincoln body panels used, although most of them had to be substantially modified.

The X-400 was delivered to Mrs. Ford in early 1953. How much she liked the car is probably shown by the fact she used it as her primary transportation, until it was retired in 1970 after 92,000 chauffeur-driven miles.

Dietrich closed his shop shortly after Mrs. Ford's towncar was completed. The X-400, which was probably the last car built by Raymond Dietrich, is now owned by the Henry Ford Museum. It is displayed in the Gatehouse at the Edsel and Eleanor Ford Estate in Grosse Point Shores, Michigan. ■

Mrs. Ford's towncar was one of the first Ford products with factory air conditioning. The air intake is below the backlight. The continental kit was molded into the rear of the body, which meant that the trunk lid had to open from the top like the original Continental.

Henry Ford Museum and Greenfield Village

(below) Mrs. Ford's X-400 towncar during construction at Dietrich's shops.

Henry Ford Museum and Greenfield Village

Suzanne Malta

The X-100 at Pebble Beach in August 1996. The car was given to the Henry Ford Museum and Greenfield Village by Ford Motor Co. in 1958. It has been maintained in a state of arrested decay since then.

X-100 (Continental 195X)

The Continental Nineteen Fifty X, as the X-100 was first called, was created between 1949 and 1953. It was first introduced to the press and public at the Chicago Auto Show on January 10, 1952 as a "pilot model." The official press release said the car was "being studied towards future development as a four-passenger sedan." The Continental Nineteen Fifty X shown at the Chicago Auto Show was a pushmobile that had no engine, running gear or gadgets and was probably made of fiberglass and plaster.

At first, Ford's public relations department wanted the public to believe the Continental Nineteen Fifty X might just be the successor to the still popular Lincoln Continental that had been discontinued after the 1948 model year. Because that's how the car was represented, the public and the automotive press began to believe it. When the Continental Nineteen Fifty X was first introduced, even Henry Ford II hoped it would become the next Continental.

Building a concept car that "looked like a rocket ship" was the idea of designer Joe Oros. The irony is that at the time he made the proposal, Oros was not even a Ford employee. He and Elwood Engel were employees of design consultant George Walker, who were assigned to work at Ford full time; Oros as a consultant in the Ford car and truck studio and Engel as a consultant in the Lincoln-Mercury studio. (In May 1955, Walker was hired as vice president of Ford design and, at that point, Oros and Engel became full time Ford employees.)

Almost immediately after he returned to Ford in 1949, Oros began working on the design of the 1952 Ford. Oros was impressed with Gil Spear's advanced designs and felt they represented the wave of the future. He was particularly interested in one of Spear's 3/8-sized models called the Cutlass, which had big round taillights with exhaust pipes exiting through the middle of them. One of the design themes Oros adopted for the '52 Ford was the round taillights, which he saw as a focal point Ford could probably use on future models.

Oros was already hearing from the GM designers hired by Ford that GM had several concept cars under way, so he suggested to Walker that Ford build its own full-sized concept car. He proposed modeling it after several of Spear's 3/8-sized models with "big rocket-tube sides and afterburners coming out of the bumpers." Oros told Walker that, if Ford wanted to preserve the big round taillights as an identifying theme for future Fords, the way to do it was to incorporate them as a predominant feature on a concept car, so no one else would dare use them without looking like they copied Ford. To Oros, it was only

Joe Oros

Joe Oros was born in Detroit to Romanian immigrant parents. English was a second language in the Oros home. As a child, Oros loved cars, but his father discouraged him from becoming an automobile designer. (His parents did not have a car, and Oros did not learn to drive until he worked for GM) Half-way through high school, his parents moved the family to Cleveland, where Oros attended the Cleveland Art Institute on a scholarship. After he completed his schooling, he went to Sweden on a fellowship but returned home early because of the start of World War II.

Ford Design Center

After spending a year at GM's Styling Section school, where he met Elwood Engel, Oros spent the war years engineering army tanks at GM. After the war, Oros took a job at George Walker's industrial design firm. At Oros' suggestion, Elwood Engel applied and was hired by Walker a few months later. Oros wanted to eventually open his own non-automotive industrial design business but was soon deeply involved in automobile design as a Walker employee.

In 1946, after a design contest with Bob Gregorie, Walker's firm was hired to design the 1949 Ford. The design work on the '49 Ford was done by Oros and Engel. (Because of Oros' engineering background, he was placed in charge of preparing Walker's entry for the '49 Ford design contest.) Once the '49 Ford was finished, Walker, Oros and Engel left Ford, only to return in less than a year as long-term design consultants. Oros was assigned as design consultant to the Ford studio.

In May 1955, when Walker was made Ford's vice president of design, Oros became a Ford employee. In March 1956 and, at least partially as a result of his design of the production 1957 Ford, Oros was appointed head of the Ford studio, a position he kept until 1968. While at Ford, Oros was responsible for the design of the X-100, the Quicksilver and the crisp design style of Ford production cars of the 1950s and '60s. Oros is also responsible for the general design of all Ford cars and trucks from 1952 to 1970. He proposed round taillights on Ford products as a general design theme. Oros is the designer primarily responsible for the resurgence of Ford design in the 1950s, when Ford almost overtook Chevrolet in sales.

In 1961, when Walker retired, he wanted Oros or Engel to replace him, but Gene Bordinat was given the job. Thereafter, Oros and Bordinat continued to work well together until Oros' retirement. In 1968, Oros was appointed vice president of design, Ford of Europe. He returned to Dearborn after nine years in Europe and was appointed executive stylist for all Fords, Lincolns and Mercurys. He retired in 1981. ■

Joe Oros with the 1957 production Ford.

Joe Oros

logical that if Ford didn't build such a car, GM almost certainly would, and Ford would lose what Oros thought was a great design theme they could lunch off on future cars. Walker agreed, and the project began immediately. The Cutlass and possibly other 3/8-sized concept cars designed in Spear's new Advanced Studio became the primary design influence for the X-100.

Art Tarabusi

As it was, General Motors introduced two concept cars in 1951, the Buick LeSabre and the Buick XP-300, both of which were futuristic factory hot rods full of gadgets. Both GM concept cars were heavily influenced by modern jet aircraft as was the X-100. Luckily, the X-100 went in a little different design direction than the GM concept cars.

Spear's 3/8-sized models were moved into the Lincoln studio, where the X-100 was built. Because those models had already been completed, there was no need for scale drawings, renderings or additional 3/8-sized clay models. There were, however, full-sized elevation drawings of the X-100 made and placed on the wall of the studio, so they could be followed in building a full-sized clay. Charles and Leonard Stobar, two modelers and brothers, who had recently come over from GM, were in charge of the full-sized clay model, although Werner Framke did most of the actual clay modeling work on the car. As with most concept cars, many deviations were made from Spear's models as the full-sized clay was built, some of them substantial. The taillights on the X-100 were taken directly from the Cutlass, but the sides of the car, the hood and front-end were changed. Oros remembers the hood and front bumper of the Spear model as more conventional than the hood and integrated front bumper on the X-100 which, for the times, were quite innovative. The sculpted lines on the rear-quarter panels were also changed as the full-sized clay model took shape, and short fins were added to give the car an aircraft feel. Oros and Spear conferred often, and Oros specifically remembers going over changes made to the front end and hood of the car with Spear.

At the same time the exterior design of the X-100 was being completed, an interior design group, including Bob Maguire, Ted Hobbs and Herman Brunn, began work on colors and the interior trim of the car. David Ash supervised the building of the full-sized

Frank Bianchi

Frank Bianchi grew up on Long Island, New York. After graduating from Pratt Institute in 1939, he began his design career in the Buick Studio at GM's Styling Section. Three years later, he enlisted in the Navy and, during World War II, he was a combat pilot flying fighters off aircraft carriers in the South Pacific. He is credited with nine enemy kills and survived 16 major crashes. Bianchi and the pilots of two other dive bombers sank a Japanese battleship at Guadalcanal. As a result, he was awarded the Navy Cross.

After the war, he was hired as a designer in Bob Gregorie's Design Department. After George Snyder created separate studios, Gil Spear became the first head of the Ford studio, and Bianchi became his assistant. When Spear took over the Advanced Studio in 1948, Bianchi became head of the Ford studio. Bianchi left Ford to take a job at the Chrysler design department in 1952.

Before he moved to Chrysler, Bianchi did renderings for the X-100 and also worked on the XL-500 and the beginnings of the Futura with Bill Schmidt.

Bianchi later worked with International Motors, George Walker and Associates and for Sundberg and Ferrar before returning to Ford in 1964.

He retired from Ford in 1982. ■

As the Continental 195X, the X-100 and the 1952 new-car lineup were shown to the automotive press at A Day With Ford Engineers, January 18, 1952.

Henry Ford Museum and Greenfield Village

(from left to right) William and Benson Ford with Henry Ford II enjoying the X-100.

Henry Ford Museum and Greenfield Village

(below) Every feature of the X-100 worked including the built-in hydraulic jacks operated by levers on the instrument panel.

Henry Ford Museum and Greenfield Village

clay model. John Najjar was primarily responsible for the instrument panel, including the aircraft-type levers, the hooded instrument pod, the steering wheel and the sliding roof he worked out with Hans Matthias, who had also been recently hired away from GM and was chief engineer on the X-100 project. Martin Regitko was the design engineer.

The designers who worked on the X-100 are all quick to point out that the car was the product of the ideas and efforts of many. Although the original idea for the X-100 came from models designed and built by Gil Spear, Oros supervised the design part of the project, including the many changes, through to completion. From time to time during the design process, Elwood Engel, who was assigned as a consultant in the Lincoln-Mercury studio, shared his ideas on the new concept car with Oros and the other designers working on it.

By mid-1951, the X-100 was taking shape, and everyone was getting pretty excited about it. There was a lot of talk going around, albeit wishful thinking at first, that this was just the type of car a revived Continental should be. Word spread to the Administration Building about the fantastic looking car being built over at the EEE Building, and more and more executives began dropping by to see it. Senior executives usually saw finished and painted clay models of proposed car designs only at formal design presentations, and it was unusual for any of them to just drop by to see design proposals in progress. It was even more unusual for the boss to show up in the studio to see a partially-completed design project. Henry Ford II showed up one day when the designers were still working on the car, and spent about 45 minutes going over it from stem to stern with Oros. He also asked a lot of detailed questions. When he left, it was obvious he was impressed, and very excited about what he had seen.

The X-100's engine had five carburetors operated by progressive linkage. The radiator, firewall and inner-fender panels were painted bright turquoise.

Soon after Henry Ford II's visit to the Design Department, the name of the car was changed from the X-100, the name it started life with, to the Continental Nineteen Fifty X. When the name was changed, it was because Henry Ford II thought he had found the new Continental.

Framke recalls no major design changes when the full-sized clay model was built. He remembers making the round taillights more oval-shaped, but all other design changes were minor. Because of an approaching deadline, clay modeler Vic Clark was brought in to help Framke finish it up. Although Joe Oros vaguely recalls that a fiberglass and plaster model of the X-100 was made before the metal-bodied car was built, the first completed model of the X-100 may have been made from plaster and wood. In any case, a full-sized plaster plug would have been made from which the fiberglass model was cast. As the plaster was being poured, the bracing holding the mold slipped, and Framke remembers spending several days shaving and hand-sanding the plaster model back into shape.

A four-barrel carburetor was located in the center of the intake manifold with four two-barrel carburetors at the outside corners of the intake manifold.

In January 1952, the same month the car was introduced as the Continental Nineteen Fifty X, Henry Ford II asked Ford's retired sales manager, John R. "Jack" Davis, to form a committee to make recommendations as to what Ford Motor Co. could do to better compete with General Motors. By June 1952, the Davis Committee recommended, among other things, that Ford produce a new low-volume luxury car. That meant the revival of the Continental. Bill Ford, the youngest of Edsel's sons, jumped at the chance to build it. In early 1953, Bill Ford asked Doug McClure, later sales manager at Continental Division, to interview prominent people around the country thought to be

Ford Industrial Archives

(above) The X-100 at the Texas State Fair where an estimated 80% of the fair's 2 1/2 million visitors saw it.

Ford Industrial Archives

(above) The X-100 at the Michigan State Fair, where it was the feature attraction at the Lincoln-Mercury display.

Ford Industrial Archives

(above) The X-100 on display at Hively Lincoln-Mercury, Cincinnati, Ohio.

Ford Industrial Archives

(left) 4400 people viewed the X-100 while it was at Ohio Motors Lincoln-Mercury in Columbus, Ohio.

Ford Industrial Archives

(above) The Duke of Edinburgh and M. R. McIntosh, manager of Lincoln Ltd., watch a demonstration of the X-100 by Ralph Kimbrough and Paul Adams of Ford's Research and Engineering Department at the International Motor Exhibit, Earl's Court, London, England.

Ford Industrial Archives

(above) More than 1000 Canadian Ford dealers watched the X-100 driven over a specially-constructed stage at the Coliseum Building at the Canadian Exposition Grounds during Ford of Canada's three-day national conference.

Ford Industrial Archives

(above) George Cozal, general sales manager of Ford of France, describes the X-100's engineering features to Mr. Auriol, president of the French Republic, Baron Periet, president of the Paris Salon and members of the French government.

Ford Industrial Archives

(right) Engineers who accompanied the X-100 to the Paris Auto Show gave two-a-day public demonstrations of the X-100's features. The sign placed beside the car indicates the time of the next performance.

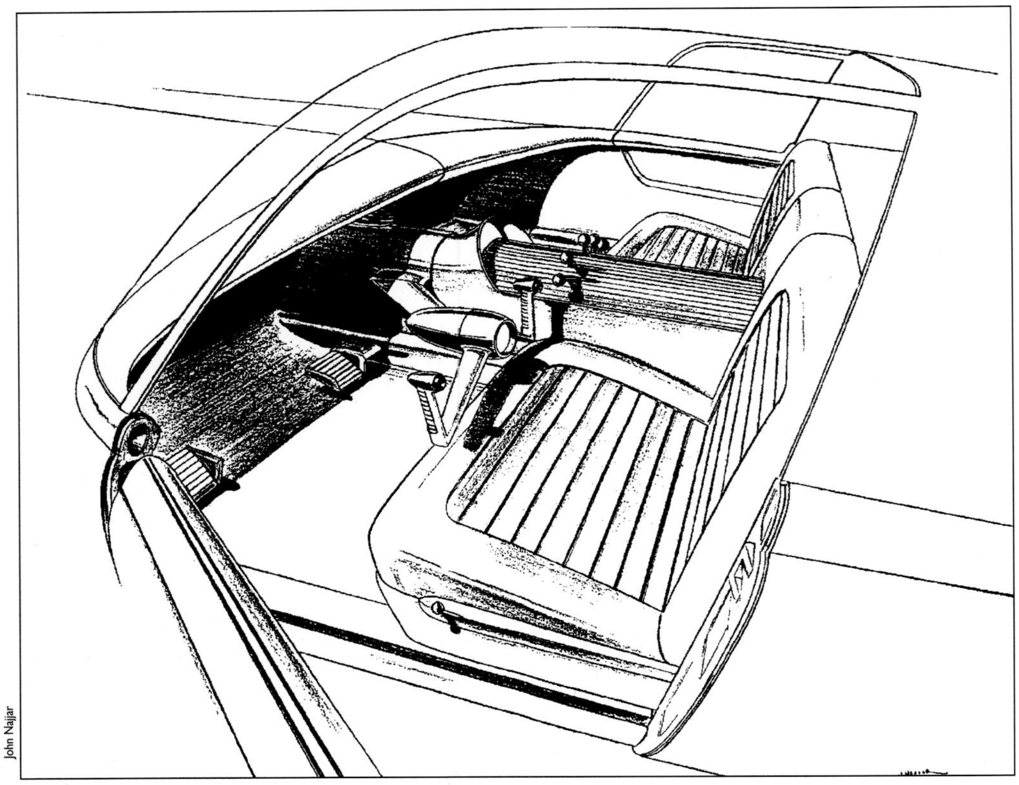
John Najjar

John Najjar's proposal for the interior of the X-100.

likely buyers of a new Continental. Those interviewed were shown pictures of several cars, including the Continental Nineteen Fifty X. The majority of those contacted said they preferred a super-luxury car with more formal lines than the Continental Nineteen Fifty X. Bill Ford decided that the new Continental he was to build -- the Continental Mark II -- would be the visual and design successor of the original Continental. The decision to build a "modern formal" Continental Mark II eliminated the Continental Nineteen Fifty X. When it was decided that the upcoming Mark II would be designed and built from scratch, the Continental Nineteen Fifty X was already at Experimental Engineering being fitted with a steel and aluminum body, an engine, and just about every gadget the engineers could dream up. A hot-rodded overhead-valve Lincoln engine and all kinds of electronic gadgets were added to the Nineteen Fifty X in an effort to match GM. Ford tried to include more gadgets on the Nineteen Fifty X than the GM concept cars had.

By May 6, 1953, the car that had been introduced the year before as the Continental Nineteen Fifty X was fully operable. That was the day it was re-introduced at Ford's 50th Anniversary press conference, and photographs were taken of it being driven around the Dearborn test track by Ernest R. Breech. The car looked almost the same as the Continental Nineteen Fifty X, but the name had been changed back to the X-100 in deference to Bill Ford's Continental Mark II project. The change in name was significant enough for a Ford photographer to record the removal of the Continental script from the car in a photograph taken June 16, 1952. On each side, where the X-100 script now is, there are signs of patched holes in approximately the same locations where the Continental script was attached, when the car was first introduced in 1952.

The X-100 had its public re-introduction on a national television show *50 Years of Progress* broadcast on station WWJ on June 15, 1953. The next day, the X-100 was at the reopening of the Ford Rotunda, which was also a part of the Ford's 50th Anniversary celebration. A new press release pointed out that the X-100, a "laboratory on wheels," was not going to become a production car. As the X-100 was taken around the world over the next several years, those who asked were assured that it was no longer a Continental, and that it would not be produced. Soon, the automotive press was also reporting that a new Continental was under construction, and then it became obvious why the name was changed from the Continental Nineteen Fifty X to the X-100.

When the Continental Nineteen Fifty X was transformed into an operable car and became the X-100 again, there were a few modifications to its exterior looks. Those changes included the installation of back-up lights, the rain cell on the roof, a non-retractable aerial on the rear deck, the air-intake slit at the bottom of the front bumper and a power dome and ornament on the hood.

Today, many people think the name X-100 was an afterthought, a name quickly picked and hung on the car because the name Continental was no longer available. Nothing could be farther from the truth. In 1953, the first of the Century Series jet fighters, the F-100 Super Sabre, took to the air. It was the first in a series of military jet fighters that could routinely surpass the sound barrier, and it captivated the imagination of the public as did the experimental rocket plane, the X-2, which, in 1953, was the first plane to fly at twice the speed of sound. There was hardly an adult or child alive in the early 1950's who was not conversant with the newest speed and altitude records that were set almost weekly by military planes. Who suggested the name X-100 for Ford's first postwar concept car is now lost to posterity, but it was named after military aircraft of the day.

Charles T. Stobar

Charlie Stobar was a Detroit native. He and his brother Leonard, also a clay modeler at both GM and Ford, grew up working in their father and uncle's pattern business. Stobar graduated from Cranbrook Academy and attended Detroit's Center for Creative Studies (at that time called the Society for Arts and Crafts). He was an accomplished oil painter and sculptor and majored in fine arts. After graduation, Stobar made the molds for the cast-bronze doors when Detroit's Fisher Building was built.

In 1932, Stobar began his clay-modeling career at General Motors. During World War II, Stobar worked as an engineer and checker for Grumman and Eastern Airlines. In 1944, at Joe Oros' request, Stobar began clay modeling George Walker's proposed industrial design products in the evenings and, in 1946, he was persuaded by Oros and Walker to move from GM to the Ford Design Department to replace Dick Beneicke as head of Ford's clay modelers. He held that position until 1954, when the chief designer in each studio was given responsibility for the clay modelers in their studios, and Stobar was placed in charge of the Ford shops replacing Jimmy Lynch, who retired. He remained as head of the shops until his retirement in 1970.

In addition to his work at Ford, the Stobar brothers started their own business making models of Ford products, from which molds were made by MPC and AMT and other promotional model companies.

Stobar was the clay modeler on Snyder's proposed 1949 Continental and on the X-100. He also built or supervised many of the other 3/8-sized and full-sized Ford concept cars. ■

Ford Design Center

Hiram Pacific, who started with Ford in 1941, was in charge of making the X-100 a fully-operable car. Pacific was assigned to Experimental Engineering in late 1952, when he returned from 27 months of military duty during the Korean Conflict. He got the job of completing the X-100, which had already been under way for about ten months by the time he got out of the service and returned to civilian duties. Working directly under Hans Matthias, who was chief engineer on the X-100 project, Pacific was given orders to finish it up on an accelerated basis. He finished the X-100 in eight months, just in time for Ford's 50th Anniversary celebration. George E. "Bud" Adams was chassis engineer on the X-100.

The interior of X-100 featured a phone, electric shaver, power seats, seat warmers and an automatic transmission controlled by a shift lever located on the instrument panel.

Ford Photomedia

The various experimental components used to make the X-100 an operable car were built by the different Ford departments or by outside vendors and were installed and made to work by Pacific and those working in his department. Because the engine and transmission were ready to install by the time he got them, Pacific does not know what was done to them internally. What he does recall is that he had a "heck of a time" tuning the engine, so it would run smoothly in parades and at highway speeds. The most time-consuming project, however, was making the X-100 a quiet-sounding and a smooth-riding car.

Almost every system on the X-100 had to be redone at least once. Everything was so advanced, many of the systems had never been used on a Ford car before. Trial and error was the rule of the day. Pacific recalls redesigning the roof system, the windows, the memory seat, and the way the microprocessors opened and closed the doors, the hood and deck lid at least once and maybe twice. Because of all the systems installed in the X-100, it turned out to be so heavy, Firestone was asked to make special tires that could handle the extra weight, especially at speed. The X-100 was the first Ford car to be equipped with radial-ply puncture-proof tires. One of the experimental radial tires blew out because of a manufacturing defect when Pacific was driving the X-100 on the Ford test track. Pacific says that, as far as he knows, he drove the X-100 the fastest — 90 mph. Although he says the car could have gone faster, he kept remembering the blowout! The tires now on the X-100 are 8.20x15 Firestone 500 brand and appear to have very little mileage on them.

Special springs and shocks had to be made to carry the car's extra weight. Because of the 300-hp engine and, because it was anticipated the X-100 would be driven a lot at slow speeds in parades, a bigger radiator and fan were used to cool the engine. One of the problems Pacific spent a lot of time on was making the fan quiet no matter what speed the engine was turning. The fan had six aluminum blades fastened to a hub with rivets in the same manner later used on the Continental Mark II.

Ford Photomedia

The Plexiglas panel over the front seats retracted into the rear section of the roof. At the time this photograph was taken, the X-100 had not yet been equipped with a rain sensor that closed the Plexiglas roof section at the first drop of rain.

Many of the special parts built for the X-100, including the steering wheel, had to be redesigned. As originally designed, the deeply-dished spokes on the steering wheel hit the driver's knees when the car was being driven. The replacement steering wheel mold and two steering wheels cost $3600. A stainless steel exhaust system was also used on the X-100. The exhaust system now on the car, including the two stock Ford mufflers, appears to be original. The exhaust inlet pipes were also wrapped in asbestos.

The operable X-100 was built at Ford. It was not built by an outside contractor as became common practice with Ford's later show and concept cars. Pacific remembers Ford's budget for the X-100 being upwards of $2,000,000, which would make it one of the most expensive Ford concept cars even by today's standards. Because many of the parts used on the X-100 were experimental and involved development of new technologies, and because they were also planned for use on future production cars, many of them were developed and manufactured by outside suppliers at their cost or gratis. When the X-100 was built, duplicates of all trim pieces and unique parts were also made.

Ford Industrial Archives

(above) Part of the estimated 140,000 people who visited the Canadian National Motor Show in Montreal waited in long lines to see the X-100.

Ford Industrial Archives

(above) Press, radio, TV and civic leaders were given an advanced showing of the X-100 at the William Penn Hotel in Pittsburgh, Pennsylvania. The car was also featured on many regional TV newscasts.

Ford Industrial Archives

(above) The X-100 was exhibited in Plellersdorf Germany at the American Club. Guests included members of the diplomatic corps, the ambassadors of Denmark, Italy and Canada, ministers of Switzerland, Sweden and Syria and junior officers from other missions.

Ford Industrial Archives

(left) The X-100 at the September 1954 Ford Motor Co. Greenbrier Management Conference, held at White Sulphur Springs, West Virginia.

(left) The rain cell on the roof of the X-100 was located just behind the Plexiglas section..

(above) idiot lights were located directly in front of driver. The clock is located in the center hub of the steering wheel.

(below) Four wipers cleared the X-100's curved-glass windshield.

The X-100's hood ornament is similar to one used on the 1956 Ford.

Pacific was the person who did all of the test-driving of the X-100 and, when it was finally completed, he remembers driving it to Henry Ford II's estate, so it could be demonstrated to Mr. Ford's guests, who were there especially to see the car. On one of the many occasions the car was at the Ford estate for a gathering, the micro-switch that operated the deck lid wouldn't work. Pacific was called and asked to come solve the problem, which turned out to be nothing more than an adjustment problem. Pacific remembers that Mr. Ford was very proud of the car. During the 50th Anniversary celebration, a new research and engineering complex was opened. To the applause of all present, Mr. Ford drove up in the X-100 and, from the front seat, waived a remote-control wand that turned on all power in the building.

On occasion and, when it was available, Mr. Ford used the car for his personal transportation. When the X-100 was not on the road or at Mr. Ford's estate, it was stored at the old Administration Building, at 3000 Schaefer Road, next to the wash rack.

After the X-100 project, Pacific continued in Experimental Engineering. He was the first to use sound-deadening fiberglass insulation in a Ford product, and helped develop Ford's air-conditioning system. Although the X-100 did not have air conditioning, Pacific remembers it as pioneering a lot of the advanced systems used on new Ford, Lincoln and Mercury automobiles over the next 15 years. The two chrome grilles used for the rear-seat heaters, which are located below the back-seat bottom on the risers, are the same design and are located in the same place as those used to let cold air into the passenger compartment on the first air-conditioned Lincolns introduced in May 1953.

The X-100 was built on a modified 1952 Lincoln frame and had a much-modified overhead-valve V-8 engine with the equivalent of "12 single venturi carburetors" and a new "Turbo-dyne" high-compression "Multi-plex throat" intake manifold. So the engine could be set lower, the oil sump was cut and welded back together, so that it was more

The X-100 also had a rear-window wiper.

Clay modelers Werner Framke (left) and Vic Clark (right) are completing last minute preparations for the introduction of the Continental 195X on January 18, 1952.

squat and widened-out on the sides as was later done on the Continental Mark II. In the middle of the special intake manifold, was a four-barrel carburetor and, at all four corners of the intake manifold there were two-barrel carburetors operated by progressive linkage. The X-100 had an electric gear selector operating an "entirely new torque converter" automatic transmission. The transmission was an experimental Fordomatic-type unit similar to those used on production Lincolns starting in 1955. The shifting mechanism was a lever on the dash with a standard PRNDL quadrant.

Henry Ford Museum and Greenfield Village

Turn signals were operated by built-in chromed buttons on the back side of the steering wheel. On the front side of the steering wheel were built-in black horn buttons. The seats in the X-100 were upholstered in a pattern similar to the production 1955 Thunderbird, and tinted glass was used on all windows. The flat glass is marked Solex, but the special curved windshield and backlight are unmarked. On the transmission hump, in front of the console, is a black leather pouch attached to the rug. It contains a brass pump-style fire extinguisher.

The X-100 has a wheelbase of 123 inches, an overall length of 220.9 inches, is 56.9 inches high, 81.25 inches wide and weighs in at a whopping 5900 pounds, even though it has over 300 parts made from light-weight aluminum for a 250 pound weight savings. There are 665 pounds of electrical equipment in the X-100 connected by eight miles of wiring. Although the press releases claimed eight miles of wiring, the engineers who built it maintain more like ten miles of wiring were used!

Paul Adams, of the Engineering and Research Department, was chief electrical engineer on the X-100 and shares the credit for all the gadgets on it. Paul Wagner, an electrical engineer who worked for Pacific in Experimental Engineering, was responsible for making much of the electrical system work and remembers it as a very difficult and involved project. Wagner still has the electrical manual that was made for use with the car. If the electrical system went down, everything on the car went dead, and the manual was needed for instructions on how to gain access to the engine compartment or trunk, so the car could be fixed.

Henry Ford Museum and Greenfield Village

A publicity picture taken before the Continental 195X was introduced to the press and public.

The X-100's frame was beefed-up with extra-thick plates on the bottom of the frame rails and X-member. Plates on the bottom of the frame rails and X-member were first seen on Lincoln convertibles in 1955 but were still not as thick as those used on the X-100. The frame and all suspension parts of the X-100 were painted 1952-54 Lincoln engine green as were the inner fender panels, the firewall and the radiator. The engine is painted slate blue, another standard color used on 1952-54 Lincoln engines. The underbody and gas tank were undercoated. The X-100 also had special body mounts and lots of gusseting on the frame. Radiator splash shields (skid plates) were used in strategic places to protect electric motors, wiring, the bottom of the radiator, the bottom of the generator, etc. Because the car weighed almost a ton more than the standard Lincoln convertible, the extra horsepower and the beefed-up frame were probably necessities.

John Najjar

Edelgard (Mrs. Gene) Bordinat worked as a model. She and the X-100 pose in the Design Department courtyard in 1954.

Ford also boasted of some 50 automotive innovations that were built into the X-100, including the following:

- *A sliding non-glare roof panel over the front seat.* A Plexiglas-roof panel can be retracted by a switch on the instrument panel or by pushing a little harder on the outside pushbuttons used to open the doors. A key inserted in the center of each outside pushbutton lock unlocks the doors. Pushbuttons are also used to open the doors from the inside. On the inside top rear of each door panel, there are sliding door-lock buttons. Sliding buttons were first used on production 1957 Lincolns. There is a rain cell on the roof that automatically closes the roof panel and rolls up the door windows at the first drop of rain. A rain cell was advertised as an option on 1955 Ford, Lincoln and Mercury convertibles. The Plexiglas-roof panel retracts into the stationary roof section over the back seat. The stationary part of the roof is covered on the outside with white leather. When the Plexiglas panel is retracted, spring-loaded covers close off the channels on either side at the front of the stationary part of the roof.
- *Independent front and rear suspension.* The X-100 has a deDion rear axle and inboard rear brakes as part of its independent rear suspension. The front suspension is standard 1952 Lincoln ball-joint suspension, except that the coil springs have been cut to make the car lower.
- *Power brakes, power steering and an electric power-assisted emergency brake.* The front brakes are cooled by thermostatically-controlled fans that turn on when the brake temperature exceeds 150 degrees. The fan motors are located behind the headlights and connected by rubber tubes that reduce from two inches down to one-inch in diameter, where they connect to metal elbows welded to the brake backing plates. The Eaton power steering unit has the large reservoir and was first used on 1953 production Ford products.
- *Seat belts on six-way, adjustable, memory power seats that have electric seat warmers.* Front seat belts are built-in and retract into the console area. The buckles fasten to two small U-shaped attachments on the outside of each front seat. The front seats are placed in recessed areas in the floor boards and seat the driver and front passenger close to the floor, like the 1955 Thunderbird.

The Continental 195X at the Design Department showroom in January 1952.

Henry Ford Museum and Greenfield Village

Ford Photomedia

The deck lid on the X-100 was opened electrically by a switch on the instrument panel. Much of the X-100's special electrical equipment was located in the trunk.

- *Power operated hood and trunk lid, that are opened, closed, locked and unlocked by switches on the instrument panel.* There is no mechanical release or lock to open either the hood or trunk if the car is without power.
- *Built-in jacks at all four wheels.* The jacks can be operated separately from the instrument panel, which also has a warning light to indicate when they are not fully retracted. When the X-100 was built, there were rubber boots on the base of each jack.
- *Four overlapping windshield wipers and hot and cold windshield washer sprays.* Because of the relatively short length of the wiper blades, the extra two wipers are used to clear the curved part of the windshield. There is also a single wiper used to clear the backlight. The wiper arms are painted car color (gloss black). There is a rear-window defroster that consists of small rectangular openings in the package tray area all along the bottom of the backlight.
- *A center console that functions as an armrest and, inside, contains a Dictaphone, telephone and an electric shaver with a pop-up mirror.* Neither the console nor the glovebox have locks. Controls on the driver's side of the console, on what was called the control heart of the car, are used to operate the signal-seeking ten-tube radio, electric shaver, phone, power seat and the heating system. There are also controls on the passenger side of the console that operate functions used by the front-seat passenger. At the back of the console, there were separate radio controls and a radio speaker.
- *The electrical system is 12 volt, with 24 electrical motors, 44 electronic tubes, 50 light bulbs, 92 control switches, 29 solenoids, 53 relays, 23 circuit breakers and 10 fuses.* There is a built-in battery charger that can be plugged into an outside electrical receptacle, so that the gadgets on the car can be operated even when the engine is not running. An extra-large generator, unlike any used on Ford's passenger cars, was installed on the X-100.
- *A soft horn for the city and a loud horn for country driving.*
- *A rearview mirror that was optically ground for a panoramic view of traffic to the rear.* The rearview mirror looks like it is attached to the windshield, but is actually on an arm attached to the upper-windshield trim.

The X-100 at the 1953 Paris Auto Show where more than 1 1/2 million people saw it. Ford engineers are shown activating the rain cell with an eye dropper of water.

Ford Industrial Archives

- *The radio is out of sight below the dash, but has a prismatic mirror that can be lowered by the press of a button to show the dial.* Like the Continental Mark II, the underside of the instrument panel is completely enclosed by upholstered metal panels.
- *A fuel gauge that has three built-in lights.* A green light shows a full tank (20 gallons), yellow means eight gallons are remaining, and red warns that there are only three gallons until empty. As on production 1952-54 Ford products, the gas filler is located behind a centered, spring-loaded back license-plate frame.

After the 50th Anniversary festivities, the X-100 was shipped to Paris, where it was displayed in a Ford showroom on the Champs Elysees, September 21-29, 1953. Then it was off to the Paris Auto Show, October 1-10. After that, it was shipped back across the Channel, by ferry, to the Earl's Court Auto Show (October 22-31) in London. Then it was taken to Germany, where the X-100 made appearances in Bonn (November 4) and Cologne (November 5). When it was shown in Cologne, the X-100 had to be driven through French doors with less than half-an-inch to spare. At each one of these showings, Ford representatives Paul Adams and Ralph Kimbrough, another one of the engineers who helped build the car, accompanied it and demonstrated all the gadgets. The demonstrations included the use of a drop of water from a eye dropper to show the audience how the automatic window and roof sensor worked to close the car up at the first sign of rain. On its final trip, the X-100 ran out of gas in the middle of the night, six miles from the port of LeHavre. After an hour's search, Adams and Kimbrough were able to scrounge up just enough gas to make it to the ship on which the X-100 was taken home.

The X-100 returned to North America for appearances in Canada, a national sports car show in Dearborn, January 9-11, 1954, and then to various shows and displays across the eastern part of the country. On May 14, 1954, the X-100 was shipped to Hollywood, California, where it spent over a month starring in a Twentieth Century-Fox movie entitled *Woman's World*.

Woman's World was a three-star movie starring Clifton Webb as the head of a large American automobile manufacturer (Gifford Motors), who couldn't make up his mind among three equally well-qualified executives, played by Van Heflin, Fred MacMurray and Cornel Wilde, to replace his recently-deceased general manager. In the movie, Webb decides to make his choice based on which candidate's wife is best suited to be a general manager's wife. The executives' wives, played by Arlene Dahl, Lauren Bacall and June Allyson, are a conniver, a woman who is faking a happy marriage and planning a divorce and a klutzy wife who doesn't want her husband to take the job. The Gifford summer home, where many of the scenes were filmed, is, in reality, the Edsel Ford Estate. The Ford family pictures were left on the walls during filming. The movie even produced a hit song, *It's A Woman's World*, which was sung by the Four Aces. Four other Ford Motor Co. concept cars, the XL-500, the XM-800, the Ventura and the Valiant also have bit parts in *Woman's World*. Movie people told Ford that over 80 million people saw the X-100 in the movie. While the X-100 was in Hollywood, newsreels were also made that were shown all over North America and Europe.

From late June 1954 until mid February 1955, the X-100 was displayed at various Ford dealers, auto shows, conferences, freeway openings, fairs, centennials, and just about everywhere else where it would be seen by lots of people. More people saw the X-100 than any Ford concept car built since.

Today, it's almost unheard of for a concept car to be driven from one show to the next, but that's exactly what happened with the X-100. When the X-100 was in Europe, it was driven more than 6000 miles. Surprisingly, it got 12 miles per gallon on high-octane gasoline. In the United States and Canada, it was driven over 12,000 miles between 1953 and early 1955. Today, the X-100 shows 33,182 miles on its odometer. In November 1954, while the X-100 was on its way back from the Texas State Fair in Dallas, headed for a dealer's showroom in Pittsburgh, Pennsylvania, it was forced to make an unexpected stop in Memphis, Tennessee because of a blown condenser. The driver of the car, Arthur Modder, simply pulled into the nearest Ford dealer, had the car fixed, displayed it to the public for a while and then drove off on his way to Pittsburgh. Modder claimed in an interview with a Memphis newspaper that, although the X-100 could go much faster, it had not yet been driven over 100 mph! Every time the X-100 was displayed, at least one Ford engineer was there to answer questions, hand out brochures about the car's special features and demonstrate those features to an eager public.

During its heyday, the X-100 was a titled and registered vehicle. It had Michigan plates and a Michigan title with Ford Motor Co. listed as the owner. The X-100 was last registered in 1956. The title, to this day, lists the X-100 as a Lincoln. The car's identifica-

John Najjar

Frank T. Francis

Frank Francis grew up in Detroit. Half way through high school, he enrolled at the Henry Ford Trade School and eventually graduated from both the Trade School and the Apprentice Schools with honors. In 1934, he was the Trade School representative at the Chicago World's Fair. In 1938, he was one of five Henry Ford Trade School students hired by Edsel Ford as Design Department apprentices. Francis became Gregorie's secretary. His job was to take care of the administrative details, so Gregorie and the other designers could spend their time designing. Francis, whose given name was Francis Zawacki, legally changed his name after he joined the Design Department.

As more designers were hired, Francis became a timekeeper. His job was to determine the time allocated to each design project and to keep projects moving towards completion. Most designers recognized that Francis was the one called on the carpet if things ran behind schedule, and that he was under the constant pressure of deadlines. Francis also worked as a clerk, a draftsman and a staff assistant. In World War II, he served in the Army and, in 1949, he became a scheduling supervisor. In 1966, he became supervisor of the Fabricating and Transmittal Section.

When the pressure was on, designers often worked a lot of overtime – sometimes all night. During times of the greatest pressure, the number of paper airplanes and other creative diversions increased at the Design Department.

At a particularly stressful time in the early 1950s, Francis became annoyed at the number of paper airplanes being produced. At the time, Gene Bordinat was assistant manager of the Design Department. At the end of one particularly frustrating day, Francis gathered up all the paper airplanes from around the studio and stormed into Bordinat's office. He unloaded them on Bordinat's desk and indignantly asked "What do you think of these?" Bordinat looked at the paper planes for a couple of minutes, picked up one and calmly told Francis, "Well, I think this one will go pretty good." Nobody remembers what happened next but, the next day, even Francis was laughing at himself.

Francis retired in 1974. ■

John Najjar

One of the drawbacks of driving the X-100 from show to show was the risk of damage. As shown in this photograph, the X-100 sustained damage to the door and rear quarter panel.

Werner K. Framke

Werner Framke was born in Berlin, Germany. He came to the United States with his family and started as an apprentice at Fisher Body in 1925 earning 50 cents an hour. After learning to assemble car bodies, he became a clay modeler almost by accident. When clay models were sent from GM's Art and Colour Section, Framke's boss, Joe Engelhart, assigned him to build similar clay models with modifications necessary to make the design feasible to build. Later, while at Fisher Body, Frank Hershey arranged to borrow Framke for several months as a clay modeler in the Art and Color Section. During the war, Framke built airplane nacelles at Fisher Body.

In 1945, Herman Sommers and Framke, both dissatisfied at GM, applied at Chrysler, Briggs and Ford and were hired by Ford as clay modelers. Framke's first project at Ford was clay modeling Ford's post-war panel delivery. Because he had worked with John Oswald during the time he was at GM, Framke was instrumental in George Snyder's termination at Ford, when Snyder took retribution on clay modelers Henry Ford II said he couldn't fire.

In 1950, Charlie Stobar assigned Framke to be the primary clay modeler on the X-100. Framke also helped clay model the FX-Atmos, the Ford Sports Car (Thunderbird) and the LaGalaxie. During his career, Framke was assigned exclusively to the Ford studio.

He retired in 1969. ■

John Najjar

tion plate is located on the left hinge-pillar face, and is the same style as was used on 1952 Fords. It lists the VIN number as "X-100," and the body data as "Ford Experimental Eng."

By early 1955, Ford thought that everybody who wanted to had already seen the X-100, and they were probably right. By 1955, Ford also had several newer concept cars to take its place. After February 1955, the X-100 was displayed only sporadically and, in late 1957, it was retired while Ford Motor Co. tried to figure out what to do with it. In August 1958, the X-100 was given to the Henry Ford Museum along with a lot of the car's extra parts and the engineering drawings. In December 1958, the title was changed to list the Museum as the X-100's new owner, and it was placed in storage.

In 1986, the X-100 was removed from its long storage, cleaned up and placed on a pedestal in the new *Car In American Life* display at the Henry Ford Museum, where it is today. Since 1986, some of the worn trim pieces have been replaced from the supply of spare parts that came with the car but, otherwise, it is kept in a state of arrested decay.

From time to time, the X-100 is loaned out for special events. In 1993 and, again in 1996, it was displayed at the Pebble Beach Concours in California. In 1995, it was displayed at the Meadow Brook Concours in Michigan.

No other car has been as influential on subsequent Ford Motor Co. products as the X-100. It heavily influenced the design of the 1955 and the 1961 Thunderbirds. Its influence can also be seen on the integrated front bumper and grille on the 1952 Mercury and in 1956 Lincoln front fender line, the power dome on the hood of 1955-57 Thunderbirds, the 1957 Thunderbird hooded instrument panel, the 1956 Ford hood scoop and in the concept and production glasstops of the 1950s. In retrospect, the gadgets on the X-100 were important, but not nearly as important as the beautiful flowing lines that became the inspiration for future Fords, Lincolns and Mercurys. Most importantly, the X-100 inspired a whole series of fantastic Ford Motor Co. concept cars that would not have been made without the genuine enthusiasm shown for the X-100 by the millions of people who saw it. ■

XL-500 (Phantom)

In 1953, for the first time since 1941, there was an auto show in Detroit. That year, Lincoln-Mercury Division decided to build an exhibit hall at the old Lincoln plant in Detroit and hold their own Detroit Auto Show. A total of 85,194 people attended the show, which was held April 26 through May 3. Lincoln-Mercury Division's new concept car, the XL-500, was displayed there along with many 3/8-sized concept cars from Gil Spear's Advanced Studio and several trim and color 1953 Lincoln and Mercury showcars. Afterwards, Detroit area Lincoln and Mercury dealers reported a dramatic increase in showroom traffic, and they credited it to the auto show. Over the next year, arrangements were made to continue the show as an annual event and, the other Detroit-area automobile manufacturers, encouraged by Ford's success, soon decided to revive and sponsor an industry-wide Detroit Auto Show.

The 1953 auto show was the idea of George O. Hackett, a Lincoln-Mercury marketing department public relations employee. He was assigned to manage the Lincoln and Mercury concept and showcars and did so until he retired in 1969.

Building the XL-500 for Ford's 50th Anniversary was the idea of Benson Ford who, at the time, was head of Lincoln-Mercury Division. He and Bill Schmidt, Lincoln's chief designer, had a good working relationship, and got to know each other well during the many times Benson Ford visited Schmidt's studio. The XL-500 got off to a late start, and there was insufficient time to make it a fully-operable car. In fact, there was barely enough time to finish it. Whether it would have been fully-operable if there had been more time is unknown.

In June 1952, when the Continental Nineteen Fifty X was changed to the Ford X-100, Benson Ford and Bill Schmidt decided that Lincoln needed its own 50th Anniversary showcar. After that decision was made, Bill Schmidt went to Gil Spear's Advanced Studio to look for ideas. He chose an unnamed 3/8-sized concept car, upon which to base his Lincoln showcar. The 3/8-sized model Schmidt chose was moved to his studio and, although substantially changed, it became the inspiration for the XL-500. Schmidt's renderings of the XL-500 are very close to the finished car, but they were much different than the 3/8-sized model borrowed from the Advanced Studio. The broken character line on the side of the car shown in the renderings was simplified during construction, but otherwise the XL-500 is almost identical to those renderings.

The XL-500 was the first postwar concept car to carry the Lincoln name. The exterior design of the car was done by Bill Schmidt, Don DeLaRossa, Jake Aldrich and Vic Frolich in the Lincoln studio. George Barbaz was temporarily assigned to the Lincoln studio to help with the design of the car. Barbaz reworked the front end of the XL-500 to its present configuration. The car's interior and instrument panel were designed by Chuck

The XL-500 on display at the Chicago Auto Show. Although the car looked fully operable, it was a pushmobile, and only the exterior and interior lights worked.

J. R. Samsen

William M. Schmidt

Bill Schmidt graduated from the Henry Ford Trade School in 1940 and took a job at Ford as a tool designer. When World War II started, Schmidt was transferred to the Air Frame Building and eventually to the Willow Run bomber plant. He began by designing tools for use in building the B-24 bomber. Because plans coming from Consolidated Aircraft in San Diego were difficult for people not trained to build aircraft to interpret, Schmidt drew exploded drawings, so he would understand what the plans meant. His exploded drawings speeded up production, and he was soon authorized to hire 35 artists to illustrate all B-24 plans that way.

Although he had never designed cars before, in 1944, Schmidt was transferred to Bob Gregorie's Design Department, where he helped complete the design of the Light Ford that later became the French Vedette. In 1947, John Oswald named Schmidt head of the Lincoln studio, where he redesigned the 1950 and '51 Lincolns. He also led the team that designed the 1952-56 Lincolns, the XL-500 (Phantom), the Futura and the 1953 Lincoln Golden Anniversary and the 1954 Lincoln trim and color show cars. From 1947-54, Schmidt was head of the Lincoln studio. During his last year at the Design Department, he was Lincoln's chief designer, but Gene Bordinat was in charge of the studio.

In early 1955, Schmidt was recruited by Jim Nance of Studebaker-Packard and, for a substantial increase in salary, left Ford to become vice president and head of Studebaker-Packard's design department. While at Studebaker-Packard, Schmidt, Dick Teague and Stan Thorwaldsen designed the Packard Predictor concept car, which shared many design features with the concept cars Schmidt designed at Ford.

In late 1955, when Chrysler's head of design, Virgil Exner, had a heart attack, Schmidt moved to Chrysler, where he stayed for two years. After Schmidt left Chrysler, he continued to do automobile design consulting work at his own firm, which he started as a side business in 1940. ■

Henry Ford Museum and Greenfield Village

(above) The cars shown with Bill Schmidt are one-eighth sized plaster models used to determine annual colors and proposed color schemes.

Henry Ford Museum and Greenfield Village

(above) An early rendering of the XL-500 then named the Phantom.

(below) The full-sized clay model of the XL-500, when it was still named the Phantom.

Henry Ford Museum and Greenfield Village

(below) A three-eighths sized model of the XL-500 (Phantom). The character lines on the side of the car were changed when the car was built.

Henry Ford Museum and Greenfield Village

Mitchell of the Lincoln interior studio. Don Beyreis was head of the Lincoln interior studio and, Mitchell, one of his designers, won a studio-wide competition to come up with a futuristic yet safety conscious design for the interior of an advanced car.

Until just before the time it was first shown, the XL-500 was named the Phantom. Phantom was a name that had been used on several prior concept cars produced by other companies, and on a one-off car built for Rust Heinz in 1938. It is unknown why the name, a favorite of Schmidt's, was changed, but Barbaz believes it may have been because the name Phantom had already been reserved by another automobile manufacturer.

The XL-500 was built at Creative Industries in Detroit. It was made entirely of fiberglass and did not have opening doors, hood or trunk, although creases in the fiberglass made the XL-500 look like it did. There were no door or rear-quarter windows on the XL-500, and access to the interior of the car was accomplished by crawling over the doors and through the "open" windows. Even the seats in the XL-500 were made by upholstering over fiberglass forms.

The XL-500 was built on a stock 123-inch Lincoln frame, was 216 inches long and 57 inches high. It supposedly had power brakes, power steering, and was the first Ford concept car to have air conditioning. It predated air conditioning on Ford production cars by less than a year; factory air conditioning was first offered on late 1953 Lincolns. There were no clear-plastic tubes protruding from the package tray or going up into the headliner as on production models. If the air conditioning on the XL-500 had worked, cooled air was designed to come from louvered vents on the package tray.

The horn on the XL-500 was operated by a foot pedal on the floor. Predicting something that was to come on the Thunderbirds and Mark III's of 10-15 years later, safety and engine warning lights were located on the header bar, above the windshield and in a more-direct line with the driver's vision. Unfortunately, none of the special features on the XL-500 were functional, except for the lights, which were battery-operated.

Except for the 1949 Continental proposal designed by George Snyder, the XL-500 was the first Ford Motor Co. car with an automatic transmission controlled by pushbuttons located in the middle of the steering wheel, like later Edsels. The speedometer went to 150 mph and was located above the instrument panel and centered in a hooded cluster that also housed a tachometer, compass and an electric calendar. The ignition key was located on the side of the steering column, a feature re-adopted by cars 25 years later.

All other controls were located in a center console, called the control pedestal, which started at the base of the instrument panel and continued through to the back seat. Towards the front of the control pedestal were throttle-type controls for air conditioning, lights, wipers and the radio. Farther back on the control pedestal, there were toggle switches for the telephone, Dictaphone, automatic jacks, air conditioning and automatic hood and deck releases. The telephone was located on a cradle at the base of the control pedestal. When it was activated, an aerial at the front of the windshield swung upright. At the back of the control pedestal were rear-seat heater vents.

Ford Design Center

Don R. DeLaRossa

Don DeLaRossa grew up in Indianapolis, Indiana. He attended the John Herrin Art Institute as a night student and sold insurance during the day. During World War II, DeLaRossa was a pilot in the Army Air Corps. In 1946, he moved to Detroit, where his uncle was in the insurance business. Although DeLaRossa's plans were to get a job in GM's insurance department, a friend got him a job interview at Ford's Design Department with Tom Hibbard, but he was not hired. The same friend introduced DeLaRossa to GM's Jules Andrade, who hired him as an apprentice designer in GM's Buick and Cadillac studios. When George Snyder went to Ford, he asked DeLaRossa to join him. DeLaRossa started at Ford, in September 1947, working in the Lincoln studio as executive stylist under Bill Schmidt.

In 1952, DeLaRossa was offered a job as chief designer on the Continental Mark II project. At Gene Bordinat's request, DeLaRossa turned down the offer. In 1952, DeLaRossa became head of the Mercury studio. In 1955, DeLaRossa was transferred to the Ford pre-production studio. When the Futura and the XM-Turnpike Cruiser were being constructed at Ghia, DeLaRossa, who spoke fluent Italian, acted as design liaison. In late 1957, DeLaRossa replaced John Najjar as head of the Lincoln studio.

In addition to many production cars, DeLaRossa worked on the XL-500, Mrs. Ford's X-400 Lincoln towncar, the 1949 Lincoln Presidential parade car and the 3/8-sized Dual Cowl Phaeton.

DeLaRossa was later head of the team that designed the Lincoln Mark V. Before he retired from Ford in 1978, he was president of Ghia, which was purchased by Ford in 1971.

He later came out of retirement to accept an offer from Lee Iacocca to become the head of Chrysler's design department. ■

Ford Photomedia

The Plexiglas roof on the XL-500 inspired the 1954-56 Ford Skyliner, the 1954-55 Mercury Sun Valley, and the roofline of the 1955-56 Ford Crown Victoria. The taillights from the XL-500 were similar to those used on the 1955-56 production Mercurys.

Charles R. Mitchell

After a year studying fine arts at the University of Akron, Chuck Mitchell spent the next four years flying B-17s in the Army Air Corps during World War II. On his 13th mission, he was shot down and spent the next eight months avoiding capture and working for the Belgian underground. With the underground, Mitchell participated in the liberation of Paris. After the war, Mitchell returned to the University of Akron, until he was able to enter the Cleveland Institute of Arts. Upon graduation in 1949, Mitchell and several other beginning designers, including Alex Musichuk, Al Mueller and Starkey Hoot, were hired by Ford. During the next three years, Mitchell worked as a designer in the Lincoln and Mercury exterior and interior studios. In 1951, Mitchell designed the soft trim and instrument panel used in the XL-500 showcar.

In 1952, Mitchell took a job as a staff designer at Packard but soon left there to work for the W. B. Ford industrial design firm. While there, Mitchell and Harold Fetty designed the W. B. Ford entry in the Continental Mark II design contest. In 1953, Mitchell took a job at Briggs Manufacturing as a designer. When Briggs was purchased by Chrysler later that year, Mitchell remained as a designer at Chrysler, until he retired in 1980. While at Chrysler, he worked in all interior and exterior studios. Mitchell was a manager in the studio that designed the second-generation Dodge Charger. He was named chief designer in the Dodge exterior studio, the compact and intermediate car studio and the standard car and truck studio.

After retirement, Mitchell returned to school, earned a Bachelor of Arts degree with distinction from Detroit's Center for Creative Studies and pursued graduate studies in sculpture in Italy. Since 1984, Mitchell has worked as a figurative sculptor. ■

Charles Mitchell

The body was painted bright scarlet, and the car was upholstered in red and white leather with matching door panels. So the body could be positioned lower on the chassis, the rear fenders had raised arches over the wheels. The rear bumper was integrated into the body and was in three pieces, with the taillights integrated into the exterior sections.

The XL-500 influenced as many Ford Motor Co. production cars over the next ten years as the X-100 did. Bill Schmidt transferred the entire front end of the car, with minimal changes, to the 1956 Lincoln. The basket handle separating the tinted Plexiglas-roof panels was copied on the 1955-56 Ford Crown Victorias, and the Plexiglas roof showed up on the 1954-56 Ford Skyliners and the 1954-55 Mercury Sun Valleys. The transmission pushbuttons in the center of the steering wheel were copied on some models of the 1958 Edsel, and taillights similar to the ones on the XL-500 appeared on the 1955-56 production Mercurys.

George Hackett and his assistant, Jim Fulton, recall that in about 1954, one of the Lincoln-Mercury showcars broke loose when the rope broke, while it was being ramped-up into New York City's Madison Square Garden for the New York Auto Show. The car careened back down the ramp and through the window of a restaurant across the street, landing inside. The car and the restaurant were badly wrecked. Unfortunately, neither Hackett nor Fulton remember what car was involved in that mishap. Bill Schmidt's son, John, and several former Ford designers recall that the XL-500 sustained substantial body damage, while it was on the show circuit. They believe the damage was repaired but are not sure. The car that was wrecked while being ramped into the New York Auto Show may have been the XL-500.

Both Hackett and Fulton also recall that one other Ford Motor Co. showcar was lost. Although they do not know the name of the car, it was supposedly in route from Italy with the Chrysler Norseman concept car on the Andrea Dorea, when it sank in 1956. No records can be found at Ford confirming Hackett and Fulton's recollection, and no designers remember such a car.

The XL-500 was not on the show circuit very long. No photographs can be found showing it with the engine compartment open, and it probably had no engine in it. There are unconfirmed reports that, sometime after it was first introduced, a 1955 Mercury engine was installed for show purposes only. There are also unconfirmed reports that the XL-500 did survive, and that it's now in Minnesota, where it is being restored by someone who wishes to remain anonymous. ■

The XL-500 at the Chicago Auto Show. Crowds gathered every place the car was shown.

J.R. Samsen

Ford Photomedia

Jack Davis (left), Lewis Crusoe (4th from left), Ernie Breech (8th from left), Henry Ford II (10th from left), Benson Ford (4th from right) with Ford retirees and the XL-500 on the night before the 50th Anniversary celebration at the Ford Rotunda.

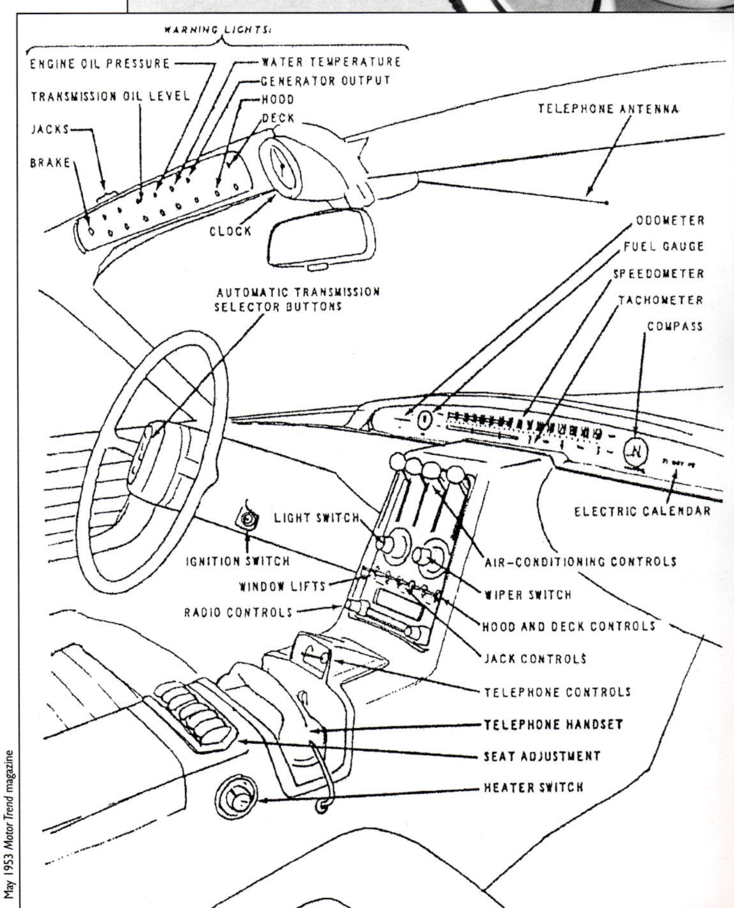

May 1953 *Motor Trend* magazine

(above and right) The instrument panel and center console of the XL-500 contain controls for many accessories, including air conditioning, which was first offered by Ford in May 1953 on the production Lincoln.

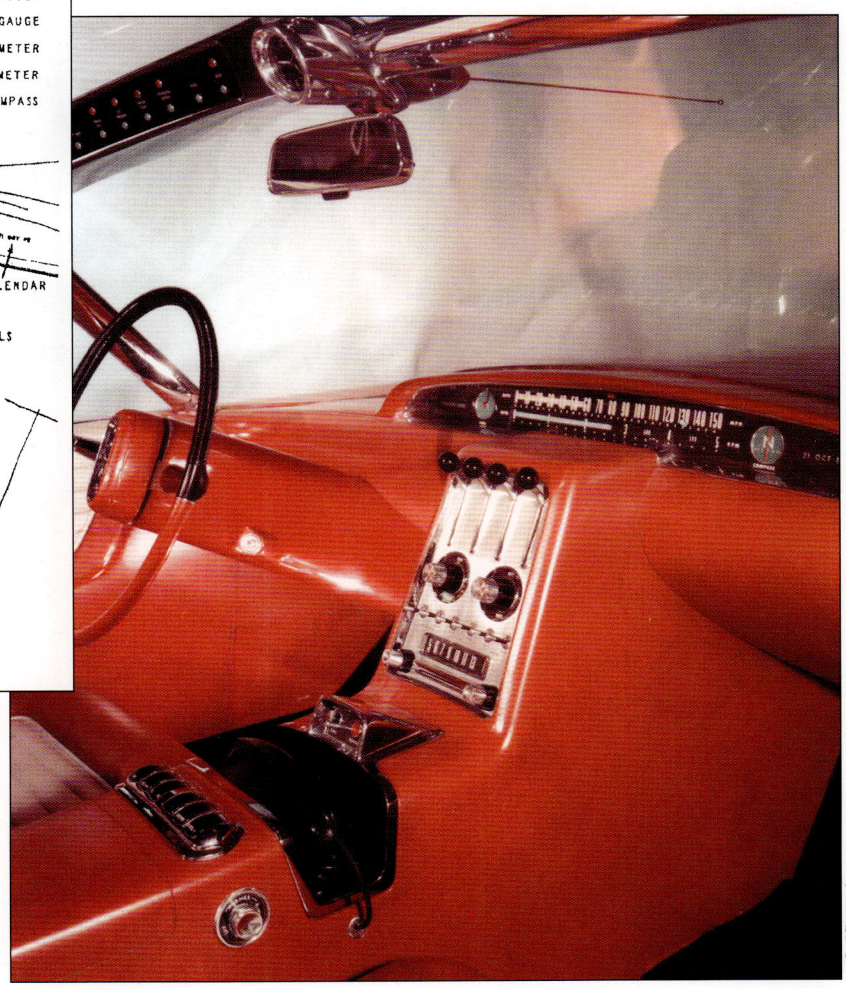

Ford Photomedia

Donald H. Beyreis

Don Beyreis' father was a commercial artist in Detroit. At a young age, Beyreis was hand-painting and lettering billboards for his father. In the 1930s, he worked in the display department at the J.L. Hudson department stores, while he attended night school at Wayne State University studying art and lettering.

In 1940, Beyreis was accepted as a student at GM's design school. When World War II started, the school was closed, and Beyreis was employed at Fisher Body as a technical and product illustrator in their aircraft production facilities.

After the war, Beyreis returned to GM as a designer. In 1949, he left to take a job at Ford. Beyreis was placed in charge of the Lincoln-Mercury interior studio. While in the Lincoln-Mercury studio, Beyreis supervised the design of the interior of the XL-500 including the instrument panel. He also was in charge of the design of all the 1953 Golden Anniversary Lincolns and Mercurys and the 1954 trim and color Lincoln and Mercury showcars.

In 1955, Beyreis left with Bill Schmidt, head of the Lincoln studio, to become a designer at Studebaker-Packard. In late 1956, he took a job at Chrysler, when Studebaker-Packard consolidated its design facilities in South Bend, Indiana. Beyreis had just built a new house and did not want to move his family.

From 1962-69, Beyreis was assistant manager of all interior design at Chrysler under Colin Neale. From 1969, until he retired in 1974, Beyreis was in charge of Chrysler's trim and ornamentation studio. ■

Stan Thorwaldsen

Ford Photomedia

(above) Except for the basket handle, the top of the XL-500 was all Plexiglas, giving the car the appearance of an open convertible.

The XL-500 at the Chicago Auto Show. Because the doors of the XL-500 did not open, access to the interior was gained by crawling through the open windows.

J.R. Samsen

Ford Photomedia

The Ford Rotunda was decorated with 50 electric candles as part of Ford's Golden Anniversary celebration.

GOLDEN JUBILEE (Pace Car)

Ford Motor Co.'s 50th Anniversary was June 16, 1953. The celebration, however, was two years in the making and held during all of 1953. In April, a book celebrating the diversity of Ford and the people who worked there was published. The book, called *Ford at Fifty*, included a section on the Design Department, including pictures of the Advanced Studio filled with concept cars. A half-million copies of the book were printed. Artist Norman Rockwell was commissioned to paint six pictures for a commemorative calendar, and 2 1/2 million copies were distributed. Historian Allan Nevins was asked to write a three-volume history of Ford Motor Co., the Ford Archives were dedicated, and a special movie, called the *American Road*, was released. It was narrated by film star Raymond Massey. Commemorative coins were also struck featuring Henry Ford, Edsel Ford and Henry Ford II.

And if that wasn't enough, as the anniversary date approached, President Eisenhower traveled to Dearborn to dedicate Ford's new Research and Engineering Center, a $50 million complex that included a new $11.5 million design building. The Ford Rotunda was also reopened to the public for the first time since World War II and, for the anniversary celebration, it was decorated like a big birthday cake with 50 lights around the top made to look like candles. On June 14th, Ed Sullivan dedicated his Sunday TV variety program to Ford's anniversary and, the next night, CBS and NBC broadcast a $500,000 prime-time special honoring Ford and featuring Edward R. Morrow, Oscar Hammerstein II, Judith Anderson, Ethel Merman, Rudy Vallee and Mary Martin. For all the entertainment that was offered, Ford didn't forget that cars were the reason they were celebrating, and many special cars were created to help celebrate their 50th Anniversary.

On January 26, 1953, it was announced that for the fourth time in its history, a Ford Motor Co. product would pace the forty-second running of the Indianapolis 500 race, and William Clay Ford would drive it. (In 1932, Edsel Ford paced the Indianapolis 500 in a Murphy-bodied Lincoln roadster. In 1946, Henry Ford II drove a 1946 Lincoln Continental Cabriolet and, in 1949, Benson Ford did the honors in a 1949 Mercury convertible.) The Pace Car was built February 2, 1953 at Ford's Rouge plant and delivered to the EEE Building, the location of the Design Department, where it was turned into the Golden Jubilee Pace Car. The car was painted Pace Car White, and had a white orlon top with a matching boot. The upholstery was done in Anniversary Gold, white pearlescent and black. The instrument panel, garnish moldings, cluster inserts and top bows were also painted Anniversary Gold.

Henry Ford Museum and Greenfield Village

Bill Ford behind the wheel of the Golden Jubilee Pace Car on the Ford test track before the car and Bill Ford went to Indianapolis for the 1953 500 mile race.

The Pace Car was designed by Ford studio interior designer Alex Musichuk, who worked closely with trim and color specialists Ted Hobbs and Andy Olinik to pick a new pearlescent-white paint and a gold vinyl to go with the contrasting white and black patent-leather-look vinyl used for the interior upholstery. According to Musichuk, the gold paint used had a tendency to tarnish rapidly and, to prevent it from doing so, it was necessary to do dry runs of the interior assembly of the car, so the interior could be finished just before it was to appear at the Indianapolis 500 race.

The Golden Jubilee Pace Car was accessorized with a Fordomatic transmission, fender skirts, rocker moldings, a Coronado deck, trim below the tail lights, dual back-up lights, dual spotlights, radio, heater, clock, special crossed flags and Kelsey-Hayes wire wheels. On the sides of the Pace Car, gold and black hand-painted lettering told everyone it was a Ford and a Pace Car -- and the date of the race.

For the first time ever, 2000 Pace Car replicas were made, so each Ford dealer could have one in his showroom for the open house all Ford dealers hosted during the entire month of June. Except for the lack of wire wheels, the replicas looked identical to the Golden Jubilee Pace Car and, if the dealers wanted it, they were lettered like the Pace Car. The replicas, which were built at all Ford plants, were identified on their data plates as "Color SS," or "Color RR," and "Trim SS," or "Trim S."

As the race date approached, the Pace Car was photographed at the Ford test track, and then it was trucked to Indianapolis, where Bill Ford, as had his father and two older brothers, helped start that year's Indianapolis 500 Race. During the pace lap, Bill Ford drove the Golden Jubilee Pace Car around the 2 1/2-mile oval at about 80 mph. After the Memorial Day race, Bill Vukovich, the race winner, was given one of the replicas, but the original pace car was delivered to the Henry Ford Museum, where it can be seen on display today.

On September 7, 1953, a Pace Car replica was used as the official car at the Pike's Peak Hill Climb in Colorado.

Also included in the 50th Anniversary celebration was the X-100, which was re-introduced as a "laboratory on wheels." The XL-500 and seven Lincoln and Mercury showcars, together with the X-100, took part in the reopening of the Ford Rotunda on June 16th. By the time plans were completed for the Pace Car and the 1953 Lincoln and Mercury showcars, the budget was pretty well depleted, and funds were very limited. ■

Theodore C. Hobbs

John Najjar

In the 1920s, Ted Hobbs worked for coachbuilders Holbrook and Brewster. In 1927, he was one of the first designers hired by Harley Earl at GM's newly-formed Art and Colour Section. While at GM, Hobbs was in charge of interior trim and exterior colors.

During World War II, GM's design studios were largely shut down, and Hobbs took a job at Fisher Body as a steel expediter on war contracts. In 1944, Hobbs left GM to work at Ford, because he was offered a job designing cars.

Hobbs became head of the Ford studio's trim and color department. His job was to pick the colors and fabrics used on all of their production, concept and showcars and to forecast industry trends.

While at Ford, Hobbs worked on the X-100 and the 1953 Golden Jubilee Pace Car. He also selected colors and fabrics used on many other concept cars.

Hobbs retired in November 1966 for medical reasons and died two weeks later. ■

Henry Ford Museum and Greenfield Village

The Golden Jubilee Pace Car in front of the EEE Building before wire wheels were added.

Andrew F. Olinik

Andy Olinik grew up in Hamtramck, Michigan and attended the Detroit College of Applied Science and the Detroit Business University before enlisting in the Navy during World War II, where he served as a Petty Officer First Class. After he was discharged in 1946, he applied for a job at Ford's Highland Park plant. He was told there was a job opening at the Design Department for someone with shorthand, typing and business skills. He interviewed and started there in September 1946. His first assignment was preparing informational drawings for the interior of the new Fords, Mercurys and Lincolns planned for 1948.

Olinik was soon made an apprentice to Ted Hobbs and Hermann Brunn in the trim and color department and spent the next 34 years there. He also took design and clay modeling classes for advancement.

Designers in the trim and color studio select materials and paints to be used on production, show and concept cars including the latest products for use on the concept and showcars. By working with paint and materials manufacturers to develop new pigments and fabrics and doing market research, trim and color specialists are able to forecast trends and tastes.

Olinik, together with the other designers in his department, chose the fabrics and colors used on the 1953 Ford Golden Jubilee Pace Car, the 1953-54 Lincoln and Mercury trim and color showcars and most subsequent Ford Motor Co. concept and showcars. Olinik retired in 1980. ■

Ford Design Center

L:ike his father and his older brothers before him, Bill Ford drove the Pace Car at the 1953 Indianapolis 500 Race. He is shown here on the Ford test track in the Golden Jubilee Pace Car next to his grandfather's 999 race car.

Henry Ford Museum and Greenfield Village

Ford Photomedia

The Lincoln Anniversary featured pearlescent-white paint and $4000 in gold plating on the trim and bumpers.

Anniversary

In addition to the X-100, the XL-500 and the Golden Jubilee Pace Car, there were also seven trim and color 1953 Lincoln and Mercury showcars created to help celebrate Ford's 50th Anniversary. The cars were given their own names, but they differed from their production counterparts only in the exterior colors used and in their upholstery and interior trim. Each showcar proposal was approved by Benson Ford, head of Lincoln-Mercury Division. After the colors, upholstery and interior trim were settled on, designer-illustrator Vern Breitmeyer prepared renderings of the interior of each car, some of which were displayed when the cars were shown.

The Anniversary was a 1953 Lincoln Capri convertible, that was painted a pearlescent white. It also had $4000 worth of gold plating instead of chrome trim. The Anniversary was upholstered in pearl-white leather and even had gold trim on the inside. ■

The Lincoln Anniversary at the 1953 Detroit Auto Show held at the Lincoln plant.

Detroit Public Library National Automotive History Collection

J.R. Samsen

Even the inside of the Lincoln Anniversary had gold-plated trim.

Robert H. Weiland

Bob Weiland was born near Pittsburgh, Pennsylvana, and attended Penn State University, where he studied business. From early childhood, he loved to draw cars. During college, he decided to become an automobile designer rather than go into business for himself. In 1939, he accepted an offer to work as a designer at GM, where he was assigned to the Buick and Cadillac studios. After World War II, Weiland was passed over for a promotion. As a result, in 1947, he left to accept a job at he Ford Design Department.

Weiland was executive designer in the Mercury exterior studio from 1947-56 and was later transferred to the corporate advanced studio. He was primarily responsible for the exterior design of the 1952 Mercury, including its integrated grille-bumper. During the time he was in charge of the Mercury studio, Weiland supervised the design of the 1953-54 trim and color Mercury showcars. When the Mustang was being developed, Weiland was one of the principal designers of its interior.

In 1967, Duncan McRae replaced Wes Dahlberg as head of design at Ford of Germany, and Weiland became McRae's executive designer. Weiland remained in Germany for 2 1/2 years. Upon his return to Dearborn in 1972, Weiland took early retirement. After he retired, he moved to Florida, where he operated an art supply business for many years. ■

Ford Design Center

The Anniversary was displayed on the show circuit in front of a freeway collage.

Ford Photomedia

Ford Photomedia

Ford Photomedia

(above and right) Except for the color, from the outside, the Maharajah looked like any other Lincoln four-door Capri, until the doors were opened to show the red and white brocade seats and the plush Bengal Red carpets.

Maharajah

The Maharajah was a 1953 Lincoln four-door Capri that was painted a gold essence and had a pearl-colored roof. It was designed by Vern Breitmeyer. The headliner was upholstered with white silk, and the rugs were done in Bengal Red plush. The seats and door panels were upholstered in red and gold brocade cloth with white leather covering the center pillars, the rear-seat heel boards and the package tray.

After the 50th Anniversary celebrations were over, the Golden Anniversary show-cars, with the exception of the Pace Car, were shown at various auto shows around the country. When the show season was over, all but the Pace Car and the Maharajah were sold to VIP's at the employees' sales lot, which was located near the Ford Rotunda. The Maharajah was actually sold before the show season was over, although the new owner wasn't able to take possession until it was off the show circuit.

The Maharajah was advertised in Lincoln-Mercury publications as "a vehicle that could more than hold its own in the garage of any Indian prince." According to Jim Fulton, then of Lincoln-Mercury public affairs, and one of the those in charge of the cars when they were on the show circuit, the Maharajah's purchaser was a real maharajah or far-eastern potentate, and the car eventually returned home with him. Where it went and what happened to it after that are unknown. ■

Henry Ford Museum and Greenfield Village

After the auto show circuit, the Maharajah was sold to a real maharajah who shipped the car overseas.

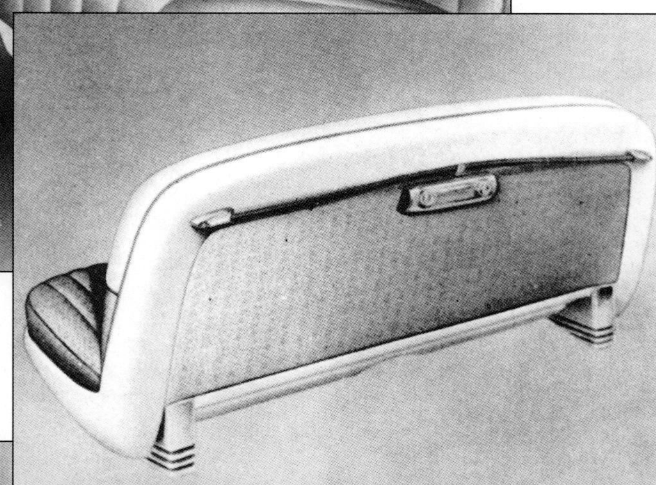

(below) The Maharajah on display at the Lincoln plant before the start of the 1953 Detroit Auto Show.

Detroit Public Library National Automotive History Collection

Midshipman

Trim and color cars were created by the Design Department on a regular basis but, except for those produced for the show circuit in 1953-54, they were usually prepared for members of the Ford family and Company executives. Although sheetmetal and running gear changes were unusual, special colors and interior trims were common. Bill Ford often ordered cars painted Honolulu blue, his favorite color, and upholstered in blue and silver, which just happen to be the team colors for his Detroit Lions. For many years, there was a special studio in the Design Department which was assigned responsibility for coordinating the colors and upholstery used on these special cars.

The Midshipman was a 1953 Lincoln Capri two-door hardtop, with a two-tone body painted a dark turquoise with an off-white top. It was designed by Stan Thorwaldsen. The interior was upholstered in matching colors and featured anchors in contrasting colors on the seats and door panels and stars on the headliner. At the 1953 Chicago and Detroit Auto Shows, and possibly others, the Midshipman was displayed in front of a large photograph of midshipmen from the United States Naval Academy on their parade ground.

All Lincoln and Mercury showcars built between 1948 and 1955, including the 1953-54 trim and color showcars, were built at Benson Ford's request. During the years he headed Lincoln-Mercury Division, he listened carefully to his dealer network and, surprisingly, the dealers told him they didn't really like true concept cars at their dealerships, because of the adverse comparison customers made to that year's production models -- but dealers loved the trim and color cars! That type of encouragement produced a total of 14 of them in 1953 and 1954. ■

Ford Photomedia

The Lincoln Midshipman was designed to honor the United States Naval Academy.

Ford Design Center

Vernon Breitmeyer

Vern Breitmeyer came to Ford in 1945. He was hired as an illustrator. Throughout his career at the Ford Design Department, Breitmeyer worked mostly in the Lincoln-Mercury studio. He generally illustrated Lincoln and Mercury interiors and occasionally designed car interiors.

Breitmeyer prepared many of the interior illustrations shown with the 1953-54 Lincoln and Mercury trim and color cars and designed the interior of the 1953 Lincoln Maharajah.

Breitmeyer reached retirement age in 1956. John Najjar, head of the Lincoln studio at the time, was able to secure a three-month extension, so that Breitmeyer could qualify for retirement benefits under Ford's new retirement plan. After he retired, Breitmeyer moved to Mexico City. ■

Detroit Public Library National Automotive History Collection

(above) The Lincoln Midshipman on display at the 1953 Detroit Auto Show. Behind it are the Mercury Contemporary and Kentucky Colonel.

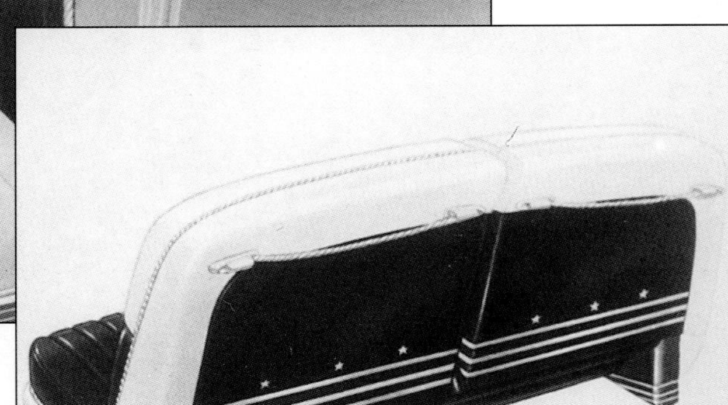

Henry Ford Museum and Greenfield Village

The Midshipman's upholstery was described as having anchors in contrasting colors on the seats and door panels.

Cadet

Although Ford's concept cars were usually the sportier convertibles and two-door hardtops, production four-door models were traditionally the biggest sellers. Three of the 1953 and at least one of the 1954 trim and color showcars were four-door sedans.

The Cadet was a 1953 Lincoln Capri four-door sedan painted a two-tone medium gray with a dark-blue top. The upholstery and color combination were selected by Vern Breitmeyer. The interior of the Cadet was gray and white, and it had a chevron pattern on the seats and door panels. The blue and gray colors on the Cadet matched the colors used on the uniforms worn by West Point cadets.

Both the Midshipman and the Cadet were designed during the Korean Conflict, when patriotism was particularly high. Designers have a vague recollection that the son of a Design Department employee was attending one of the military academies at the time.

The Cadet was displayed at the 1953 Chicago and Detroit Auto Shows and possibly others. It was usually displayed in front of a mural showing the U. S. Military Academy at West Point. ■

Henry Ford Museum and Greenfield Village

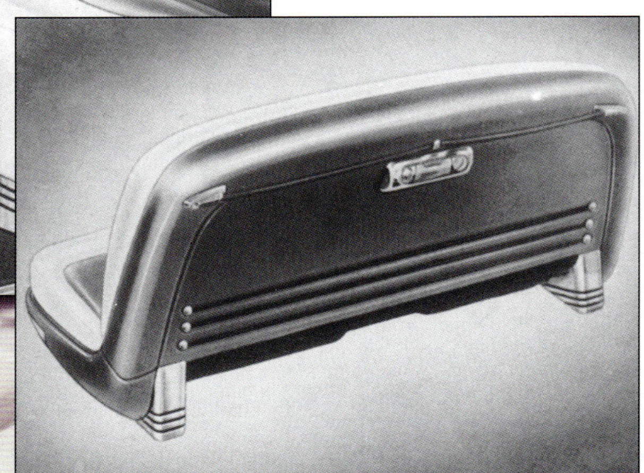

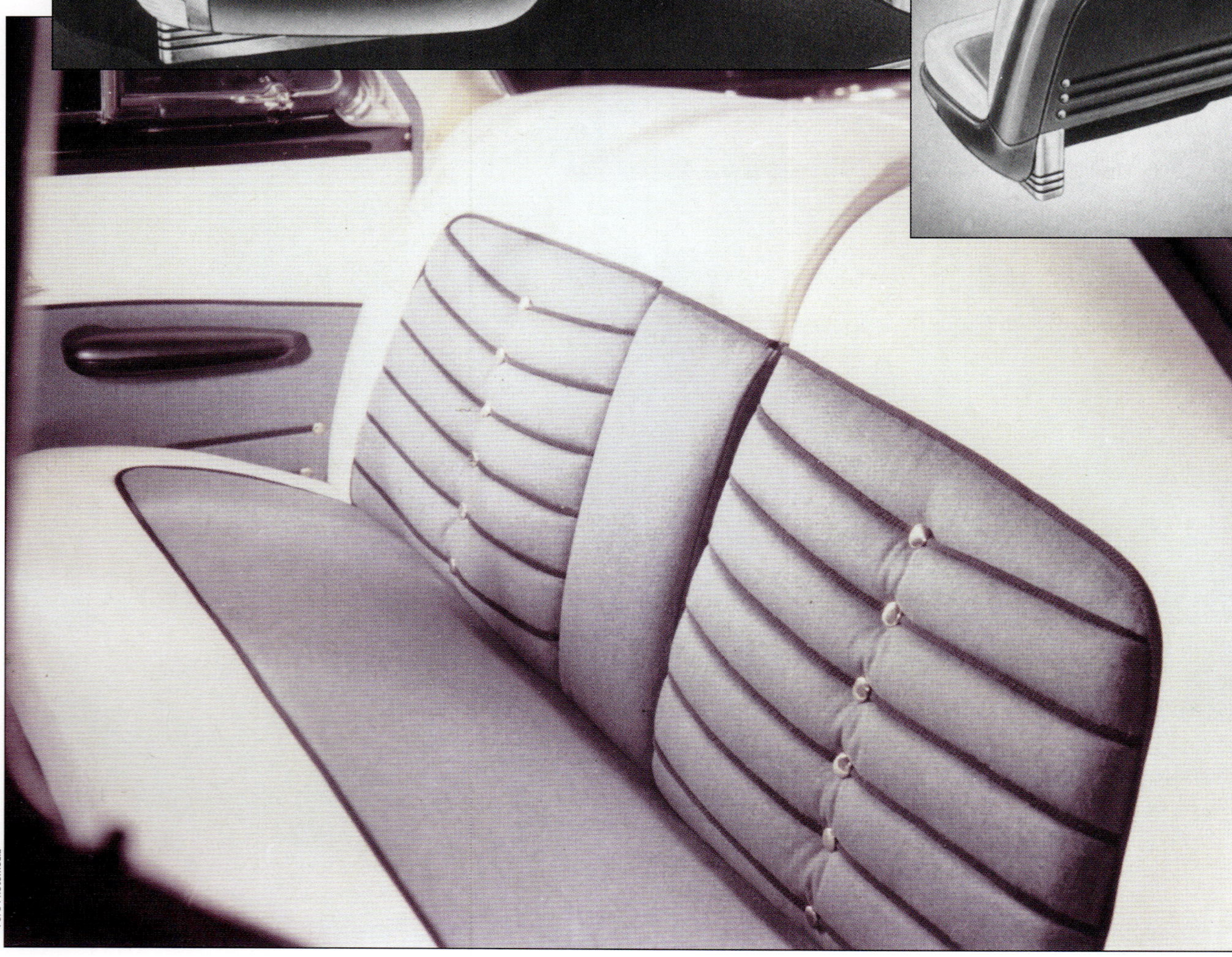

Ford Photomedia

(above and left) The Lincoln Cadet was upholstered in gray, dark blue and white in honor of the U.S. Military Academy.

Ford Photomedia

The Kentucky Colonel was first shown at the 1953 Detroit Auto Show held at the old Lincoln plant.

Kentucky Colonel

The Kentucky Colonel was a 1953 Mercury Monterey four-door sedan that was painted a bright kelly green and had green and white upholstery. Although 1953 Mercurys could be ordered in four different shades of green and many two-tone combinations, the kelly green color used on the Kentucky Colonel was much brighter than the other shades of green offered on production Mercurys. ■

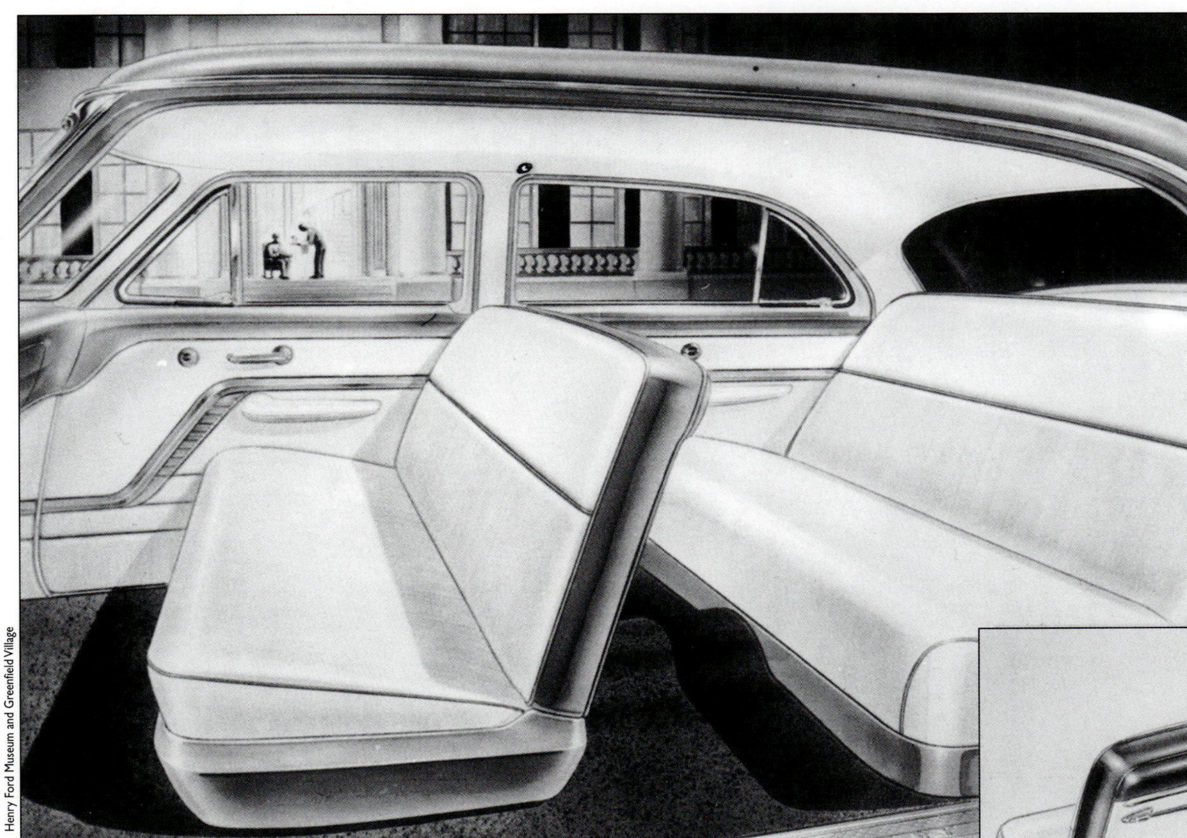

Henry Ford Museum and Greenfield Village

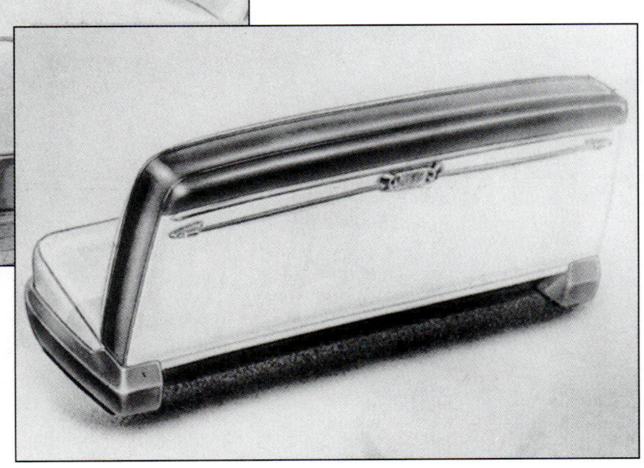

(top and above) The kelly green color on the exterior of the Kentucky Colonel was carried over to the inside of the car and was complimented by white seats and door panel inserts.

Contemporaire

The Contemporaire was originally called the Mercury Contemporary. After this particular 1953 Mercury Monterey convertible was shown at the auto show held at the Lincoln plant and at the Chicago Auto Show, the name was changed to the Mercury Contemporaire. Designers in the Lincoln-Mercury interior studio at the time were partial to French-sounding names and were apparently successful in persuading their bosses that the name Contemporary was too ordinary sounding. The Contemporaire was painted jet black and had black and white upholstery. In addition, the white areas of the upholstery had a pattern of interlocking rectangles.

The Public Relations Departments took many photographs of showcars for their own purposes. For internal use, and to make a permanent record, the Design Department took their own eight-point photographs (from eight different positions around the car) of the exterior of each design. Because photographs did not always show sufficient interior detail, renderings like the one at left were often made and then photographed. ■

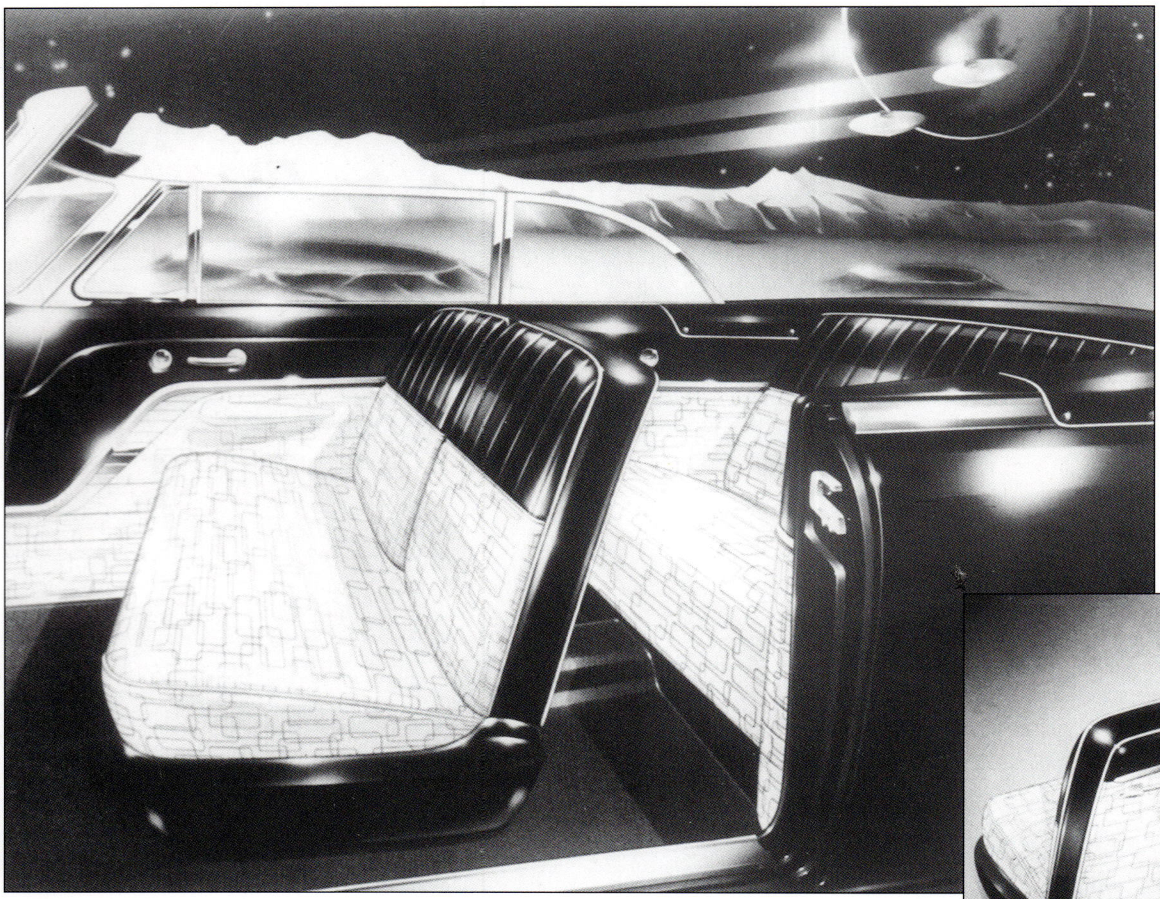

Henry Ford Museum and Greenfield Village

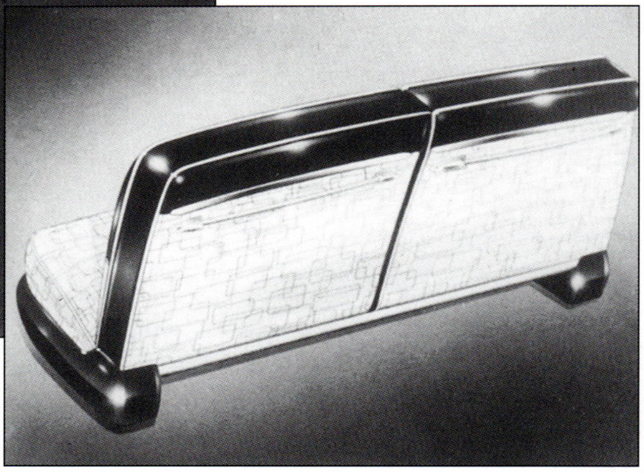

(above) Designer Vern Breitmeyer prepared the renderings of the1953 Lincoln and Mercury color and trim showcars including this one of the Mercury Contemporaire.

(below) The Mercury Contemporary at the 1953 Detroit Auto Show.

Ford Photomedia

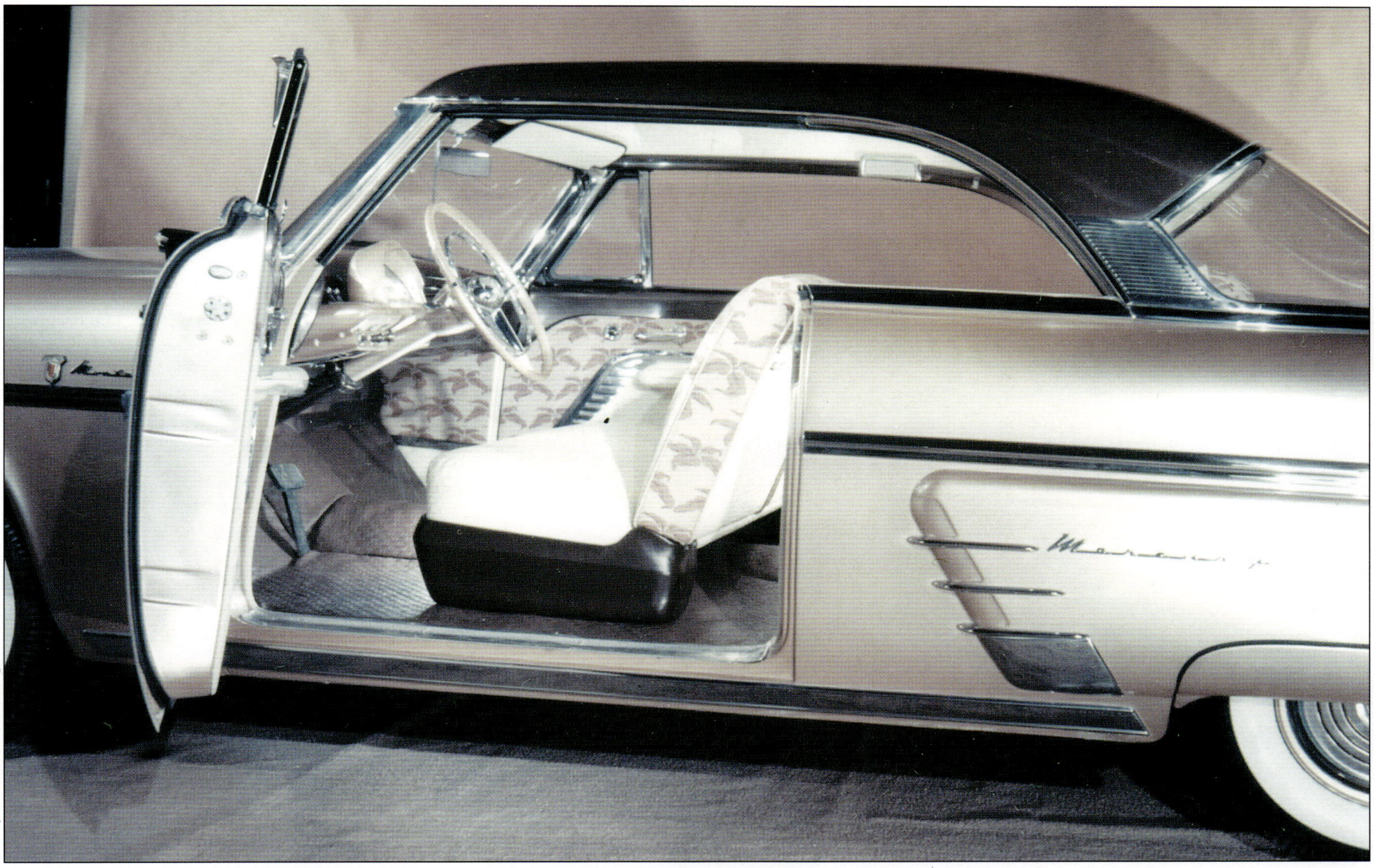

Henry Ford Museum and Greenfield Village

The Bahamian was a 1953 Mercury Monterey two-door hardtop with a suntan theme. It was painted tan and had a dark Bahamian Brown roof. The Bahamian's exterior color scheme was only slightly different than what was offered on production 1953 Mercurys, but the interior was dramatically different. The interior of the Bahamian was upholstered in a brown, tan and cream tropical motif. The door panels and portions of the seats were upholstered in cloth with large palm fronds on them. ■

Bahamian

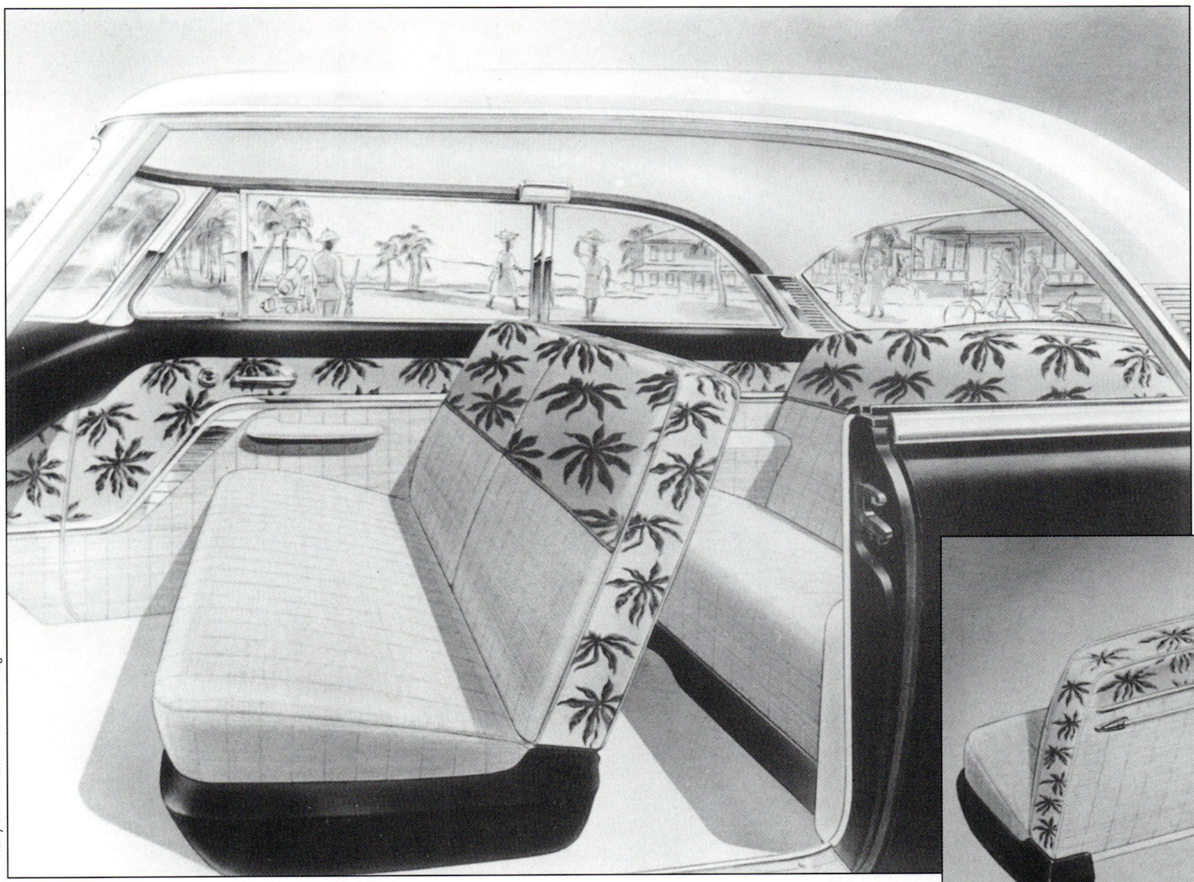

Henry Ford Museum and Greenfield Village

(above and left) The Mercury Bahamian featured a tropical motif with large palm fronds on the seats and door panels.

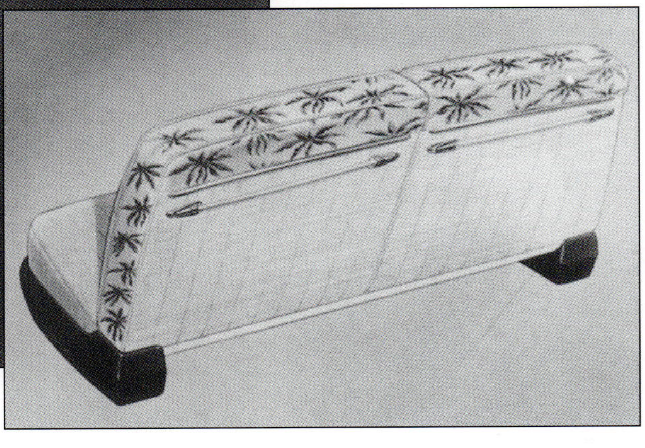

FX-Atmos

Many of those who attended the 1954 Chicago Auto Show on opening day, March 18, 1954, thought that Buck Rogers had landed. If Ford was trying to convince the public that it was laying claim to number one in advanced styling, the FX-Atmos did just that at the expense of General Motors and their Firebird I, which was also new for the 1954 show circuit. The Firebird I was a one-passenger vehicle and, except for the wheels, it looked like a jet plane. The FX-Atmos looked like a cross between a car and a flying saucer, which it was.

The designer of the FX-Atmos was John Middlestead, a young designer in the Ford studio with an fascination for spacecraft and cars with tri-hulled designs. At the time the FX-Atmos was conceived, Frank Hershey was head of the Ford studio, Bob Maguire was Hershey's executive stylist, and Dave Ash was Middlestead's design manager. Ash, Maguire and Damon Woods, who was in charge of Ford's body development studio, all contributed to the final design of the FX-Atmos, but Hershey did not like the design and, except for looking in on the project from time to time, made no significant contribution to it.

Ford Photomedia

When the FX-Atmos was introduced at the Chicago Auto Show in March 1954, many show-goers thought Buck Rogers had landed.

The FX-Atmos was the first of four super-futuristic concept cars styled at Ford in the 1950s, including the 1955 Futura, the 1955 Mystere and the Gyron. To many stylists and to a large part of the public, the 1954 FX-Atmos represented the future, and it told people that space travel was just around the corner. As fantastic as the FX-Atmos was, people believed it represented *their* future, and they loved it. The name Atmos came from the word atmosphere, and FX meant Future Experimental or Ford Experimental. The name was coined by executive stylist Bob Maguire. According to Middlestead, the name Atmos was a typical Maguire-ism and, as far as he knows, it was the only name considered for the car.

Most designers, including Middlestead, acknowledge that the design of almost all cars, whether they be concept cars or production models, is a team effort. The FX-Atmos, however, was less of a team effort than other Ford concept cars. Because the XL-500 and the X-100 brought Ford much favorable publicity and, because GM was continuing to build concept cars, Ford's management decided that they needed to produce more of their own, but it was decided there was as much publicity to be gained from inoperable show-cars as from fully-operable ones, which were more expensive to build.

In fall 1953, Bob Maguire called a meeting of the designers under him. He told them they had a green light from Lewis Crusoe, head of Ford Division, to build a new concept car for the upcoming Chicago Auto Show, but he could assign only one of the approximately ten Ford studio designers to work full-time designing it. From the beginning, the FX-Atmos was planned as a fiberglass pushmobile without any mechanical or engineering components. Middlestead was selected and was the only designer assigned to work on the new concept car, because Maguire did not feel he could divert more than one designer from production work, which had already been expanded to include the Ford Sports Car later introduced with the FX-Atmos as the Thunderbird.

Middlestead was given one corner of the Ford studio, and began by preparing different pencil sketches of cars that he meant to look as much like flying saucers as he could make them. Ash and Maguire decided Middlestead should make full-sized color side-view renderings on black paper of the proposals they thought were the best of the ones he showed them. As Middlestead was finishing one of his drawings, Crusoe stopped by on one of his regular visits. Crusoe was well respected by designers, and it wasn't unusual

Ford Photomedia

A.J. Middlestead

for him to look at what individual designers were doing. Crusoe saw Middlestead's drawing, liked it and immediately authorized construction of the car Middlestead had drawn as Ford's new showcar. Middlestead made three slightly-different renderings as the design work progressed. Of the three renderings Middlestead made, the one closest to the completed car eventually wound up on the wall in Crusoe's office. The second rendering was sent to *Popular Mechanics* magazine, although Middlestead is unsure if they ever published it. Middlestead kept the third rendering, and it is still in his possession.

The FX-Atmos was clay modeled in the Ford body development studio. No 3/8-sized clay model of it was built before the full-sized clay model was built, because time was short. Leonard Stobar was in charge of the clay modeling, which took approximately two months to complete. The most significant change made in the design of the FX-Atmos during the clay modeling process was moving the fins from the middle of the trunk to the top of the rear fenders.

Middlestead acknowledges that several suggestions from Woods were incorporated into the final clay model of the FX-Atmos. The original design of the FX-Atmos, when first approved by Crusoe, was more tri-hulled looking than the final car turned out to be. During the building of the full-sized clay model, Woods suggested that the tri-hulled look be de-emphasized in favor of two outer-hulls, that looked more connected in the center. The tri-hulled design still came out looking a little too radical, so Woods decided he wanted a light bar across the front of the car, which Middlestead thought was inconsistent with a true tri-hulled design.

A. J. Middlestead

John Middlestead grew up in Dearborn. He began drawing at age five and was drawing cars by age seven. Middlestead got his artistic training at Cranbrook Academy in Bloomfield Hills, Michigan, where he majored in fine arts and design. His studies were interrupted by World War II, during which he was a combat engineer in the European Theater. After the war, he studied at Meinzinger's Art Academy. When he graduated in 1949, Bob Jones, a designer at Ford, suggested he revise his portfolio and apply at the Ford Design Department. In December 1949, Middlestead was hired by Bill Schmidt to work in the Lincoln-Mercury Studio. His first assignment, however, was a three month stint in Gil Spear's Advanced Studio, where he worked on the rear-end design for the X-100.

He was then assigned to the Lincoln-Mercury studio working on interiors under design manager David Ash. Later, Middlestead was assigned to the Ford exterior, and then the Ford interior studio, where he again worked under Ash and then Bob Maguire. Middlestead was one of the many young designers who looked upon Maguire as a mentor and father-figure.

As a young designer in the Ford studio, Middlestead designed the FX-Atmos and later worked on the Mystere. In the early 1960s, while working as a manager in the advanced interior studio, he helped design the interior of the Aurora concept car and the gas-turbine truck, Big Red.

Middlestead later worked in the Lincoln-Mercury interior and exterior studios, in advanced interiors, in the Special Projects Studio and, in 1968, he was made a design executive in the Lincoln-Mercury exterior studio. He also worked as a design executive in the medium-sized car exterior, color and trim and in the luxury car studios.

He retired in 1981. ■

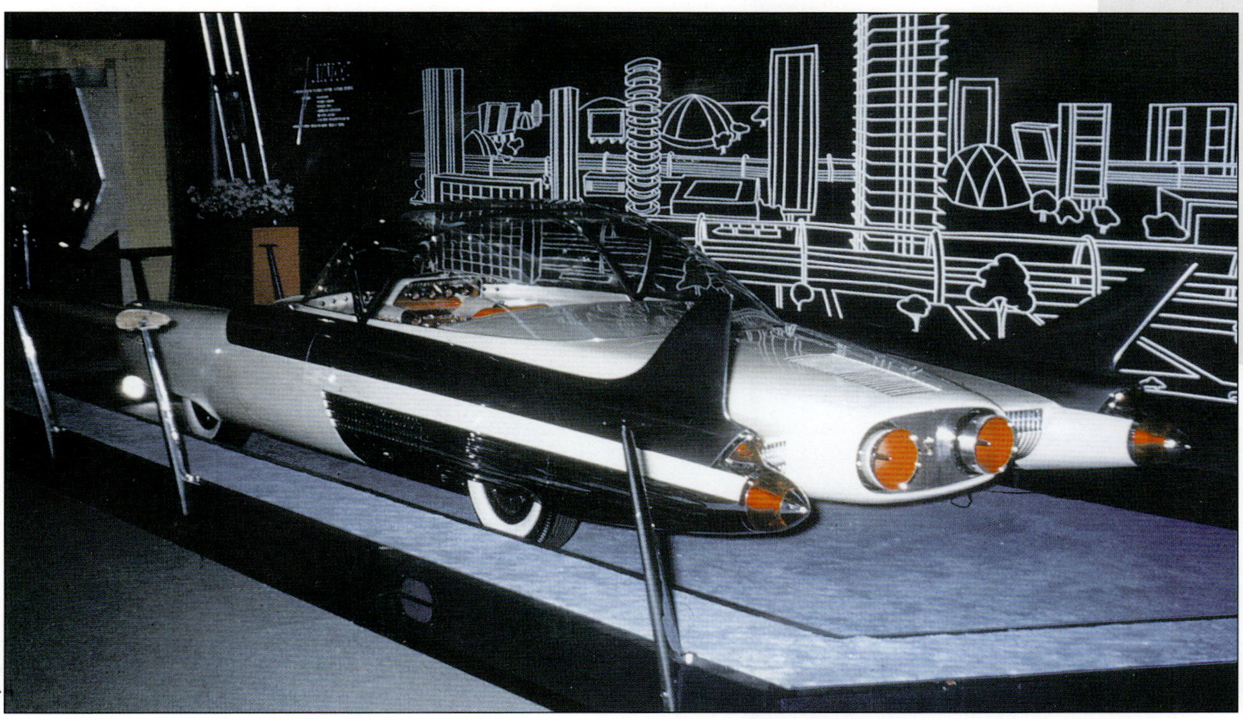

J.R. Samsen

(above left) The automotive press was as shocked by the FX-Atmos as everyone else.

(left) The FX-Atmos at the Chicago Auto Show.

Although Middlestead prepared the initial renderings for the interior of the FX-Atmos, several other Ford studio designers, including Art Querfeld and Al Mueller, assisted on the interior design. A full-sized interior buck was constructed, and both it and the full-sized clay model were trucked to Creative Industries, so the car could be built from fiberglass. The seats of the FX-Atmos were molded fiberglass with upholstery placed over the top of them. The FX-Atmos was designed with no opening windows, because it was intended to have full-time air conditioning and heating.

The FX-Atmos at the Chicago Auto Show. The car was steered by wrist twisters from the middle seat, which was forward of the other two seats.

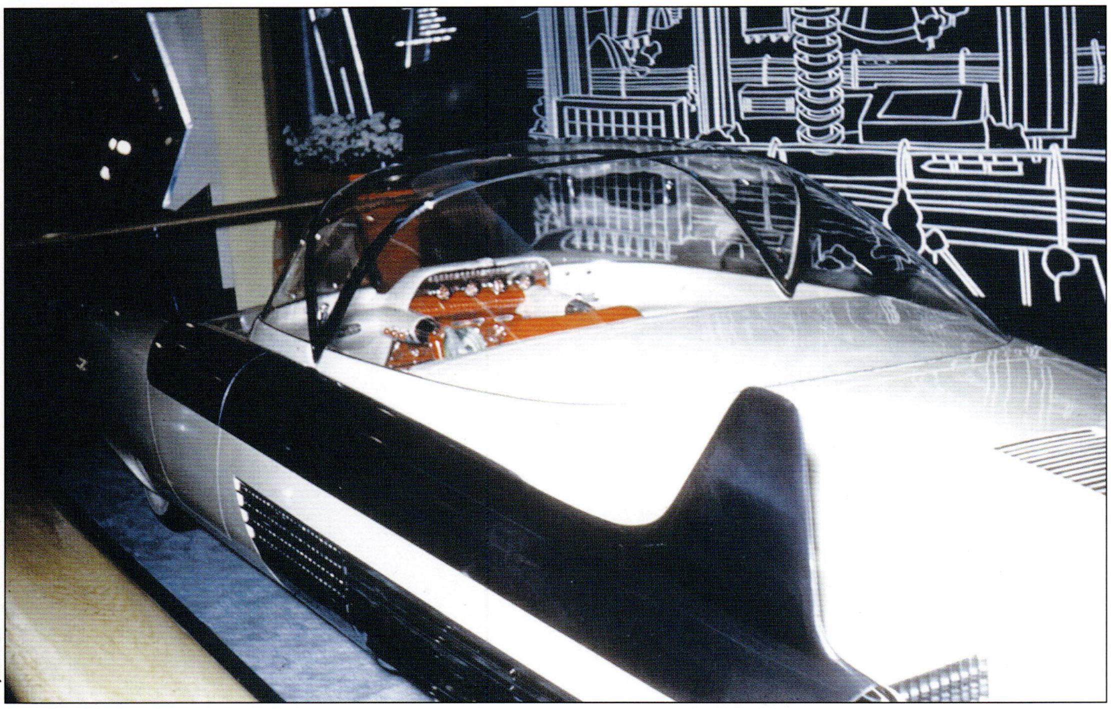
J.R. Samsen

Because the FX-Atmos was planned for a rear engine, the air-conditioning unit was supposed to be located in the front, so the cold air vents were incorporated into the instrument panel. The rest of the instruments were clustered in the center in front of the driver. They included a speedometer and four round gauges below it. Middlestead says that the gauges used in the FX-Atmos were conventional and were the same ones generally found on Ford passenger cars. Below the instrument cluster, there was a small roadair screen to show the driver what was ahead of him on the highway.

Although Frank Hershey later claimed that the FX-Atmos was designed to be powered by a gas-turbine engine, Middlestead insists that information is incorrect. The FX-Atmos had no running gear, and Middlestead is the first to acknowledge that not much consideration was given to the type of running gear it would have. However, Middlestead designed the FX-Atmos with the idea that its power plant would be a conventional reciprocating engine located in the rear, and the venting in the rear-quarter panels and on the rear deck was designed to provide cooling for a reciprocating engine. The two jet pods at the rear of the car were intended as part of the design only and not intended as exhaust outlets. On the completed car, they were used for lights. None of the contemporary press releases claimed that the FX-Atmos was to be powered by a gas turbine. The Mystere, not the FX-Atmos, was the first Ford concept car designed to be powered by a gas turbine. In contemporary car magazines, there was some specula-

One of John Middlestead's alternate renderings of the FX-Atmos.

A.J. Middlestead

tion that cars like the FX-Atmos might someday use atomic power, but that was not something encouraged by the car's designers.

Building the fiberglass FX-Atmos was supervised by Dave Margolius of Creative Industries and took about three months. It was finished shortly before the March 1954 Chicago Auto Show. Before it was sent to Chicago, the completed FX-Atmos was returned to the Design Department, where it was shown to the press, and publicity photographs were taken of it with Ash and Middlestead. At the same time, publicity photographs were also taken with other people inside the car. The woman in the passenger seat in the publicity photographs is not 1950s pinup girl Betty Page, as has been reported on the Internet but, rather, she was a Design Department secretary who resembled Betty Page and now wishes to remain anonymous.

Ford Design Center

A.J. Middlestead

One of John Middlestead's alternate renderings of the FX-Atmos.

Middlestead and Leonard Stobar made many trips back and forth to Creative Industries to supervise the construction of the FX-Atmos. No major changes were made to the design of the exterior or interior of the car while it was at Creative Industries, although the color used on it was changed several times while the car was being built. At the time the full-sized clay model was taken to Creative Industries, the colors were suppose to be a red body with black upper-body sides and fins.

Creative Industries molded the FX-Atmos body as a solid unit with doors, hood and trunk that did not open. The Plexiglas canopy, which was also molded at Creative Industries, was a separate piece. It was fastened to the car, and only the center section could be lifted off. Although the center section was designed to slide back to provide passenger access, as built, the center section had to be removed, so people could be lifted up by a crane and lowered into the car. They were removed the same way.

Arthur L. Querfeld

Art Querfeld's forebearers purchased a French land grant and settled in the Dearborn area in the 1830s. Much of what is now Ford Motor Co. property, including the location of the Ford Division building, was originally part of the Querfeld family farm.

Querfeld didn't want to work in the mortuary business his family had established in Dearborn in the 1920s. Instead, he attended Pratt Institute, where he studied industrial design and advertising. He graduated in 1941. Later that same year, he applied for a job at the Ford Design Department, but they were not hiring because of the war. He was hired at the Rouge plant by Tom Hibbard, who was working there as a body engineer, and put to work preparing interpretive drawings of tank parts.

After he was drafted, Querfeld became a part of the military occupational government in Germany, where he stayed until 1947. In Germany, he was placed in charge of the Mercedes-Benz automobile manufacturing plant.

In January 1947, Querfeld returned home and was hired at Ford's Design Department by Tom Hibbard, who was, by then, Ford's chief designer. His first assignment was working in Bill Wagner's truck studio.

Bob Maguire saw Querfeld's freelance drawings and arranged for him to move to the Ford interior studio. He also worked for Maguire in the Ford exterior studio. While in that studio, he helped design the interior of the FX-Atmos. Querfeld later worked on the interior designs for the 1961 and 1964 Thunderbirds, and in Buzz Grisinger's Lincoln-Mercury studio.

Early in his career, Querfeld and Maguire developed the design of the 1950 and '51 Ford Crestliners. Later in his career, Querfeld was executive designer in Dave Ash's studio, when the 1969 Mark III was designed. He was also one of the designers of the 1980 Lincoln Town Car and the 1982 Lincoln Continental. He retired in 1985. ■

A.J. Middlestead

The FX-Atmos is shown in the Design Department with a model in the driver's seat. Dave Ash (left) and (right) John Middlestead are standing behind it. Ash was Middlestead's studio manager at the time the FX-Atmos was designed.

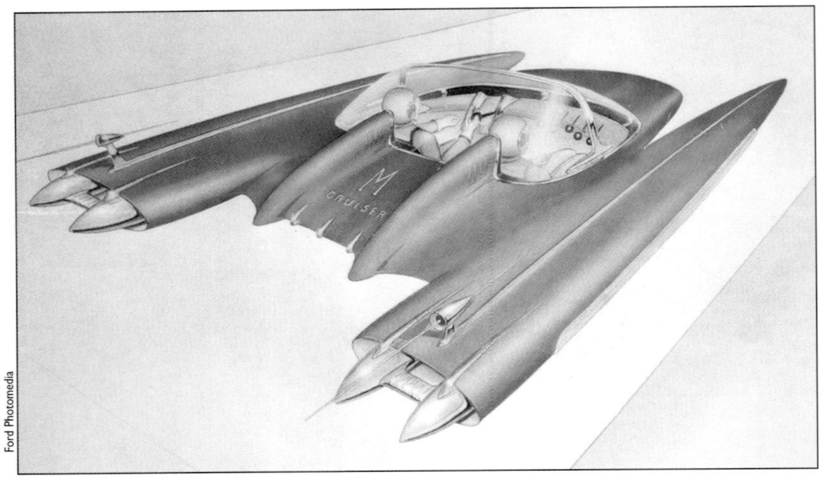

Ford Photomedia

Tri-hull designs were popular in the early 1950s. The designer of this rendering is unknown.

Middlestead does not recall whether the FX-Atmos was ever painted red and black, or who picked the final colors that were used on the car. The colors finally selected were a pearlescent-white, with a red band on the front luggage compartment lid and along the lower rear-quarter panel. The upper part of the body sides, from the windshield to the rear stabilizers, was painted ice blue. At the time, pearlescent-white was a new color, and Middlestead believes it was selected for its impact. Middlestead claims that the red, white and blue color scheme was not necessarily picked for its patriotic impact, although that was the assumption made by many who saw the car.

The most innovative feature on the car was the roadair scope, which was intended to give the driver a constant picture of the road ahead via aerials sticking straight out from the front fenders. The driver's seat was in the center, with two additional seats for passengers on either side of the driver's seat and a little to the rear. There was no center console between the two most rearward seats. All of Middlestead's designs for the FX-Atmos included a three-person seating arrangement with the driver forward of the two passengers, because that was how he thought a flying saucer-type car should be designed. The car had no steering wheel as such, but was steered by two handgrip wrist-twisters situated at elbow height in front of the driver's seat. The wrist-twister was also Middlestead's idea and, although it was not functional on the FX-Atmos, several years later he developed a working prototype of the wrist-twister steering, which was installed on a driveable two-passenger Thunderbird.

The FX-Atmos featured a concave-shaped light bar across the front grille area instead of conventional headlights. For show, the light bar, the interior lights, the rear tail-lights and the lights in the jet pods could be illuminated by a battery hidden under the car. Middlestead recalls that when the light bar and the lights in the jet pods were illuminated, it gave the car a very dramatic look.

The FX-Atmos was built on a short, extra-wide frame, and it had a 105-inch wheelbase, was 220.5 inches long, 4 feet high and 6.5 feet wide. The dimensions of the FX-Atmos were not taken from a production car, but were taken directly from the full-sized blackboard drawing prepared by Middlestead and approved by Crusoe.

From the back, the FX-Atmos looks like it has jet exhausts, but they were intended only as lights.

Ford Photomedia

A. L. Mueller

Al Mueller was born and raised in Cleveland, Ohio. After serving in the Army during World War II, he attended the Cleveland Institute of Arts. Mueller majored in industrial design with a minor in sculpture. When he graduated in 1949, he was hired as a designer at the Ford Design Department.

His first assignment was in the Ford interior studio, where he worked on the interior of the FX Atmos. After three years in the Ford interior studio, Mueller was promoted to design manager in Bill Wagner's truck studio, where he helped design the 1955 Ford tilt-cab (COE) truck. Several years later, Mueller was transferred back to the Ford, and then to the Thunderbird studio as a design manager, where he worked on the interior and instrument panels of the 1955-57 Thunderbirds.

After he was reassigned to the Ford exterior studio, Mueller also worked on the redesign of the 1958 Ford, which was meant to have a family resemblance to the Thunderbird. He is credited with the 1958 Ford taillight design. Mueller was also a manager in Bob Maguire's Advanced Studio and worked on the interchangeability studies (L'Avant and Xemplar). He was also on the team that designed the interior of the 1961 Thunderbird, and was lead designer on the team that designed the exterior of the 1964 Thunderbird.

In 1961, Mueller was named an executive stylist in the Lincoln-Mercury interior studio and, later, held that same position in the Ford interior studio, the Corporate Projects Studio and the international studio.

For three years in the early 1970s, Mueller was a liaison with Ghia in Italy. He was later executive designer in Buzz Grisinger's Lincoln-Mercury studio. Mueller retired in 1987, after serving as head of the Industrial Design Studio and as director of the advanced studio. ■

Ford Design Center

Middlestead and several other designers recall that the FX-Atmos was around until sometime in the mid-1960s. Because it was built only as a showcar and was without a supporting structure, after several years, the fiberglass began to sag and the Plexiglas roof began to discolor and check. By the mid-1960s, with no place to store it, the FX-Atmos was broken up. When the FX-Atmos was built, Ford gave Creative Industries permission to build a small model of it for their showroom. That model still exists and is currently owned by Creative Industries.

Although the press releases about the FX-Atmos emphasized that its features and design were fanciful and only represented one of the many directions future Ford styling might take, current model new cars are only now becoming available with map screens in their instrument panels and light bars across the front. ■

The FX-Atmos at the time of its press preview. Frank Hershey (third from left) and Bob Maguire (fifth from left) answer questions from the automotive press.

Detroit Public Library National Automotive History Collection

Gil Treweek

The full-sized clay model of the Ford Sports Car is just outside the EEE Building in early 1953. From left to right are Leonard Stobar, Lloyd Kamm (with back turned), Bob Maguire (motioning), Bill Ripking, Al Kellum, Ernie Gregory, Dick Mazur, Bill Fuhrman and Gene Froberg.

Thunderbird

As with many great automobiles, several designers claim credit for the two-passenger Thunderbird. George Walker, then an outside consultant to Ford's Design Department, Frank Hershey, head of the Ford studio, and Bill Boyer, a young designer hired away from GM by Hershey specifically to help design a Ford Sports Car, have all claimed credit for the Thunderbird. Although the design of all cars is a team effort, there has been much confusion about who really designed the Thunderbird. Several of the Thunderbird's designers have been forgotten, probably because they left the Ford Design Department, before or just after the design of the Thunderbird was finished.

Most histories that chronicle the birth of the Thunderbird claim that it was a four month project, rushed through to completion during the first half of 1953 after George Walker fibbed to Lewis Crusoe, head of Ford Division, by telling him that Ford designers had already designed just such a car. The real story is of a car that was designed in a back room without official authorization, and then put on hold until Walker could steer management to the conclusion he, Frank Hershey and other designers and engineers had already reached — that Ford must build a car to compete with GM's new Corvette.

The first time the four month time schedule was put in written form was in a July 1953 internal Design Department report summarizing the increased work load and advances the Design Department had made since 1950. The chronology of work on what that report called the 195X Ford Sports Car was as follows:

"On February 9, 1953, the Sports Car project was approved by the Product Planning Committee. Full-size layouts were started on February 10, 1953 and actual work on the clay buck began on February 16, 1953. The model was completed and shown during the months of April and May and basically approved for exterior appearance. The interior and instrument panel were designed, modeled and shown in conjunction with the exterior model. Upon review, a new instrument panel containing the "astra-dial" treatment was requested. The new panel and revised interior design were shown and approved in July 1953. This program consumed considerable man hours not in the original budget."

The most incredible part of the Thunderbird's history is that the car was already a finished full-sized clay model, and a mechanical prototype had already been built by the time Ford's official records say it was started. When Walker told Crusoe a Ford Sports Car had already been designed, it was the truth, but it was also less than the whole truth. As Boyer puts it, what Walker told Crusoe was with "malice aforethought." Walker knew the car had already been secretly built but, to go any farther with it, he also knew he had to convince Crusoe that Ford needed something Crusoe may not have known existed. In classic Walker fashion, he let Crusoe lead himself to the conclusion Walker wanted him to make. While it is difficult to believe that Crusoe was unaware that a Ford Sports Car had already been designed, it is consistent with what has been Ford's official position since July 1953. One of the problems with the official story of the Thunderbird has always been the sequence of events. It's all wrong. The Corvette wasn't publicly shown until the New York City Motorama, in January 1953 and, yet, the Ford Sports Car project supposedly wasn't begun until that same month. Given the January starting date, there simply wasn't sufficient time to design, engineer, build a finished clay model and build a finished mechanical prototype of the car by May 1953, the acknowledged date of its exterior design approval. More lead time than the four month period claimed was needed to complete the design and engineering process necessary to produce a new car -- even at Ford Motor Co.

J.R. Samsen

One of Dick Samsen's 1952 renderings of the proposed Ford Sports Car.

Before the Thunderbird came along, Frank Hershey was an automobile collector and historian, but not the sports-car buff the press later claimed him to be. He and other designers and engineers at Ford, and elsewhere, were generally aware of the resurgence of the sports car genre in the United States after World War II but, before the Thunderbird, Hershey had been partial to classic American cars, not European sports cars. There was at Ford, however, a small group that were passionate advocates of true sports cars. In late 1951 or early 1952, a group of young engineers and designers at Ford, influenced by the sports cars many of them had first seen during World War II service in Europe, formed the Ford Motorsports Club. Ford designers who belonged to that club included Dave Ash, Jake Aldrich, Dick Samsen, Chuck Mitchell and Ken Nelson.

Sometime in early 1952, Dick Samsen recalls asking his boss, Frank Hershey, why Ford didn't build an American sports car. Hershey responded by telling him that Ford wasn't interested, because it couldn't make money on a sports car. That all changed several months later, when word leaked out that arch-rival Chevrolet was going to build a sports car. At that point, Hershey became a certified sports-car fanatic, and even bought his own Jaguar XK-120 roadster.

It had only been five years since Hershey worked for the GM Styling Section. He still had many friends there and, from time to time, one or more of those friends shared information and pictures with him. In February or March 1952, over dinner at Hershey's

Members of the Ford Motorsports Club in 1952. Designers Ken Dunn (fourth from left), Dick Samsen (fifth from left), Ken Nelson (eighth from left), Jake Aldrich (ninth from left), Charles Mitchell (twelfth from left), and Dave Ash (fourth from right) were members.

J.R. Samsen

Frank Hershey

Franklin Q. Hershey

Frank Hershey was born in Detroit but considered California his home. When he was a youngster, his family moved to Beverly Hills, California. Shortly after that, his parents divorced. Hershey lived with his mother, who became a part-time movie actress. In November 1928, his mother's financial advisor got Hershey a job working for the Walter M. Murphy Body Co., a Southern California builder of custom-bodied cars.

When Murphy Body went out of business, Hershey moved to Detroit to take a job designing Hudsons for Frank Spring, his former boss at Murphy Body. He soon left Hudson for a job at GM as a designer in the Pontiac studio.

In 1939, after spending several years at GM's European design center, Hershey returned to Detroit as head of the Cadillac advanced studio, where he remained until he left to serve in the Navy during World War II. In 1944, Hershey returned to GM and was appointed head of the Cadillac studio. He is responsible for designing the fins on the 1948 Cadillac.

In 1947, while employed at GM, Hershey started a side business making reproduction brass ashtrays. Moonlighting was prohibited, and he was terminated. Until the Korean Conflict cut off his supply of metal, Hershey successfully made and sold his reproduction brass ashtrays. He then took a job as chief designer for Packard but, in early 1952, he was recruited by Charlie Waterhouse to head the Ford studio.

Hershey headed the team that designed the FX-Atmos, the Mystere and the Thunderbird. He also supervised the design of the production 1955-57 Fords. By early 1955, Hershey was at odds with George Walker and Robert McNamara. In May 1955, after Walker was made vice president of design, Hershey left for California to accept a job as head of Kaiser Aluminium's design department. He stayed at Kaiser until 1964, when he left to accept a position as a designer at Rite Autotronics in Southern California. After retiring in 1976, Hershey built a vacation house in Puerto Vallarta, Mexico, raced motorcycles and was active in the old car hobby. ■

house, a GM designer, who was one of his sources, showed him a picture of a proposed sports car Chevrolet Division was planning to build. It was a picture of the Corvette, which wasn't introduced until the New York City Motorama in January 1953 and in Chevrolet show rooms late in the 1953 model year. As Hershey put it, he "flipped" when he found out the competition was building a sports car!

Harley Earl, GM's design czar, first decided to build the car that became the Corvette in late fall 1951. It was done in secret and under the code name Project Opel. Earl and designer Robert F. MacLean copied the dimensions of the Jaguar XK-120. By April 1952, a plaster model of the Corvette was shown to GM management and, by mid-1952, the decision was made to include the Corvette in the 1953 Motoramas. Although public enthusiasm shown at the 1953 Motoramas encouraged production of the Corvette, Earl and Chevrolet's chief engineer, Edward N. Cole, had already decided that the Corvette would likely be produced.

When Hershey saw the picture of the proposed Corvette, there was no doubt in his mind that Ford must build its own sports car and outdo the Corvette. In the early 1950s, everyone at Ford, from Henry Ford II on down, was on a holy war to beat GM. There were even signs hung in Ford's assembly plants urging just that. Hershey lost no time. In addition to buying his own Jaguar XK-120 roadster, he studied all other available sports cars, until he arrived at what he thought was the optimum package for a Ford Sports Car. Although the inspiration for the Ford Sports Car was the Corvette, as with the Corvette, the Jaguar XK-120 was its model.

At a church social, shortly after he decided Ford should build its own sports car, Hershey and a young GM designer named Bill Boyer found themselves comparing notes about the hard lifestyle at GM's Styling Section. Hershey, sensing that Boyer was unhappy at GM, asked to see his portfolio. Hershey then accompanied Boyer to his house, where Boyer gave him his portfolio, which included a rendering of a proposed Cadillac he had recently completed. Although Boyer thought Hershey might offer him a job, he did not know that Hershey was looking for someone to design a Ford Sports Car to compete with the Corvette. Because Gene Bordinat was always looking for good designers to hire away from GM, Hershey showed Boyer's portfolio to him. Bordinat was impressed and expressed surprise that the drawing of the proposed Cadillac was done by someone out of school only one year. Bordinat suggested that Hershey offer Boyer a job designing Ford's answer to the Corvette.

In March 1952, Boyer received a call from Hershey. Hershey's call was made directly into the GM design studio, where Boyer was working at the time. Hershey told Boyer that Ford was going to build a sports car to compete with the Chevrolet sports car, and that he had been authorized to offer him a job as one of its designers. While Hershey was talking to Boyer, Harley Earl was literally standing right in front of him, so Boyer told Hershey he had a problem, and that he would have to call him back. During his lunch hour, Boyer took the tunnel under Grand Boulevard to a pay phone and returned Hershey's call. Hershey offered Boyer the job. Boyer took it, gave his two-week notice at GM and started as a Ford employee the next day. GM's Styling Section had already lost several of its designers to Ford, and had decided there was no reason to let those jumping ship stay longer than about two hours after they gave their notice.

Boyer's first assignment at Ford was working on the production 1955 Ford but, by mid-summer 1952, he was assigned to the Ford body development studio, which was in a small room off of the main Ford studio. Hershey also assigned two other young designers to the body development studio, J. R. "Dick" Samsen, who had excellent rendering skills and a background in aeronautical engineering and Allan Kornmiller, a young designer and a recent graduate of the Cleveland Institute of Arts, who had shown promising design talent while in Gil Spear's Advanced Studio.By the time Samsen arrived at the body development studio, Boyer was already there, Kornmiller came a few weeks later.

After work had been completed on the production 1955 Ford in the body development studio, that car was moved into the Ford exterior studio for ornamentation. After that, there was some slack time before body development was to begin work on the 1956 production Ford. It was during this lull, that Boyer, Samsen and Kornmiller and their superior in the body development studio, Damon Woods, began work on the Ford Sports Car. Boyer, Samsen and Kornmiller each recall being told that their sole assignment was to design a Ford Sports Car to compete with what was to become the Corvette, and that the job was to be done under strict security. They were told that the body development studio was off-limits to most people in the Design Department, and they all sensed that security was much tighter than it normally was.

Henry Ford Museum and Greenfield Village

Three full-sized paper and plywood Ford Sports Cars on display for the Product Planning Committee in early 1953.

Bob Todd, whose father had been Henry Ford's personal mechanic and had worked on the flathead V-8 engine with Henry Ford, was the layout man in Woods' body development studio. Just before Boyer, Samsen and Kornmiller began designing the Ford Sports Car, Todd was given a grainy photograph of GM's proposed sports car, and he was asked to make a full-sized side-view drawing of it to scale. Todd's drawing of the Corvette was done on heavy manila-colored paper. When it was finished, the drawing was pasted on a piece of plywood that was trimmed to resemble a real car. Once the drawing was mounted on plywood and trimmed, wheels and tires were placed in the wheelwell cutouts, a Plexiglas windshield was added, and a chair was placed on the backside of the drawing, so that someone could sit behind the steering wheel. The process actually provided a very realistic view of the Corvette for the designers to work from.

At the Design Department, what was called an EJN number was required to go beyond the rendering stage on any proposal. In other words, to build a clay model or begin a mechanical prototype, authorization from management was needed, and that meant justification and approval of the project. At the time, Ford new-car sales were down and money was tight. According to Boyer, one of the reasons the body development studio was declared off-limits in 1952 was because the Ford Sports Car being designed there didn't have authorization or an EJN number. The primary reason, however, was that Hershey knew that management was unlikely to allocate scarce development funds for a proposed limited-market sports car, so he didn't ask.

Henry Ford Museum and Greenfield Village

A full-sized clay model of the Ford Sports Car in early 1953.

There is general agreement that design work done on the Ford Sports Car during 1952 was without any formal authorization. It is not clear how unusual that was, but most designers at Ford at the time, while they question whether Hershey would have risked his career on an unauthorized sports car project, were not aware that a Ford Sports Car was being designed until January 1953, and then they acknowledge that the clay model just showed up one day. There is a dispute about which managers were aware of what Hershey, Woods, Boyer, Samsen and Kornmiller were designing. Boyer believes that George Walker, then an outside consultant to the Design Center, general manager Charlie Waterhouse, assistant manager Gene Bordinat, Joe Oros, at that time an outside consultant working for Walker, Frank Hershey, Ford studio chief stylist, Bob Maguire, executive designer in the Ford studio, product planners Thomas B. Case and Chase Morsey, Bill Burnett, an engineer who built the mechanical prototype of the Ford Sports Car and Henry Grebe, chief body engineer, were the only people aware a Ford Sports Car was under way. If there was a conspiracy it was widespread.

This photograph was taken from a program on the Thunderbird shown on the History channel. This is the original armature used to clay model the full-sized Ford Sports Car in 1952.

Ford Photomedia

As Boyer recalls it, Burnett, although enthusiastic about the project, expressed the most concern about proceeding without authorization. Maguire and Oros were busy with the 1955 Ford, which was behind schedule, so they didn't pay much attention to what was going on behind the body development studio's closed door. Hershey and Woods checked on the car's progress daily and, on many occasions, they became directly involved in its design. Bordinat looked in on the Ford Sports Car project often. Kornmiller, although he confirms the tight security,

Damon C. Woods

Damon Woods grew up in Dijon, France, where his father was a U.S. consular official. Woods spoke fluent French and, from an early age, developed a keen appreciation of classic European automobile design. Woods attended college in Canada and later studied mechanical engineering at the University of Michigan. Although he had an engineering background, Woods also had a natural design talent he was able to use to become a self-taught designer.

During World War II, Woods worked for Curtis-Wright as an aeronautical engineer. After the war, Woods was hired as a designer by Kaiser-Frazer, where he worked with Buzz Grisinger. In 1951, Woods was hired as a designer at Ford. Woods worked mostly in the Ford studio. In 1953, he became design executive for the Ford car and truck exterior studios. At the same time, he was also placed in charge of the Ford body development studio, where the two-passenger Thunderbird was being designed. Woods also supervised the design of the FX-Atmos and the Mystere concept cars.

In the mid-1960s, Woods was appointed vice president of design, Ford of Europe. While on home leave in 1968, he was killed in a freak car accident on I-94 near Ypsilanti, Michigan. Most designers remember him as an extremely talented and innovative designer with superior diplomatic skills. ■

believes everyone in Ford's management was aware of what they were doing. So many designers and engineers knew about the project that Samsen doubts it could have been a secret from anyone of Ford's higher management, who were regularly at the Design Department, including Crusoe.

The only other employees with daily access to the body development studio were engineers Frank Pinkham, a senior layout man (design engineer), who was responsible for such things as making sure the angle of the windshield was correct, the windows dropped and the engine fit under the hood, and John Zimmerly, who was responsible for making sure the full-sized clay model stayed within the package dimensions previously set.

Hershey recalls that, during this period, it was common for Robert McNamara, then assistant manager of Ford Division, to walk through at unexpected times even on weekends, and he had a habit of canceling or changing whatever he didn't like. Crusoe, who was McNamara's boss, would come by a few days later and override what McNamara had changed or cancelled. Needless to say, designers didn't appreciate the problems those type of disputes caused for them and sought ways to avoid having their projects subject to the conflicting orders. Sometimes, that included hiding work in progress.

Once plans for the Corvette became common knowledge in the industry, Ford's Product Planning Committee initiated a market research study to find out how large the sports car market was, and what features potential buyers of sports cars wanted. Based partially on that study and, partially on his own instinct, Hershey recommended that the Ford Sports Car's dimensions be based on the Jaguar XK-120. The package Hershey, Bordinat and the engineers agreed on was for a two-passenger car that weighed less than 2500 pounds, was 175 inches long, had a wheelbase of 102 inches, was 51 1/2 inches tall, 70 inches wide, had a road clearance of 5 1/2 inches and had a V-8 engine with at least 160 horsepower. The price of the car was hoped to be less than $2500. The production Thunderbird weighed in at 2833 pounds, sold for a little more than $3000, and the engine was actually a bored-out Mercury block with 292 CID and 193-198 horsepower.

While Boyer, Samsen and Kornmiller were making renderings of the proposed Ford Sports Car, engineers were working on the car's chassis and running gear. At first, Engineering proposed a simple chop-job to make a car similar to the 1953 Cadillac Eldorado or Buick Skylark. Hershey and Bordinat felt that to do the job right, a special frame was needed that would permit the engine and the passenger compartment to be set back as far as possible. They felt that a special frame was needed to permit for the short deck and long hood that were being designed on the Ford Sports Car. Hershey and Bordinat got their way and their special frame but, as a trade off, Grebe insisted that the body be made from steel instead of fiberglass. Making the body from steel meant that about 15,000 cars per year would have to be sold to break even. That meant the Ford

The full-sized clay model of the Ford Sports Car at the Product Planning Committee showing held in April 1953.

Sports Car had to appeal to as wide a market as possible. Grebe insisted on roll-up windows, rather than the Corvette's side curtains, but there wasn't much resistance to that. Everyone agreed that to sell in even modest numbers, a Ford Sports Car had to be more refined and appeal to more buyers than the XK-120 or the Corvette. For a Ford Sports Car to be economically feasible, they also agreed it had to lunch off the 1955 Ford passenger car as much as possible. In other words, to limit complexity and tooling cost, it had to use as many parts from its production cousin as possible.

Kornmiller remembers that all three designers spent weeks preparing renderings using the 102-inch package of the Jaguar XK-120 and the Corvette. He recalls that the biggest initial design question was whether the Ford Sports Car should have a hopup in the belt line, or whether it should have a straight-through belt line. That question was not initially resolved, and many renderings were prepared both ways. Each designer made multiple 3/8-sized side-view renderings on blueprint paper of a car that was very structured, given the package and the practical restrictions imposed by Engineering.

After a sufficient number of renderings were made, Woods, Hershey and Bordinat picked one rendering by each designer, and asked each to make them into full-sized side-view renderings that were to be mounted on plywood, the same as Bob Todd had done with the Corvette. Three additional full-sized side-view renderings were also made of current production sports cars. As Samsen remembers it, one was of a Jaguar XK-120, one was an MG-TD and the other was a Nash-Healey. In addition to adding wheels and tires, the five profiles were painted different colors, and Plexiglas windshields were added. When completed, all six profiles were taken to the showroom. No one remembers who selected from the profiles in the design showroom but, after it was over, Boyer, Samsen and Kornmiller each recall being told that his rendering had been chosen, and that a full-sized clay model of the selected rendering was to start immediately. It is likely that features from each designer's rendering were chosen for the clay model and, as happened on occasion, each designer was encouraged with approval.

No 3/8-sized clay model of the Ford Sports Car was built. Instead, one full-sized clay model was authorized. Boyer, Samsen and Kornmiller all worked on the full-sized clay model, although Kornmiller acknowledges that Boyer and Samsen were more involved than he was. At least six clay modelers were assigned to build the full-sized clay model, including Walter Amrozi, Leonard Stobar, John Snider, Clyde Trombley, Werner Framke and Mike Nowicki. They were all cautioned not to talk to others about the project and, as far as Design Department records were concerned, the clay modelers were all still assigned to work on the production 1955 Ford.

When the Ford Sports Car program was started, in mid-summer 1952, Hershey assigned a designer from the Ford studio to design the interior soft trim for the proposed Ford Sports Car. That designer was Alex Musichuk. Musichuk was hired as a designer at Ford in 1951. In the short time he had been at Ford, Hershey had promoted him to head up design projects involving Ford interior soft trim. At the time, Musichuk was assigned the sports car project, it was in addition to his other duties.

Musichuk was assigned to work with Gordon Buehrig. Although Buehrig was a legendary designer, in 1951, he was assigned to work in Lincoln-Mercury Engineering. (In 1953, Buehrig became chief body engineer on the Continental Mark II.) Buehrig was borrowed from Lincoln-Mercury Division to set the package for the interior of the Ford Sports Car. Using Buehrig's package measurements, it was Musichuk's job to start the interior layout and design of the car's soft trim. When the sports car project was stopped in the fall of 1952, Musichuk does not recall being very far along. When the project was restarted in January 1953, Musichuk was in the process of changing jobs, so designing the soft trim for the Ford Sports Car was reassigned to Alden "Gib" Giberson.

Giberson had been hired at Ford in 1952 and was pulled out of the Ford interior studio to work on the Ford Sports Car. Art Querfeld also later worked on the interior of the Ford Sports Car. The full-sized clay model was made with the exterior and interior on the same buck and, because it was a convertible, the designers and modelers were able to work on the interior as well as the exterior at the same time. The clay seat buck was removable, so Giberson and the modelers could take it out, put in small stools and then work on the instrument panel.

John Najjar

Robert J. Todd

Bob Todd's father was Henry Todd, an engineer who worked on the development of Henry Ford's flathead V-8 engine. Henry Todd was also Henry Ford's personal mechanic. From the seventh grade, through his first year in college, Bob Todd attended school at Greenfield Village. In December 1941, one week after Pearl Harbor, he left school and started work as a foreman on the B-24.

Because many unskilled war workers could not read blue prints, Bill Schmidt and others were assigned to prepare exploded drawings to assist in assembling the B-24. Todd took a class from Bill Schmidt in perspective drawing and spent the rest of the war preparing drawings for the B-24.

When the war was over, Todd's dad suggested that he apply for a job at Ford's patent office preparing patent drawings. After he completed his interview, Todd stopped to see his father at the EEE Building. When Todd arrived, Tom Hibbard, who was visiting with his father, saw the drawing Bob Todd had used in his interview at the patent office, and suggested that he come to work in the Design Department working with Martin Regitko as a design layout man. (A layout man had to determine how to make a design work from the engineering standpoint, so the clay modeler could start building it. Much of the layout man's function is now performed by computer.)

One of Todd's first assignments at the Design Department was working on the Light Ford. He was again working for Bill Schmidt in an advanced studio, which was located in the old Airport Terminal Building. Later in his career, Todd was the layout man (later called a PED or product design engineer) on the 1955 Thunderbird. He prepared the original full-sized drawing of the Corvette that was glued to plywood and used during the design of the '55 Thunderbird.

Todd retired in 1979. ■

The high front bumper was not approved because of objections from Engineering.

Henry Ford Museum and Greenfield Village

The instrument panel as originally proposed for the Ford Sports Car in May 1953. This proposal was eventually used on the 1956 production Ford and on the 1957 Thunderbird.

Henry Ford Museum and Greenfield Village

While the full-sized clay model was under way, engineer Bill Burnett began construction of a mechanical prototype he cobbled from a production 1953 Ford. It was nicknamed the Burnetti. According to Boyer, Burnett built the mechanical prototype in a remote corner of the old Air Frame Building, where he knew it was unlikely anyone would notice what he was doing. Apparently no one ever did.

Sometime during the early fall of 1952, Boyer recalls that at least one unauthorized designer came into the body development studio. Robert H. Gurr dropped by to borrow a current publication from Boyer. As Gurr recalls it, after getting the publication and, as he turned to leave, he saw a full-sized rendering of a very striking-looking sporty car on the far wall. (The clay was probably covered.) That car Gurr later recognized as the Thunderbird. Before Gurr could make his exit, Hershey came in, chewed him out for being somewhere he wasn't supposed to be, and told him never to return! Gurr was puzzled by Hershey's anger as he felt he and Hershey had always gotten along quite well. Gurr assumed there must have been a "hush-hush" project in Boyer's studio he wasn't supposed to see. Shortly after that incident, Gurr left the Design Department to work for Walker's independent design firm on its design proposal for the Continental Mark II, which became one of the unsuccessful competitors in the Mark II design contest. (Gurr later became chief designer for Walt Disney Productions.)

(below) This May 1953 clay model of the Ford Sports Car had a blister over the left rear wheel. The blister did not make it into production.

Henry Ford Museum and Greenfield Village

In late 1952, just as the full-sized clay model was being completed, work was abruptly stopped. Samsen recalls being told by Hershey that it had been decided that anticipated sales just didn't justify building a Ford Sports Car. Boyer recalls being told that someone in management had found out about the Ford Sports Car, and that it was too risky to continue. Hershey claimed that, on one of McNamara's visits to the Design Department, he stumbled on a young design engineer laying out the unauthorized chassis for the Ford Sports Car. McNamara apparently then told chief engineer Earl MacPherson that he had caught someone working on an unauthorized project. When MacPherson confronted Hershey, Hershey told him he had no knowledge of the unauthorized work and blamed it on the over-enthusiastic young engineer. Hershey apparently felt that McNamara and MacPherson's discovery of the special chassis was too close for comfort and stopped work on the Ford Sports Car before the true extent of the unauthorized work was discovered.

(below) A cobbled 1953 Ford, called the Burnetti, was built by engineer Bill Burnett as a mule or mechanical prototype of the Ford Sports Car.

Henry Ford Museum and Greenfield Village

The almost-finished clay model of the Ford Sports Car was taken to the basement and placed in covered storage. The clay modelers were sent back to work on the production 1955 Ford, which was behind schedule anyway. Boyer was left alone in the body development studio with not much else to do, and Samsen and Kornmiller were assigned to other projects. All three designers were deeply disappointed that the sports car project had been canceled.

The New York City Motorama took place in January 1953. Ford Division head Lewis Crusoe and design consultant George Walker went to see what new concept cars GM had on display. Walker probably knew the Corvette was going to be there and had undoubtedly seen pictures of it. Crusoe was probably aware that GM was building a sports car but may not have seen pictures of it yet. Once there, Walker made sure he steered Crusoe and their conversation towards the Corvette. Before long, Crusoe wanted to know why Ford didn't have anything like the Corvette. Although Crusoe apparently didn't know it, and Walker didn't tell him the whole story, Walker did tell him that Ford's designers had already finished the design of a Ford Sports Car. As soon as he got the chance, Walker called Bob Maguire back at the Design Department and told him "for God's sake, get that car finished! I just told Crusoe we have the car finished back in styling!" Boyer remembers being told of Walker's call by a worried Maguire. Boyer also recalls that the call was made in January 1953, just before work started up again on the Ford Sports Car.

The next few weeks after Walker's call were hectic at the Design Department. Hershey called Woods, Boyer, Samsen and Kornmiller together, told them that the Ford Sports Car program had been revived, and that they had to finish it in a hurry. The clay model was retrieved from the basement, and Boyer, Samsen and Kornmiller started work on it again.

Word in the Design Department was that Henry Ford II had decided that no matter what the economics were, if Chevrolet was going to build a sports car, Ford had to have one too. In January 1953, an EJN number was assigned to the project and, by mid-February 1953, the Product Planning Committee had formally approved building the Ford Sports Car. To save costs, in January 1953, it was also decided that as many parts as possible, including the running gear, the headlight bezels, the taillights and finials above the taillights from the production 1955 Ford had to be used on the Ford Sports Car. Not only would it save money, the '55 Ford passenger cars could then be advertised as having sports car styling.

After the Ford Sports Car project was revived, Samsen was assigned to refine the design of the front of the clay model and Boyer the rear. Kornmiller worked with both Samsen and Boyer and prepared additional full-sized renderings. Boyer and Samsen collaborated on the sides of the car. The hopup over the back wheels was ordered removed from one side of the car, so Samsen and Boyer took a long stick, put it along one side of the car and scribed a straight knife line. Using that line, the clay modelers took off the blister to create a straight-through fender line on one side, while leaving the other side with the blister.

At this point, Boyer and Samsen became directly involved in the clay modeling process working on the clay right along with the clay modelers. After incorporating the 1955 Ford's round taillights onto the Ford Sports Car, Boyer carried the character line forward along the sides of the car. According to Samsen, he and Boyer came up with the idea of angling the character line upwards to end it on the front fender, and adding louvers just behind its termination. Samsen proposed real louvers but, after the car was moved to the main Ford exterior studio for ornamentation, diecast hash marks were added by Dave Ash as a cost savings measure. (Some reports credit Joe Oros with suggesting the upturned character line and the louvers.)

The grille Samsen designed extended below his split-bumper design. Engineering disapproved the split bumper, so Samsen connected the round bumperettes he had designed at the inside ends of the bumper with a straight bumper section. Engineering then decided that the bumper had to be redesigned and lowered, so that there was no sheet metal showing below it. (It wasn't until the Mustang that Engineering okayed a car with high bumpers, and then it was required that the exposed sheetmetal below the bumpers be .090 instead of .040 thick sheetmetal.) Because the low bumper made the grille vulnerable, Samsen went to a Mercury dealer, bought the bumperettes used on the 1953 Mercury, had the metal shop cut them down, put lights in them and installed them above the bumper, so they looked like they were coming out of the grille. According to Samsen, Boyer bought the same '53 Mercury bumperettes, also had them cut down in the metal shop and installed them on the back bumper. The exhaust exits were then routed through the middle of the back bumperettes.

After the car was moved to the Ford exterior studio, the Ferrari-like egg-crate grille Samsen designed was changed to a less-expensive stamped grille. For cost reasons, the front bumperettes were also redesigned with spinners and without lights. The front and the back bumpers were eventually redesigned, so they had the same bow and peak, and thus could be made with the same tooling.

Originally, a unique instrument panel cluster was proposed for the Ford Sports Car, but the cost of tooling it for just the Ford Sports Car proved too expensive. Instead, the production 1955 Ford astra-dial instrument panel cluster, designed by John Najjar, was substituted and used on the 1955 Thunderbird. The instrument panel cluster originally designed for the '55 Thunderbird was later used on the production 1956 Ford and then became available for use on the 1957 Thunderbird.

According to Hershey, after the full-sized clay had been moved to the Ford exterior studio, he and Oros agreed that Oros should hurriedly build a second full-sized clay of the Ford Sports Car, so there would be an alternative clay for the product planning people to

J.R. Samsen

John R. "Dick" Samsen

John Samsen (then known as Dick Samsen) graduated from Purdue University in 1950 with a degree in aeronautical engineering. His first job was at McDonald Aviation in St. Louis, Missouri working on jet fighters. After two years at McDonald, Samsen decided he liked the aesthetic end of design rather than the engineering part of it. He left McDonald and took a job as an engineer at Studebaker working on jet-engine design. At the same time, he learned to design automobiles from Bob Bourke, who was in charge of the Loewy-Studebaker design studio located in the same building.

After nine months at Studebaker, Samsen's job ended, and he moved to Dearborn, where he was hired by Frank Hershey as a Ford designer. Because of his engineering background, Samsen was assigned to Damon Woods' body development studio. Several months later, he, Bill Boyer and Allan Kornmiller were assigned to work on the Ford Sports Car program. Samsen worked on the front end, grille, front fenders and hood design of the car that became the Thunderbird.

After the Thunderbird full-sized clay was moved out of the body development studio, Samsen was transferred to the special projects and international studios to work under Gil Spear. Samsen designed the exterior of the D-523 and the interior of the D-524. He was transferred to the Ford exterior studio to work on Ford station wagons, just as the clay models of the D-523 and D-524 were completed, and just before Spear left to set up Engineering's Special Vehicles Department design studio.

In November 1955, Virgil Exner, head of Chrysler design, offered Samsen a substantial increase in salary to leave Ford. Samsen helped design many of Chrysler's muscle cars. In 1976, Samsen left Chrysler to open his own independent design business. ■

Allen Kornmiller

Allan E. Kornmiller

Allan Kornmiller grew up in Cleveland, Ohio. From an early age, his mother encouraged him to develop his artistic talent. He attended the Cleveland Institute of Arts on a scholarship. At the end of his second year, one of Kornmiller's teachers saw his renderings and encouraged him to change his major from fine arts to industrial design. After graduation, he traveled the western United States for a year on a painting scholarship.

In 1951, he applied for jobs at both Ford and GM and was hired by Ford. His first assignment was in Gil Spear's Advanced Studio, where the Syrtis was being completed. Two months later, he was assigned to Damon Woods' Ford body development studio. For the first two weeks there, he worked on the design of a new model of the French Vedette. He and the other designers in Woods' body development studio were then told that they had been selected to design Ford's new sports car, which was eventually built as the Thunderbird.

Kornmiller worked on the design of the Thunderbird for approximately six months but, in late 1952, he left Ford to accept a job at American Motors. At American Motors, he helped design the 1955 Hudson, and then worked with Italian designer Pininfarina on a proposed Nash design. After 3 1/2 years at American Motors, Kornmiller moved to Chrysler, where he was assigned to one of their exterior design studios. He became a design manager in 1964 and was promoted to chief of exterior design in 1968. While at Chrysler, Kornmiller worked on the 1965 Chrysler, several Imperials and the Cordova. When Richard Macadams was appointed head of Chrysler design, Kornmiller became a design manager. He retired from Chrysler in April 1987. ■

Ford Photomedia

(above) Dick Samsen with the full-sized plaster and wood model of the Ford Sports Car, just before it was named the Thunderbird.

Bill Boyer

(above) Bill Boyer prepared this rendering of a racing version of the yet unnamed Ford Sports Car in June 1953. At the time, there were no official plans to race the car, but Boyer thought it only a matter of time before someone decided to do so.

Allen Kornmiller

This rendering of the Ford Sports Car was prepared by Allen Kornmiller in 1952. It incorporated many of the design cues used on the Thunderbird, including the front fender and hood lines, the shape of the grille, and the positioning of the headlights and parking lights. The hopup in the belt line was not eliminated until quite late in the design process.

consider. According to Hershey, Ford studio designers were also asked to make full-sized renderings of variations of the proposed Ford Sports Car. When completed, five full-sized renderings were mounted on plywood, so that it would look like alternative designs had been considered before selecting the two designs reflected in the Boyer-Samsen-Kornmiller and the Oros full-sized clays. Hershey claimed that Walker favored Oros' clay and recommended it to Crusoe, but that Chase Morsey, head of corporate planning, went to bat for the Boyer-Samsen-Kornmiller design and eventually sold both Crusoe and McNamara on it. Oros does not recall building a full-sized clay alternative of the Ford Sports Car, and believes that Hershey's memory is incorrect. Other designers and photographs confirm the full-sized cut-outs but not a second full-sized clay model

During construction of the full-sized clay model in the body development studio, plug points were placed inside the clay model to let the designers know where the top of the engine was supposed to be. In the design process, they kept getting hammered down, so the hood could be made lower. At one point, Engineering agreed to come up with a special depressed intake manifold to accommodate the low hood on the Ford Sports Car. Unfortunately, the prototype performed poorly. The volumetric efficiency was bad, causing the test car to develop acceleration problems and a rough idle, so they decided to go back to the original intake manifold and carburetor, which then stuck up above the original plug points in the clay. According to Boyer, once the full-sized clay model was moved into the main Ford Studio, David Ash remembered the power dome he had designed for the X-100 and suggested it be used on the Ford Sports Car. Ash has generally been credited with the power dome used on the Ford Sports Car but, as Samsen points out, his original 1952 renderings include both a shaker-style hood and the power dome. Because no one liked his shaker hood proposal, he changed it to the power dome. The clay model of the Ford's Sports Car was completed on May 1, 1953 and, by May 18, Crusoe made the formal recommendation that production proceed.

Hundreds of names were considered for the Ford Sports Car, as it was still being called. Most of the renderings had the Ford crest on them, because no one had yet come up with a name for the car; names most often used on renderings by the designers were Sportsman and Sportliner. Crusoe proposed the name Savile, after the Savile Row street of haberdasheries in London. No one else liked that name and, eventually, Crusoe dropped it. Crusoe also proposed the name El Tigre. David Ash proposed the name Denab. Someone else proposed the name Cougar, but that name was rejected. Other names in serious contention were Coronado and Lightening Rod, which was proposed by Ken Nelson.

Giberson, who had once worked in New Mexico and was familiar with southwest Indian legends, proposed the name Thunderbird. Giberson had a coffee cup he brought back from New Mexico with a two-headed Thunderbird logo on it. He copied the original Thunderbird logo from that coffee cup. At another designer's suggestion, Giberson redesigned the logo so that it had only one head, but it too had to be further redesigned, because it came out looking too much like Roosevelt's National Recovery Act eagle. Giberson then made a turquoise and silver crest that had the Thunderbird script on it and presented it to Boyer, who had copies of it made in the metal shop. Boyer then attached them to the car. Other designers, including Hershey, Woods and Samsen were enthusiastic about the name Thunderbird, but management still couldn't make up its mind. At one point, the full-sized plaster model of the Thunderbird was hauled over to the Administration Building for a special showing to Henry Ford II, Ernest Breech and Lewis Crusoe. When it left the Design Department, it had the Thunderbird script on it; when it returned, the script was gone, and the designers were told Henry Ford II didn't like that name. Crusoe, who was getting desperate to find a name, offered a $250 suit to whoever came up with an acceptable name for the car. Bill Ford later told Boyer that the Executive Committee spent hours agonizing over a name for the Ford Sports Car. It was not named until February 15, 1954 -- just days before its press preview. No one could come up with another satisfactory name, so the name Thunderbird won by default. Giberson got a $95 suit and a $45 pair of slacks. He couldn't find a $250 suit he felt comfortable wearing.

The Thunderbird was originally meant to come with only a convertible top. Late in the project, it was decided to also offer a removable hardtop. Boyer and Samsen designed several proposed hardtops, most of which were rounded in what they considered to be true sports-car fashion. One day Joe Oros brought in sketch of a squared-off top for the car. Oros' sketch was clay modeled on an armature by Boyer and Samsen and accepted as the Thunderbird's removable hardtop. Only later did Boyer and Samsen learn that Oros' hardtop copied the design of the Mark II. Chase Morsey had suggested that Walker

Alex Musichuck

Alexander Musichuk

Alex Musichuk grew up in Youngstown Ohio. He served in the Army during World War II. After the war, he attended the Cleveland Institute of Arts, finishing in three years by attending classes year round. In his last term at school, he was hired as a designer at Ford. Musichuk was assigned to the interior studio, and after it was divided into the Ford and Lincoln-Mercury interior studios and then further divided into the soft trim and instrument panels, he was placed in charge of the studio where Ford soft trim was designed.

In 1952, Musichuk designed the special gold, black and white interior for the 1953 Indianapolis Pace Car (Golden Jubilee). He also worked with Ted Hobbs and Andy Olinik of the trim and color department to pick the materials and colors used on that car.

In 1952, in addition to his regular production work, Musichuk was assigned to work with Gordon Buehrig on the interior soft trim for the Ford Sports Car project that eventually became the Thunderbird. Buehrig, who by that time was a Lincoln-Mercury body engineer, established the package, and Musichuk began the design of the interior soft trim. In 1953, before the Thunderbird interior was completed, Musichuk left to take a job as a designer at Nash-Kelvinator.

After four months at Nash-Kelvinator, Musichuk accepted a job at Smith and Shearer, a consulting industrial design firm in Akron Ohio. While there, he taught part-time at the Cleveland Institute of Arts.

In 1955, Gene Bordinat asked Musichuk to return to Ford as head of the Edsel interior studio. At Bordinat's request, Musichuk came to Dearborn to consider the offer but decided not to accept the position. Later, he became head of MTD products design department. He retired from MTD in 1986.

Musichuk is the founder and the current director of the non-profit Ohio Boys Choir. ■

Walt Amrozi

Walter J. Amrozi

Walt Amrozi was born in Chicago, but his family moved to Detroit when he was 12 years old. Beginning at age 11, Amrozi was constantly drawing cars. After he graduated from high school, he worked as a sailor on a Great Lakes freighter. In 1936, he was hired at Ford's Rouge plant as an apprentice tool and diemaker. He worked as a tool and diemaker until 1950, when he applied for and was accepted in Bill Leverenz' first clay modeling class. After approximately six months of evening classes, Amrozi was hired by the Design Department as a clay modeler.

Throughout his career at the Design Department, Amrozi regularly worked more overtime than most other clay modelers, who generally worked 68-hour weeks. Although Amrozi was assigned to various production studios, he and Jim Doyle were often included on the clay modeling teams assigned to model the concept and showcars. Amrozi helped clay model the FX-Atmos, the Thunderbird, the Futura, the Quicksilver and the Mustang among others.

Amrozi worked in most production studios at the Design Department. Before he retired in 1970, he became a modeling supervisor in the Ford studio. ■

take a look at the roof line of the proposed Continental Mark II. Walker did so and, when he returned, he suggested that the Ford Sports Car use the Continental Mark II's blind C-pillar on the removable hardtop. Walker justified borrowing a design cue from the yet to be introduced Continental Mark II, because he felt it emphasized continuity between Ford Motor Co. products. Bill Ford, who was head of Continental Division, wasn't so charitable.

The Thunderbird and the Mercury Monterey XM-800 were introduced at the Detroit Auto Show, February 24, 1954. The Thunderbird shown at the Detroit Auto Show was not a prototype as the public assumed. Although it looked like the real thing, it was made of plaster and wood and painted a turquoise color. It also had 1955 production Ford wheel-covers on it.

Several changes were made to the Thunderbird just before the prototype was built. A floor-shifter was suggested to make it more like a true sports car. By working all night, the engineers at Experimental Engineering were able to adapt a floor-shifter from a Triumph sports car. The next morning they took the floor-shifting prototype Thunderbird back to the Design Department, and showed everyone the floor-shifter they had cobbled. It was a big hit, and all two-passenger Thunderbirds were produced with floor-shifters.

In late summer 1954, while Hershey was on vacation, Walker convinced Crusoe the Thunderbird needed more kinship with the 1955 Ford line. Walker thought they could get it by adding the Fairlane stripes from the top-of-the-line production '55 Ford to the sides of the Thunderbird. (Publicity photographs, taken August 20, 1954, show Fairlane stripes on a light-colored Thunderbird. Crusoe later had Fairlane side moldings installed on his black Thunderbird prototype.) When Hershey returned from vacation, he exploded when he saw side moldings on *his* Thunderbird. When he found out that the publicity brochures had already been printed using a Thunderbird sporting the side molding, Hershey accused Walker of interfering with his design and of trying to create more work for Trim Trends, a business owned by Walker and his son. (Trim Trends made most of Ford's pressed moldings including the Fairlane side molding.) A very angry Hershey went to MacPherson, who went to Henry Ford II, who concurred with Hershey's very strongly expressed belief that Fairlane stripes did not belong on a sports car. The Fairlane stripes came off, and new brochures were printed.

Budd Manufacturing of Philadelphia was selected to build the bodies for the Thunderbird. Budd was looking for work at the time, and the engineering workload involved in producing the 1955 production Ford precluded building another body at Ford's facilities. Completed bodies were delivered in white to a second-floor production line at the Rouge plant, where they were finished and then dropped through a hole in the floor to the '55 Ford production chassis line on the first floor, where the Thunderbird was mixed in. In September 1954, the Thunderbird went on sale. It sold for $400 less than the 674 Corvettes sold that year. In 1955, Ford sold 16,155 Thunderbirds.

Initial press releases reported that the Thunderbird would have a power convertible top. It came with a manual top. There was also some confusion about the removable hardtop. Press releases said it would be steel. It was made from fiberglass. Press releases also said the Thunderbird would have an elapsed-time clock. It came with a standard 12-hour clock, but it was not an elapsed-time clock.

The interior of the wood and plaster model of the Thunderbird at the Detroit Auto Show, February 24, 1954.

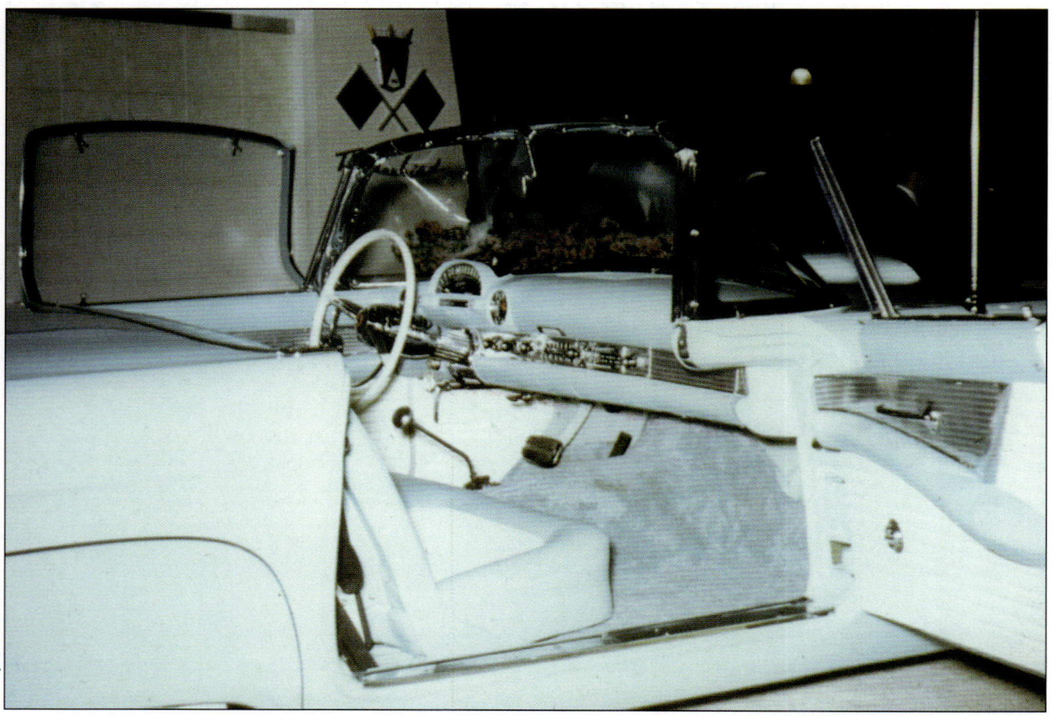

J.R. Samsen

Originally, a hardtop with quarter windows was proposed, but none of the designs Boyer and Samsen came up with for a removable hardtop looked right with quarter windows, so it was built without windows. Crusoe was enthusiastic about the Thunderbird and drove one back and forth to work daily. On his way to work he had to drive through an oblique intersection located at Silvery Lane and Outer Drive. Every time he approached that intersection, he was visibly reminded that he couldn't see traffic through the blind quarter of the removable hardtop. Crusoe insisted that windows of some sort be included on the removable hardtop. Boyer and Samsen spent days looking for window ideas, but none of them had the right look. One day, when looking through a book on old-English carriages, they noticed a picture of a carriage with a porthole-type window in it called the booby hatch. They copied it for the Thunderbird, and it became optional on all removable hardtops starting in 1956.

Crusoe also had a continental kit installed on his prototype Thunderbird. Most designers thought the splash pan on Crusoe's continental kit was overdone and a bit garish, but Crusoe liked the looks of the Thunderbird with it. A continental kit had originally been suggested as a way to create a location for a spare tire that took up almost the entire trunk in the 1955 Thunderbird, but it didn't make it into production until 1956. Crossed flags were also used on the '55 Thunderbird, but they were not carried over on either the 1956 or '57 Thunderbirds. In late 1955, Corvette started using crossed flags in its advertising. According to Boyer, that was followed by a directive from Ford management to discontinue their use on the Thunderbird.

Frank Hershey, as head of the Ford studio, is entitled to much of the credit for the design of the Thunderbird. He is the one who approved and directed the design process. Damon Woods, who was in charge of the day-to-day design work on the Thunderbird, also deserves credit but, if there was a disagreement, it was usually Hershey who prevailed. Boyer, Samsen and Kornmiller are the actual designers responsible for the aesthetic look of the finished product -- the Thunderbird looks like it does because of their efforts. Boyer, Samsen and Kornmiller did the initial sketches, the full-sized renderings and supervised the clay modeling of the first full-sized clay of the Thunderbird. Boyer, Samsen and Kornmiller each acknowledge that the car's package was set by Hershey, and that Hershey and Woods approved what they designed. There are many others who are also entitled to credit for the Thunderbird including the clay modelers, interior designers Alex Musichuk and Gib Giberson, and Frank Pinkham, who made sure that what Boyer, Samsen and Kornmiller designed and Hershey and Woods approved would work from an engineering standpoint.

As a result of the run-in between Walker and Hershey over the Fairlane stripes on the Thunderbird, Ford's management became even more convinced that the decentralized structure in place at the Design Department wasn't working, and that one strong overall boss was required. Walker also started looking for a way to get rid of Hershey, who was seen as an excellent designer but a bit of a maverick. Whatever Walker found resulted in Hershey receiving a summons to McNamara's office. According to Hershey's account of the meeting, McNamara suggested that he might be better suited to running "an antique shop somewhere." According to Hershey, after he compared McNamara to a certain part of a horse's anatomy, he quit. Hershey's departure also coincided with Walker's appoint-

Ford Design Center

Leonard Stobar

Leonard Stobar was the younger brother of Charlie Stobar. Both brothers grew up working in their father's pattern shop. As a teenager, Charlie taught Leonard to clay model. Leonard followed his brother to GM's Styling Section as a clay modeler in 1936. During World War II, both brothers were assigned to GM's Eastern Aircraft, where they worked on the GBF dive bomber and the Grumman Hellcat. Before the war was over, both were also working as part-time clay modelers for George Walker in his independent design firm.

Leonard was offered a job at the Ford Design Department and left GM in 1947, about eight months after Charlie started at Ford. In 1953, Leonard became clay modeling supervisor in the Ford studio. Towards the end of his career as a clay modeler, Leonard was transferred to the Ford interior studio.

Leonard worked on many concept and showcars, including the X-100, the 1955 Thunderbird, the Futura and the LaGalaxie. In 1970, he was offered and took early retirement. Starting in 1955, and with Henry Ford II's permission, Charlie and Leonard started a side business making small promotional models of Ford cars and trucks. That business led to contracts with MPC and AMT making prototype models of Fords, Lincolns and Mercurys that became the basis for their plastic promotional models. ■

Henry Ford II (left foreground) and Bill Ford (right foreground) admire the Ford Sports Car at the Styling Rotunda shortly before the announcement was made that the car would be called the Thunderbird.

Henry Ford Museum and Greenfield Village

Henry Ford Museum and Greenfield Village

Ford Division head Lewis Crusoe had one set of Fairlane stripes saved and installed on his personal Thunderbird. He also had a continental kit fabricated and installed on his Thunderbird in an effort to gain trunk space.

Alden "Gib" Giberson

Ford Design Center

An interest in airplanes led Gib Giberson to study aeronautical engineering at Detroit's Cass Technical High School. After service in the Army Air Corps during World War II, Giberson studied architecture at the University of Michigan and then worked at the Ross Roy Advertising Agency, where his father, an architect, was employed designing instruments.

In 1947, his father was hired as a town planner in Los Alamos, New Mexico, and Giberson followed. At that time, Los Alamos was a government town, where atomic and hydrogen bombs were being developed. While in New Mexico,. Giberson worked as an architectural designer and illustrator. In 1952, he returned to Dearborn, married, went back to work at his father's old firm and took art courses at night at Meinzinger's Academy and the Allen Air Brush Academy. A Ford Design Department employee saw Giberson's airbrush work and suggested that he submit his portfolio to Ford. He did and was hired by Frank Hershey in June 1952 as an entry level designer, at a salary of $325 a month.

After working on the 1954 Ford glasstop and the 1955 and '56 Fords, Giberson was transferred to the body development studio, where the 1955 Thunderbird was being designed. Giberson worked on the interior of the first Thunderbird and also gave the Thunderbird its name.

He later worked as a manager in Alex Tremulis' Advanced Studio, and then worked as a manager in the Ford, advanced and interior studios. In 1960, he was assigned to Ford of England as a design executive.

He helped design the Volante, LaGalaxie, Gyron and Technia concept cars and the 1971 Mustang, the Granada, Escort and Topaz production cars before retiring in 1981. ■

ment as a Ford vice president and general manager of the Design Department. When Hershey left, he took a job at Kaiser Aluminum and asked Boyer to come with him. Boyer caught an all-night flight to California, took a look, thought about it and decided to stay at Ford. Kornmiller left Ford for a job at Chrysler, just as the Ford Sports Car was moved into the Ford exterior studio. Samsen also left for a job at Chrysler, just as the Thunderbird was being finalized in the Ford exterior studio.

In a Society of Automotive Engineers paper Hershey delivered in late 1954, he traced the history of the American sports car, and the Thunderbird in particular, from the Model T Ford speedster and the Stutz Bearcat to the resurgence of interest in sports cars brought back to this country by G.I.s returning from Europe after World War II. Hershey singled out the Mercer Raceabout as the first example of a refined American sports car like the Thunderbird. Although his SAE presentation only alluded to it in passing, the Thunderbird owes a lot less to the Jaguar, MG and other true sports cars, and more than anyone has yet been willing to acknowledge to the hot-rod culture of Hershey's California, where chopped and channeled custom cars were built by the thousands. European sports cars and the Corvette may have influenced the Thunderbird but, it was, in every sense of the word, a chopped and channeled American original in the style of the first Continental.

Even though he thought the Thunderbird should have been designed as a true sportscar in the tradition of the Jaguar XK-120 or his MG, Samsen placed an order for one, even before the dealer knew what it was. After educating the dealer, Samsen made arrangements to get the first Thunderbird off the production line. When his Thunderbird came in, Samsen was already at Chrysler. The dealer told him that his car was the number two production car, because Bill Ford had asked for the first production Thunderbird. Samsen was also told he would have to pay more for the car than originally quoted, because he was no longer a Ford employee and was not entitled to an employee discount. Samsen has always regretted that he did not buy the car anyway. ■

Collectible Automobile

The Premiere predicted the 1955 production Lincoln. The rear half of the roof was covered with white vinyl.

Premiere

The 1953 trim and color cars shown during the Golden Anniversary were such a hit, especially with the dealer network, that Benson Ford's Lincoln-Mercury Division decided to develop new ones for the 1954 show circuit. Because there were three full-size concept cars built for the 1954 show circuit, the FX-Atmos, Thunderbird and the Mercury Monterey XM-800, there was very little money left in anyone's budget to work with. The Lincoln-Mercury Studio also had three Lincolns and four Mercurys altered for the show circuit. Except for the Lincoln Premiere, these 1954 showcars were only color and trim variations of their production counterparts.

The 1954 Lincoln Premiere, which was actually a prototype of the 1955 Lincoln body, was given the top-of-the-line name Lincoln would adopt in 1956. The Premiere had a half-Plexiglas roof over the front-seat area, just like the new 1954 Ford Skyliner and the Mercury Sun Valley. It was planned to offer a Lincoln model in 1956 with a Plexiglas half roof, but it never made it into production, because the Skyliner and Sun Valley did not sell in the numbers initially anticipated and, by 1955, it was felt that a Lincoln glasstop would only dilute the Ford and Mercury glasstop market.

The Premiere showcar followed a theme of a party-goer or firstnighter and, at one time, it was also referred to by the names Firstnighter and Boulevardier.

The exterior of the Premiere was designed by Bill Schmidt, and the interior was primarily designed by Don Stumpf. The body was painted black and it had a padded off-white vinyl top. As on the Skyliner and the Sun Valley, the glasstop on the Premiere had a green tint to it. The seat cushions were upholstered in worsted broadcloth with gray, black and white stripes and had a wide white leather boarder around the edges. The headliner on the Premiere was finished in white cloth, and the floor had thick black carpets. In a design cue not followed on any production car, the white leather in the middle of the door panels wrapped around the door jams. Interior moldings were done in black leather and chrome. On the inside, at the top of the rear window and down each side to the package-tray area, was an arch of small lights that lit up when the doors were opened, or the inside lights were turned on.

As with the interior design of the Premiere, the 1954 Lincoln and Mercury trim and color showcars were a joint effort of many of the designers assigned to the Lincoln and Mercury interior studios. On some cars it's possible to single out a designer most responsible for a particular car but, in general, the Lincoln trim and color cars produced in 1954 were designed by Lincoln-Mercury interior stylists Don Beyreis, Art Querfeld, Art Tarabusi, Art Miller, Don Stumpf, Jake Aldrich, Vern Breitmeyer, Regg Bennett and Stan Thorwaldsen, and the Mercurys were designed by Lincoln-Mercury interior stylists Dave Ash, Ernie Gregory, John Middlestead, Earl Bunge, Bob Stoffel and maybe others. ■

Don Stumpf

Donald A. Stumpf

Don Stumpf practically grew up at his father's upstate New York Ford dealership. As a child, he entered the Fisher Auto Body Craftsman League design contest and, on his second try, his car won. That encouraged him to attend the Art Center College in Los Angeles, California.

After he graduated in 1951, Stumpf was hired at the Ford Design Department. For the first year, he was assigned to the Advanced Studio working under Gil Spear. During his time in the Advanced Studio, Stumpf helped design the instrument panel eventually used on at least one production Lincoln.

In 1952, Stumpf was transferred to Don Beyreis' Lincoln interior studio, where he helped design the 1953 Golden Anniversary Lincolns and the 1954 trim and color Lincoln showcars. Stumpf worked on the Premiere showcar and is credited with coming up with the name and designing the arch of interior lights used inside the car. Stumpf is also credited with the design of the steering wheel on the 1956 Lincoln, and coming up with the name for the light-purple exterior color called Wisteria also used on the '56 Lincoln.

In 1957, Stumpf left Ford to take a job in the interior studio at American Motors, but soon left American Motors to take a job at Chrysler. While at Chrysler, he spent a year in France working on the redesign of the Simca. Stumpf retired from Chrysler in 1969. ■

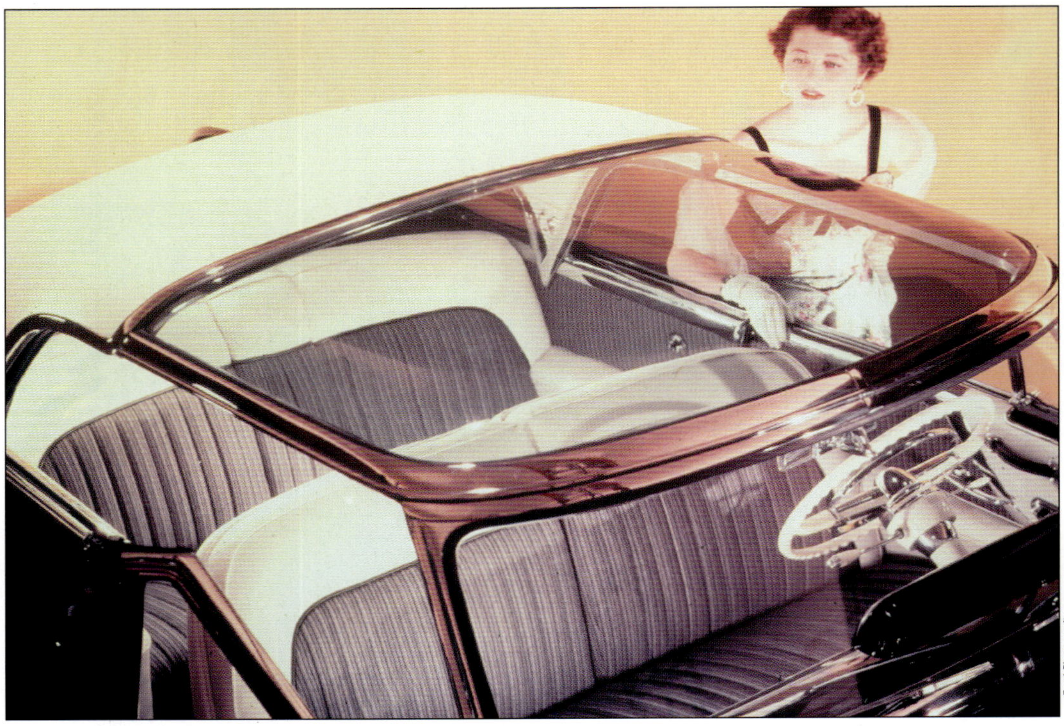

Ford Photomedia

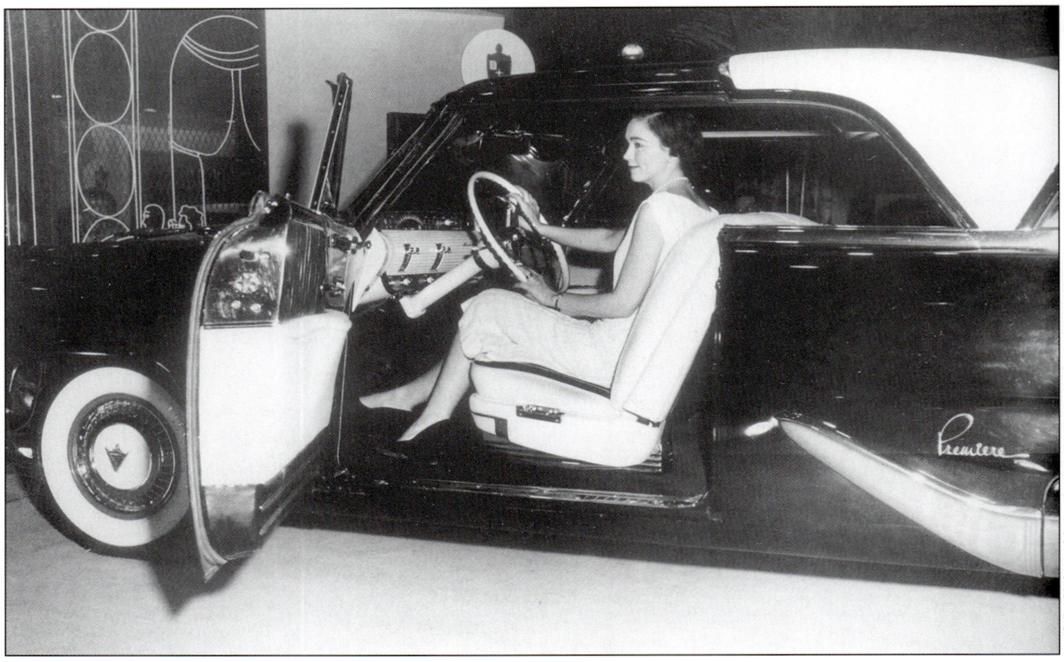

Collectible Automobile

Henry Ford Museum and Greenfield Village

(top) The interior of the Premiere featured an arch of small lights above the backlight and contrasting striped seat inserts.

(middle) The Premiere was originally planned as a production Lincoln glasstop. Poor sales of Ford and Mercury glasstops discouraged production of a similar Lincoln model.

(bottom) This early rendering of a proposed 1956 Lincoln featured a glasstop model.

Ford Photomedia

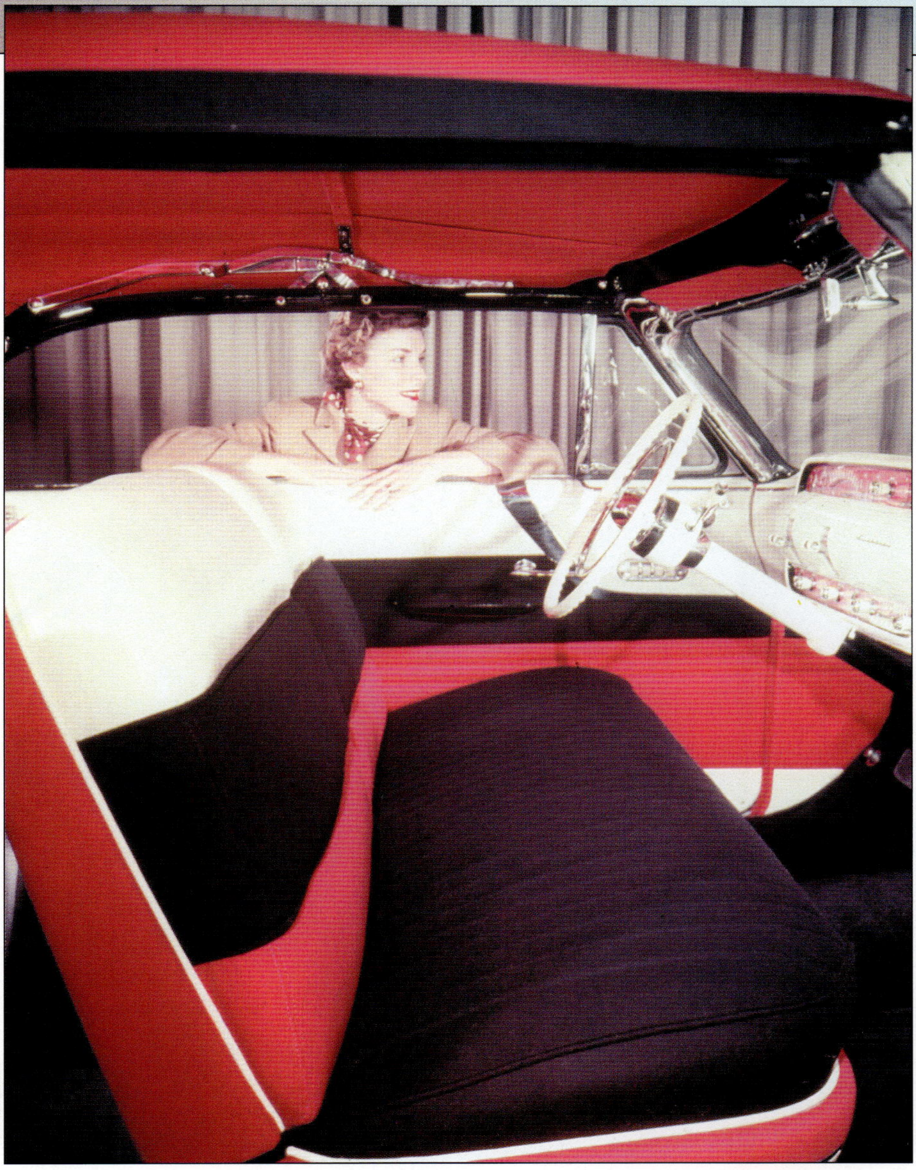

Ford Photomedia

(above and left) The Lincoln Huntsman was a 1954 Lincoln Capri convertible painted white with a red fabric top and special red, white and black upholstery.

Huntsman

The Huntsman was an otherwise stock-bodied 1954 Lincoln Capri convertible, that was painted ermine white and had a red convertible top with black binding. The inside of the Huntsman was upholstered in ermine white and black and red broadcloth and was painted to match. The floors were covered with a new synthetic uncut loop black-colored material called Fibre E. The trim and colors of the Huntsman and its name are credited to designer Stan Thorwaldsen. ■

The Lincoln Mardi Gras at the 1954 Chicago Auto Show.

Collectible Automobile

Mardi Gras

Although they were not as well publicized, there were several other trim and color showcars produced in 1954. One of them was a 1954 Lincoln Capri two-door hardtop called the Mardi Gras. The Mardi Gras was shown at the 1954 Chicago Auto Show, but there is little evidence it was shown anywhere else. The Mardi Gras was painted pearlescent white with a gold top, and it had a white, light purple, green and gold interior. A patterned light purple and white brocaded insert was also used on the door panels. The selection of the colors and trim for the Mardi Gras is attributed to designer Art Miller. ■

Vern Breitmeyer's rendering of the Mardi Gras.

Dave Beyreis

Resorter

The Resorter was a 1954 Lincoln Capri four-door sedan, that had a black body with an off-white convertible top-type material on the roof. The interior was done in black and white leather in a pattern similar to that used on the Huntsman. The carpet in the Resorter was very thick off-white plush. Like the other Lincoln and Mercury trim and color show cars for 1954, the brake arm, brake pedal and accelerator were color-coded to match the carpeting. The Resorter was displayed at the 1954 Chicago Auto Show and possibly at other auto shows that year. The Resorter was displayed in front of a mural that had a Southern California or Palm Springs theme. The selection of colors and trim used on the Resorter is attributed to Don Beyreis. ■

Ford Photomedia

(left) The interior of the Lincoln Resorter featured leather seats and plush off-white carpeting.

(below) Palm Springs was the theme of the Resorter.

Dave Beyreis

Carnival

The Carnival was a 1954 Mercury Monterey two-door hardtop with a circus theme. The body was painted Yosemite Yellow, and it had a Siren Red roof. On the inside, the seat cushions and backs were upholstered in Siren Red vinyl with confetti dots on the door panels and on portions of the seats. The headliner and steering wheel were Yosemite Yellow, and the rugs, instrument panel and steering column were Siren Red.

Although Yosemite Yellow and Siren Red were standard colors offered on 1954 Mercurys, Siren Red was only offered on convertibles, and was not a part of the many two-toned color schemes '54 Mercurys normally came in.

There were two other trim and color 1954 Mercurys created for the show circuit. Recollections are in conflict about the colors and upholstery used on the Mercury Fantasy and the Mercury Debonair, but they were also displayed at the 1954 Chicago Auto Show and possibly elsewhere. They are described as only slightly different from their production counterparts. No photographs of either car have yet been located. ■

Will Fox

(left and below) The Mercury Carnival featured red and yellow paint with confetti-dot seats and door panel inserts.

Henry Ford Museum and Greenfield Village

Stan Thorwaldsen

Designers, modelers and studio engineers from the Lincoln-Mercury interior studios in spring 1951. Standing, from left to right are: Art Tarabusi, Bill Leverenz, Dave Ash, Alex Musichuk, Art Querfeld, Riley Quarles, Bob Stoffel, Ernie Collette, John Middlestead, Vern Breitmeyer, Regg Bennett, Don Beyreis and George Spiroff. In the front row, from left to right are: Ernie Gregory, Dave Barr, Bob Tarran, Don Hillier, Earl Bunge, Dick Hill, John Grebe and Art Miller. (This photograph was taken behind the Dearborn Inn.)

Reginald O. Bennett

Regg Bennett was a true son of the west, who was born and raised in Oklahoma. He was known as more of a fine arts painter than an automobile designer. In his early years, he studied art under Charles Russell and Frederic Remington.

In 1935-36, Bennett was employed at GM to run Harley Earl's apprentice design school, which he did for many years. While at GM, Bennett was also responsible for selecting the themes, preparing the decor, painting the murals and planning many of Earl's annual Styling Section rodeos.

Bennett came to Ford in 1946-47. He worked on the 1953 and '54 Lincoln and Mercury trim and color cars and, in the mid-1950s, he was assigned teaching duties in Alex Tremulis' Advanced Studio. He retired in the early 1960s. ■

Chrissie Tremulis

Jake Aldrich at his drawing board.

Henry Ford Museum and Greenfield Village

John G. (Jake) Aldrich

Jake Aldrich grew up on Long Island, New York. His first job in the automobile industry was as an illustrator for Brewster and Co. In 1931, he was hired as a designer in GM's Art and Color Section. (He later told others that one of the reasons he was hired at GM was because he could draw wheels and tires in perspective.) While at GM, Aldrich worked in the Oldsmobile studio. Shortly before World War II, Aldrich left GM for a position as a designer with Hudson. During the war, he worked for Raymond Loewy and Associates designing Studebakers in South Bend, Indiana.

In 1946, Aldrich was hired at the Ford Design Department, where he worked in the Lincoln production and, later, in the Mercury pre-production studios, before he retired in 1960. Aldrich was one of the designers of the XL-500.

During his design career, Aldrich was known by the nickname Tiger, probably because he owned and drove exotic vehicles. At his retirement, Aldrich acknowledged having owned 88 foreign and one domestic car during his life. While at Ford, Aldrich drove a Citroen and a Mercedes 300 SL Gullwing. Aldrich was also one of the founders of the Ford Motorsports Club. He was quiet but was recognized for his prolific design ideas, which he kept on 3"x5" index cards. He was also an accomplished watercolor artist and an avid photographer. ■

Ford Photomedia

The iridescent color of the Futura did not photograph well. In most pictures, it appears either white or a green–blue.

Futura

The Futura was the brainchild of Bill Schmidt. As Lincoln's head stylist from 1947 to 1955, Schmidt was responsible for the update of the 1950 and '51 Lincolns, the design of the 1952 Lincoln, the 1953 Lincoln XL-500 concept car, the 1956 Lincoln and the 1956 Lincoln Dual Cowl Phaeton concept car. Gene Bordinat was placed in charge of the Lincoln-Mercury studio in January 1954, but Schmidt retained the position of Lincoln chief designer. Schmidt was a talented and progressive stylist, who got his start at the Henry Ford Trade School, worked on the B-24 bomber during World War II and, beginning in 1944, was assigned to Bob Gregorie's Design Department. Schmidt's first assignment was in Gregorie's small advanced studio, which was run by Eric Ramstrum. Three years later, Schmidt was placed in charge of the Lincoln studio.

The Lincoln-Mercury public relations department claimed that Schmidt got the idea for the Futura as a result of a 1952 diving encounter with a shark, while he was vacationing in the Bahamas. That's true, but there's more to the story. In 1952, during a cold Michigan winter, Bill Schmidt and Bill Mitchell vacationed together in the Bahamas. Mitchell worked under Harley Earl as a GM stylist and, when Earl retired in 1958, he became head of the GM Styling Section. Both Schmidt and Mitchell were fascinated by the tropical sea life they saw and especially by the movement and the shimmer of the fish they observed under water. (Mitchell also found inspiration during that vacation, which in the early 1960s, resulted in the Corvette XP-755 Shark concept car and, in 1963, the production Corvette Stingray.)

Schmidt was responsible for the concept and initial design, and he also selected the name Futura from a list of type faces. He acknowledged that a Mako shark and a Manta Ray got some of the credit for the design of the Futura, but that, too, is only part of the story. The gills on the deck lid of the finished Futura are the most predominant feature that looks like it was copied from the anatomy of a shark. Interestingly enough, the original drawings of the Futura, done by Schmidt in 1952, have no gills or other features, except possibly the hooded headlights, that are attributable to a shark or any other sea creature. The gills first appear on the 3/8-sized clay model of the Futura, which was completed in 1953. Schmidt's original 1952 drawings of the Futura show a car with taillights much like those that appeared on the 1955 Lincoln and with rear bumper-exhaust exits much like those used on the 1956 Lincoln.

The predominate influence on the design of the Futura was modern jet aircraft. The twin canopies, the front and back fins, the grille, the air intakes at the front of the rear fins, the rear-end treatment, the interior and the ventilation system were all inspired by the aerodynamic, sophisticated jet fighters developed for the American military. Since other automobile manufacturers were also taking their inspiration from jet fighters, it may well be that Ford, by emphasizing the influence of a shark in the design of the Futura, was trying to be just a little different from GM or Chrysler who, at that time, were turning out lots of their own concept cars.

Benson Ford took great pride in the Futura, the first fully operational concept car from his Lincoln-Mercury Division.

Arby Schuman

Schmidt, who usually spent his spare time sketching, drew three or four 18" x 24" renderings showing his ideas for the Futura concept car. At that time, about 10% of the Design Department budget was available for discretionary projects, so Schmidt showed his design proposals for the Futura to Lincoln-Mercury Division head Benson Ford, who especially liked show and concept cars. Benson Ford asked how much it would cost to build a fully-operable car, and Schmidt estimated that it could probably be done for about $100,000 plus the man hours involved. General approval was given by Benson Ford, and the Futura became one of the discretionary projects. The Engineering Department also had other discretionary money in its budget for experimentation, and some of that was also earmarked for the Futura project. Because the Futura was planned as a full-sized concept car, it was necessary to have concurrence between styling and engineering that what the stylist wanted to build could actually be done in practice.

The team Schmidt assembled to work on the Futura included John Najjar, who was in charge of body development throughout the project up through the finished full-sized plaster model, Martin Regitko, who was in charge of engineering, Roy Brown, who supervised the design work, Ken Spencer, who was the designer in charge of the clay models, Stan Thorwaldsen, who did the interior design and instrument panel and Ron Perry, who designed the Plexiglas top.

The Futura was Brown's first project at the Ford Design Department, after he had been hired by Gene Bordinat. Brown was hired to design the Edsel, and the Futura was his first opportunity to demonstrate his design and leadership skills. Thorwaldsen recalls that most of the interior design innovations on the Futura, including the roll-top doors that closed-off the instruments, were inspired by nothing in particular, but were ideas he had been considering for some time.

In addition to the 3/8-sized clay, a full-sized clay, a full-sized plaster model and at least one partial clay buck of the interior of the Futura were made. Spencer recalls that while the full-sized clay model of the Futura was being completed late one evening, everyone got a little silly, and he ran over the top of the fresh clay on the Futura leaving rather deep shoe prints. Schmidt came in the next day, saw the hardened footprints and called a security officer, who came and made plaster casts of the footprints. Spencer says he was able to smooth things over with Schmidt before everyone in the studio was required to compare shoes with the plaster casts.

Henry Ford Museum and Greenfield Village

The Futura's speedometer, odometer and four idiot lights were in the center hub of the steering wheel. Turn-signal controls were hung from the steering-wheel spokes.

By late 1954, a full-sized plaster model of the Futura, a complete set of plans and blueprints, and a chassis and running gear were sent to Ghia in Turin, Italy to be used for construction of the fully-operable Futura. Although bids were obtained from other firms, Ghia was selected to build the Futura, because their bid of $75,000 for the finished product was the lowest one. Ghia's craftsmen were famous for their workmanship, and they preferred to work in metal rather than fiberglass. During construction, the only problem that developed was with the design of the wheelwell openings, which had to be slightly modified.

Stan Thorwaldsen

Stanley E. Thorwaldsen

Stan Thorwaldsen grew up in Brooklyn, New York and attended Pratt Institute. He graduated in 1949 and began looking for a job. After several months, he got a call from classmate Ken Spencer, who told him Ford was hiring designers. Thorwaldsen began his design career in the Lincoln-Mercury interior studio, under Don Beyreis. He remained in that studio until he left Ford with Bill Schmidt in 1955, to work as Schmidt's head of interior design at Packard.

When Thorwaldsen started at the Design Department, Schmidt was just finishing up the XL-500. Beyreis assigned Thorwaldsen to the team that designed the 1953 Golden Anniversary Lincolns, the 1954 Lincoln Premiere and the 1954 Lincoln Huntsman. Thorwaldsen also designed the interior of the Lincoln Futura. Although he spent his entire career at Ford in the Lincoln-Mercury interior studio, one of his drawings was used for the taillights on the production 1954 Mercury.

Thorwaldsen worked with Bill Schmidt for a total of thirty years. While at Packard, he designed the interior and Schmidt and Richard Teague designed the exterior of the Packard Predictor concept car. Thorwaldsen recalls that the Ford Styling Center loaned Packard its most experienced painter to touch up damage done when the Predictor was shipped from Ghia. The Predictor was painted with a fish-scale paint, similar to that on the Futura.

In 1989, Thorwaldsen left Schmidt's employment to become an industrial design teacher at Detroit's Center For Creative Studies. He retired in 1994. ■

The chassis used for the Futura was from an unused mechanical prototype built by Hess and Eisenhardt of Cincinnati, Ohio for the Continental Mark II project. John R. Hollowell, chief engineer for the Continental Division retractable project, arranged to sell the extra chassis, which was cobbled from a 1953 Lincoln, to Lincoln-Mercury Division for $17,000. The engine in the Futura was rated in press releases at 330 hp, and was "an advanced version of Lincoln's present overhead valve V-8." The Futura's engine had an unusual oil bath-type air cleaner that, because of the low profile of the car, was mounted to the side of the engine. The engine also had twin fans and modified carburetion. The chassis Hollowell sold Lincoln-Mercury Division for use on the Futura included stock 1953 Lincoln running gear. What was done to achieve a horsepower rating of 330 is unknown. The Futura was advertised as a "$250,000 laboratory on wheels," which probably represented the total cost, including all of the design and engineering man hours involved.

Because Italian carrozzerias had a well-justified reputation for doing things their own way, once they started a project, Don DeLaRossa was assigned responsibility for follow-up and reportedly visited the Ghia factory several times to make sure there were no surprises. Mid-construction, Schmidt also went to Italy to check on the progress of the Futura. At that time, the Ghia factory had an open courtyard with trees in the middle. Some of those trees had died or been cut down, and Schmidt noticed that the stumps had odd shapes. It seems that Ghia's craftsmen, used to doing things the way they were taught, were using the stumps to shape the Futura's metal panels before welding them together.

The Plexiglas canopy and the complicated interlock mechanism that operated it were constructed in-house at Ford, under the direction of Martin Regitko, and sent to Ghia for installation on the Futura.

For the color of the Futura, Schmidt had his heart set on the iridescent color of the fish he had seen in Caribbean waters, and he got it. The color of the car was referred to in press releases as "pearlescent frost blue-white." The finish had an iridescent quality which presented each of the viewer's eyes with a slightly different color. The conflict of color between both eyes causes what is known as retinal rivalry and creates an impossible third color in the mind of the viewer. The colors are also affected by the lighting being used. This is why color photographs of the Futura look flat and don't capture the iridescent shimmer, and it also accounts for the reason no two eyewitnesses seem to agree on what color the car really was. To get the iridescent color Schmidt wanted, Ghia ground and pulverized the scales from thousands of fish to mix in with the paint. In contrast, Schmidt's first renderings of the Futura are of a black car with red upholstery and a blue car with gold upholstery.

It only took three months for Ghia to complete the Futura and, in early January 1955, the finished Futura was delivered to Chicago, where it was unveiled at the 1955 Chicago Auto Show, January 8-16, 1955. The second showing of the Futura was at the Detroit Auto Show.

The third showing of the Futura was slated for the New York Auto Show. On March 3, just before the New York Auto Show opened, Benson Ford and Bill Schmidt took the Futura for a drive, starting at the United Nations and then through Central Park to the *Tavern On The Green*. The public was not told, however, that in this sojourn, the air-conditioning and ventilation system failed, turning the Plexiglas-covered, windowless passenger compartment into a sauna. For only a part of their drive through New York City

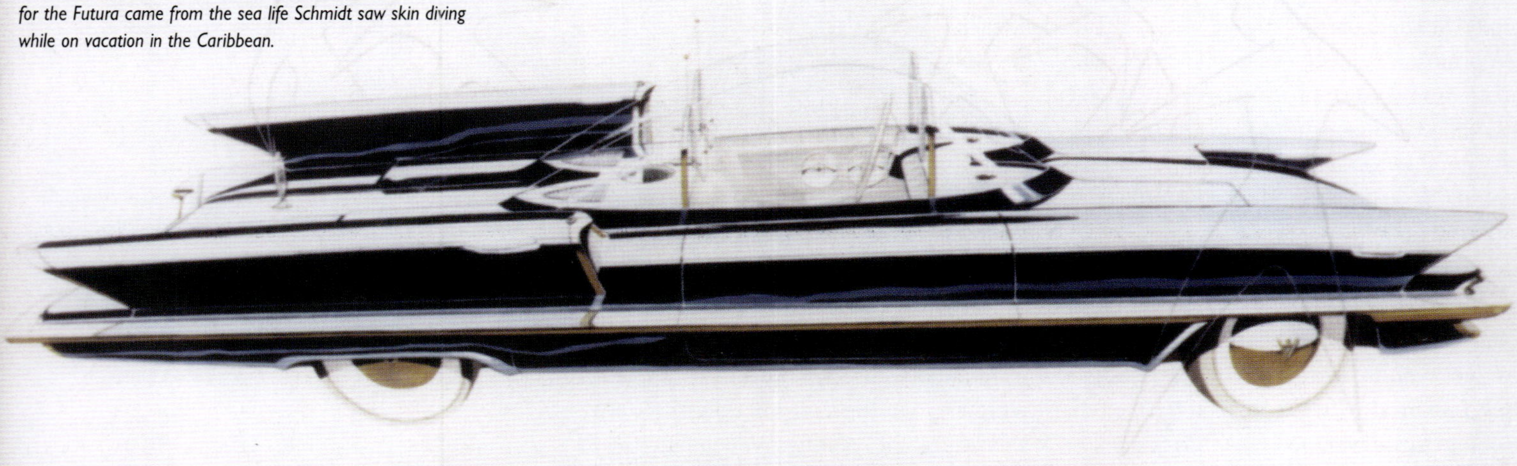

An early rendering of the Futura by Bill Schmidt. The inspiration for the Futura came from the sea life Schmidt saw skin diving while on vacation in the Caribbean.

Henry Ford Museum and Greenfield Village

were Benson Ford and Bill Schmidt able to override the safety relays that prevented the roof section and the doors from opening when the car was moving. Most of the time, especially when the press was around, the roof and doors were closed, even though the car's two passengers were sweating profusely. In the photographs, at least, Benson Ford and Schmidt looked like they had the time of their lives.

On the Futura, everything except the nonexistent windshield wipers worked. In addition, there were some unusual and innovative features, including the following:

- When in motion, the interior was supposed to be sealed off from the outside. To make sure sounds outside the car could be heard inside the passenger compartment, there was a microphone located in the center of the aerial on the trunk. The microphone looks like a black ball. The speaker inside the passenger compartment was located on the vertical chrome panel between and just behind the seat backs.
- Instruments were located in what was called a binnacle in the center of the steering wheel. The steering wheel turned but not the binnacle. In the center of the binnacle was a disc with the odometer. Towards the bottom of the binnacle was a 150-mph speedometer and a tachometer. At the top of the binnacle were five chrome-bezeled idiot lights for fuel, oil, battery, temperature and high beam indicator. Transmission gear selection was made by pushing one of five pushbuttons located near the driver's right side on the center console. As a safety feature, the transmission pushbuttons were

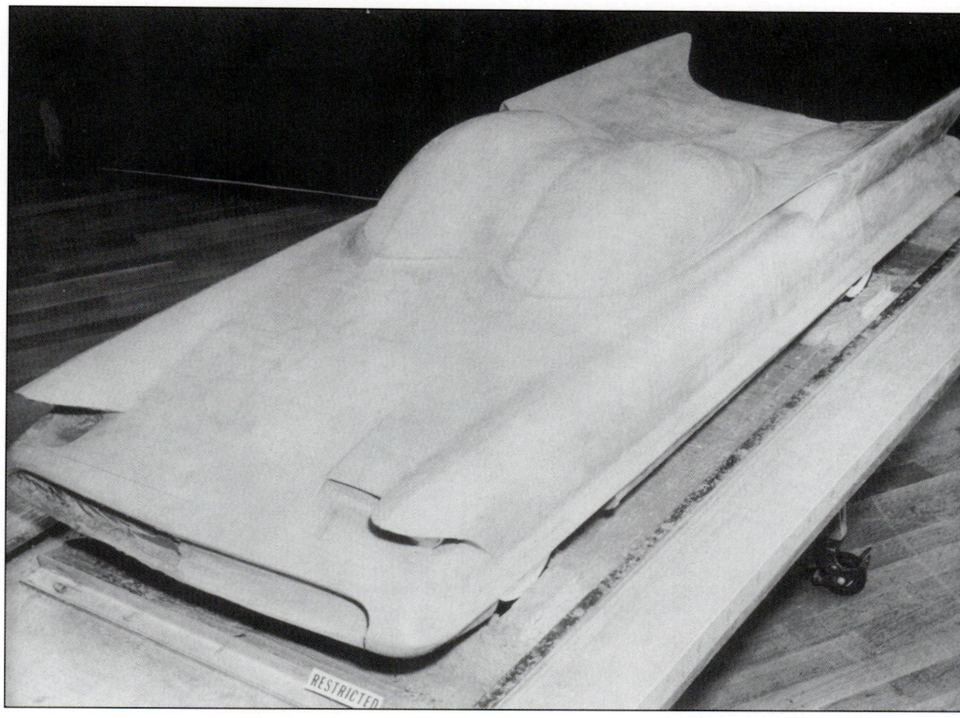

Henry Ford Museum and Greenfield Village

The three-eighths sized clay model of the Futura.

One of Schmidt's later renderings of the Futura.

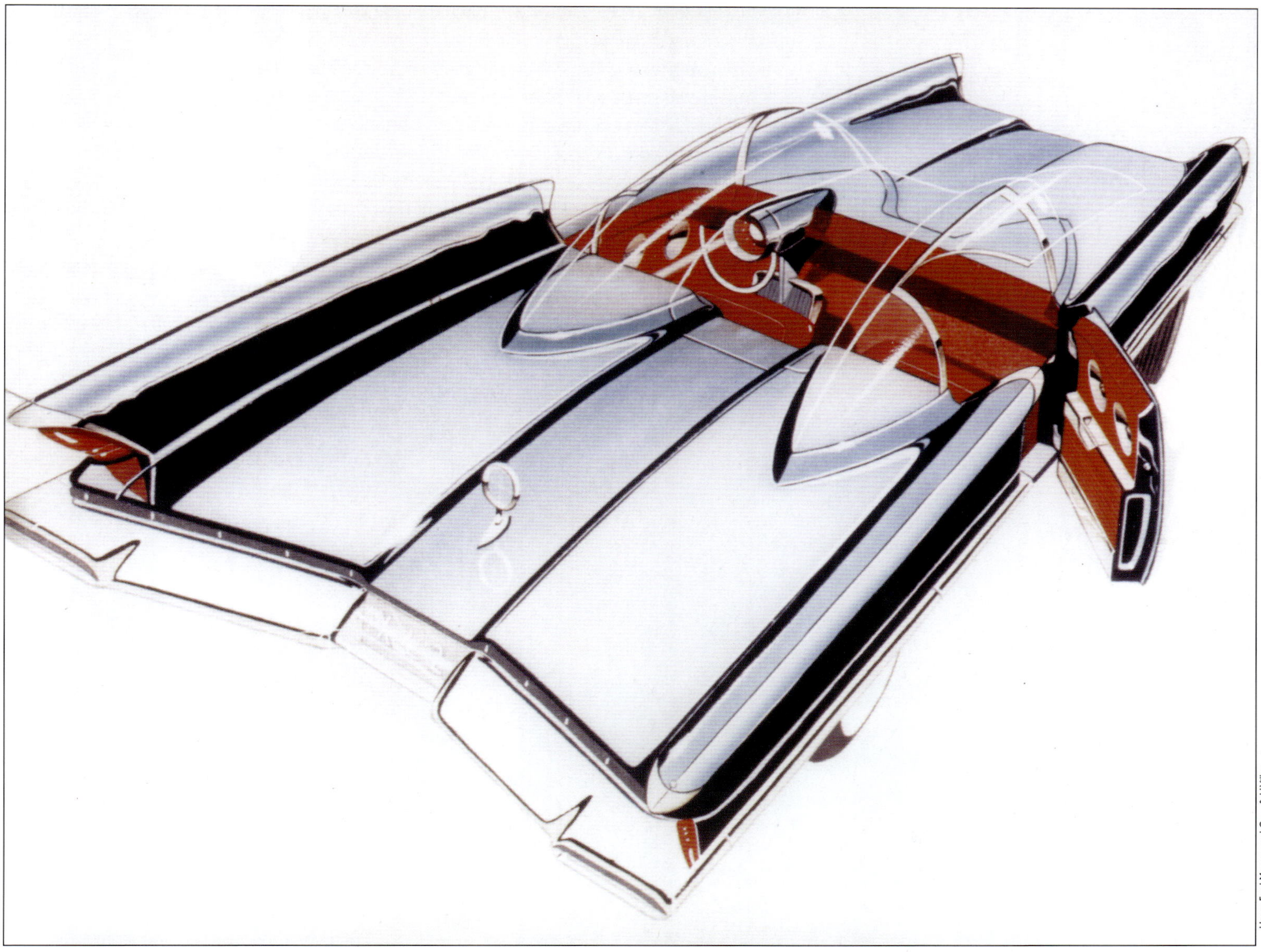

Henry Ford Museum and Greenfield Village

The full-sized clay model of the Futura in the Design Department showroom.

This full-sized plaster model of the Futura was shipped to Ghia in Turin, Italy, and was used with a chassis and running gear from a Continental Mark II mule to build the fully-operable Futura.

different sizes for each gear, and the parking gear-selector button was tied into the interlock system, so the doors and canopy could only be opened when the gear selector was in park.

- The wheelcovers, although similar to those on the 1957 Lincoln, have no valve stem holes in them; the valve stems were on the inside of the modified Lincoln 6L 15" wheels and accessible only from under the car. The tires were Firestone gum-dipped 8:00x15 tubeless deluxe Champions, with 3 1/2 inch wide whitewalls, and a spare tire was in the trunk.
- The gas filler is behind the back license plate, as was standard Ford practice beginning in 1952. The Futura had a full-sized gas tank mounted beneath the trunk floor.
- The hood of the Futura was hinged at the front, and the hood release was inside the passenger compartment much like the production 1957 Ford.
- There was a console between the two bucket seats. Behind the console, and in line with the seat backs, was a phone.
- The outside door handles were in indentations at the tops of the doors. Interior door handles were levers with chrome balls on the ends. On the door jambs were cam-type devices with microswitches on them. As the doors opened, the interior lights went on, and the complicated interlock system raised the top. When the doors were closed, the top automatically closed.

- On the inside of the Futura, in the coved area close to the windshield on the driver's side, was a compass and on the passenger side, there was a clock. In front of the compass on the driver's side were five small round lights that indicated the gear that had been selected. On either side of the gear-selector lights were small triangular lights that illuminated when the turn signals were on. The directional signals were activated by buttons located in chrome pods hung from the spokes of the steering wheel
- The instrument panel could be closed-off by five roll-top desk-type chromed metal doors that, when down, hid the controls and closed-off the glovebox. From driver's side to passenger side, the roll-down compartments were for: a) air-conditioning controls; b) four knobs for the lights; c) cigar lighter, temperature control and dummy wiper switch; d) AM radio; and e) glovebox. All but the air-conditioning controls were to the right of the steering wheel,
- The horn was activated by a pedal on the floor.
- The thin vents inboard of the headlights were used to help cool the engine compartment. Neither the grille opening nor the vents near the headlights did a very good job of cooling the Futura, and it always had a tendency to overheat.
- At the top center of the windshield, there was a vent used to let fresh air in. It didn't work very well, even when the car was moving. On the package tray area there was an air-exhaust vent about three inches high extending from one side of the passenger compartment to the other.
- The air-conditioning system, when originally installed, had two clear-plastic funnels coming out of the package tray area. The evaporator for the air-conditioning system was located in the trunk. (The gills on the deck lid were nonfunctional and just for looks; they had no connection with the air-conditioning system.) The vents at the front of each rear fin were used to cool the rear brakes and as the source of outside air for the air-conditioning system. Sometime after the Futura was introduced, the plastic funnels on the package-tray area were removed. Whether the air-conditioning system remained operable after that is unknown.
- The only other known change made to the Futura, after it was introduced, was the addition of 1956 Lincoln script on each side at the front of the car.

Lincoln standard features on the Futura included power steering, brakes and seats, ball-joint suspension and dual exhausts. For some unknown reason, Lincoln-Mercury press releases listed an automatic starter as standard equipment. Whether that meant the Futura was started any differently than other Ford products is unknown.

The interior design of the Futura was functional and integrated. The upper part of the instrument panel was covered with black leather, and the lower part was in blue-white pearlescent leather. The area underneath the instrument panel was a continuation of the upholstery extending almost to the floor before flaring out into heater openings on both sides. The door panels were extensions of the instrument panel design and included built-in armrests. The power-seat switches were on the center console. The seats were high-back bucket-type, padded with foam rubber and covered in blue-white pearlescent leather

Ford Photomedia

Roy A. Brown

Roy Brown was born in Hamilton, Ontario, Canada. At age 15, his family moved to Grosse Point. He attended the Detroit Art Academy (later Meinzinger's Art Academy) on a scholarship. Upon completion in 1937, he was hired by GM and worked in Bill Mitchell's Cadillac studio, where he met Gil Spear and Frank Hershey. Both Spear and Brown were impressed with the vertical grille of the proposed 1941 LaSalle, which Brown later used as inspiration for the grille design of the Edsel. At GM, Brown also worked with Joe Oros and Elwood Engel, who were several years his junior. In 1939, Brown left GM to work in Harry Pack's private industrial design firm but, when it was disbanded in 1941, he was rehired by GM as a senior stylist in the Oldsmobile studio. Brown had a hard time handling the politics at GM and soon left again for a job in Bob Bingman's independent industrial design firm.

After serving in the Army during World War II, Brown returned to Bingman's firm. By 1946, Brown was tired of designing cars, so he left to accept a job at PPG's Ditzler Paint Division as a color consultant. In 1953, after six years at Ditzler and a year as an independent paint and color consultant, Brown called Gene Bordinat and asked for a job. Brown was hired, and his first assignment was in Bill Schmidt's Lincoln studio, where he directed the building of the clays models of the Futura.

After the Futura was completed, Bordinat recommended that Brown be appointed chief designer for the Edsel. Brown is responsible for the design of the Edsel, including its LaSalle-like front grille. In the spring of 1958, because of the Edsel's poor sales, George Walker, general manager of the Styling Center, transferred Brown to the truck studio, and then arranged to send him to Ford of England, where he was chief stylist from September 1959 until March 1964.

After five years in England, Brown returned to the Design Center, where he became executive designer in the Lincoln-Mercury, and then the Ford interior studio.

He retired in 1975. ■

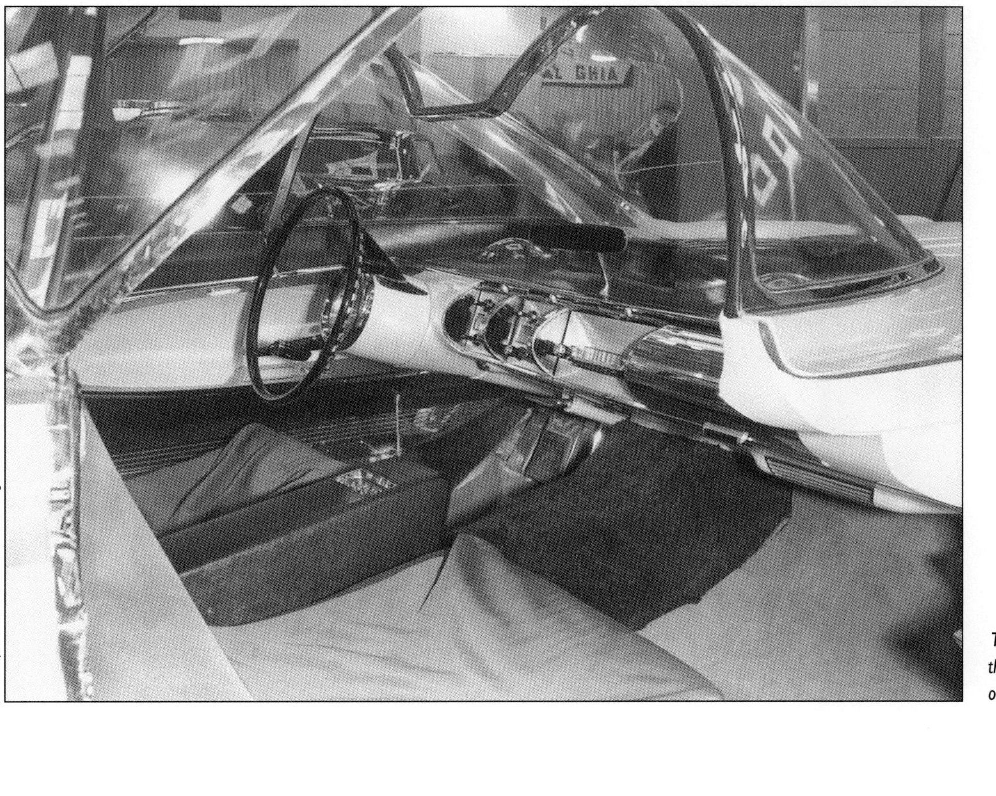

Henry Ford Museum and Greenfield Village

The Futura's instrument panel featured five roll-down doors that could be used to close-off the instruments. When not on display, the Futura was fitted with slip covers over the seats.

Art Tarabusi

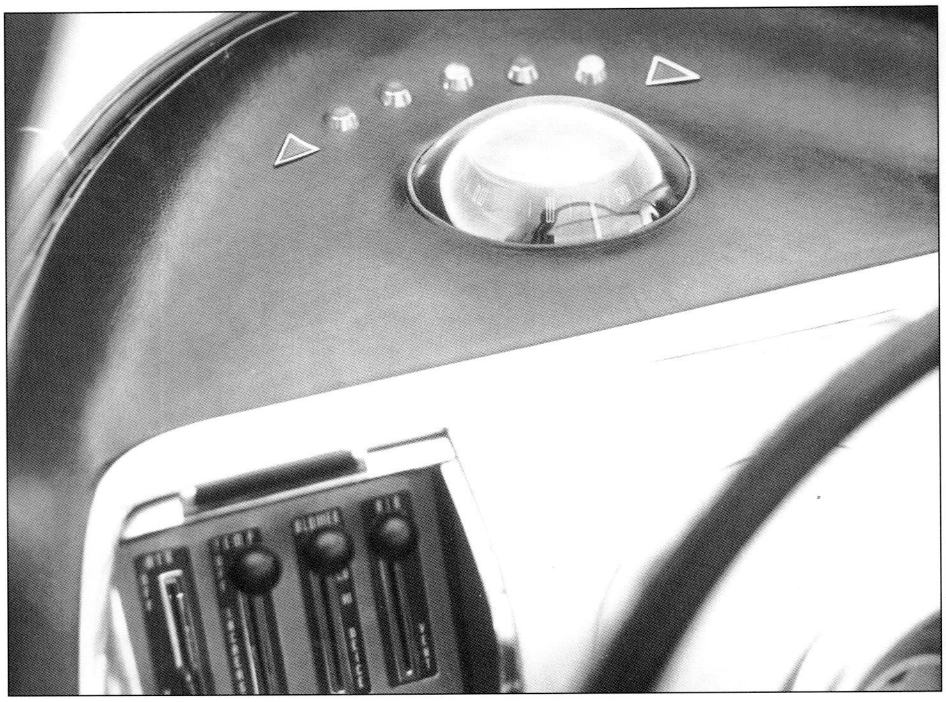

Ford Photomedia

Arthur F. Tarabusi

Art Tarabusi grew up in the Detroit area. During World War II, as a Navy fighter pilot, he was shot down over Japan but was able to ditch at sea and was recovered by an American ship. After his release form the service in 1946, he and several friends operated an aircraft ferry service. In late 1947, he was recalled by the Navy. Just before his discharge, his friend, former Navy pilot Frank Bianchi, told him how to prepare a portfolio. He spent three months doing so, and then applied at the Ford Design Department and was hired. Tarabusi had no formal training as a designer but loved cars and loved to draw.

Tarabusi's first assignment was in the Advanced Studio, where he worked on many of Gil Spear's 3/8-sized models of advanced cars. He was then transferred to the Lincoln studio, where he did some of the initial sketch work on the Futura, but he generally worked on production Lincolns.

In 1953, Tarabusi was again recalled by the Navy. After he was released in 1955, he returned to Ford for a few months, but was soon invited to come to Chrysler by Frank Bianchi, who had moved from Ford to Chrysler in 1952. From 1955-60, Tarabusi worked as a designer in the Chrysler and DeSoto studios.

In 1960, he accepted a job at Reynolds Aluminum as their liaison with the design departments of all domestic automobile manufacturers. Just after he started at Reynolds, Tarabusi was recalled by the Navy again and assigned to an anti-submarine squadron. In 1962, he returned to Reynolds Aluminum, where he remained until his retirement in 1990. While at Reynolds, Tarabusi borrowed many of the Ford, General Motors and Chrysler concept cars for display at Reynolds Aluminium's headquarters building, some of which he got to drive. ■

Ford Photomedia

(above) A compass was located in front of the driver on top of the instrument panel. In front of the compass were five gear-selector lights. On either side of the gear-selector lights were the turn-signal lights. The controls to the left of the steering wheel were for heating and air conditioning.

(left) The controls on the face of the instrument panel could be closed-off by roll-down chromed doors. When closed, the doors resembled the glovebox door at the right of the instrument panel.

(below) Pushbutton gear-selector controls were located on the center console. The large indented coves in the door panels gave added arm room.

Ford Photomedia

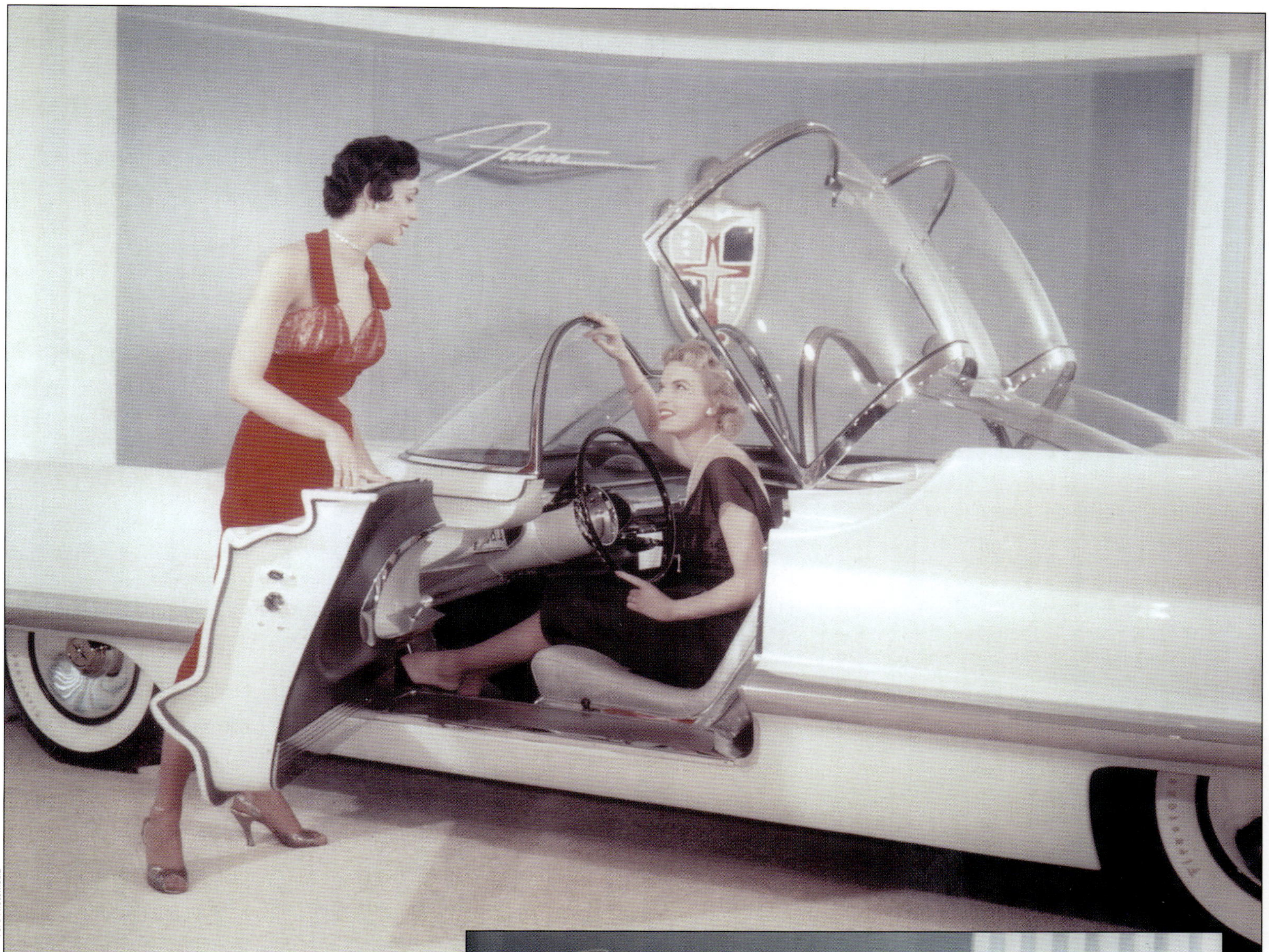

Ford Photomedia

with pleated black leather inserts. The steering wheel was black, as was the carpet. The rear part of the console was a raised armrest.

After the 1955 show circuit, and for the next several years, the Futura continued on a very busy schedule appearing at auto events, fairs, races, selected dealer showrooms, etc. When not on tour, it was usually displayed at the Ford Rotunda.

By 1959, the Futura was four years old, and considered a little long in the tooth. Automobile customizer George Barris claims that it was at his urging that, in 1959, the Futura wound up in Hollywood, where it starred in the movie *It Started With A Kiss* co-starring Glen Ford and Debbie Reynolds. The Futura played a one-off concept car won by a military noncom serving in Europe. When the Futura was screen tested for the roll, its iridescent color came out flat. The only filming system then available that could have recorded the iridescent quality of the Futura's paint, with some degree of accuracy, was 3-D, which was never seriously considered. With Ford's permission, the studio decided to have the Futura repainted red, a color that filmed better. The Futura was repainted at Barris Kustom City in North Hollywood, California. After the movie, the Futura, still repainted red, was returned to Dearborn, where it was featured in a 1963 Ford film about the Design Center. It was then returned to Southern California, where it sat in the MGM studio lot and then outside Barris' yard next to his shop until about 1965. During that time, the tires went flat, the paint faded, one or two of the wheelcovers were lost and the car generally deteriorated.

Ford Photomedia

(top and above) The Plexiglas roof canopy rose automatically when the doors were opened and automatically closed when the doors closed.

Richard Steiger

Richard W. Steiger

Dick Steiger grew up in East Cleveland, Ohio. After he was discharged from the service in 1946, he enrolled at the Cleveland Institute of Arts and graduated in 1949. His first job was as a designer in the Kaiser-Frazer exterior studio. In 1952, as a result of a lay-off at Kaiser-Frazer, Steiger took a job with the independent Detroit design firm of Sundberg and Ferrar. In 1953, he was rehired by Kaiser-Frazer. Three weeks later, he was interviewed and hired by Gene Bordinat and Frank Hershey as a Ford designer.

Because he had prior design experience, Steiger was placed in the Ford interior studio working under Art Querfeld. That same year, he was reassigned to the Lincoln interior studio to work for Don Beyreis. While in the Lincoln studio, Steiger worked on the 1953 and '54 Lincoln trim and color cars and on the interior of the Futura.

In 1955, Steiger was promoted to manager in the new special projects (Edsel) interior studio, where he stayed until 1958, when he was reassigned to the Mercury-Edsel-Lincoln pre-production studio. While in the Lincoln pre-production studio, Steiger developed a front swivel seat, similar to the one that was later used on production Chryslers products.

In 1960, Beyreis, who had already moved to Chrysler, recruited Steiger, and he spent the next 20 years as a manager in the Plymouth, Chrysler, Dodge and the Safety interior studios. In 1980, Steiger left Chrysler, but returned two years later. He retired from Chrysler in 1986. From 1982 until 1989, Steiger also worked as an independent industrial designer. ■

There are many stories about how George Barris acquired ownership of the Futura, and some of them have even achieved the status of folklore. The most common of these stories is of Barris foreclosing a storage lien against the Futura, buying it for $1, and then being sued by Ford. In these stories Barris always wins the lawsuit, sometimes after a trip to Dearborn to make his peace with Ford executives. Barris dismisses such stories as totally untrue and, in fact, there are no records of a lien foreclosure or a lawsuit involving Ford and Barris in Los Angeles County, where such records would most likely be if they existed. What really happened is not as exciting as folklore, but it probably makes a lot more sense.

During the 1960's, Ford and Barris worked together on several projects. In 1964-65, Barris was involved in the Lincoln-Mercury Caravan Of Cars, during which Lincoln and Mercury showcars and California custom Ford products were displayed around the country on behalf of Ford Motor Co. Barris contracted to build the Super Marauder, and David Ash was assigned to oversee the project. Barris showed him the Hollywood scene and introduced him to a number of Hollywood celebrities. It was during this time that Barris says he bought the Futura from Ford for $1. The Futura has never been a titled vehicle, so there are no motor vehicle records to check. Ford had no further use for the Futura by that time, and Barris says that if he had not purchased it, Ford intended to destroy it because of liability concerns. It is unclear whether, at the time the Futura was sold to Barris, Ford knew of or Barris already had plans to convert it into the Batmobile. Ford did know of and encouraged Barris' efforts to get the Futura on TV and in the movies.

Sketches of the Batmobile were originally made by Barris. At the time, Barris was a partner in Barris Kustom City. His other two partners were Lester E. Tompkins and Irvin Kuns. Barris' drawings were submitted to Twentieth Century Fox Television and Greenway Productions. Studio artists revised the drawings. The studios included the revised drawings, the original drawings by Barris and a list of requirements for the Batmobile in a 13-page written contract signed September 1, 1965. The list of requirements for the Batmobile (in the contract the Futura was referred to as the "proto-Lincoln chassis") included a "Bing Bong Warning Bell," a "Batscope," something called a "Bat-Ray Projector Mechanism" and nine other special effects. The contract required that the Batmobile be completed and delivered by October 11, 1965, and the studios agreed to pay Barris Kustom City $5000 to convert the Futura into the Batmobile for use in the *Batman* TV show pilot plus a promise of $5000 to $9000 more depending on merchandising rights and completion of the *Batman* pilot. If the series went into production, there was also a provision in the contract guaranteeing the studios the right to rent the Batmobile from Barris during production for $150 a day. Barris also agreed to build a second Batmobile for $10,000-$12,000 if the studios needed it. In the contract, Barris retained the design rights and exclusive right to publicly display the Batmobile.

The influence of the Futura on the 1956 Lincoln was noted in Lincoln advertising.

Ford Photomedia

Benson Ford is behind the wheel of the Futura. He and Alex Rhodes, an electrical engineer and troubleshooter, took the Futura for a test drive through New York City.

Henry Ford Museum and Greenfield Village

Bill Cushionberry, an auto-body craftsman noted for his metal work, had a customizing shop several blocks from Barris in

North Hollywood. Barris took the trim off the Futura, then subcontracted the metal work necessary to convert the Futura into the Batmobile to Cushionberry. Cushionberry describes the Futura as very solid but weathered. Cushionberry was given no written plans, but was told by Barris what he wanted done. He was then left to his own imagination to figure out how to do the job. According to Cushionberry, extending the fins into the doors was the hardest part of the job. Cushionberry also remembers the original metal work on the car as having been extremely well executed. After the metal work was finished, the Batmobile was returned to Barris' shop, where it was painted the gloss black it is today. In making the Batmobile, the center section of the top was removed and a new Plexiglas cover was fabricated, leaving the window area on both sides open. In 1956, Bob Adomeit, a young design student working at the Styling Center for the summer, got a chance to drive the Futura on the Ford test track. He recalls that engine noise from the car reverberated throughout the Plexiglas-double canopy, making it feel like he was driving inside a constantly ringing bell. Interior noise may be one of the reasons Barris removed the canopy.

Bill Schmidt in the driver's seat of the Futura.

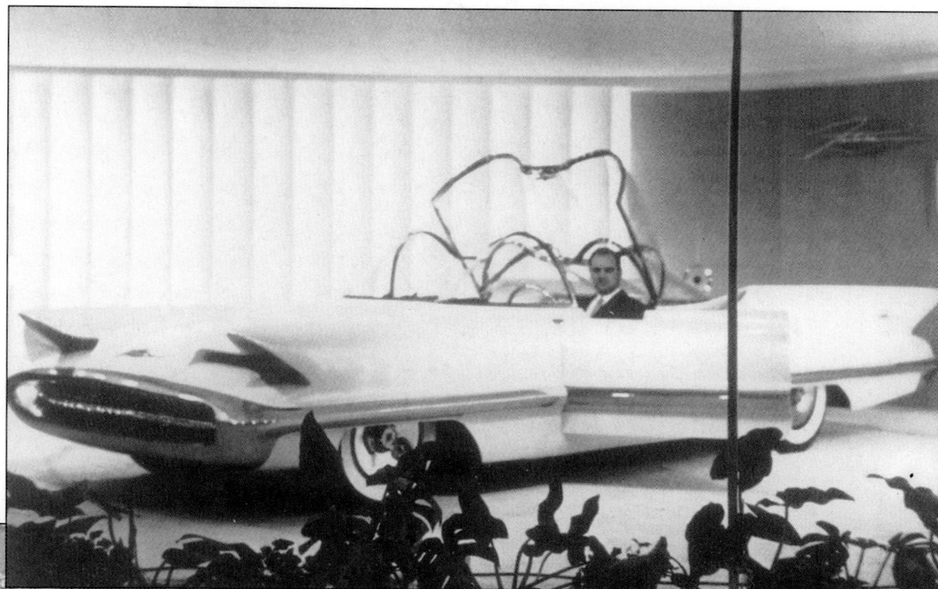

Ford Photomedia

Later, the original Plexiglas center section that had been removed from the Futura was cut into two sections by Barris. One-half of it was used on a custom rear- engined speedster built for Richard Boone of the *Paladin* TV series fame. The other half is now owned by a collector in Wisconsin. When the Futura was converted to the Batmobile, the steering wheel was cut so that it was U-shaped with handles at the top of each side. Adam West, star of the *Batman* TV series, disliked the steering-wheel configuration so much that Barris removed the modified Futura steering wheel, binnacle and all, and replaced it with a standard Edsel steering wheel which, to this day, remains a part of the Batmobile.

Arby Schuman

Hard use during filming of the *Batman* TV series, which ran between 1966-68, took its toll on the car's frame and running gear. After cracks developed in the frame and troubles developed with the transmission, Barris replaced the frame and the running gear as a means of avoiding the risk of further breakdown during filming. A Ford Galaxie frame that was lengthened 11 inches, a 390 CID Ford engine and a Ford automatic transmission replaced what had originally come on the Futura. Barris complains that hobbyist constantly ask him for information about the Batmobile's frame and engine, which were hauled off to the junkyard immediately after he removed them from the car.

Henry Ford Museum and Greenfield Village

(middle) The Futura at a race track in Wisconsin.

(bottom) The Futura being pushed up a ramp at an auto show prior to clean up.

In March 1966, Barris applied for a patent on his "new design" Batmobile and, on October 18, 1966, the U.S. Patent office granted him patent number 205998, which was good for 14 years.

Barris built five replicas of the Batmobile, which were displayed and later sold. The replicas were fiberglass made from molds taken from the original Batmobile. The molds are supposedly stored in the rafters at Barris' shop. Stretched Galaxie frames were used for the replicas. One of the replicas got a 429 CID Ford engine, and was apparently used for drag-racing events. The other replicas got Ford 390 CID engines.

Benson Ford, Bill Schmidt and the Futura in New York City.

Ford Photomedia

After the *Batman* series ended in 1968, the Batmobile sat for several more years at Barris' shop gathering dust. In 1973, Barris held an auction to dispose of many of the cars he had built for the movies. Sale prices were not what Barris expected. Although the Batmobile, formerly the Futura, was included in the auction, it failed to garner any bids and went unsold. Barris still has the Batmobile and, with the revived interest as a result of the new *Batman* movies, the original Batmobile is constantly on tour. ■

Ronald T. Perry

Ron Perry

Ron Perry grew up in Urbana, Ohio. As a child, he loved to draw planes and cars. After high school, he took a correspondence course in aeronautical drawing and was later accepted as a student at the Dayton Art Institute. While at the Dayton Art Institute, Perry worked summers as a draftsman for a manufacturer of airplane lighting. He graduated in 1953 and was hired by Charlie Waterhouse as a Ford designer. While at Ford, Perry also received a diploma from the Chicago Technical Institute.

Perry's first assignment was in Gil Spear's Advanced Studio. He was next assigned to the Lincoln exterior studio. Perry and other Lincoln exterior studio designers were asked by Bill Schmidt to come up with a unique roof treatment for an upcoming concept car. Perry's double-bubble canopy design was selected for use on the Lincoln Futura.

Perry later worked in the Ford exterior, the Lincoln-Mercury interior, the Lincoln-Mercury exterior studios and in the international studio. He became a supervisor and was on the team that designed the Maverick, the Designer Series Lincolns, the Tempo, Topaz, Taurus and Sable before retiring in August 1985. ■

Gilbert Treweek

Gil Treweek

Gil Treweek is a Detroit native. After serving in the Army Air Corps during World War II, he attended the University of New Mexico. After graduating in 1951 with a degree in fine arts, he returned to Detroit and spent a year doing post-graduate work at Cranbrook Academy studying car design, sculpture, metal-smithing and clay modeling. Treweek's friend, John Reinhart, who had just been hired to design the Continental Mark II, got Treweek a job at Ford. He started as a B-grade clay modeler at the Ford Design Department on July 1, 1952. (Entry level clay modelers began as B-modelers, advanced to A-modelers, then to sculptor modelers, who were responsible for a front end or body side, master modelers, who were in charge of 5-6 other modelers, and then to modeling supervisors, who were in charge of 20-30 modelers. The highest ranking clay modeler, a modeling manager, was in charge of all clay modelers in a studio.)

During most of the early 1950s, Treweek worked in the Lincoln-Mercury body development studio on production cars, including the 1956 Lincoln. Throughout his career, he worked in all studios, except the truck studio. He was promoted to master modeler in the Edsel interior studio. Treweek was one of the modelers who worked on the interior of the Mystere and the exterior of the Futura. He later worked on the Aurora show car. He became modeling manager upon Fred Hoadley's retirement in 1978.

In 1979, design director Gene Bordinat asked Treweek to start a Design Center newsletter and, until it was made a part of the Company wide Ford News, Treweek was the editor of Highlight. After he retired, he helped build the modeling staff at the Ford-sponsored Concept Center in California and worked there as a consultant until 1993 ■

At 11:30 am, on Wednesday, October 5, 1955, the curtain was pulled back in the Styling Rotunda, and out walked George Walker, who introduced a breathtaking new concept car. It was called the Mystere, and it was painted a fuchia-pink magenta and black, with a narrow pearlescent-white band over the top of a huge transparent canopy that covered the passenger compartment. The Mystere had a compartment in the rear for a gas-turbine engine. The front part of the glass canopy was hinged at the cowl (supposedly), and it swung up and forward as the doors opened, allowing passengers to enter and exit an airplane-like cockpit. Ford press releases said the Mystere "was built as an experimental car to consider future styling innovations and primarily to study styling themes, particularly in new contours of sheetmetal."

Mystere

The Mystere was originally built as a Ford show car for the 1955 show circuit and as the stablemate for the Lincoln Futura. They were both intended to make their first appearances at the 1955 Chicago Auto Show. By the time the Mystere was finished, in December 1954, many of its design features had already been borrowed for the production 1957 Ford, so it could not to be publically shown until fall 1955. The distinctive sculptured swoosh marks on either side of the car, the fins and the entire front end were transferred, albeit toned down, to the '57 Ford. Concept cars, by their very nature, are meant to be copied, and their designers hope they will influence future production vehicles, but the Mystere was so successful, showing it during the 1955 show season would have disclosed features of the upcoming '57 Ford in time for the competition to copy them, so it was held back for the better part of a model year.

Ford Photomedia

The Mystere was introduced to the automotive press at the Styling Rotunda on October 5, 1955.

The Mystere did not influence the design of the 1955 or '56 production Fords as is often reported. Those designs were already locked up by the time the Mystere was started in the summer of 1954. If anything, it was the other way around. The Mystere was designed to show continuity between the 1955 and '56 Fords *and* future model Fords.

The design of the Mystere was completed by Bill Boyer, after he finished work on the 1955 Thunderbird, and while it was being readied for production. Boyer's immediate supervisor at the time was Damon Woods, who was executive designer in the Ford car and truck exterior studio and ran the body development studio. Bob Maguire was executive stylist of the Ford studio under chief designer Frank Hershey. Hershey was busy with production cars at the time the Mystere was built and, although he occasionally looked in, he took no part in its Mystere's design.

In July 1954, Hershey asked Woods, Boyer and the other designers in Woods' studio to come up with a design for a showcar for the fast-approaching 1955 show season. Hershey wanted a car that would emphasize the swoosh that was designed for the 1957

Ford Photomedia

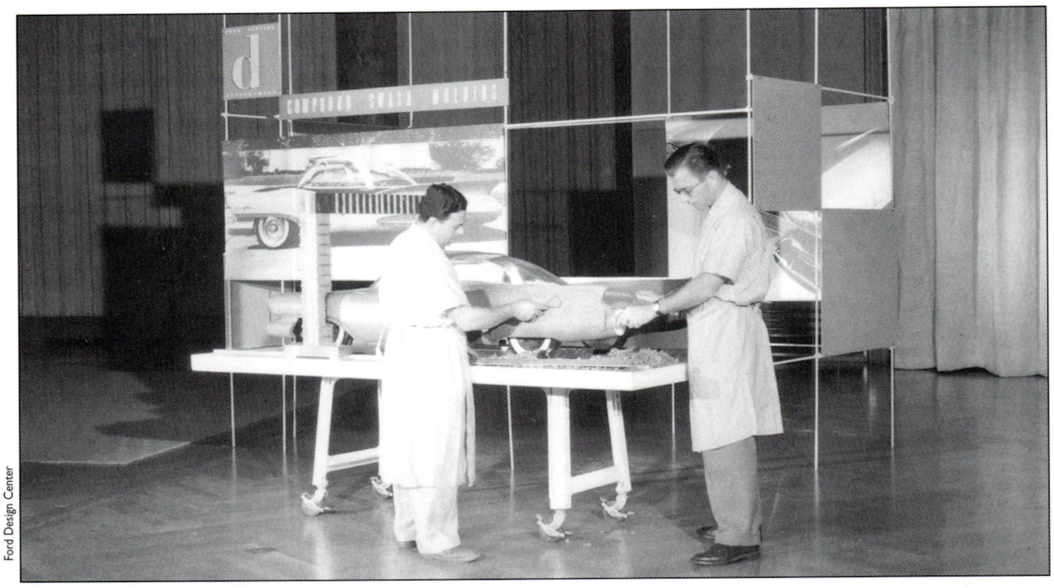

Ford Design Center

(above) Clay modelers Anton Vella and Gil Treweek working on a three-eighths sized model of the Mystere. This clay model was built after the Mystere was completed.

William P. Boyer

Bill Boyer's design career grew out of a life-long love affair with airplanes. In 1944, Boyer enlisted in the Navy intending to become a Naval Aviator. He completed preflight school and was in flight training at Memphis Naval Air Station, when the war ended. Because there were 40,000 other pilots fighting for rank in the service, he took the Navy up on its offer of an early discharge in April 1946.

Boyer then enrolled at the University of Maryland intending to become an aeronautical engineer. After two years, he woke up one morning, saw all the drawings of cars and planes he had made in his calculus notebook, and knew he was in the wrong field. He left to enroll at Pratt Institute to study car design. At the end of his second year at Pratt, Boyer was interviewed by Tom Christianson, Ned Nickels and Art Ross from GM's Styling Section. They offered him a scholarship and a job at GM when he graduated. In 1950, Boyer graduated from Pratt and started at GM as an entry-level designer. His first job was illustrating maintenance manuals for the LeSabre concept car. Later, he was assigned to the B-studio designing Oldsmobiles and Buicks. Even though his work was more than satisfactory, Boyer didn't feel he fit in at GM.

In March 1952, Frank Hershey offered him a job at Ford designing the Ford Sports Car. Boyer was one of the designers of the 1955 Thunderbird and all Thunderbirds through 1964. He was also responsible for the design of the Mystere concept car.

In June 1961, Boyer was named design executive for all Ford Division cars. He later worked on the 1967 Thunderbird, before he was transferred to the Lincoln-Mercury studio, where he helped design the 1970 and '71 Mark IIIs and the 1972 Mark IV. Boyer was then transferred to Ford of Australia as their chief designer. He remained in Australia for three years. After he returned to Dearborn, he was on the team that designed the 1983 Thunderbird and Cougar, and then became design executive in the truck exterior and interior studios, before he retired in 1985. ■

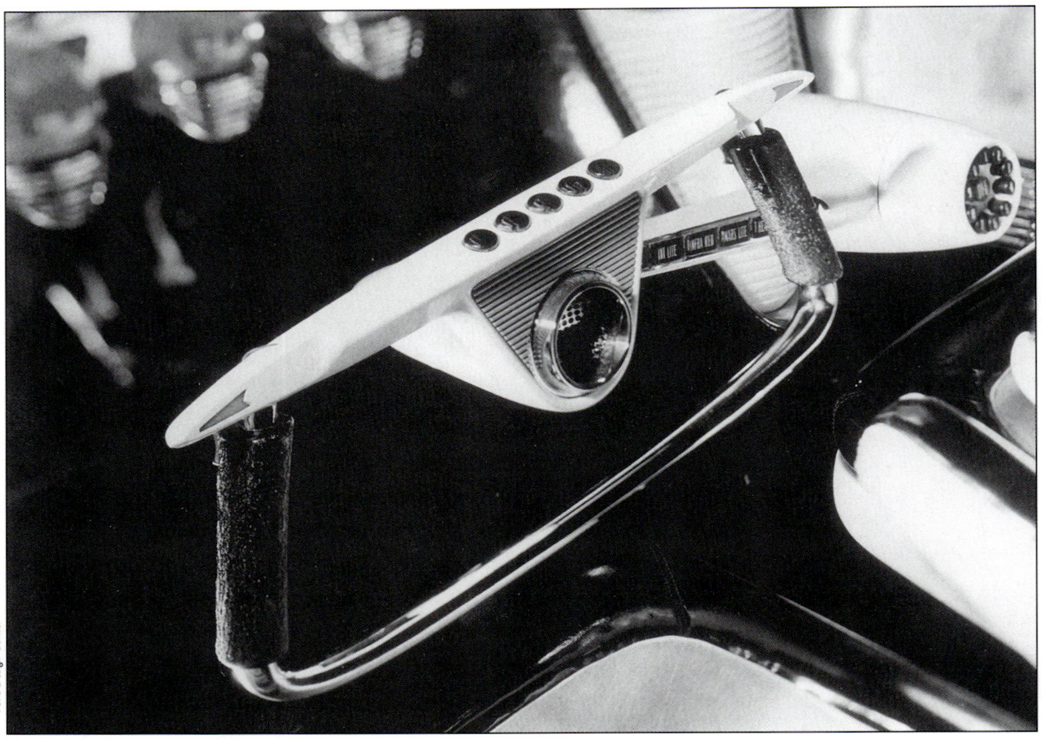

Ford Design Center

(above) Pushbutton automatic-transmission controls were on top of the Mystere's aircraft-inspired steering wheel.

(below) The Mystere shortly before it was introduced to the automotive press.

Ford Design Center

Ford. Engineering had recently determined that the compound curves necessary for the shape of the 1957 Ford's fins could be built by using multiple rollers to form the metal as it went through the die and had given the go-ahead for the proposed design.

Boyer prepared and submitted one design, and Woods fell in love with it. As Boyer puts it, "it had flair." Woods said the design reminded him of the custom-bodied French Delahayes he remembered from his youth growing up in France. The name Mystere was suggested by Woods after the French Mystere jet fighter. When Hershey saw the proposed design, he approved it, commenting that "this is it!" Boyer took the theme for the Mystere from the Fairlane side moldings Ken Nelson originally designed for the 1955 Ford Fairlane, but made them three-dimensional and a part of the sculpture of the car. Boyer's original sketch of the Mystere now hangs in the Michigan studio of his daughter Lynn.

Ford Design Center

Herb Tod (left) and Bob Genslak (right), with a three-eighths sized clay model of the Mystere.

The other designers in the studio, at that time, were given two days to complete a full-sized rendering based on the package set by Frank Hershey, Bob Maguire and Damon Woods. The package measurements were put on a large blackboard, and designers Bill Braathan, Ralph Ortiz, Pierre Crease, John Van Tilburg and John Vanchovie, along with student-designer Ed Jaquet, began drawing out the car full-sized on the blackboard. It took awhile to set the main lines of the car. After that, as Jaquet remembers it, the rest of the design went together very fast, although they had to work many hours of overtime to meet Hershey's deadline. Once the full-sized drawing was completed, a full-sized blueprint was made, the highlights were drawn in on it, and Bill Braathan and several other designers air brushed a rendering of the car right over the blueprint. Everyone was relieved to have the design done in time for a Friday showing to the department's design executives, until word came back that they had decided to increase the package size including the wheelbase. After working most of the weekend, the studio designers were able to finish another full-sized black-paper rendering of the Mystere based on the new package measurements.

There was no money in the budget, so the entire Mystere project was done on the cheap. No engineering was done for the car; it was just designed and built. Changes from what was shown in the blackboard design were incorporated directly onto the full-sized clay.

The more Woods looked at the car on the blackboard, however, the more he doubted that the front end, as it was designed, could be executed. He asked that a quick 3/8-sized clay model be prepared, so he could see the proposal in three-dimensional form. After the model was completed, Boyer and Woods agreed that the front end shown on the rendering had to be changed. Boyer's sketch did not include an interior, so a 3/8-sized clay interior buck was also made and, when approved, a full-sized interior buck was started. A full-sized clay model of the Mystere was quickly started. Only after the Mystere was completed were several 3/8-sized models of it made for display.

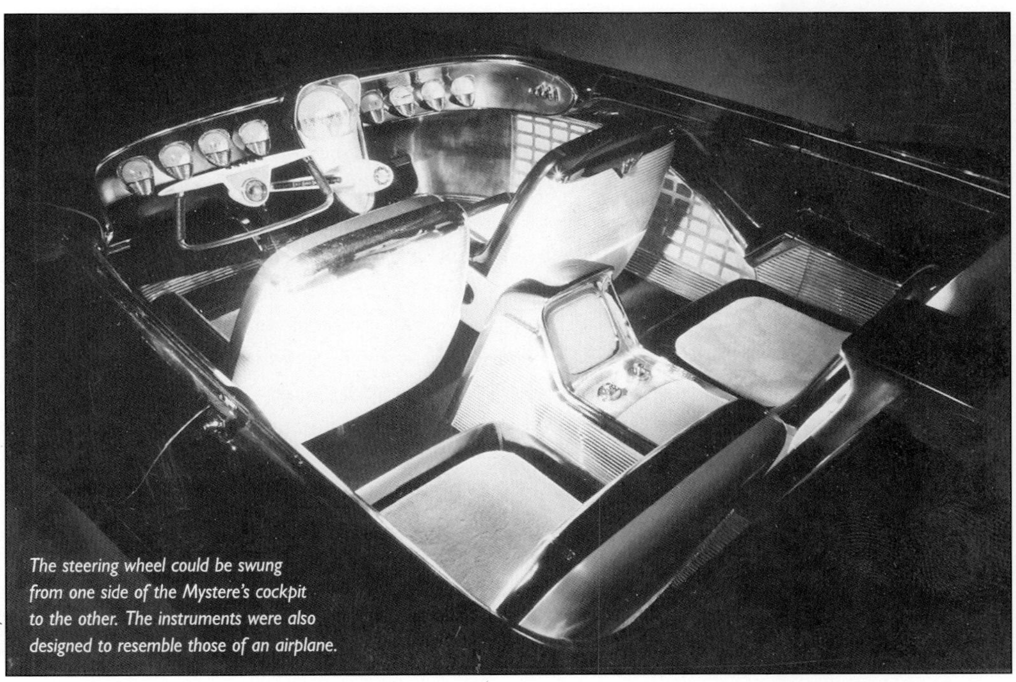

Bill Boyer

The steering wheel could be swung from one side of the Mystere's cockpit to the other. The instruments were also designed to resemble those of an airplane.

Chrissie Tremulis

Ralph G. Ortiz

Ralph Ortiz grew up in the car culture of prewar Southern California. In high school, he developed an interest in art, but his post high school education was interrupted by World War II. During the war, Ortiz flew B-17s over Europe. In 1945, he left the service and took a job back in Southern California at Douglas Aircraft. In1948, he prepared a portfolio and applied at the Art Center College. He was accepted and graduated in 1951. Upon graduation, he was offered a job at the Ford Design Department but decided to accept a job as an industrial designer in Buffalo, New York. Several years later, Frank Hershey again offered him a job at Ford and, this time, he accepted.

After several months in Gil Spear's Advanced Studio, Ortiz was transferred to the Ford pre-production studio, where he worked on the two-passenger Thunderbird and then on the design of the Mystere. Ortiz was later transferred to the Ford production studio and is credited with the design configuration of the fins and rear quarters of the 1957 Ford passenger car. Later, he was promoted to a supervisory position in the truck studio, where he worked on the design of the 1957 Ranchero with Jim Darden.

In 1957, Ortiz left Ford to accept a job with American Motors but, instead of starting there, he decided to accept an offer from Markel Electric in Buffalo, New York, designing appliances and other products. He retired from Markel in 1980. ■

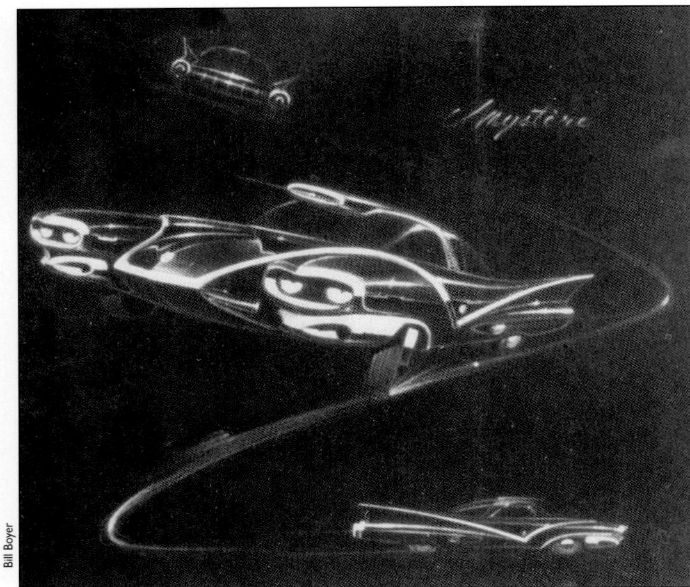

Bill Boyer

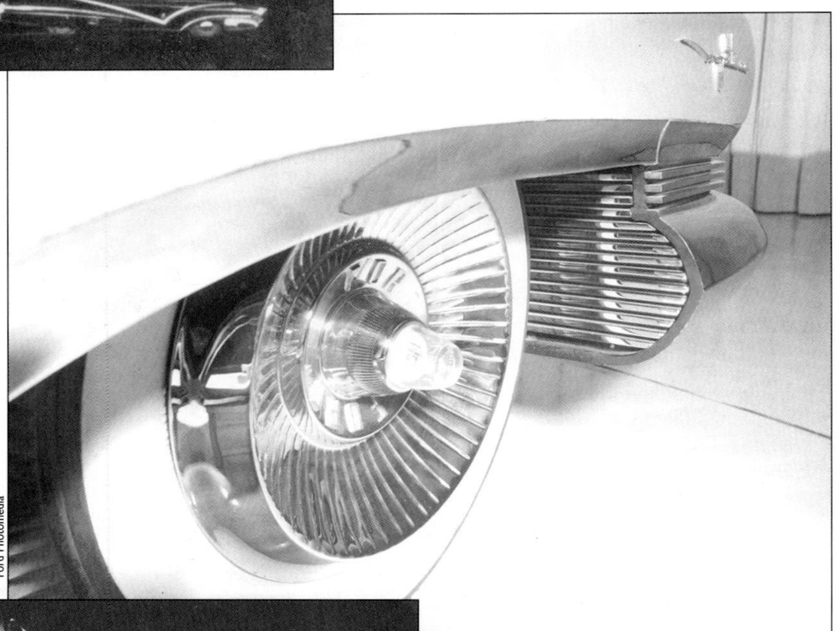

Ford Photomedia

(top) Designer Bill Boyer's original rendering of the Mystere.

(middle) Wheelcovers for the Mystere were designed by Gale Halderman. They influenced the design of the wheel-covers used on the production 1957 Ford.

(below) The Mystere's taillights were designed to look like jet exhausts.

Jim Quinlan, a classmate of Boyer's at Pratt Institute, was hired on Boyer's recommendation to help design the interior of the Mystere. He was assisted on the interior design of the car by Van Tilburg and Jaquet, before he returned to school. Gale Halderman, who later became head of the Lincoln-Mercury studio, also worked on the Mystere. In 1954, Halderman, fresh out of the Dayton Art Institute, was hired by Gene Bordinat to work in the Lincoln-Mercury studio. Bordinat sent him upstairs to Woods' studio to help out for a couple of weeks. While there, Halderman designed the wheelcovers used on the Mystere and also worked on the clay model under Boyer's supervision. Halderman made the wheelcovers on the Mystere as big as possible and in a star-burst pattern to emphasize the Mystere's wheels. The clear-crystal cap in the center of the wheelcover came from an idea Halderman developed while in school.

Because the new Styling Center shops were not yet fully staffed after the move from the smaller EEE building, the full-sized clay model of the Mystere was trucked to Creative Industries in Detroit, where a fiberglass body was molded from a plaster plug. The Plexiglas canopy was also made at Creative Industries by placing felt over the green-house part of the plaster plug and heat forming the Plexiglas over it. Creative Industries also made the chassis for the Mystere from steel I-beams. No suspension or any running gear were included on the Mystere. Boyer made many trips to Creative Industries, and worked closely with Gary Hutching supervising the building of the finished car. About once a week, Hershey visited Creative Industries to check their progress.

Ford Photomedia

The Mystere had a wheelbase of 121 inches, was 220 inches long and 80 inches wide. With its 52-inch height, which was the same as the Thunderbird, press releases said the Mystere had a particularly low center of gravity. The clear-Plexiglas roof on the Mystere was said to permit an almost unlimited view from any direction, except for what the designers called the cummerbund, or roll bar, that formed a meeting bar for the two roof sections. At the front-top of the roof, extending forward from the cummerbund, was a chrome-plated fresh-air intake used for interior ventilation and the air conditioner/heater. When the front roof section was closed, press releases said the Mystere, which had no windows, relied solely on air conditioning for passenger comfort.

Although Ford's press release said the car was built to accommodate either a gas-turbine or a conventional engine in the rear, there were oil-cooler reservoirs at the front of the car, and air intakes on the rear quarter panels of the car specifically designed for a gas turbine. The Mystere was the Design Department's first concept car with four headlights, and the rear bumper assembly included not only the taillight/exhaust outlets but, in the recessed center section, there was a series of exhaust-discharge holes for the gas-turbine engine.

The interior of the Mystere was designed to look like an airplane's cockpit. The ignition switch was a set of buttons, much like today's keyless-entry systems, that required punching in a combination to start the car. On the grip-type U-shaped steering wheel were buttons and triggers that allowed the driver to control many of the car's features without removing either hand from the wheel. Spherical instruments were recessed under a padded safety crash pad at the top of the instrument panel and included gauges for bearing temperature, fuel and oil pressure, a tachometer, inclinometer and an altimeter. In the center console there was a two-way radio that could be used by passengers in both the front and back seats. For the driver, there was a microphone in the steering wheel. The front and rear bucket-type seats pivoted and, for passengers in the back seat, there was a combination radio-television set located on the console, just aft of the front-seat backs. The exterior design of the Mystere rolled over into the inside, where the same trim and color scheme was used. The colors used on the Mystere were picked by Boyer. Even though the design of the Mystere proved a popular one, there was never any talk of building the car as an operable vehicle.

The press releases, newspaper and magazine articles about the Mystere referred to its many features, and some even described what it was like to sit in the cockpit. Not one article, however, mentioned that the doors, trunk and hood did not open, but were really just crease marks in the fiberglass body. The clear-glass canopy didn't open either. The entire canopy was one piece, and it had to be lifted off to gain access over the dummy doors into the cockpit. The only features that actually worked on the Mystere were the headlights, taillights and some of the interior lights, which were all powered by a battery.

Ford Design Center

William E. Braathan

Bill Braathan grew up in the Detroit area and attended Wayne State University. He loved cars, taught himself to be a good airbrush artist, and decided he wanted to become a car designer. In an effort to eventually secure a job at the Ford Design Department, in the late 1940s, Braathan took a job as a guide at Greenfield Village. By 1950, he was working in planning and scheduling in Gil Spear's Advanced Studio. While working at that job, Spear helped him prepare a portfolio, and he was hired as a beginning designer in 1951.

His first assignment was in Gil Spear's Advanced Studio. Braathan was next assigned to work in Bob Maguire's Ford studio and was later reassigned to Damon Woods' body development studio. Braathan was one of the designers who worked on the original full-sized blackboard design of the Mystere and turned it into a full-sized airbrush rendering.

In 1955, when Bill Schmidt left Ford for Studebaker-Packard, Braathan followed. In 1957, he was hired as a designer at Chrysler, where he worked until retirement in 1985. While at Chrysler, Braathan helped design the 1960 Valiant. ■

Phyllis Braathan

James L. Quinlan

Jim Quinlan grew up in Dearborn and attended Henry Ford Community College. He later attended Pratt Institute, where he studied Industrial Design. After graduating from Pratt in 1951, he spent two years in the Army. After he was discharged, he was hired as an architectural display designer at New York's American Museum of Natural History. He also took courses at Wayne State University in Detroit.

In August 1954, Bill Boyer, a classmate at Pratt, asked Quinlan to come to work at Ford. One of his first jobs at Ford was working on the interior design of the Mystere concept car.

Quinlan later designed interiors on the 1956, 1958, 1961 and the 1964 Thunderbirds. He also worked on the interior design of the first Mustang and, in 1969, was on the team that designed the exterior of the full-sized Ford. In 1981, Quinlan was appointed a senior stylist in the interior studio and, in September 1966, he became a manager in the Ford studio.

He retired in 1989. ■

(below) The Mystere in the Styling Center courtyard with design engineer Robert Todd's Model T Ford. Todd bought the Model T from Frank Hershey.

Bill Boyer

Ed Jaquet

Edwin F. Jaquet

As a boy growing up in Clovis, New Mexico during World War II, Ed Jaquet saw all the fancy cars owned by the B-29 pilot-trainees stationed near his home, and decided he wanted to become a car designer. A teacher at Amarillo Junior College, where Jaquet attended school, recommended that he further his artistic education at the Cleveland Institute of Arts.

In the Summer of 1954, between his junior and senior years at the Cleveland Institute of Arts, Jaquet became the first student-intern to work at the Ford Design Department. While a student-intern, Jaquet worked on the design of the Mystere. In June 1955, after he graduated, Jaquet was hired as a designer at Ford. As the winner of the Gunn scholarship, Jaquet interrupted his work at the Design Department to study in Europe for 8 1/2 months beginning in September 1955. (Joe Oros, a former winner of the same scholarship, helped arrange Jaquet's itinerary.)

Jaquet returned to the Ford Design Department in 1956 and was drafted five months later. After two years in the Army working as a graphic designer at the Pentagon, he returned to Ford and worked in the the corporate advanced studio, where he, Bob Thomas and others designed the Allegro. Promoted to management under Joe Oros, he worked on updates for the 1965 Thunderbird and the Mustang, styled the 1966 Fairlane and Falcon and the 1968 Torino line. Assignments followed in various interior, advanced and the Ford exterior studio, where he created the 1974 1/2 Gran Torino exterior. Jaquet later worked as a manager in the Mustang and Falcon studio and in the Fairlane studio. Jaquet also worked on the 1964 Falcon, the Torino fastback, and the Granada and Monarch. He was also on the team that first developed image boards, now adopted industry wide.

Since retiring from Ford in 1991, Jaquet has been a part-time instructor at the Cleveland Institute of Arts. ■

Almost every feature of the Mystere was aircraft-inspired. According to Boyer, the oil coolers in the front of the car were inspired by the jet pods on the B-47 bomber. The canopy and the fins on the car were also aircraft-inspired, as were the jet-pod taillights, which had, by 1954, become a design theme on all Fords. The interior of the Mystere used aircraft-type controls on the center console and in an overhead console. The way the center console wrapped up, over, and towards the driver made it look like an aircraft cockpit. The steering wheel, which could be flipped from one side of the front-passenger compartment to the other, was taken directly from the Beech Bonanza, a light plane popular at the time. Even the shape of the gauges on the Mystere's instrument panel were made to look like a whiskey compass on a plane. Gary Hutching recalls great difficulty installing the Mystere's gauges, which were ultimately floated in alcohol, so they would look like they were suspended.

The Mystere's gas-turbine engine was also fanciful. No studies were done to see if one would fit, nor was there a specific space provided for it in the Mystere. Although the car looks like it was designed in a windtunnel, there wasn't time. In school, Boyer used windtunnels and was generally familiar with the principles of aerodynamics. The Mystere looks like it was designed in a windtunnel, because Boyer and Woods made some pretty good guesses when it was being designed.

The Mystere influenced the design of more than just the 1957 Ford and the 1957 Thunderbird. It also influenced the 1958 and the 1964 Thunderbirds, including the console and the ball-like instruments on the '64 Thunderbird. The roof vent, in smaller form, was used on the 1957 Mercury Turnpike Cruiser. The wheelcovers inspired those used on the production '57 Ford and '57 Thunderbird.

After it was introduced to the press, the Mystere was turned over to George Hackett of the Lincoln-Mercury marketing department, and it went on the show circuit. After the 1955 show season, the Mystere was returned to the Design Center and put in storage, occasionally to be removed, dusted-off and shown to new designers and sometimes to Ford executives at their request. In 1960-61, designers in Elwood Engel's corporate advanced studio repainted the Mystere a pearlescent-white. As Gale Halderman remembers it, the car just didn't have the same impact after it was repainted. Several designers report seeing it as late as the mid-1960s. In 1970, after many years as head of the Thunderbird studio and some time in the Lincoln-Mercury studio, Boyer was transferred to Ford of Australia as chief designer. The Mystere was still intact when he left, but he was later told it was destroyed while he was in Australia. ■

The rear end of the Mystere featured dual aerials and perforated panels below the deck lid designed to exhaust heat from a gas-turbine engine.

Ford Design Center

Ford Design Center

(above) *The Mystere was predictive of many design features used on the 1957 Ford.*

Ford Design Center

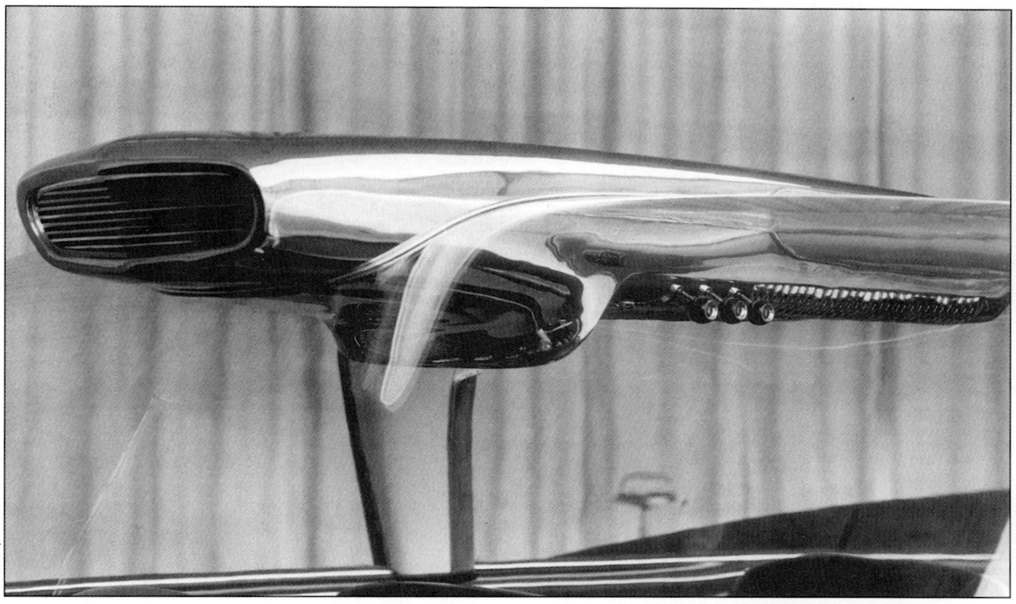

Bill Boyer

(above) *The Mystere's heating and air-conditioning system and their controls, were located in the car's roof console.*

(below) *The side swoosh on the Mystere was intended to demonstrate design continuity between the 1955-57 production Fords.*

Bill Boyer

C. Kenneth Nelson

Ken Nelson grew up in Oregon. After flying B-24s during World War II, he studied engineering at the University of Oregon for two years and then attended the Art Center College in Los Angeles, California. After graduating in 1950, Nelson took a job at GM as a designer. In 1952, he left GM to work at Ford's Design Department.

Nelson started in the Ford studio, and remained there for ten years. He designed the swoosh molding used on 1955-57 Fords and on the Mystere concept car. After days of trying to design an acceptable side molding for the upcoming '55 Ford, Nelson, one of several designers working on the project, drew the swoosh on a side-elevation drawing as a joke. Dave Ash came over to see what the other designers were laughing at and, to Nelson's surprise, selected it as the final design. Nelson also did some of the presentation renderings of the mid 1950s concept and showcars designed in the Ford body development studio.

Nelson was later assigned to the Lincoln interior and then the truck interior studios. In the late 1960s, he was named manager of the advanced studio and, in 1971, he replaced Gil Spear as chief designer at Ford of England.

In 1975, Nelson returned to Dearborn, and was named executive designer in the advanced studio, where he was in charge of all concept and showcars and special vehicles for the Ford family. Before he retired in 1980, Nelson spent several years in the truck interior studio, working for Gale Halderman. ■

(above and right) Bill Schmidt liked the look of the more formal towncars and limousines. The fins on the Dual Cowl Phaeton were predictive of the 1957 Lincoln. The Futura's influence is also unmistakable, even with the top on the car

Dual Cowl Phaeton

Bill Schmidt had been the head of the Lincoln and then the Lincoln-Mercury studio since 1947. He disagreed with Earl MacPherson's decision to downsize the 1952 Lincoln and, when it came time for the next-generation Lincoln, he felt it should be a bigger car to compete head-to-head with Cadillac and not Oldsmobile as was the case with the '52 Lincoln. A dark-horse design done by Schmidt of the type and size car he thought the new Lincoln should be won Ernie Breech's approval and became the 1956 Lincoln.

While the Futura was being built at Ghia in 1954, Schmidt and Don DeLaRossa designed a proposed Lincoln towncar based on the 1956 Lincoln sheetmetal but with many of the design characteristics of the Futura. The proposed '56 Lincoln Dual Cowl Phaeton was built as a 3/8-sized clay model, but went no further. A separate top was built that could be placed on the model to turn it into a towncar.

Ford usually chose to go in different directions than GM did with its concept cars but, many times, Ford concept cars were a reaction to what GM was rumored to be coming out with. The 1953 Pontiac Parisienne and, especially the 1956 Cadillac Eldorado Brougham Town Car, were GM concept cars that probably influenced Schmidt to design and build a Lincoln towncar as Ford's response.

Schmidt's departure from Ford in 1955 and the failure of the super-luxury Continental Mark II, probably doomed any thought of offering a production Lincoln towncar, assuming a production towncar was seriously being considered by Ford Motor Co. GM's showcars did not immediately lead to a Ford towncar show vehicle, but Lincoln did produce two full-sized towncars for the show circuit in the mid-1960s.

During his years at the Design Department, Bill Schmidt thought Ford Motor Co. ought to build a production Lincoln Dual Cowl Phaeton or towncar, and he designed many of them throughout his career. Even after he left the Ford Styling Center and, after he retired from Packard and Chrysler, he continued to design proposed Lincoln towncars and limousines. Some of those designs are in the design collection at the Henry Ford Museum. In the 1970s, some of the special Lincoln limousines Schmidt designed were privately built under his supervision. No one knows what happened to the Dual Cowl Phaeton, but it was probably destroyed shortly after Schmidt left Ford. ■

6

Continental Division Design Studio

Bill Ford is the youngest of Edsel and Eleanor Ford's four children. He is most like his father in his appreciation of fine automotive design. Like his father, he is also a genuine car enthusiast. As a young man, Bill Ford drove all of his father's special cars, including the first Continental. During World War II, Bill Ford was in training to become a Naval Aviator. When the war ended, he enrolled at Yale University and graduated in 1948 with a degree in engineering. He started work at Ford Motor Co. that year and, in June 1948, he was appointed a Company director.

Between the end of World War II and 1960, Ernest R. Breech, executive vice president, Lewis Crusoe, head of Ford Division and George Walker were primarily responsible for Ford's corporate design policy. The exception is the Continental Mark II, which was designed, built and marketed under Bill Ford's direction.

After the Continental Mark II project was authorized in 1952, Bill Ford established a separate Ford Motor Co. Division at the recently-vacated Henry Ford Trade School building. He hired his own engineers and designers to build a car that was a direct descendent of his father's Continental. Bill Ford was totally devoted to and involved in the design and development of the Continental Mark II and the new facilities the Continental Division eventually occupied. He championed the retractable-top concept and developed it as the Mark II Retractable. Continental Division also developed and built the prototypes of the Ford Skyliner retractable after the Mark II Retractable was canceled.

Ford Design Center

William Clay Ford

The Continental Mark II was only built during the 1956-57 model years. In 1958, after the Mark II had been discontinued, and a proposed Continental Division Mark III Berline had been rejected for cost reasons, Bill Ford was assigned responsibility for Ford's corporate design and, in May 1973, he became Ford's vice president of product design. His other avocation is the Detroit Lions football team, which he owns.

Starting with the 1961 Lincoln, Bill Ford's Continental Mark II was adopted as the design inspiration for all Lincolns up through current production. The Continental Mark II also inspired the design of many other Ford products, including the original Mustang.

Bill Ford, like his father, is modest, unassuming, well respected and well liked. Although he was trained as an engineer, he has inherent good design sense. In 1989, when he retired, his full-sized portrait was placed next to the one of his father in the foyer of the Design Center. Those two portraits now symbolize the influence father and son have had on Ford design over the past 65 years. ■

Continental XC-1500R Retractable

The open-model Continental Mark II was supposed to be a retractable hardtop. Three years of research and development and over two-million dollars were spent on the Mark II Retractable, and one prototype was built. It was a fully-operable car known as the mechanical prototype #5 or MP#5 for short. Because there was no room at the old Henry Ford Trade School building, where the rest of the Special Products Division (later Continental Division) was initially located, the team put together to build MP#5 occupied its own separate quarters located in the front garage building (Building 19) of the old Lincoln assembly plant at Warren and Livernois streets on the western edge of Detroit. At its peak, the team working on the retractable project had 150 employees including its own engineers, mechanics, body and paint people, etc.

Ford Motor Co. was not the first automobile manufacturer to consider building a retractable, nor was it the first to actually build one. By the 1930s, efforts were actively underway to develop a car with a fixed hardtop that could be collapsed either on top of or into the trunk. In the United States, retractable prototypes were proposed at Chrysler, American Motors, Kaiser-Fraser, GM, Ford and probably other manufacturers. Some one-offs were built, but none were mass produced. In 1934, a British Grimston Hartley retractable prototype was built with a top that retracted to a position on top of the trunk. In 1934, Peugeot built an experimental car with a top that retracted into the trunk and, in 1936, Peugeot displayed a retractable hardtop-convertible model at the Paris Auto Show. From 1936-39, Peugeot produced Types 402 and 601 with optional steel tops that electrically retracted into the trunk and, in 1940-41, Chrysler produced several Thunderbolt showcars with hardtops that retracted into the trunk.

Michael Schultz

The fully-operable Continental Mark II Retractable on the turntable at the Styling Rotunda, after it was shown to Ford's board of directors. When the top made its first full cycle, everyone in the room broke into a spontaneous round of applause.

In the early 1950s, Gil Spear, head of Ford's Advanced Studio, championed the retractable concept at Ford Motor Co. and, in 1953, Spear sold Bill Ford and Harley Copp on the idea of a retractable hardtop for the new Continental Mark II. Financial support to the tune of a $2.19 million allocation for development of a retractable model Continental was approved by Ernest R. Breech, Ford's executive vice president, Lewis D. Crusoe, vice president and head of Ford Division and Earle S. MacPherson, Ford Motor Co.'s chief engineer. The proposed retractable budget was confirmed by Ford's executive committee as part of the overall Continental Division budget.

The decision to develop a Continental retractable model was made early enough in the design process of the Mark II, so that the retractable mechanism could be custom-fit within the Mark II's existing design, which was one of the initial requirements of the project. Some Special Products Division designers, including chief designer John Reinhart, initially opposed inclusion of a retractable model in the Continental lineup as too gimmicky. There was also a fear in the Continental design studio, that a retractable model Continental could not be built without compromising the overall design of the car. Nevertheless, the decision to go ahead with the retractable as a second model in the new Continental Mark II lineup was made by Bill Ford, head of Special Products Division and

chief engineer Harley Copp. Those who worked on the retractable project remember Reinhart as very cooperative and helpful, even though he was still concerned that a retractable model might compromise the design.

Harley Copp hired his brother-in-law, John R. Hollowell, to head the team assigned responsibility for developing an operable Continental Mark II retractable. Because Hollowell had expertise in linkage, which was the heart of the retractable project, he was recruited to head the project and started work on it July 1, 1953. Hollowell, in turn, recruited a 30-year-old engineer named Ben Smith, who he had earlier met while working at Ford. Smith was recruited from GM's Fisher Body Division to help design and build a working full-sized retractable.

Smith, early in his career, worked with Hollowell on the Light Ford project, but left Ford in 1949 and, since then, had worked for Nash and GM. At GM, Smith was an engineer working on convertible tops, and that expertise was the reason he was recruited to work on the Mark II Retractable.

(left to right) Ben Mills, assistant manager of Continental Division, Harley Copp, Continental Division chief engineer and John Hollowell, engineer in charge of the retractable project, are inspecting the retractable's linkage, before they took the prototype for its first test drive.

Because it had already been decided that the Continental Mark II would be developed separately from other Ford car lines and in secret, when Hollowell first offered Smith the job, he couldn't even tell him what he would be doing. It took four or five meetings and a full disclosure before Smith, finally intrigued by the idea of being one of the engineers selected to actually build a retractable vehicle for production, agreed to take a cut in pay to come to work on the project. When he left GM, Smith told his boss he was leaving for a higher-paying job at Ford. Nobody would have been the wiser, except that Smith's boss at GM sent a nasty letter to Bill Ford claiming that recruiting Smith away from GM by offering him more money than he was worth was an unethical recruiting practice. When Smith explained that he couldn't very well have told his boss at GM that he was leaving for a job at Ford that paid less money, Bill Ford got a good chuckle out of it, and the complaint was quickly round filed.

At Smith's suggestion, Hollowell hired Roy Butler, one of Smith's co-workers at GM, as Smith's assistant. John Hollowell, Ben Smith and Roy Butler produced the working Mark II Retractable prototype, and they did it within budget and within one week of being on time.

Special Products Division gave Hollowell 18 months to produce a working Continental Retractable. The retractable project got started in earnest on July 1, 1953, and it was to be completed by January 1, 1955.

The Continental Mark II Retractable was badged as the XC 1500 R.

Although the Mark II Retractable's sheetmetal was similar to the production Mark II, it was made from a cobbled 1953 Lincoln mule.

Michael Schultz

(above) Ford's board of directors asked to see the top go up and down so many times it almost discharged the car's battery.

Mark Hollowell

John R. Hollowell

John Hollowell attended the Edison Institute, now Henry Ford Community College, from 1941 to 1943 and then transferred to the University of Michigan, where he graduated in 1944 with an engineering degree. In 1946, after serving in the Navy during World War II, Hollowell took a job at GM for a short time. He then worked at Ford on the Light Ford (Vedette) project. After the Light Ford project was canceled, Hollowell transferred to Ford's Cincinnati transmission plant, where he helped develop Ford's automatic transmission. In July 1953, he was recruited by Harley Copp, his brother-in-law, as chief engineer on the Mark II retractable project.

Hollowell's expertise was linkage, which was the heart of the retractable project. He continued as head of the retractable project until October 1955, when Continental Division was awarded a contract to develop a Ford retractable. Hollowell then moved to Lincoln Division as advanced chassis engineer. Hollowell soon moved to Ford Division as a chassis engineer, and was in charge of developing the unibody 1958 Thunderbird. Hollowell then became advanced light car engineer on the Cardinal and Comet programs, but left Ford in 1960 to accept a job at Fairchild Stratus developing a proposed infinitely-variable-hydraulic automatic transmission.

In 1962, Hollowell returned to Dearborn and opened his own engineering firm. One of his first jobs was developing the chassis for the new Bronco 4-wheel drive. Hollowell's firm also developed military vehicles for Kaiser-Jeep and AMT-General and many other automotive products. Hollowell's engineering firm is now operated by his son. ■

Michael Schultz

(above) The package tray rotated upwards to allow the top to retract into the trunk. When the roof was in the trunk, the package tray returned to rest.

Michael Schultz

(left) Unlike a standard convertible, the back seat in the Continental Mark II Retractable was full-sized.

Ford Photomedia

The deck lid on the original three-eighths sized model of the retractable opened from both ends like the Syrtis. Engineers later determined it was impractical and dangerous to build a trunk lid that opened from both front and back.

The rear-seat area of the Continental Mark II Retractable was full width with the top in the up or dawn position. XC 1500 R script is on the sail panel.

By the time Hollowell's engineers started work on the retractable, a full-sized clay model of the standard Mark II had already been approved. The instructions Hollowell's team were given required that the retractable top be fully automatic, the roof be the same size and shape as the one already designed for the standard Continental Mark II, no other exterior dimensions of the vehicle were to be changed, the trunk lid had to open from the back and from the front, and the backlight had to roll down.

According to Hollowell, the problems that had to be solved and the difficulties that had to be overcome to build a production-ready, working retractable prototype were monumental. In some cases, entirely new technologies had to be developed from scratch. It was hard enough just to develop the complicated mechanical linkage that ran the top up and down and opened and closed the trunk. The complicated linkage that moved the top from the up position to the trunk had to be designed to not only fold into the trunk, but to travel high enough so the passengers would not be decapitated in the process. Hydraulic technology as a way of operating the top was quickly rejected in favor of new, smaller and more powerful electric motors Hollowell and Smith developed with the help of outside suppliers. A system of screw-jacks and locks was developed that, at the trunk end of the car, moved and collapsed the top into the trunk and, at the other end, fastened the leading edge of the top to the header by electric motors. This new method of attaching the convertible top to the header by screw locks was first used on the 1956 production Lincoln convertible.

Michael Schultz

In the beginning, Hollowell had constant personnel problems. Continental Division was set up with six separate departments; design and engineering, purchasing, finance and budget, program and timing, and shop prototype build. The personnel working on the retractable project were from all six separate departments. When there were disputes, and employees didn't like what Hollowell told them to do, they went to their department heads and, more often than not, Hollowell was overruled. Hollowell finally did some bargaining with Copp on something extra Copp wanted done within the existing budget. The only way Hollowell said that what Copp wanted could be done was if Hollowell had the final say on the project instead of the various department heads. Within days, an order came down, signed by Bill Ford, naming Hollowell as the final authority on what happened on the retractable project.

Halfway through the development process, an alternate study was initiated with a different group of engineers to see if a way could be devised to articulate the top in the middle and thus take up less trunk room. As part of the alternate proposal, which came after the budget had been set, Copp also proposed that two mechanical prototypes, each with a different style retractable roof, be built. Copp was quite taken with the new articulated-top proposal someone not associated with the project had made, and he wanted it built as an alternative retractable prototype. The alternate proposal would have involved putting the front section of the roof on rollers, and retracting it on top of the back section, as the entire top retracted into the trunk. A supplemental budget was initially approved but, before the money was made available, someone with more authority than whoever approved the supplemental appropriation canceled it. Hollowell and Smith said it couldn't be done. After spending about $140,000 on the alternative proposal, Hollowell and Smith's judgment was confirmed.

In the meantime, two mechanical prototypes had been ordered from Hess and Eisenhardt of Cincinnati, Ohio, for both variations of the retractable project. The mechanical prototypes were cobbled 1953 Lincolns, but they had frames that were custom-built to the expected dimensions of the proposed Continental Mark II. Only one of the mechanical prototypes was used, because only one retractable prototype was built. Hollowell later sold the leftover mechanical prototype to Lincoln-Mercury Division for $17,000, and it was used to build the Futura concept car.

Because the Continental Mark II was designed as a four-passenger vehicle, with front and back seats, the top was too long to fit in the trunk in one piece. At first, Smith proposed that the front section of the top remain permanently attached to the header as a kind of a stylish sunshade. That proposal was rejected and, eventually, Hollowell, the expert in linkage, worked out a roof design, where the front section of the top, or the flipper as it came to be called, was articulated so it folded under as the top moved towards the trunk. The design Hollowell's team came up with required a three-inch concession in the body sizing of the proposed retractable, so that the top, even with the flipper folded under, would fit in the trunk. The practical effect was that, on MP#5, the back seat was moved slightly forward, and the rear deck was made slightly higher and longer than on the standard Continental Mark II. Proposals for a retracting backlight and a trunk lid that opened both ways were eventually abandoned. The retracting backlight was just not workable, and the two-way trunk lid was eliminated after a small working model showed the two-way trunk lid to be dangerous if something went wrong.

Because of all the switches and relays necessary to make the top retract and go back up in the right sequence, and the trunk open and close at the right time, a new way of controlling and sequencing all of the separate functions had to be developed. That led to production and the first use of electrical multiconnectors, which are now in common use in the automobile industry and elsewhere.

As work progressed, there was a lot of talk about what to call the retractable. It was already being referred to by those working on the project as everything from the flip top to a hardtop convertible. At one point, someone proposed that the system be named the "fully automatic retractable top." That name was almost adopted, until Bill Ford realized

Ben Smith

Ben J. Smith

Ben Smith grew up in Detroit. At the beginning of World War II, he took a job at Ford Motor Co. as an apprentice tool and diemaker. He attended the Henry Ford Engineering School and became its first graduate. Starting in 1943, Smith transferred to Willow Run as a die designer. In 1946, after 17 months in the Navy, Smith returned to Ford as a chassis draftsman on the Light Ford program. In 1949, Smith took a job with Nash and worked on the Nash Rambler. He then moved to GM's Fisher Body Division as a body engineer, and worked in Fisher Body's hardtop and convertible section.

In 1953, because of his experience with convertible tops, Smith was recruited by Ford's Special Projects Division to work on the Continental retractable program. Later, as a Continental Division employee, Smith engineered the 1957 Ford Hide-Away-Hardtop.

In March 1956, Smith was promoted to Lincoln advanced engineer. From late 1959 until early 1964, he was chief engineer at Ford of Argentina, where he helped redesign the U.S. Ford Falcon for Argentine production. In 1964, Smith was appointed executive engineer for Ford International and, in 1965, at Bill Ford's request, he was placed in charge of advanced package department at the Design Center. In 1966, Smith developed a manually-retractable hardtop for the Mustang, but it was not produced.

From 1967, until his early retirement in 1968 because of his wife's declining health, Smith was product director at Ford of Brazil. ■

Ray Smith's three-eighths sized model of the Continental Mark II Retractable was built to show what the production Mark II Retractable was expected to look like. Smith's job was to make the retractable model look as much like the standard Mark II as possible.

Jesse Richards

Jesse Richards

From the rear, Ray Smith's three-eighths sized clay model of the proposed Mark II Retractable shows that it would not have looked all that much different than the standard two-door hardtop model. The greenhouse on the Retractable was two inches shorter and the deck lid two inches higher.

what the acronym for that set of words was. The problem of a name for the mechanism disappeared when the name Retractable came into general usage.

On October 13, 1953, the decision was made to build a 3/8-sized model of a Continental Mark II with a retractable top that worked. Contemporaneously with the construction of the 3/8-sized model and a static buck, a full-sized mechanical prototype Continental Retractable was started. Because there had already been four full-sized operable mechanical prototype Continental Mark IIs built to test other development work being done elsewhere in Special Products Division, the retractable prototype was called Mechanical Prototype #5. It was designed and built to be as close as possible in looks and engineering to the Continental Mark II at that stage of its design development. The mechanical prototype came from Hess and Eisenhardt with front fenders, grille, bumpers, hood, doors and part of the back fenders. The rest of the back fenders, deck lid and the trunk area sheetmetal had to be fabricated or refabricated by Hollowell's team. MP#5 was built and rebuilt by trial and error until the retractable mechanism worked the way it was supposed to. The design work on the 3/8-sized models and on the full-sized mechanical prototype was done by Continental Division designer Ray Smith.

The Continental Mark II Retractable was also given the code name XC 1500 R. The original code name for the Continental Mark II was X 1500. The "C" was added to signify the Continental and the "R" was added to identify it as a retractable model. At one point, it was proposed to put XC 1500 R script on the C-pillars of the retractable top, but that was never actually done. XC 1500 R script was placed on the inside of the sail panels and at the top of the grille.

In April 1954, a survey was sent out to prospective Mark II customers to see how many of them would buy a Mark II Retractable. The proposed price listed for the Mark II on the survey was $8,000, and prospective buyers were asked if they were willing to pay an extra $2500 for a retractable model. Although a response of about 20% was expected, 90% of prospective buyers expressed a willingness to spend the extra $2500 for a retractable Mark II. The engineers working on the project were not prepared to guarantee they could build that many Mark II Retractables, or that there would be no problems. Ultimately, the retractable's anticipated success was one of the factors that killed it. By September 1954, at Hollowell's direction, Smith and Butler and possibly others, began work on a bootlegged copy of the design and engineering studies of the proposed 1957 Ford to see if it could be made to accommodate the retractable mechanism. With some modifications, they determined it could.

By mid-May 1954, a retractable top on a static buck had been constructed that worked. It was tested by three engineers riding the top up and off the header and halfway into the trunk. By mid-September 1954, Ford Division became aware that Hollowell's crew would likely succeed in producing a working retractable -- and they also became aware of the results of the market survey. Ford Division executives requested and were given a briefing on the project. Word soon spread, and the other Ford Motor Co. divisions were trying to figure out if and how they could market their own retractable. By October 1954, the retractable top mechanism was perfected to the point that it had cycled 2400 times on the static buck without a problem. By this time, the full-sized MP#5 was also about 80% complete.

On September 20, 1954, Hollowell, Smith and Butler filed for a patent for a Vehicle With Retractable Rigid Top. Patent number 2,860,004 was granted to them on November 11, 1958. Other patents were granted for the header mechanism, the retractable mechanism itself and the saddle gas tank that was located behind the backseat and straddled the driveline. As was customary, all of the patents were assigned to Ford Motor Co..

A continental spare tire outline was added during the design process of the Mark II Retractable, but after MP#5 was built.

Michael Schultz

On November 11, 1954, Hollowell and Smith made their first formal presentation of the retractable top mechanism on the static buck to Benson Ford, Ben Mills, Lewis Crusoe, Robert McNamara and other executives. McNamara was so enthusiastic, he left the presentation convinced that the retractable might completely replace the regular convertible. There is an often repeated story that, during the November 11 presentation, while the retractable top was being cycled, the trunk lid came off, flew through the air, was caught by one of the embarrassed engineers and quickly reinstalled on the static buck, after which the presentation was continued. Roy Butler, in addition to being a talented engineer, liked to write poetry. The story of the flying trunk lid was included in one of Butler's poems and has since been taken as fact. It is not.

By December 9, 1954, MP#5 was done except for a few pieces of trim. By January 7, 1955, only one week behind schedule, MP#5 was finished. It was painted pearlescent light silver-blue, with mother-of-pearl flakes in the paint. The top was painted a darker blue. The seats were upholstered in light-blue and white Bridge of Weir leather, and the instrument panel was painted a medium-blue. Light-blue 1 1/2 inch-thick Mouton carpet was installed on the floor.

On January 7, 1955, Hollowell and Smith showed MP#5 to Bill Ford, Ben Mills, who was by then Bill Ford's assistant at Continental Division, and Harley Copp. The top worked flawlessly. Hollowell then invited Ford, Mills and Copp to accompany him on the maiden voyage in MP#5. As they were getting into the car, Copp, the voice of experience, asked Hollowell if there was grease in the rear end. Hollowell assured Copp that there was. Soon, the four of them were riding through the streets of Detroit in MP#5. Every once in awhile, they pulled over to the curb to operate the retractable top. After driving a few miles, they started to hear a howling sound coming from the back of the car. Hollowell quickly made a U-turn and headed back, but they didn't make it very far. A chassis engineer had forgotten to fill the rear end with grease, and it had frozen up tight stranding Bill Ford, Mills, Copp and an embarrassed Hollowell with Ford's top secret car on a public street in the middle of Detroit! Soon, help was summoned and car and passengers were returned to the plant. By the following Monday, the problem was fixed with no apparent adverse consequences to car or crew -- except for the chassis engineer, whose ears were probably still ringing from the chewing out he got.

Hollowell was then asked to show the car at a Ford Motor Co. board of directors meeting, which was arranged at the Design Rotunda. MP#5 was placed on one of the three turntables built into the showroom floor and was slowly rotating. At just the right moment, the curtains were opened, and the directors saw a car that very few had heard anything but rumors about. When the top cycled, the room broke into applause. It took about 25 cyclings of the top to satisfy everyone's curiosity, and the car performed flawlessly. Over the next month, Hollowell's people were at the Styling Center continuously showing the car to Ford, Lincoln, Mercury and Edsel executives and engineers.

By January 1955, it was obvious that the retractable developed by Continental Division was going to be built as a production vehicle. By the first part of February 1955, the decision had been made to build a 1957 Ford retractable -- not in addition to but instead of the Mark II. It was decided that a retractable model would only increase the financial burden on the already financially overburdened Continental project. Members of Ford's executive committee also questioned the wisdom of risking such radically new and untried technology on a model meant to be Ford's luxury flagship. Between 1957 and 1959, Ford Motor Co. built and marketed the Ford Skyliner. With the pull of a knob, its 13 switches, ten solenoids, nine circuit breakers, five electric motors and over 600 feet of wire would make the top go from hardtop to convertible and back again on demand.

In reality, the retractable mechanism on MP#5 worked well when it had to, but the engineers who built it crossed their fingers every time the top went up or down. By the time the retractable became a production Ford, the problems and quirks had been solved. On MP#5, however, it was common for the circuits to work in the wrong sequence and, when that happened, Hollowell says they were always afraid the mechanism would self-destruct. Although a lot of executives and engineers drove MP#5, Hollowell and Smith were always careful that whoever drove it knew what not to do if something went wrong. Only once did things go wrong to the point of doing damage that had to be repaired.

After the Continental Mark II Retractable project was canceled, MP#5 was driven and even taken home overnight by Hollowell, Smith, Butler, other engineers who helped design and build it and by Ford Motor Co. executives. Smith regularly drove it for two years after the Continental Retractable program was discontinued. In the summer of 1957, he came back from a vacation to find that MP#5 had been scrapped. Smith saw the scrap

Roy Butler

Roy T. Butler

Roy Butler grew up in England and attended Essex Technical College. After graduating in 1938, he took a job in the blueprint department at Briggs Bodies Ltd., a company that built car bodies for Ford and other automobile manufacturers. By the time World War II started, Butler had been promoted to Briggs' design department. During the war, he was in the Royal Air Force and flew night fighters. After the war, Butler returned to Briggs' design department, where he stayed until 1950. He then moved to the United States to work at GM's Fisher Body plant as a layout man developing convertible tops.

While at GM, Butler became friends with Ben Smith, a co-worker. When John Hollowell persuaded Smith to come to work for Ford's Special Products Division to help engineer and build the Mark II retractable, Smith, in turn, persuaded Butler to follow.

Butler worked on the Mark II Retractable through completion, and then worked on the Ford retractable until 1956, when he returned to England to work for Ford of England. In 1958, Hollowell persuaded Butler to return to Dearborn as a supervisor in Experimental Engineering.

In 1964, Butler transferred from Experimental Engineering to the Design Center as a design engineer and department manager. By 1966, Harley Copp, former chief engineer for Continental and Lincoln Divisions, had moved to England, where he was in charge of Ford of England's engineering and product development. Copp asked Butler to start a design engineering group at Ford of England. Butler worked with designers Duncan McRae, Gil Spear, Damon Woods and Joe Oros.

In 1970, Butler returned to the Design Center in Dearborn, where he was placed in charge of the safety section, a job he held until his retirement in 1980. ■

Michael Schultz

(above) Left to right, shown admiring their creation, are Roy Butler, body engineer Avery Morris, chassis engineer Gene Hildebrandt, test engineer Dale Pettre and electrical engineer Baird Wcllace. The workmen in the background are unidentified.

order but has also heard persistent rumors of MP#5's continued existence. In 1994, a former Continental Division engineer reports seeing MP#5 at the home of a retired Ford employee who, he claims, took it home rather than see it destroyed. That report has not yet been confirmed. ■

Charles C. Keresztes

Charles Keresztes was part of the Hungarian aristocracy by birth. He attended the Royal Hungarian Military Academy before World War II. Keresztes was also a world-class athlete in swimming, fencing, tennis and skiing. He attended the Academy of Fine Arts in Dusseldorf, Germany and graduated with an advanced degree in sculpture. He was in training for Hungary's equestrian olympic team when World War II started. When his superior officers were killed in action, Keresztes, then age 19, assumed command and led his troops in the last horse calvary charge in history.

Hans Rockel

His family's wealth, position and property disappeared in the aftermath of the war. In 1951, he emigrated to the United States and was hired to work as an illustrator at the Ford Administration Building. In 1952, Keresztes transferred to Special Products Division, (Continental Division), where he worked as a clay modeler on the Mark II. Keresztes, a skilled metal sculptor, made many of the Continental Mark II's trim parts.

After Continental Division was disbanded in 1956, Keresztes transferred to the Styling Center, where he designed and sculpted in relief many of the insignias used on Ford's concept cars and many of Ford's most recognizable symbols, including the Mustang horse, the Cougar's head, and the Pinto script with flag.

Keresztes taught figure drawing at Wayne State University. He was also an accomplished yachtsman, sailing in 14 Mackinaw races. He retired in 1982. ■

Ben Smith

(above) The Retractable was licensed, and Ben Smith used it as his daily transportation for two years.

(below) Ben Smith is shown with the Continental Mark II Retractable. While on vacation the car was supposedly scrapped.

Ben Smith

Ford Photomedia

The Continental Mark II Cabriolet was introduced to the public at the Texas State Fair in October 1956.

Continental Mark II Cabriolet

When Special Projects Division (later renamed the Continental Division) ordered the mechanical prototypes from Hess and Eisenhardt, the order included a true convertible, and two mechanical prototypes that were to be made into the retractable prototypes. At that time, plans still may have been in flux, but the cost of the retractable program probably indicates that the convertible model was intended as a backup in case the retractable program was not successful or, at the very least, that it was undecided whether both would be offered. Whatever the case, the retractable and the convertible both depended on the success of the production Continental Mark II two-door hardtop.

By February 1955, even though Continental Division had successfully completed the first working retractable prototype, it was decided that a retractable Mark II would not be built. There were several reasons the retractable was not built as a Mark II but, as Bill Ford confirms, those reasons really come down to a lack of money. The Mark II remained a two-door with a fixed roof. although studies were undertaken to determine if Mercurys, Lincolns, Thunderbirds and even Edsels should include retractable models. Although there was a Ford retractable, no other Ford product was built as a retractable model.

Originally, when the Mark II was authorized, a four-year design cycle was anticipated. Because the Mark II depended on the production Lincoln for its running gear, when it was decided that, for 1958, Lincoln would have a unibody and use the Lincoln 430 CID engine, the Mark II's four-year design cycle became a two-year cycle, which eliminated any chance of breaking even. That killed the Mark II convertible and the Mark II .

In early 1956, Continental Division decided to build three Mark II Cabriolets, but what turned out to be the only Mark II Cabriolet was built as a showcar. According to Bill Ford, Continental Division was also putting its toe in the water to see what the response would be to the first Mark II Cabriolet. That turned out to be wishful thinking. By that time, it had already been decided that any Mark II built in 1958 would have to be a unibody Mark III, so by the time the Mark II Cabriolet was authorized, it was a last hurrah. By the time it was built, there was no plan to market a Mark II convertible and, by July 1956, Continental Division was folded back into Lincoln Division.

All Mark II bodies were made by an outside supplier, Ionia Manufacturing Co. But, on a one-time-only basis, Continental Division contracted with coachbuilder Enos Derham of Rosemead, Pennsylvania, to build three Mark II Cabriolets. The first car delivered to Derham's shop bore Mark II serial number C56 G 3190. It was delivered in May 1956. Paul Wagner, who later bought the cabriolet from Bill Ford, was told that a leftover pre-production Mark II was used to make it.

Design studies for the Mark II Cabriolet were completed at the Continental Division design studio by designers John Reinhart and Charlie Phaneuf and, at that time, it was decided the top should have blind quarters like the original Continental. Although Charlie Phaneuf prepared renderings of a proposed Mark II convertible as early as 1954, those were meant only to tempt management, and they were not the basis for the one-off cabriolet built at Derham. When it was decided to build the Mark II Cabriolet as a showcar,

William C. Ford

An early full-sized clay model of the Continental Mark II. From left to right are Bill Ford, clay modeler Hugh Bush, chief engineer Harley Copp, Bob Thomas, John Reinhart and an unidentified clay modeler.

Derham was sent line drawings and a 3/8-sized rendering of the proposed convertible. Derham was given great latitude in building the Mark II Cabriolet. Changes were made to its design by Derham's craftsmen to accommodate the stock Lincoln parts available, and the car was actually built around the convertible-top assembly.

Derham was sent a stock Mark II two-door hardtop, and they cut the top off. Stock 1957 Lincoln top bows and other parts from the 1957 Lincoln convertible, including the top well, were used. Although it is often reported that the doors on the cabriolet are thicker than the stock doors on a Mark II hardtop, inspection of the cabriolet confirms that is not true, nor is the body wider than other Mark IIs. The back-seat area has been reduced to accommodate the top bows on each side. The frame on the Cabriolet was strengthened at Derham by welding 1/4-inch steel plates to the bottom of each outside-frame rail, and anti-shake braces were bolted to the frame underneath the back-seat area. The windshield header of the Mark II Cabriolet was custom-built from aluminum as were the corresponding parts on the convertible top. The A-pillars were reinforced with rods that extend into the kick panel area. The C-pillar area of the body was also reinforced. The top motors and hydraulic system used on the Mark II Cabriolet were stock 1957 Lincoln.

Because three Mark II Cabriolets were planned, Derham ordered or made three sets of the special parts necessary to build them. Mark II cabriolets were assigned separate

Ford Design Center

production numbers, and it is likely a new serial number was assigned the Mark II Cabriolet Derham actually built. Paul Wagner claims that the Mark II Derham converted to a convertible originally had exposed air-conditioning nostrils that were covered over with lead by Derham during the rebuilding process.

After the first Mark II Cabriolet was completed, the other two cabriolets were canceled because of the financial problems being experienced at Continental Division and the lack of any orders for further custom Mark II Cabriolets. After completing the authorized Mark II Cabriolet, Derham advertised that it would build other Mark II convertibles to order. There is some evidence that Derham built one additional convertible for a private party. That evidence, however, has not yet been confirmed. Only one Mark II Cabriolet was ever authorized by Ford.

After Derham finished the authorized Mark II Cabriolet, Harry Mills, the project engineer, brought it back to the Continental Division, so all the 1956 parts could be updated, and the car changed to a 1957 model. In the process, the engine and transmission were replaced with stock 1957 versions.

(above right and below) Early renderings of a proposed Continental Mark II convertible model were prepared by designer Charlie Phaneuf.

Ford Design Center

The Mark II Cabriolet was originally painted a pearlescent-white lacquer, and had a white and red leather interior. The top bows, side rails and linkage on the top mechanism were chrome plated, and the top was made of white orlon. The top boot was red leather. In the back-seat area, the side and middle armrests were eliminated, when the rear-seat upholstery was redone, so that the seat back, now with a single biscuit, was more upright than on the standard Mark II hardtop.

The Mark II Cabriolet was first displayed to the public at the Texas State Fair, on October 12, 1956. An accompanying press release said a Mark II convertible model was being produced in limited volume but, by that time, Continental Division had already been disbanded. George Haviland, public relations manager for Continental Division and later with Lincoln Division Public Affairs, recalls that soon after the Mark II Cabriolet was shown at the Texas State Fair, a modest number of letters started arriving from prospective customers asking if they could order a Mark II convertible. By that time, Haviland had already left Continental Division and was working at Lincoln Division public affairs. He approached his manager, John Mayhew, and suggested that a limited number of Mark II convertibles be authorized. Mayhew thought it was a good idea, and suggested that Haviland make his proposal to the Lincoln marketing manager, which he did. Haviland was unsuccessful in convincing the Lincoln Division marketing people they should offer a Mark II convertible.

When it came off the show circuit in early 1957, the Mark II Cabriolet was given to Bill Ford's wife, Martha. Bill Ford's recollection is that the Cabriolet was offered to him first. Because he was satisfied with the car he was driving, he suggested giving it to his wife, who liked convertibles, as an "unexpected present." She was delighted with the gift and had it repainted a light-blue metallic, and a black top was installed. Later, at her request, the Cabriolet was again repainted a dark metallic blue-green. In about 1959, Mrs. Ford had the car redone for a third time at the Styling Center shops. At that time, the three previous coats of paint were removed, and the car was repainted with a light-blue lacquer. Seven new hides of blue and white Bridge of Weir leather were also ordered to reupholster the seats and instrument panel, the top boot was remade in blue leather, and the carpeting was redone in black mouton. One of the upholsterers at the Styling Center shops made Russian-style fur hats for many Styling Center employees from the left-over mouton carpeting. The car was then returned to Mrs. Ford and, for several more years, Bill and Martha Ford used it in good weather for their personal transportation.

Bill Ford recalls that Mrs. Ford tired of driving such a big car in city traffic, and they had nowhere to store it. In 1961, the Fords decided to dispose of the Mark II Cabriolet, and Ford transportation director Storms Arnold was assigned the task of selling it. Although the car was a present to Mrs. Ford, it was still technically owned by Ford Motor Co.

When the Mark II Cabriolet was offered for sale, Paul Wagner, then an electrical engineer in Ford Motor Co.'s Experimental Engineering Laboratory, was first there with his money. Wagner paid $7213 for the car including tax and title transfer. Mr. Arnold voluntarily agreed to reduce the price of the car by $1000, "because it didn't have a top." Wagner didn't argue, paid the money, got a receipt, drove the car home and put it in his garage. In the meantime, several Ford vice presidents also heard that the Mark II Cabriolet was for sale, and they wanted it. They each talked to Wagner, and each hinted that Wagner's job was on the line if they didn't get the car. Wagner went to one of Ford's attorneys to find out his position. The attorney who Wagner saw told him the receipt was a contract, and he would see to it that Wagner got the title. He did. When Wagner bought the car, it had 15,000 miles on it. After the fact, Wagner agreed to let Ford Motor Co. use the car when they wanted to. This condition was also passed on to the Goeppingers, the next owners. When Wagner bought the car, he also got many of the spare parts from Enos Derham, five gallons of leftover light-blue lacquer and an extra hide of leather for any future upholstery repairs.

In 1962, Wagner and his wife needed money to finish an addition to their house, so the Mark II Cabriolet was sold to Walter and Marne Goeppinger of Boone, Iowa. The sale was arranged privately by Dr. Erwin C. Ruth, then president of the Lincoln Continental Owners Club, who called Wagner and told him that if he ever wanted to sell the car, the Goeppingers wanted to buy it and would provide a good

Fred Beamish

John M. Reinhart

After studying design at the Central Academy of Commercial Arts in his home town of Cincinnati, Ohio, John Reinhart began his design career in 1936 at GM working under Bill Mitchell in the Cadillac studio. His first assignment was helping design the interior of the 1938 Cadillac 60 Special. Later, Reinhart joined Gordon Buehrig's independent design firm, but soon left because of a lack of business.

From 1939-42, Reinhart worked as a designer at Packard. He left Packard to work for Budd Manufacturing as a product illustrator. From 1943-47, he and Bob Koto worked for Raymond Loewy and Associates in England. In 1947, Reinhart was hired as chief stylist at Packard, where he designed the 1951 Packard. In 1952, Frank Hershey displaced Reinhart as chief stylist at Packard.

In 1952, when Bill Ford and Harley Copp were putting together a team to design the Continental Mark II, Copp recommended Reinhart as the Mark II's chief designer. Copp thought Reinhart was a good designer and also felt Reinhart's and Bill Ford's personalities would compliment each other, which they did.

After Continental Division was disbanded in 1956, Reinhart was transferred to Ford's Styling Center as an executive stylist in John Najjar's Lincoln studio. After Continental Division's proposed 1958 Mark III Berline design was rejected, Reinhart helped design the 1958 Lincoln-based Mark III. Reinhart also designed the Mark X for the Stylerama program and began the design of the Envoy. Although Najjar and Reinhart worked well together, Reinhart left Ford to become head of a new design department at U. S. Steel in November 1957.

In the mid-1970s, Reinhart returned to Ford and worked as a designer in an interior design studio on special projects. He retired in 1980. ■

The top on the Mark II Cabriolet was made with blind-quarter panels to resemble the top on the early Continental.

Henry Ford Museum and Greenfield Village

Charles Phaneuf

Charlie Phaneuf graduated from the Commercial Arts Department of Cass Technical High School in Detroit. He served in the Army Air Corps during World War II as a gunner in a B-24. While stationed in Denver, he studied art part-time at the University of Denver. After the war, Phaneuf finished his design studies at Meinzinger's Art Academy in Detroit, graduating in 1947.

In 1947, Phaneuf was hired as a designer at Kaiser-Frazer, where he worked under Herb Weisinger. Phaneuf was one of the designers of the 1951 Kaiser. In October 1948, Phaneuf left Kaiser to take a job in the Packard design department. There, he met John Reinhart. In August 1952, Reinhart left Packard to accept a job as chief designer at Ford's Special Projects Division (later the Continental Division). Phaneuf was the first designer Reinhart hired to work with him on the Mark II design project. Phaneuf is generally credited with designing the sides of the Mark II.

Phaneuf and Bob Thomas designed the Continental star, and Phaneuf designed the hood ornament on the Mark II, which was built under his supervision by master carver John Kruscil. (The Mark II star has since become the symbol for all Lincolns.) Phaneuf also worked on the design of Continental Division's proposed Mark III Berline.

In 1956, when Continental Division was dissolved, Phaneuf was transferred to the Styling Center, where he was assigned to the Ford studio. Phaneuf was one of the leaders of the team that designed the exterior of the 1964 Thunderbird and was one of the primary designers of the 1965 Mustang, which he modeled after the lines of the Continental Mark II.

Beginning in 1966, he was a design manager in the Corporate Projects Studio, the advanced interiors studio, and the studio where the Mark IV was designed.

Phaneuf retired from Ford in 1988. ■

Charles Phaneuf

home for it. Wagner called the Goeppingers and the terms of the sale were agreed to. The exchange took place in March 1962, at Chicago's O'Hare International Airport. Wagner took an 11,000-original-mile 1947 Lincoln Continental Coupe in trade, plus some cash. (Ford designer Jim Powers later bought the '47 Lincoln Continental from Wagner.) At the time the Goeppingers took possession, the Mark II Cabriolet had 19,000 miles on it.

Before the Mark II Cabriolet was sold to the Goeppingers, Wagner received several calls from Colonel Parker, Elvis Presley's manager, who explained that Elvis wanted the car as a gift for someone special. Colonel Parker never called back and, by the time Wagner got ready to sell the car, he had not been recontacted by Parker for several months.

Once they acquired the Mark II Cabriolet, the Goeppingers had it repainted, some chrome was replated, the stainless was polished, the engine was detailed, and a new black top was installed. On at least two occasions, the Goeppingers honored Wagner's promise to return the car to Dearborn for display at Ford Motor Co.'s request.

In 1988, the Goeppingers sold the car to their sons, Hans and Neil Goeppinger, but it is still kept at the senior Goeppinger's home. It currently has a little over 40,000 miles on it. ■

(above) The Continental Mark II Cabriolet as it appeared in the summer of 1996.

(below) Although other Mark II convertibles have since been made from production two-door hardtops, there was only one factory-authorized Mark II Cabriolet.

Gordon Buehrig's Plastic Models

In 1948, when he built the quarter-scale model of his proposed Tasco sports car, Gordon Buehrig employed a new process recently developed by U.S. Rubber. It involved vacuum forming many of the Tasco's parts from ABS plastic. The Tasco project failed and, in 1949, Buehrig was hired as a designer at the Ford Design Department. One of the first things Buehrig did was suggest to Henry Grebe, who was then Ford's chief body engineer, that three-dimensional models be used to demonstrate new construction proposals. Grebe rejected the idea because it had already been tried without success. Grebe showed Buehrig a metal model that had been made at too great of a cost to Ford. When Buehrig suggested vacuum forming the parts using ABS plastic, Grebe, who had never heard of the material or the process, declined, but he did agree to let Buehrig use a set of wooden patterns for a demonstration.

Buehrig sent the patterns to U.S. Rubber in Chicago and then flew there to supervise their construction in plastic. It took 30 minutes to vacuum form all 13 parts from ABS plastic. Buehrig then took the parts home with him to trim and assemble. When Grebe saw the finished product and the bill, Ford Motor Co. became the first automobile manufacturer to regularly make use of the process.

In 1952, Buehrig was appointed chief body engineer of the Continental Mark II project. Using the technology he developed, he made three-eighths sized vacuum-formed ABS plastic models of the proposed Mark II, which became an easy and inexpensive way of determining any body structure problems in advance and of evaluating new construction proposals. The process, although much refined, is now in use by all automobile manufacturers. The Continental Mark II was the first Ford Motor Co. product developed using this new process. ■

Gordon Buehrig with a quarter-sized model of the Tasco. This model and the full-sized Tasco are on display at the Auburn-Cord-Duesenberg Museum in Auburn, Indiana.

Kathryn Buehrig

Ford Photomedia

(above) Three-eighths sized plastic models were made from vacuum-formed ABS plastic to help validate planned construction methods. Gordon Buehrig developed the process, which saved Ford a lot of time and money.

(below) This vacuum-formed plastic model of the Continental Mark II was a first for Ford Motor Co.

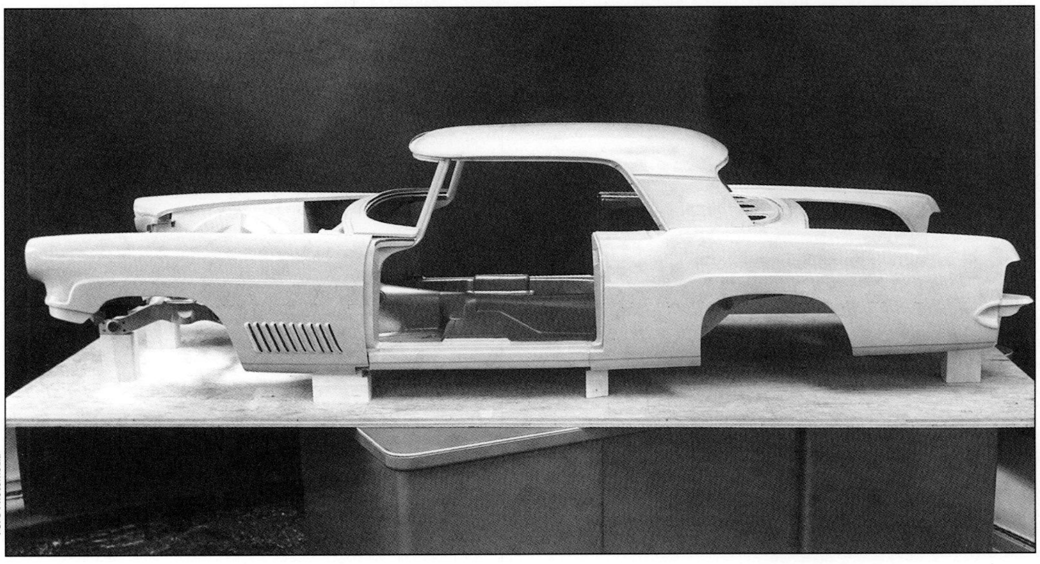

Ford Photomedia

Gordon M. Buehrig

Gordon Buehrig began his automobile career working as an apprentice body engineer at the Golfriedson body plant in Wayne, Michigan, building car bodies for Wills St. Claire, Peerless and Jewett. After four years at Golfriedson, Buehrig worked in succession for Deitrich, Packard, General Motors and Duesenberg.

In 1928, he got his first job designing cars as a GM apprentice designer. He later worked as chief designer at Cord, where he designed Duesenbergs, the 1935 Auburn boattail and the 1936 Cord. Thereafter, he worked for Budd Manufacturing, Consolidated Aircraft, Goodyear Aircraft, Raymond Loewy and Associates designing Studebakers and, in 1947-48, he formed a company to produce the Tasco, a sports car of his own design.

Unable to find financial backing to produce the Tasco, in August 1949, he was hired as a designer at Ford by John Oswald partly on the basis of the Plexiglas roof technology he had developed for the Tasco. Buehrig's first assignment was as a manager in the Ford body development studio, where he was in charge of designing the 1951 Ford Victoria two-door hardtop and the 1952 Ford all-steel station wagon. While working at Ford's Design Department, Buehrig also developed a method of vacuum-molding plastic scale models used to test automobile production methods.

In 1951, Buehrig transferred to Lincoln-Mercury Division as a body engineer and, in 1952, he was assigned to the Special Products Division as chief body engineer on the Continental Mark II project. With his assistant, Jesse Richards, Buehrig was responsible for development of the cobbled cars used as mechanical prototypes of the Mark II.

After Continental Division was disbanded, Buehrig became a product planner and researcher for Ford Division, where he helped develop light-weight, durable plastic parts to replace metal parts in all Ford Motor Co. products.

Although Buehrig is best known for the design of the 1935 Auburn and the 1936 Cord, his training and most of his career was as an engineer. After he retired from Ford in 1965, he taught design at the Art Center College in Pasadena, California, until 1970. ■

Continental Mark III Berline

Bill Ford began his automotive career in June 1948, after he finished college. He graduated from Yale with a degree in engineering, but his real interest was the design of automobiles and product planning. For the first four years he was at Ford Motor Co., he was in a training program working in various capacities to familiarize himself with all aspects of the Company. During his training program, Bill Ford let it be known that designing and building a new Continental was something he really wanted to be a part of.

In early 1952, Bill Ford was in Ernie Breech's office, for some reason he can't now recall. He had just finished his training program, and it was probably about what permanent position he was to be assigned. During their conversation, Breech mentioned that in the four years since the Continental had been discontinued, he had received a steady stream of mail asking when a new one was going to be produced. Breech indicated that it might just be the right time to resurrect it. He then asked Bill Ford if he was interested in taking charge of the program to develop a new Continental. Needless to say, Bill Ford jumped at the opportunity to resurrect his father's car. Breech then told him to do some thinking about personnel he'd need and to put together a budget for a new division to build a top-of-the-line prestige Continental, that would act as corporate advertising for the entire Ford Motor Co. product line. As Breech explained it, a new Continental was not expected to make a profit, nor would it be a mass-produced automobile. Rather, it was to be a loss leader that would bring prestige to Ford's entire car line, just as Edsel Ford's Continental had done.

By summer 1952, Special Products Operations, which was later renamed the Special Projects Division and then the Continental Division, was formed to design, engineer and build a new Continental — and the delivery date was set for the fall of 1955. Because of an industry wide shortage of personnel, Bill Ford was also told that his division would have to recruit most of its personnel from outside the Company.

Because Bill Ford was not yet 30-years old, an effort was made to fill the executive ranks of Continental Division with personalities who would compliment his and could also teach him the automobile business. Harley Copp was hired as chief engineer. Soon after that, John Reinhart was named chief stylist, and then Gordon Buehrig was transferred from Lincoln-Mercury Division as body engineer on the new Continental.

The first order of business was to find a place to establish operations. Several locations were considered, including Fairlane, Henry Ford's Estate. Consideration was given to covering over Fairlane's swimming pool and setting up the design studio there. Finally, the north end of the old Henry Ford Trade School was selected as the first home for Continental Division. Because the building was still occupied, initial remodeling plans were worked out in Buehrig's basement. The engineering offices were established in the former classrooms, and the design studio was located in the gymnasium. Compared to the newly-constructed facilities at the Design Department, the old Trade School building, where the Continental Mark II was designed and built, was primitive. The garage area had

Although Ford executive vice president Ernest R. Breech wanted the Mark III Berline produced as the 1958 Continental Mark III, the cost of tooling made production of a separate-bodied Mark III prohibitive.

Michael Schultz

to be shared with the gardener and, if an engine had to be removed from a car, a chain-fall was hung from a nearby tree. The rural Trade School location, however, proved ideal, and soon became known throughout the rest of Ford Motor Co. as the "country club" supposedly because of its wooded location. Due to Bill Ford's persistence, Continental Division finally moved into a new building of its own in Ecorse Township in April 1955.

Bill Ford surprised many at Ford Motor Co. by working long hours right along with the engineers and the designers, who developed the Continental Mark II. His relationship with co-workers was informal and friendly, and he spent many a late evening with Reinhart personally developing the styling direction of the Mark II. It was also common for Continental Division designers to arrive at work in the morning to find that, after everyone else had gone home the night before, Bill Ford had restyled the clay model they had been working on. There was usually a note on the clay model, apologizing for changes he had made, but he wanted them to at least consider them. Designers came to appreciate Bill Ford's understanding of design principles and his good taste, and some of his proposed changes actually made it into production

Michael Schulz

(above) The Continental Mark III Berline proposed for production had suicide doors that locked to the floor. The B-pillars were entirely eliminated.

Bill Ford's childhood friend, Doug McClure, was named sales manager for the new Continental Division and was sent around the country to find out what kind of prestige car the public wanted. Ultimately it was decided that the new Continental would be a "modern formal" reincarnation of Edsel Ford's first Continental, and that its design would feature a long hood and a short deck, with the outline of the spare tire on it. From the beginning, it was Bill Ford's assumption that the Mark II should be built as a two-door, like his father's Continental. Because of that assumption, a four-door Mark II was planned as a later addition. According to McClure, the decision to build a two-door Continental, instead of a four-door car, was a fatal flaw in their initial planning, but it was not discovered until after the Mark II went into production. Even Bill Ford acknowledges that the Mark II "was a two-door car in a four-door market." Unfortunately, by the time they discovered they should have built a four-door Mark II first, sales weren't strong enough to support another model.

Michael Schulz

(above) Several full-sized clay models were continually changed throughout the Continental Mark III Berline's design process.

(below) The slabsides, the hopup and the suicide doors later became design features on the 1961 Lincoln Continental.

Bill Ford's Continental Mark II, although attractive, did not sell well. That was probably because it priced itself out of the market. The Mark II, however, faced additional problems and, even before the first production model came off the assembly line, "the rules of the game were changed," as Bill Ford puts it. After initial planning for the Mark II was completed, Ford Motor Co. planned its first-ever public stock offering for 1956, and Ernie Breech was loath to go to the first stockholders' meeting with one of Ford's divisions losing money. To this day, Bill Ford believes that stock offering became the insurmountable obstacle, and if the Mark II had only been produced two years earlier, it probably would have survived.

Original projections for Continental Division were that it would lose $1.6 million a year, assuming sales of 1600 cars a year. When Breech decided that each of the Company's divisions had to make a prof-

Michael Schulz

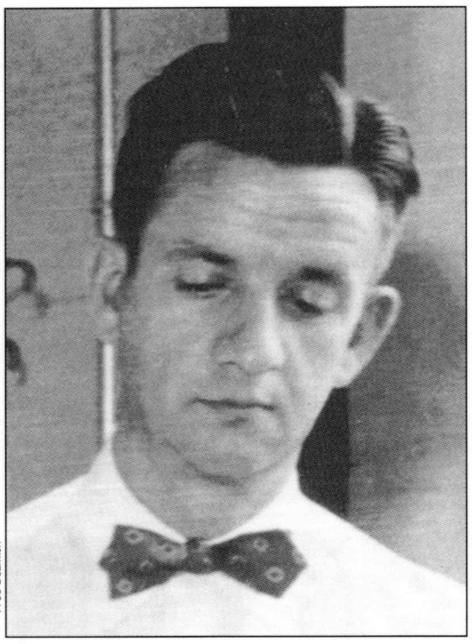

Fred Beamish

Raymond Smith

Ray Smith was born in Bad Axe, Michigan. His father died when he was nine-years old, and it was arranged for him to attend the Henry Ford Trade School in Dearborn. After Smith completed the Trade School, he studied drafting at the Laurence Institute and was employed at Greenfield Village working in the pottery shop. He worked there until Henry Ford caught him chewing tobacco, something Ford disapproved of. Smith was transferred to the Rouge assembly plant, where he worked until he entered the service.

During World War II, Smith was in the Seabees serving overseas. After the war, Smith was hired by Nash as their chief clay modeler and eventually became a designer. While at Nash, Smith built the quarter-sized clay model of the Metropolitan. In 1952, his friend and co-worker at Nash, Bob Thomas, was hired as executive stylist on the Continental Mark II project. Thomas suggested that Smith come to Ford as a designer on the Continental Mark II. Because of Smith's clay modeling and mechanical abilities, he was selected to design the proposed Mark II Retractable, so that it looked as close as possible to the fixed-roof Mark II model.

When Thomas was reassigned to the Styling Center in 1955, Smith replaced him as executive stylist at the Continental Division design department. When Continental Division was disbanded in 1956, Smith was transferred to the Styling Center, where he worked as a design analyst until he retired in 1972. Smith became head design analyst, when Bob Thomas was promoted to a design manager in 1961.

While at the Styling Center, Smith was responsible for developing the parallel-action door system, the swing-away steering wheel, the vertical-action windshield wiper, the pivot-back front seat and many of the innovations used on the Mustang I and the Allegro show cars. Smith also patented several of his own inventions including the collapsible camper. ■

it, the only way the finance people at Continental Division determined they could do that was by selling 10,000 cars over a four-year period at $10,000 per car, which was something that proved impossible. Approximately 3000 Mark IIs were sold during the two years it was produced. Breech assigned "Whiz Kid" Ben Mills to help stretch Continental Division's dollar, and Mills made some headway by selling Continental Division's retractable technology and contracting with Ford Division to engineer and produce the Ford retractable prototypes but, in the end, what was planned as a loss leader could not be made profitable.

In 1955, Ford Motor Co started building the Wixom Assembly Plant. It was one of the expansion projects planned in Ford's all-out effort to overtake General Motors. Ford's chief engineer, Earle MacPherson, himself a former GM engineer, decided that Wixom would build only unibody Thunderbirds. MacPherson believed that unibody cars were the wave of the future, and the only way the four-passenger 1958 Thunderbird could be built even lower in height was as a unibody car. The only problem was that Ford Division didn't think enough Thunderbirds could be sold in 1958 to devote a whole plant to building just that one car. Their solution was that Lincolns would also have to be built at Wixom, and they too would have to be unibody, because Ford Division engineers convinced MacPherson it was impossible to build unibody and separate body and frame cars on the same assembly line.

In early summer 1955, Harley Copp, Continental Division's chief engineer, transferred to Lincoln Division as Lincoln's chief engineer and was replaced at Continental Division by Harold Johnson. Copp, and most other Lincoln Division engineers, argued that the 131-inch wheelbase Lincoln planned for 1958 was too big to build as a unibody car. Even though they were correct, they lost the argument. Eleven years later, Lincolns returned to separate body and frame construction.

Because the Continental Mark II relied on the production 1956-57 Lincolns for its mechanical components, the Mark II's fate was tied to what Lincoln Division planned to do in 1958. That meant a unibody Continental starting in 1958. The decision to build a unibody Lincoln meant that the Mark II had to be completely retooled for 1958, and it had to become a 131-inch wheelbase unibody car. Harold Johnson and his Continental Division engineers came up with a way to build unibody and separate-frame cars together at the new Wixom assembly plant, but their plan was rejected. They then made plans to graft a body designed by Continental Division onto the unibody platform of the 1958 production Lincoln and to use the 1958 Lincoln's running gear. While that car was being designed, Continental Division engineers were determining how to mate the car to the 1958 Lincoln unibody platform.

Because it was originally planned to produce the Mark II for at least four years with very little design change, by late 1954, things slowed down in the Continental Division design department. Harley Copp arranged for Bob Thomas to transfer to the Design Department. Thomas went there as an executive designer, shared an office with Bill Wagner and worked with Jim Darden and Bill Boyer in Frank Hershey's Ford studio. When George Walker became a Ford vice president and general manager of the Design Department, in May 1955, he reorganized the studios. Frank Hershey left rather than work under Walker, and Bob Maguire was appointed head of the Ford studio. When it

Michael Schulz

On one of his visits to the Continental Division design studio, Elwood Engel, George Walker's assistant at the Styling Center, directed the modelers to make changes to one of the full-sized clays they had been working on. This photograph was taken of the results, which were undone as soon as he left.

became obvious to Thomas that everyone else at the Styling Center was getting assignments except him, he went to Reinhart, and arranged to return to Continental Division as head of the Continental Division advanced studio. Thomas' new assignment meant that he and Ray Smith, who had replaced Thomas as Continental Division executive designer when Thomas left, would work together on the 1958 Continental Mark III Berline.

After it was determined that Continental Division had to develop a unibody Mark III, Bill Ford decided that, if his Division was going to die, it wasn't going to be without a fight. In the meantime, and after Continental Division's engineers completed work on the Ford retractable, all 105 of them were reassigned to work with Harold Johnson developing a full engineering study for the four-door Mark III Berline. They were asked to engineer a car with innovations that would appeal to luxury-car buyers, but could be sold for approximately $6000. Gordon Buehrig is credited with naming the Mark III the Berline, which is a type of formal chauffeur-driven sedan.

In January or February 1955, design work began on the Mark III Berline, but Continental Division was initially unable to find designers needed to design it. Experienced designers were at a premium and, no matter how hard Bill Ford and Reinhart tried, they were unable to find enough experienced designers to design their car. At the time, Edward "Tink" Adams, president of the Art Center College, was trying to raise funds to build a new building. He was already indebted to Ford Motor Co. for past financial support and cautiously hopeful about its future support. Presumably, with Bill Ford's support and maybe even with his participation, John Reinhart proposed to Adams that Continental Division hire ten of the Art Center's best automotive design students to help design and clay model the new Continental. Adams agreed, and ten second and third-year students were recruited and, within days, left on a plane for Dearborn. All the students were told was that Continental Division had to design a new car in a hurry, that Ford's continued financial support for the school was tied to their recruitment, and that five of them were being hired as designers and five as clay modelers. As it turned out, the second-year students were the clay modelers.

With John Reinhart, Bob Thomas, Charlie Phaneuf, Ray Smith, Charles Russell and Fred Beamish, the Art Center students became the designers and clay modelers of the Mark III Berline. The Art Center students, who were recruited to design and clay model the Mark III Berline, included Paul Wong, Ed Albright, Dick Tagee, Jimmy Nelson, Ed Asadoorian, John Bailey, Rey Isip, Bill Davidson and Ed Haddock.

The Art Center students arrived in Dearborn in mid-winter 1955 and were put up at the Dearborn Inn. They assumed that Continental Division was paying the bill, and Continental Division assumed that, because the students were being paid the same money as other beginning designers, they were paying their own room tab. After a week-and-a-half, someone at Continental Division realized that the Dearborn Inn was billing them. Continental Division paid the bill and very soon arranged for the students to stay with families in the local area.

The students also assumed that Ford would furnish any tools they needed, so they didn't bring theirs with them. Continental Division assumed they would bring their own tools. By scrounging around, enough tools were located, and the students started off on what they saw as a great adventure. Most of the students worked at Continental Division

Fred Beamish

Frederick J. Beamish

Fred Beamish's design career began when his Detroit high school teacher saw his engineering drawings, and suggested he consider becoming a designer. An interview was arranged at GM with Jules Andrade and, in 1940, he started as one of six student in that year's class at the GM's design school. When the school was shut down at the beginning of World War II, Beamish took a job at Continental Aircraft working on aircraft engines. He was drafted in 1943, served in the Second Infantry Division and participated in the D-Day landing. After the war, Beamish returned to Continental Aircraft and then to the GM Styling section, where he worked in Bob Maguire's interior studio.

In 1948, Beamish left GM and worked for a short time as a designer at both Nash-Kelvinator and Kaiser-Frazer. That same year, he was hired as a designer by Briggs Manufacturing. At Briggs, Beamish worked on the Packard account and became friends with designers John Reinhart and Charlie Phaneuf.

In 1952, when Reinhart was appointed chief designer on the Mark II project, he asked Beamish to join the Mark II design team. Beamish designed the proposed Mark II that became the winning in-house entry in the Mark II design contest. After the design of the Mark II was completed, Beamish helped design Continental Division's proposed Mark III Berline.

In 1955, Beamish was laid-off during the cutbacks at Continental Division. That same year, he was hired by Chrysler, where he worked mostly in the interior studio. He retired as a manager in 1980. ■

Michael Schulz

The renderings on the wall above this clay model were only a few of the hundreds of proposals for the Mark III Berline prepared by studio designers. By the time this clay model was completed, studio designers had settled on an evolutionary design based on the modern-formal look of the Mark II.

Michael Schulz

(above) This was Continental Division's alternate Mark III Berline proposal. The front end was slightly different, and its doors opened conventionally, using a four-inch wide exposed half-B-pillar.

Michael Schulz

(right) The headlights of the two Mark III Berline proposals were slightly different.

for about a year. All of the students were great admirers of Raymond Loewy and, in short order, they bought a total of seven 1953 Studebakers for their personal transportation, which they all thought was the best-designed car on the road.

The first thing the students did, when they started at Continental Division, was to draw literally hundreds of 3/8-sized side-view renderings and the same number of 1/10-scale front and rear-view Mark IIIs. After that was completed, several of the students and Continental Division's seasoned clay modelers, including Charles Keresztes and John Kruscil, started work on several full-sized clay models of what were to eventually become the two separate proposals for the Mark III Berline.

Over the course of about six months, four to six different full-sized clay models, two seating and instrument panel bucks, at least four rear-end bucks and two front end bucks were built. The first full-sized clay model of the Mark III Berline was completed in late June 1955, but it did not look anything like the Mark II, or either of the Mark III Berlines eventually proposed. The Mark III Berline's designers remember a lot of false starts and difficulty in coming up with a theme for the car they were designing. Although the clay models started out looking like anything but the Mark II, as they got closer to the finished project, both proposals began to resemble the Mark II.

Several months after the design of the Mark III Berline began, Elwood Engel started showing up at the Continental Division design studio to "try to light a fire" under the designers. Engel would arrive unannounced, with a box of tools and his tennis shoes. He would put his tennis shoes on, take a clay-modeling knife from the box and start hacking up the clays. Engel encouraged the student-designers to come up with more futuristic designs they thought were inconsistent with the principals of good design they had been taught at the Art Center. One of the designs Engel encouraged had a roof that slid back on the rear deck lid on roller bearings.

Several of the students remember rolling their eyes at Engel's suggestions, and believe he slowed down the development of the clean Mark III Berline Reinhart was encouraging them to design. The final designs of the Mark III Berline apparently came almost overnight, when everything just seemed to come together. Although Proposal-B was finished first, by late August 1955, both clay models were completed, and fiberglass tops and Plexiglas windows were installed.

There were several differences in the alternate Mark III Berline design proposals, although both looked very similar. Proposal-A had conventionally-hinged rear doors and a backlight that retracted into the body of the car. Proposal B featured rear doors that were

M. Rey Isip

M. Rey Isip

Rey Isip grew up in New York. His father was a movie advertising designer. As a child, he loved to draw cars. After attending Pratt Institute for two years, he transferred to the Art Center College, because it offered a better automotive design program. In summer 1954, he was awarded a Ford scholarship and worked as an intern at the Ford Design Department. In 1955, one semester before graduating, he was one of ten Art Center students recruited to design the Mark III Berline at Continental Division. Isip worked on the proposed interior for the car. One year later, he left Continental Division, when offered a job in an advanced studio at Chrysler.

After several months at Chrysler, Isip was drafted and spent two years in the Army. For six months after he was discharged, Isip worked for Raymond Loewy as a designer in his Paris office, before returning to Chrysler.

At Chrysler, Isip worked on the interior design of two concept cars (one was the Norseman) that were built at Ghia and were being delivered to Chrysler on the Andrea Doria, when it sank. After a year at Chrysler, Isip left to work for W.B. Ford as a product designer. He left there to become an automotive consultant in Japan for Mazda, Honda, Datsun and Daihotsu. He later returned to Southern California, where he currently produces automotive artwork and is a part-time industrial design consultant. ■

hinged at the rear and no B-pillars, so that when all four of its doors were opened, there was one large opening on either side of the car. Each proposal had different taillights, headlights and parking lights, and proposal-A had a squared-off rear tire hump. One of the most unusual features on both proposals was the sharp ridgeline on either side of the roof. It was about two inches wide and located on the roofline crease. The ridgeline is not particularly noticeable in photographs, but designers who saw the car remember actually being able to get a fingerhold on it.

Although there were no interiors in either full-sized clay model, it is likely the Mark III Berline would have had pushbutton gear-shifting, with the selector buttons located on the top of the instrument panel to the left of the main instrument pod.

After both clay models were finished, they were di-noced in black, and foil was placed on the bumpers and trim. According to Bill Ford, all they were waiting for was budget approval to build the first prototype, which they planned to build in-house. Although both Continental Division Mark III proposals were shown to management on November 15, 1955, one of them had already been selected as the car Continental Division wanted to build and, in fact, the wooden diemodel from which the tooling was to be built had already been completed. Early in 1956, Bill Ford went before Ford's Executive Committee with Continental Division's proposed budget for a unibody 1958 Mark III. The full-sized clay models were trucked from the Continental Division plant to the Styling Center, so the Executive Committee could see them along side the planned 1958 Lincoln. The prestige and beauty of the proposed Mark III Berline could not overcome its questionable financial status, and Continental Division received the axe. Over the next several months, all of Continental Division's functions were transferred to Lincoln Division, and Continental Division was formally dissolved in July 1956. After it was decided to dissolve Continental Division, cost projections were run, assuming that there would be separate Lincoln and Continental models built, even if both were built by Lincoln Division, but the tooling cost was just too high to build two separate cars.

There was still one more chance for the Mark III Berline. Bob Thomas recalls that, when it was shown to Ford's Executive Committee, even Ernest Breech agreed that it was a better design than the proposed 1958 Lincoln. Breech even wondered out loud if it could become the new Lincoln. As Ben Mills, then head of Lincoln Division, pointed out, it was too late to change the design of the Lincoln, which had already been partially tooled.

Michael Schulz

(above) The alternate Mark III Berline proposal almost had this front end. At the last minute, it was decided to use the same grille on both proposals. The Mark III Berline's designers expected it to be the first production car with stacked quad headlights.

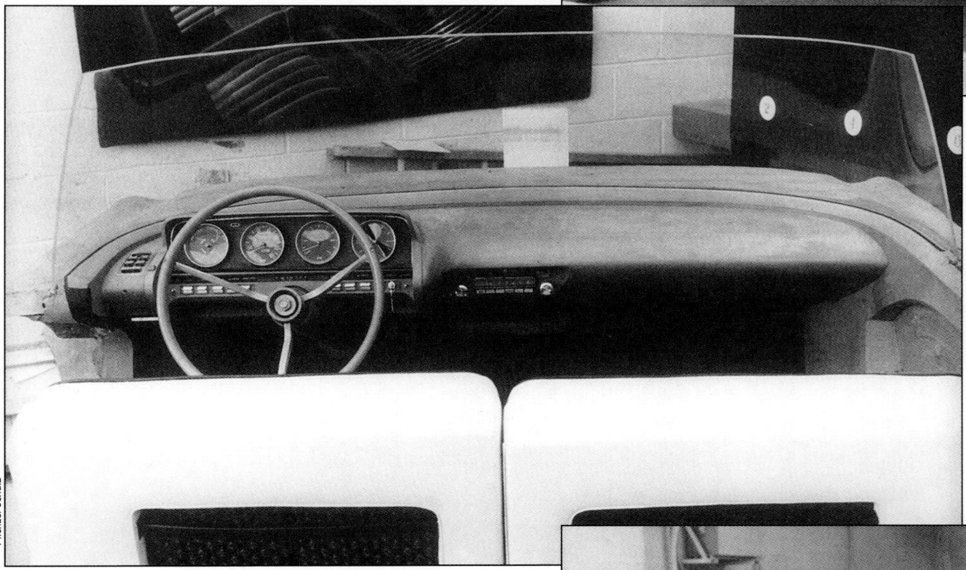

Michael Schulz

(left) This Mark III Berline interior buck included seats from a Mark II. The instrument panel is close to the production proposal, which would have probably included an instrument cluster similar to the Mark II, and pushbutton automatic transmission controls, which are located to the left of the instrument panel.

Michael Schulz

Continental Division was so sure their Mark III Berline would be produced, a wooden diemodel was built for making the tooling to be used to produce the car. According to Bill Ford, all he was waiting for was production approval, and Continental Division would have built an operable-prototype Mark III Berline.

Joseph B. Seibold

Joe Seibold was born and raised in Elwood, Indiana. As a child, he enjoyed building models. After high school, he enlisted in the Air Force. In 1952, Seibold left the military. In 1953, as a result of reading an article about the school in the Saturday Evening Post, he applied at the Art Center College. He was accepted, discovered he liked working in three dimensions, and decided he wanted to become a clay modeler. In October 1954, before he completed his schooling, Seibold was recruited by John Reinhart as a clay modeler at Continental Division. Seibold completed the current semester and left for Dearborn. While at Continental Division, he worked as a clay modeler on the Mark III Berline. When Continental Division was dissolved in June 1956, Seibold was transferred to the Styling Center, but was let go after one month.

Seibold then took a job with Newman Engraving, a job shop that built automobile prototypes. He was rehired by Ford in May 1958 and assigned to Elwood Engel's corporate advanced studio, where he was one of the clay modelers on the Levacar, the Gyron, the Lincoln Marlin, the 1961 Thunderbird/Lincoln, Big Red, the Allegro, the Aurora and the Mustang I, among others.

Seibold became a modeling manager in the early 1960s. Later in his career, he worked in Larry Shinoda's studio, the Mercury studio and in the small car studio. He was assigned to Germany for three months, and then spent 3 1/2 years at Mazda in Japan, clay modeling the Ford Probe and the Escort.

Seibold retired in 1995. ■

Joe Seibold

According to John Najjar, Lincoln studio's chief designer at the time, there was a meeting in the conference room at the Styling Center attended by Henry Ford II and key personnel from the manufacturing, engineering, finance and scheduling departments. In preparation for that meeting, Najjar assigned one of the designers in his studio, Howard Payne, to graft four headlights onto a Mark II, in an effort to see if there was any interest in continuing the Mark II during the 1958 model year. There was not. The outcome of the meeting was that, for 1958, Lincoln would be the basic car, and the Continental would be a badge-engineered model made from the basic Lincoln sheetmetal. That meant that not only were Continental Division and the proposed Mark III Berline dead, the Mark II would not get another chance either.

Elimination of Continental Division meant that its 24 remaining designers and clay modelers had to look for jobs elsewhere. Only eight of them were hired at the Styling Center, including John Reinhart, who became an executive stylist in the Lincoln studio, Bob Thomas and Ray Smith, who became design analysts, and Charlie Phaneuf, who became a senior stylist — they all had to accept demotions and cuts in pay to work at the Styling Center. Most of the engineers who developed the Mark II and the proposed Mark III Berline found jobs at Lincoln Division, including Harold Johnson, who became Lincoln executive engineer-advanced. Of the students recruited from the Art Center, only Paul Wong, Ed Asadoorian and Ed Albright were hired at the Styling Center, and Asadoorian left within the month.

Although he chose to spend more time on his other outside business interests, including the Detroit Lions football team, Bill Ford was later appointed the titular head of design at Ford. Najjar and Reinhart worked well together, but Reinhart soon left Ford to take a job as a designer for U.S. Steel. Charlie Phaneuf, Ray Smith and Bob Thomas stayed at the Styling Center until retirement. Thomas was later restored to his position as a designer, when Gene Bordinat became design director in 1961. Thomas later helped design the Allegro, Aurora and Stiletto concept cars, the 1961 Lincoln Continental and the Pinto. Ray Smith became chief design analyst, when Thomas was restored to his position as a designer. Charlie Phaneuf was one of the primary designers of the Mustang, which he modeled after the Mark II. Phaneuf believes that designers who transferred from Continental Division to the Styling Center were at a disadvantage, because the designers already there always felt Bill Ford would try to place his people in charge. George Walker saw to it that that never happened.

The Mark III Berline clay models and the diemodel were destroyed, but the car was not forgotten. It later became one of the inspirations for the 1961 Lincoln Continental four-door, and the engineering features developed for it influenced many future Ford products including the Mark VI four-door and the 1980s Lincoln Town Cars, which have more than a passing resemblance to the Mark III Berline. ■

The alternate Mark III Berline had a squared-off spare-tire hump on the deck lid and a retractable backlight. The taillights had a more forward cant to them. Although this was not the selected design, studio designers were almost equally divided as to which car they preferred.

Michael Schultz

7

Aerodynamics and Alex Tremulis

Mexico

LaTosca

Palomar

Taj Mahal

Maxima

Madam X

999

X-1000

X-2000

Wind Brake Car

Mexico

In 1935, Edsel Ford, Preston Tucker and Harry Miller built 12 Ford racecars to run at that year's Indianapolis 500 race. There was a lack of time to adequately test the cars, and all of them dropped out of the race, because engine heat cooked the lubrication out of the steering knuckles, which were placed too close to the exhaust manifold causing them to seize. An embarrassed Henry Ford forbid anyone at Ford from ever again involving the company in racing. Only after Henry Ford died in 1947 did that edict change.

In 1948, Henry Ford's grandson, Benson Ford, was elected a Ford vice president and appointed general manager of Lincoln-Mercury Division. By 1952, factory-sponsored Lincolns from Benson Ford's division were participating in the Mexican Road Races, and Lincoln's Road Race wins in 1952-54 had at least an indirect effect on Ford's designers, who couldn't wait to start designing Ford racecars. There was, however, another racing event that was to have an even more direct effect on the Design Department.

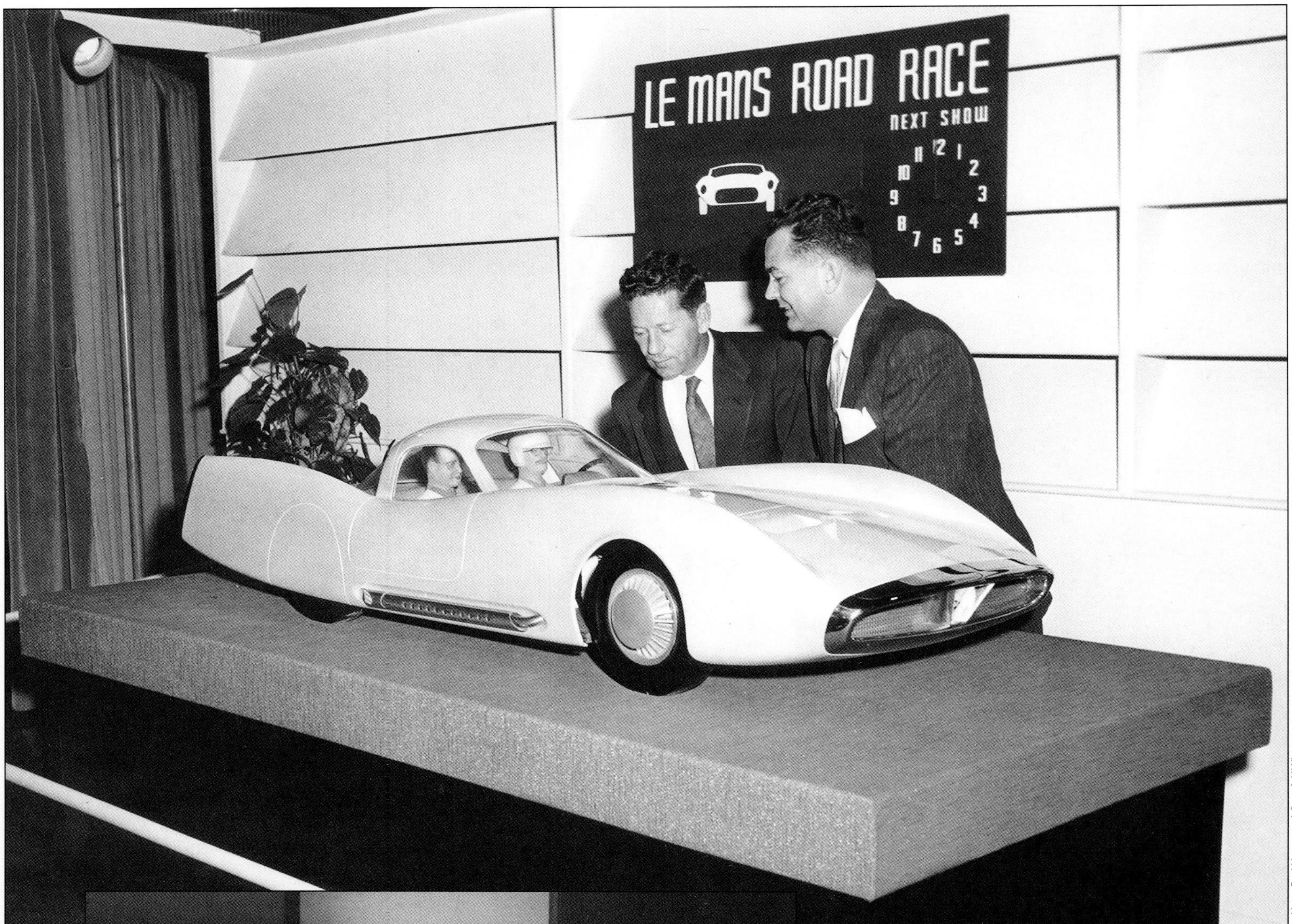

Henry Ford Museum and Greenfield Village

Ford Photomedia

(top) Mexican Road Race participants Walt Faulkner (left) and Bill Stroppe (right) admire the Mexico.

(bottom) Because of budget and time constraints, the partially completed Gran Turismo was remodeled to become the Mexico.

In the 1950s, the long, straight beaches at Daytona Beach, Florida, were where domestic automobile manufacturers took their cars to make speed runs. Zora Arkus-Duntov, the GM engineer who put the small-block Chevrolet V-8 in the Corvette, frequented Daytona Beach with his specially-prepared Corvettes. Sometime after the Thunderbird went on sale in September 1954, Henry Ford II agreed to let Ford engineers take several Thunderbirds that had been specially prepared to Daytona Beach to race against the Corvettes. During one of the speed runs, a Thunderbird hit a bump and took to the air at about 135 mph. Then it went upside down and slid down the beach for about 1500 feet, before it finally stopped. To Henry Ford II's dismay, cameras recorded the whole thing. After the wreck and back in Dearborn, Henry Ford II contacted his engineers and, they, in turn, directed him to Ford's resident expert on aerodynamics -- Alex Tremulis at the Advanced studio.

Tremulis came to Ford from Kaiser-Fraser in 1952. By that time, he had already acquired a reputation as a colorful and brilliant designer. Tremulis had been head of the Advanced Studio for only a short time when Mr. Ford came to see him. The first thing he asked Tremulis was what had gone wrong. Tremulis told Mr. Ford that the Thunderbird had developed too much lift over the rear end at the speed it was going and, when it hit a small bump, it went airborne. Tremulis also told Mr. Ford that the same thing would happen to the Thunderbird the next time it was going that fast, and there was no real way to correct the problem. The next thing Mr. Ford asked was,"What do we do?" Tremulis told him they needed a zero-lift car, and he volunteered to design it. Mr. Ford gave him the go-ahead, and Tremulis proceeded to design and build a 3/8-sized model of a car he named the Mexico after the Mexican Road Races.

The Mexico was not a planned project, so Charlie Waterhouse, the Design Department's general manager, asked Tremulis how long it would take to design. Tremulis told him he could do it in a week. Waterhouse reminded Tremulis that, since it had taken him four months to complete the LaTosca, how could he expect to complete the design of an entirely new car in one week? Waterhouse bet Tremulis a steak dinner for all those who worked on the project that he couldn't do it in a week. That's all Tremulis needed to hear, because he finished the 3/8-sized model of the Mexico in a week, and Waterhouse paid the $300 bill for a steak dinner for all designers and clay modelers who worked on it. What Waterhouse didn't know was that Tremulis had clay modeler Larry Wilson remodel Fred Hoadley's almost completed Gran Turismo into the Mexico, thereby saving a lot of time.

(top) The Mexico was designed to stay on the ground at high speeds.

(middle) The Mexico had the same wheelbase as the 1952 Ford passenger car, but because of the front and rear overhang, it looked much longer.

(bottom) The Mexico's instrument panel was designed specifically for racing. It included a speedometer in the middle of the steering wheel, an idea borrowed from the Futura.

Tremulis designed the Mexico with a full belly pan under it. Waterhouse said not to, but Tremulis and Romeyn Hammond, his assistant manager, worked all night to design and place it under the car. The belly pan was built from aluminum and had flush grooves running front to back.

After the Mexico was completed, Waterhouse made an appointment to use the University of Maryland windtunnel, and Tremulis and Hammond took the Mexico to Ann Arbor for their $200-an-hour session. Tremulis bragged that the Mexico would attain a drag coefficient of .22 or .23 in the windtunnel tests. When Bill Bowman, who ran the windtunnel, saw the belly pan with all the grooves in it, he told Tremulis it would take two hours to remove the grooves and, that if they weren't removed, it would make the windtunnel results worse, not better. Tremulis disagreed and finally talked Bowman into letting him make one test run in the windtunnel with the grooved belly pan on. Not only did the windtunnel test demonstrate that the model had negligible lift, the coefficient of drag turned out to be .22. Two more windtunnel tests confirmed the .22 result. There was one more test made in the windtunnel, this time without the belly pan, and it almost doubled the coefficient of drag.

Tremulis returned to Dearborn, and told Henry Ford II that the results of the windtunnel tests meant that a full-sized Mexico, with a belly pan, would have to go 400 mph before it developed lift problems, and that the Mexico was slippery enough to run at 200 mph, using only a 240 horsepower engine in it. Tremulis was asked to work up some preliminary plans for building a full-sized operable Mexico. He designed a full-sized Mexico with a tubular frame and a weight of 2500 pounds, including a 400 horsepower Mercury Marauder engine, which he calculated would give the car a top speed of 240 miles an hour — enough to beat the world's two leading sports cars,

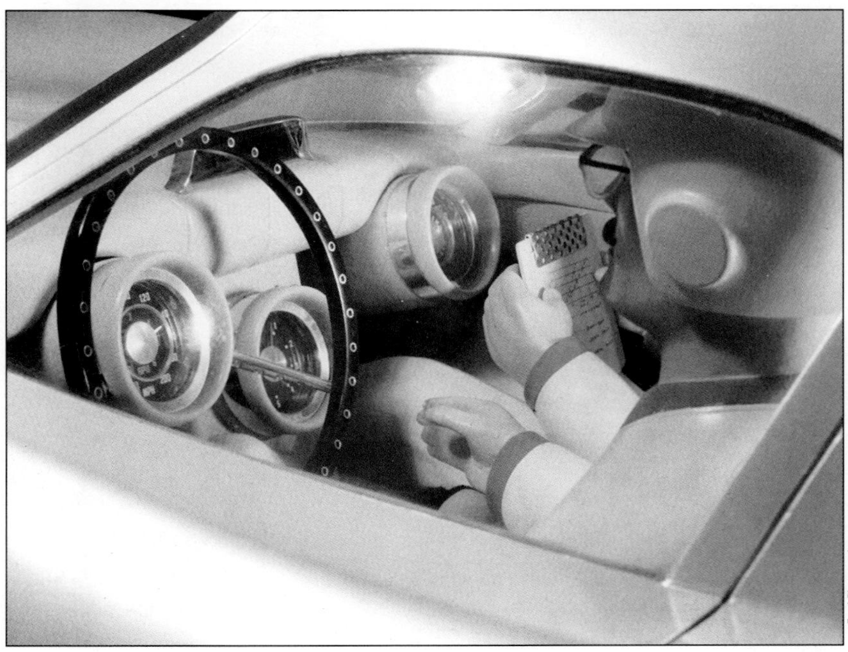

Ford Design Center

Alex S. Tremulis

In 1933, after he graduated from high school, Tremulis enrolled in an aerodynamics course given by the Army at Wright Field in Dayton, Ohio. After completing the course, he began his design career at Duesenberg and ultimately became their head stylist. He next enrolled in the General Motors Training School but, at Howard O'Leary's suggestion, he left early to take courses in watercolor painting. (Tremulis claimed to have flunked Art in high school!) Before the war, Tremulis was also a designer for Briggs Manufacturing, Custom Motors and American Bantam.

During World War II, Tremulis designed rockets and taught courses on aircraft and rocket design for the Army. In January 1947, he was hired to design the Tucker automobile. After the Tucker failed, Tremulis took a job as a designer for Kaiser-Frazer. While at Kaiser-Frazer, Tremulis and other designers were invited to an open house at Ford, where futuristic car models from Gil Spear's Advanced Studio were on display. Tremulis saw the models and thought that if they represented what Ford was all about, he wanted to work there. Tremulis was offered a job by Elwood Engel and started at the Ford Design Department in 1952.

Tremulis' first assignment at Ford was working in the Mercury studio. When the new Design Department building was opened in 1953, Spear asked Tremulis to become his assistant and to begin an in-house school to teach Ford's new designers the Ford way of designing cars.

In January 1954, Tremulis became head of the Advanced Studio, when Spear moved to the new Special Vehicles Studio. In March 1956, Tremulis' Advanced Studio was dissolved, and he became a manager in Bob Maguire's new Advanced Studio. He was later assigned to the Thunderbird and Ford pre-production studios, and eventually was assigned as a senior stylist to the corporate advanced studio. He left Ford in 1962. He is easily one of Ford's most controversial designers.

Tremulis was largely a self-taught designer, who worked for 20 automobile companies throughout his career. He was an effective teacher and an early advocate of aerodynamics. He designed the Mexico, LaTosca, Gyron, X-1000, X-2000, Taj Mahal, 999, Maxima, Seattle-ite and many other concept and showcars while at Ford. ■

(below) Tremulis designed the small Oscar in the driver's seat to resemble himself, and the passenger to resemble Charlie Waterhouse. When Waterhouse saw himself in miniature in the passenger seat of the Mexico, he told Tremulis he had wondered why the clay modelers had been watching him so intently, when he walked through the studio. Waterhouse initially expressed concern, because the Oscar of him was reading the Wall Street Journal. Because he thought others might think he spent too much time reading the paper, he asked Tremulis to remove it. This photograph was taken after the newspaper was removed.

Ford Photomedia

(above) Alex Tremulis admiring the Mexico.

Ford Photomedia

(Bottom) Both the Mexico and the LaTosca were motorized by Romeyn Hammond.

Ford Photomedia

the Mercedes- Benz and the Lancia. As Lincoln-Mercury Division was preparing for the 1955 Mexican Road Race, it was unexpectedly canceled. Even if the Mexican Road Race hadn't been canceled, it's unlikely the Mexico would have been produced, because there was no market for a second production sports car. With cancellation of the Mexican Road Race, any chance that Ford would build a limited production or even a one-off Mexico also died.

Steve Tremulis

The Mexico, just before it was tested at the University of Maryland windtunnel.

After the Mexico was built as a 3/8-sized fiberglass model, like the LaTosca, it was placed on a motorized radio-controlled platform. After it had been run and photographed, the Mexico was placed on a raised bench by the door of the Advanced Studio, where it was displayed for several years. Several years after Tremulis left Ford in 1962, he was told that about 20 of his models were gathered up, put under a tarp in a corner of the Rouge plant and forgotten. Some years later, someone else told him of seeing his deteriorated models somewhere in the Rouge plant. The last thing Tremulis heard was that the Mexico, which he considered his best model, had been saved because of its aerodynamics. Today, nobody at Ford knows where the Mexico is, or if it survived.

In the meantime, happenstance was not making Tremulis friends among those who ran the Design Department. Don DeLaRossa bought a Jaguar XK-120 with a Corvette engine in it. He arranged to have some paint work done on it at the Design Department shops. After the paint work was finished, he and Bob Maguire decided to take it out on the freeway and see how fast it would go. Maguire and DeLaRossa were finishing up something in the studio, and asked Ken Spencer if he would put the new license plates on the car. Spencer tried, but discovered that the attaching holes in the plates didn't line up with those on the car, so he took the plates back to the shops to redrill the holes. Thinking the plates were on the car, DeLaRossa and Maguire took off down I-94 with Maguire driving. They apparently didn't see the police car unsuccessfully trying to catch them -- until they were stopped by a road block near Ann Arbor about 30 miles by freeway from Dearborn. Reckless driving, speeding, no license plates and failure to stop got them a trip to jail. When they were finally allowed to make a phone call, they called Alex Tremulis to come bail them out. He was the only one they knew who lived in Ann Arbor. By that time, the banks were closed, so Tremulis had to borrow the day's receipts -- most of it in $1 bills -- from friends who owned the local movie theaters. Maguire swore Tremulis to secrecy but, when he got to work the next day, everyone at the Design Department knew he spent the night in jail. Although DeLaRossa was the one who told, Maguire always believed Tremulis was responsible for his embarrassment. The relationship between Maguire and Tremulis did not improve with time. ■

The Mexico was displayed in the foyer of the Styling Center. John Reinhart (middle) is showing the model to visitors.

Ford Design Center

Romeyn Hammond

(above) Romeyn Hammond (left), with the LaTosca, and Alex Tremulis (right), with the Mexico at the Ford test track.

Chrissie Tremulis

(left) The rear end of the Mexico had a U-shaped grille, capped on either side by taillights. Tremulis' original design had lights inside the entire "U."

Adolphus Painter

Dolph Painter attended the Henry Ford Trade School from 1946-49. After graduation, he was hired as a Ford draftsman. His desk in the Engineering Department was close to the wall separating it from the Design Department. Clay modeling manager Bill Leverenz usually left by the exit door near Painter's desk, so Painter and Leverenz got to know each other. Leverenz soon asked Painter to attend his clay modeling class. Six months later, Painter was hired as a clay modeler at the Design Department.

Dolph Painter

Painter's first assignment was working for modeler Dick Beneicke in a Ford pre-production studio. For the next 15 years (except for time in the service), he worked in the prove-out area double checking production surface drawings prepared by the engineers against the finished clay models.

Because Painter's personal dimensions were close to those of the average man, he was often asked by engineers to accompany them to car dealers, so they could compare the interior dimensions of other manufacturer's cars. In 1951, at the suggestion of Edwin C. Pickard of the engineering staff, Parke Hanner, a staff modeler, made a mannequin using Painter's exact dimensions. It took two months to make the mannequin but, from then on, other manufacturers' cars were rented and brought to the Design Department, where they were measured and compared using Oscar. Oscar was finished in March 1952, the day Painter was drafted into the Marine Corp.

All mannequins were named Oscar and, although the dimensions were later changed when the average man became larger, to this day, Oscar's features remain those of Painter.

After he was discharged from the Marine Crop., Painter resumed his career as a Ford clay modeler. In 1968, he became modeling manager of all exterior studios. He retired in 1985 but was asked to work as a contract clay modeler for several years at Ford and several other companies. ■

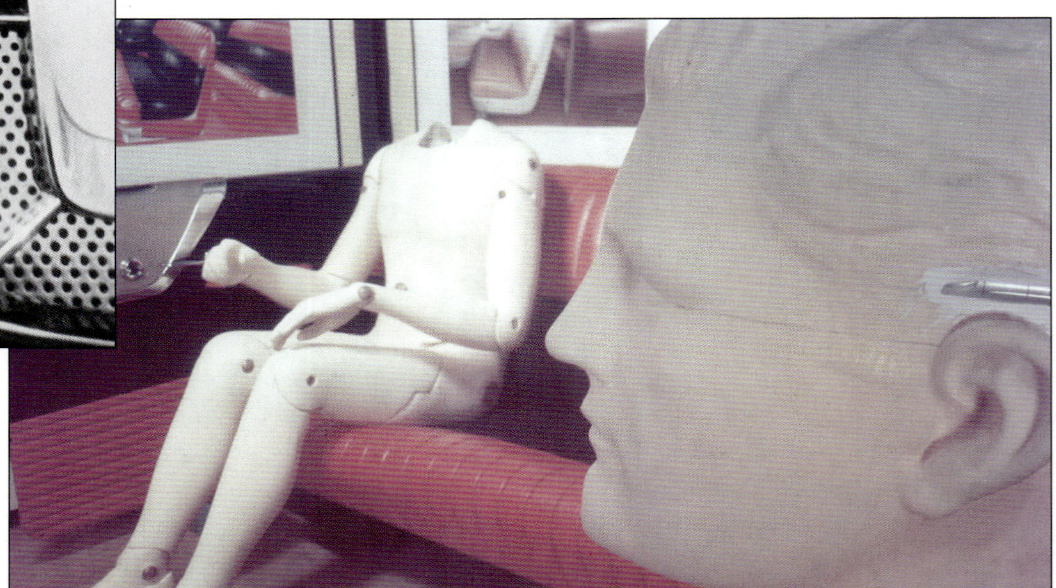

Ford Photomedia

(right) Oscars were mannequins made to approximate the average man's measurements. They were used to check the interior measurements of all proposed Ford products. Clay modeler Dolph Painter, who just happened to have average dimensions, was used as the model for Oscar. Although the Oscars in use at the Design Center today have been revised to keep pace with changes in the average man's measurements, they all still bare an uncanny resemblance to Painter, as he was in 1952.

The LaTosca was designed by Alex Tremulis and Romeyn Hammond in 1954. At the time, Hammond was Tremulis' assistant manger in the Advanced Studio. According to Hammond, the LaTosca was built as an exercise to show students in the Advanced Studio how hard it was, even for professional designers, to design a car. Tremulis said the LaTosca was the hardest design he ever did because of its radical canted fenders, and the optical illusions they created.

The LaTosca became the inspiration for the production 1958 Lincoln. When designers began working on the 1958 Lincoln, Elwood Engel asked Tremulis to "drive" the LaTosca down to the Lincoln Studio, which he did.

LaTosca

The LaTosca project was approved after Tremulis assured Charlie Waterhouse, general manager of the Design Department, that it would only take 30 days to complete. It actually took four months to design and build, and it went through many changes and revisions. The LaTosca was built with Waterhouse's knowledge, but without formal approval, because there was no money in the department budget for additional concept cars. Tremulis and Hammond had to scrounge the parts they used. Hammond estimated that it cost Ford less than $1000 in mostly used parts to design and build the motorized LaTosca.

The LaTosca was painted salmon-pink, and the trim was chrome plated. The car had a full interior, and a little Oscar sitting in the driver's seat. The sculptured fins gave the car a wider look. Headlights were located under lids that were flush with the hood, and raised up when they were on. The exhausts were jet-tubes located in the middle of the rear fenders. Taillights, stop lights and turn signals were set at the rear of each fin. Tremulis anticipated that, if a full-sized LaTosca was built, it would be powered by a conventional V-8 engine, be 195 inches long, have a wheelbase of 108 inches, be 50 inches tall and have a road clearance of about seven inches.

The LaTosca became the first 3/8-sized, battery-powered, motorized concept car that was radio controlled. Although Tremulis was one of the designers of the car, the motorized unit was the idea of Romeyn Hammond. Two Lincoln top motors were used to drive the motorized unit, the transmission was adapted from a power-window regulator, and the LaTosca's steering mechanism was adapted from a power-seat unit. The relays were from Ford power-window units, and the radio controls were the same type used on model airplanes. The power source, which was a standard car battery, could make the LaTosca travel at speeds of up to five mph, go backward, forward and turn corners.

For *Life* magazine and other writers and photographers, the LaTosca was demonstrated on the Ford test track with George Walker at the controls. Photographs and articles written about Ford's motorized concept car appeared in *Life* and many other magazines and newspapers around the country. Even though Walker got the publicity, Tremulis was the person who spent the most time "driving" the LaTosca.

Ford Photomedia

(above) According to Alex Tremulis, the LaTosca was a difficult design to execute, because of the many optical illusions it created.

A new building was occupied by the Design Department in the summer of 1953. It had an enclosed courtyard, that was primarily used to display prototypes, proposed production cars and completed concept cars. Designers felt that their cars could be better displayed, and the design critiques would be more thorough if done in a more natural setting using natural lighting. The new building was originally built so each studio had big windows opening onto that courtyard, and so cars could be easily moved from the studios to the courtyard. Even the hallways in the new building were designed wide enough to move full-sized cars through them. A big brick wall enclosed the outside courtyard, which was paved with flagstone pads with grass in between them. It didn't take Tremulis long to start driving the LaTosca over the bumpy flagstone or, if the weather was bad, around the Design Rotunda, which was a big showroom located towards the front of the building. On occasion, Tremulis would also drive the LaTosca up and down the wide hallways and in and out of the open doors to the various offices fronting on the hallways.

Tremulis would send the LaTosca out on the street on occasion while controlling it from inside the building. Designers and clay modelers could watch it through binoculars as it traveled down Oakwood Blvd., in and out of traffic, stopping at traffic signals and finally turning into the driveway of the Ford plant. In time, Tremulis discovered that the radio controls worked up to 1 1/4 miles away. Security guards at the Styling Center recall stopping traffic on Oakwood Blvd., as the LaTosca crossed the street under its own power. On one occasion, while Tremulis was "driving" the car down Oakwood Blvd, it created a traffic jam and was holding up a line of traffic that included Ford's chief engineer Earle S. MacPherson. MacPherson didn't have an excess of patience anyway and, when he learned it was Tremulis who was driving his toy on the public streets and creating traffic jams from inside the Design Department building, he made sure Tremulis and Waterhouse knew what he thought of it. Many designers believe that Tremulis' later problems at Ford started with the LaTosca, and even Tremulis himself, in later interviews, acknowledged he created many of his own problems with his motorized LaTosca.

(below) Like the Futura, some of the LaTosca's instruments were on top of the instrument panel.

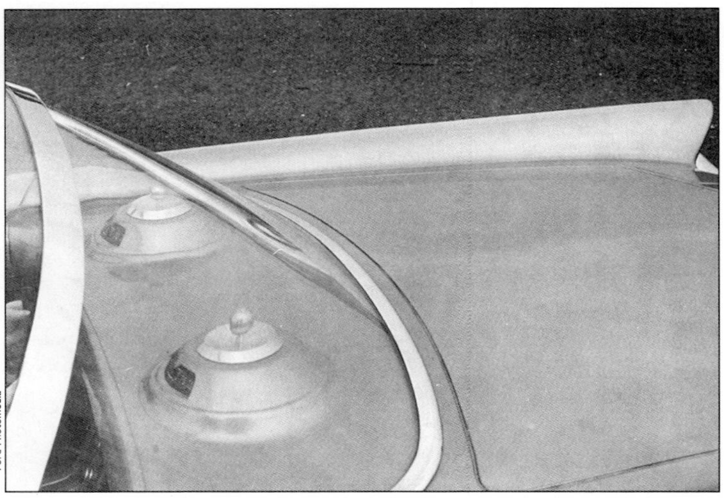

Ford Photomedia

When the car was designed, Tremulis hoped that it would lead to a small production Ford, possibly a sports car. The LaTosca, however, gets credit as the inspiration for the 1958 Lincoln. In 1954, at the time the LaTosca was designed, Alex Tremulis' influence at Ford was probably at its peak, and the designs he did during this time period were not subject to his overriding interest in aerodynamics, which came several years later. In 1955, after John Najjar was appointed head of the Lincoln studio, he and Elwood Engel decided there were lines on the LaTosca that could be used for the design of the '58 Lincoln Najjar and Engel were about to start designing. Engel went to the Advanced Studio, and asked Tremulis to have the LaTosca "driven over" to Najjar's Studio and stopped in front of Najjar. During the time the '58 Lincoln was being designed, the LaTosca was "parked" in Najjar's studio, and it was used by Najjar, Bud Kaufman, David Ash, Engel and the other studio designers as the basis for the design of the '58 Lincoln. Later, Tremulis said he was not pleased with the interpretation of the LaTosca made in designing the '58 production Lincoln.

(below) The LaTosca was one of the first serious design proposals at Ford to place the main air intake below the bumper.

Ford Photomedia

The LaTosca was designed so that the motorized unit could be used on other 3/8-sized models. For years, designers report seeing that unit in storage without the LaTosca on it, and they assume the LaTosca was chipped or broken with the passage of years and eventually destroyed. ■

(above) Romeyn Hammond (left) and Alex Tremulis (right) with the LaTosca on the Ford test track.

Ford Photomedia

Ford Design Center

(right) The motorized unit of the LaTosca with the body removed. According to Hammond, a battery-powered convertible top motor, power-window regulator and a power-seat motor were used to build it.

Ford Photomedia

George Walker operating the radio-controlled LaTosca at the Ford test track. Walker thought the idea of motorizing three-eighths sized models was an excellent way to see designs in motion.

Ford Photomedia

Romeyn S. Hammond

Romeyn Hammond grew up in New England. As a child, he developed a love for drawing cars and planes. After graduating from high school, he enrolled in a course on aircraft manufacture and repair at Quoddy, Maine, and received his diploma shortly before the start of World War II. For the first three years of the war, Hammond was an aircraft repair foreman. In 1944, he was accepted as an air cadet, but was prevented from flying because he wore glasses. When the war was over, Hammond was employed as a civilian aircraft repair inspector for the Air Force.

In January 1948, he returned to school to study industrial design and, in1950, graduated from the Vesper George School of Art in Boston, Massachusetts. From 1950-53, after completing a post-graduate course in interior design, Hammond became a partner in an interior design firm in Wooster, Massachusetts. In April 1953, he decided he wanted to design automobiles, so he prepared a portfolio and began applying at automobile manufacturers. He had an appointment at Hudson but, while on the train to Detroit, he decided he wanted to work for Ford. Without an appointment, he stopped by the Ford Design Department, was interviewed by Charlie Waterhouse and Gil Spear and hired immediately.

Hammond's first assignment was in Gil Spear's Advanced Studio. Alex Tremulis soon became head of the Advanced Studio, and Hammond was appointed his assistant. During the time Hammond worked as Tremulis' assistant, he helped design many of the advanced cars produced in that studio, including the Mexico and the LaTosca. Using his mechanical background, Hammond motorized the LaTosca and the Mexico.

In 1955, Hammond was appointed a manager in the Edsel interior studio. In 1958, he was transferred to the Lincoln-Mercury studio as a senior stylist. In 1960, he was assigned to the advanced interior studio. Later, he became a manager in the corporate projects, the Lincoln-Mercury and the corporate advanced studios.

In the early 1960s, Hammond was on the team that designed the Aurora show car, the Big Red gas-turbine truck and the interior of the Bordinat Cobra.

He retired in 1980, after working on the F-Series pickups, the 1972 Mercury, the 1977 Thunderbird and the Granada and Monarch. ■

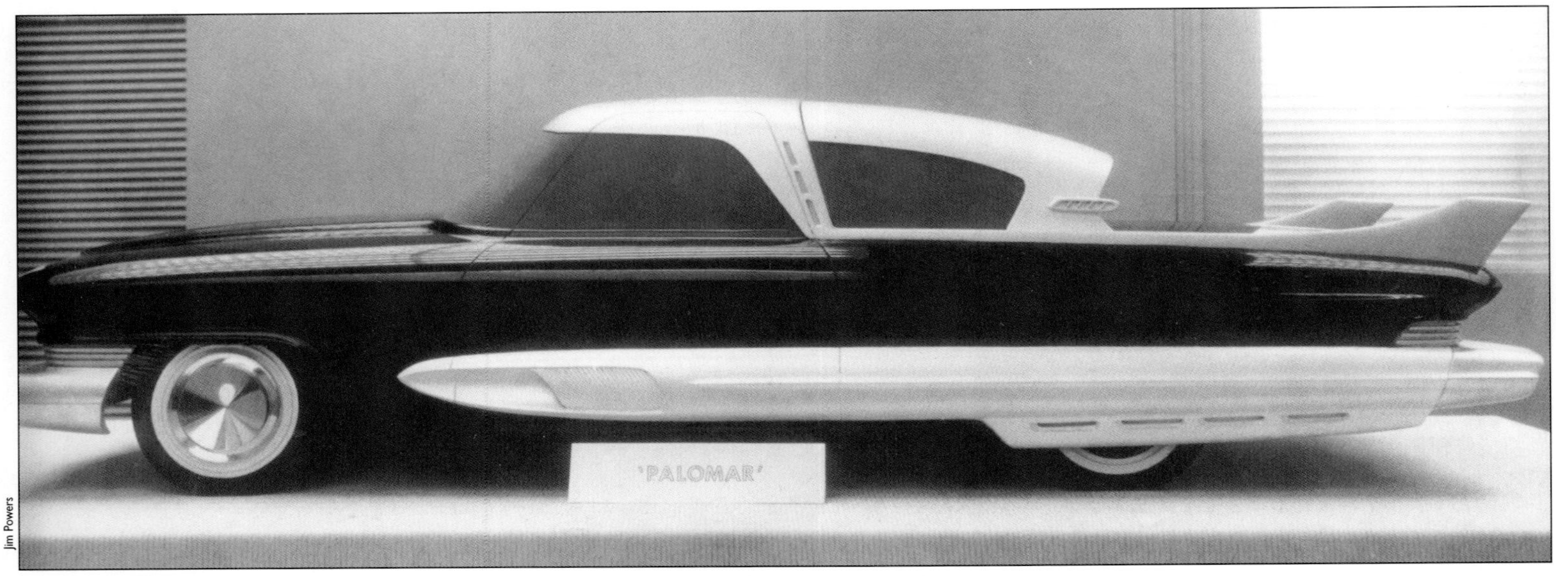

Jim Powers

The Palomar was designed by Jim Powers who, at the time, was a newly hired designer in Tremulis' Advanced Studio. The Palomar was designed without A-pillars, and the rear part of the roof slid back on top of the rear deck to create an open back-seat area.

Palomar

Jim Powers designed the 3/8-sized Palomar in 1956 while a designer-trainee in Alex Tremulis' Advanced Studio. Of the three cars Powers designed and built while he was in the Advanced Studio, the Palomar was the most conventional. Although the Nucleon and the Volante were publicly displayed, the Palomar was never shown outside the Styling Center.

The Palomar was designed as a two-door sedan, with door openings that were cut high into the roof. It also had prominent tail fins and large side sculptures extending forward from the back bumpers to just in front of the doors. The most unique features on the Palomar were its rear roof section that could be retracted onto the deck lid to make the rear-passenger compartment into a convertible and the car's lack of A-pillars.

Powers named the Palomar for the California observatory that influenced the way the roof opened. The car was painted a dark blue with a pearlescent-white top, fins and side spears. In 1961, the idea of a retracting roof section and the name Palomar were recycled on another full-sized Mercury station wagon showcar.

Powers' Palomar hasn't been seen in years, and he assumes it was destroyed. ■

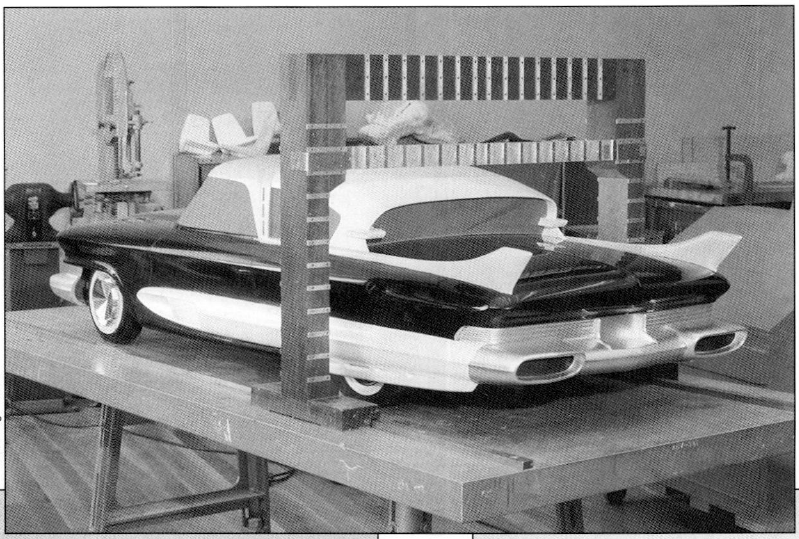

Ford Design Center

(left) Powers named the Palomar, shown here under a three-eighths sized styling bridge, after the California observatory of the same name.

(bottom left) Tremulis thought the Palomar would look better with the hidden headlight covers painted a contrasting color. This photograph was taken before they were repainted body color.

(bottom right) From the rear, the Palomar had strong 1956 Plymouth overtones.

Jim Powers

Jim Powers

Alex Tremulis started at Ford at about the same time Bob Maguire was replaced by Frank Hershey as head of the Ford studio. Maguire then became the number-two man in the Ford studio. Although Tremulis was first assigned to the Mercury studio, he and Maguire soon came to know each other very well. Maguire, who was not known for his sense of humor, became the butt of many of Tremulis' practical jokes.

Tremulis loved to make what he called his Egyptian music by blowing air from an airbrush hose through his sanded thumb and index fingers, while tapping his watercolor brushes on the glass containers he used for brush water. Tremulis' Egyptian music made a high-pitched whine that always seemed to filter into Maguire's office. Tremulis usually began making his Egyptian music late in the evening. When he did, the studio would invariably receive an irate phone call from Maguire, telling whoever answered the phone to tell Tremulis to cut out that blankety-blank Egyptian music! Needless to say, Tremulis didn't get off on the right foot with Maguire.

Almost immediately after he was assigned to the Mercury studio, Tremulis took a dislike to one of the grille designs proposed for the production 1955 Mercury. The proposal had teeth across the entire grille opening with recessed areas between them. On show day, when executives were scheduled to select the final grille design to be used, Tremulis arrived early and began painting the recessed areas between the teeth black. Nobody could figure out what he was doing but, when the executives showed up later that morning, Tremulis pulled up a modeler's stool, faced the grille, sat down, and pretended like he was playing the piano. Everyone got a good laugh, but Tremulis lost the battle -- the grille chosen for the '55 Mercury was the one Tremulis called their attention to -- painted areas and all! The antics got him noticed, but didn't set well with the very people he needed to champion his designs.

Tremulis was much admired by the young designers at Ford, even before he was assigned to the Advanced Studio, where most of them started out. He and his wife, Chrissie, invited many new designers to their Ann Arbor home on weekends for home-cooked meals and great stories -- and Tremulis' past adventures were almost always featured in those stories.

Taj Mahal

The Taj Mahal was a 3/8-sized model designed by Tremulis shortly before he created the Mexico. The Taj Mahal was designed as a Mercury sports car, which was being considered for production at the time. The Taj Mahal was a serious design proposal, and Tremulis considered it one of his most creative efforts.

When Tremulis took the Mexico to the University of Michigan for windtunnel tests, he also took the Taj Mahal and a 3/8-sized model of a production 1955 Ford. On his way to Ann Arbor, where the University is located, Tremulis stopped to pick up one of the clay modelers, who planned to accompany him. The modeler had a 200 pound plaster elephant

The scallops in back of the front wheelwells and the roof line on the Taj Mahal were design features Tremulis used on more than one of his designs.

Steve Tremulis

The Taj Mahal at the time it was tested in the University of Maryland windtunnel. Tremulis did not consider the front end of the car one of its better features, but he designed it to be aerodynamically functional.

Steve Tremulis

in his living room apparently left over from his college days. It was painted black and, underneath scratches, it was obvious that it was made from white plaster. When Tremulis expressed an interest in the plaster elephant, the modeler told him he could have it. For some unknown reason, Tremulis decided to test the aerodynamics of the plaster elephant in the windtunnel with the three cars, so they took the elephant with them. Tremulis later said he thought it would be interesting to test the aerodynamics of the plaster elephant at the same time they tested the three car models, since they were paying a flat rate of $200 an hour to use the windtunnel.

Chrissie Tremulis

Tremulis believed the fins on the Taj Mahal influenced the design of the Mystere.

Because the plaster elephant was about the same size as the car models, he planned four tests for it. To the consternation of the University personnel who operated the windtunnel, Tremulis put the elephant in the wind tunnel and *glued* its feet to the floor. They were apparently in the middle of the first test, when Earle MacPherson and Val Tallberg, MacPherson's assistant, walked in. Tremulis had not expected them to be there, and MacPherson and Tallberg had certainly not expected to find a plaster elephant in the windtunnel on Ford's $200 an hour. MacPherson stood there for a minute, and then asked why Tremulis had a plaster elephant in the windtunnel. Tremulis said he was interested in the aerodynamics of the elephant, and then he told MacPherson that on the fourth test he planned to cut-off the elephant's ears. MacPherson was quiet for a few moments, and then asked Tremulis to tell him again why the plaster elephant was in the windtunnel. Tremulis explained that he was "testing aerodynamic braking." MacPherson held his temper but, just before he walked out, he asked Tremulis of what possible interest to Ford Motor Co. could the aerodynamics of an elephant be? Tremulis didn't get a chance to respond before MacPherson noticed a pair of full-sized skis with the car models and left in a huff. Tremulis planned to place Allen Bushnell, one of the designers and a ski jumper, in the windtunnel to do drag tests on him also. No one remembers the results of the windtunnel test on the Taj Mahal.

Tremulis believed that the fins on the Taj Mahal influenced the design of the Mystere and, many years later, he recalled seeing Damon Woods and Bill Boyer in his studio studying the back end of the Taj Mahal shortly before the Mystere was designed.

There is no record of the Taj Mahal being shown at any car show, and it's unknown what happened to it. ■

Ford Photomedia

The Maxima was a three-eighths sized model designed by Alex Tremulis to commemorate the 1904 speed record set by Henry Ford in a car called the Arrow.

Maxima

In 1954, Alex Tremulis designed and built a 3/8-sized concept car he called the Maxima. It was built to commemorate the anniversary of the speed record set by Henry Ford 50 years earlier. Setting speed records was one way for early-day automobile manufacturers to generate a little free publicity. Henry Ford believed that winning races on Sunday sold cars on Monday. On January 12, 1904, he drove a car of his own design to a world land-speed record of 99 1/3 mph. The car Henry Ford drove was called the Arrow, and he drove it over a frozen and bumpy Lake Clair.

The Maxima was a rocket car designed to be propelled by a jet engine at speeds of over 500 mph. It was also the first of Tremulis' three-wheeled vehicles. The Maxima did not influence the design of any production vehicles, but it did inspire other real life three-wheeled land-speed cars, including Craig Breedlove's Spirit of America. Ford press releases said the Maxima was the "product of research into the application of radical aircraft shapes to automobiles."

Tremulis donated the Maxima to The Henry Ford Museum in 1985. It is in storage there and, apparently, has never been displayed. ■

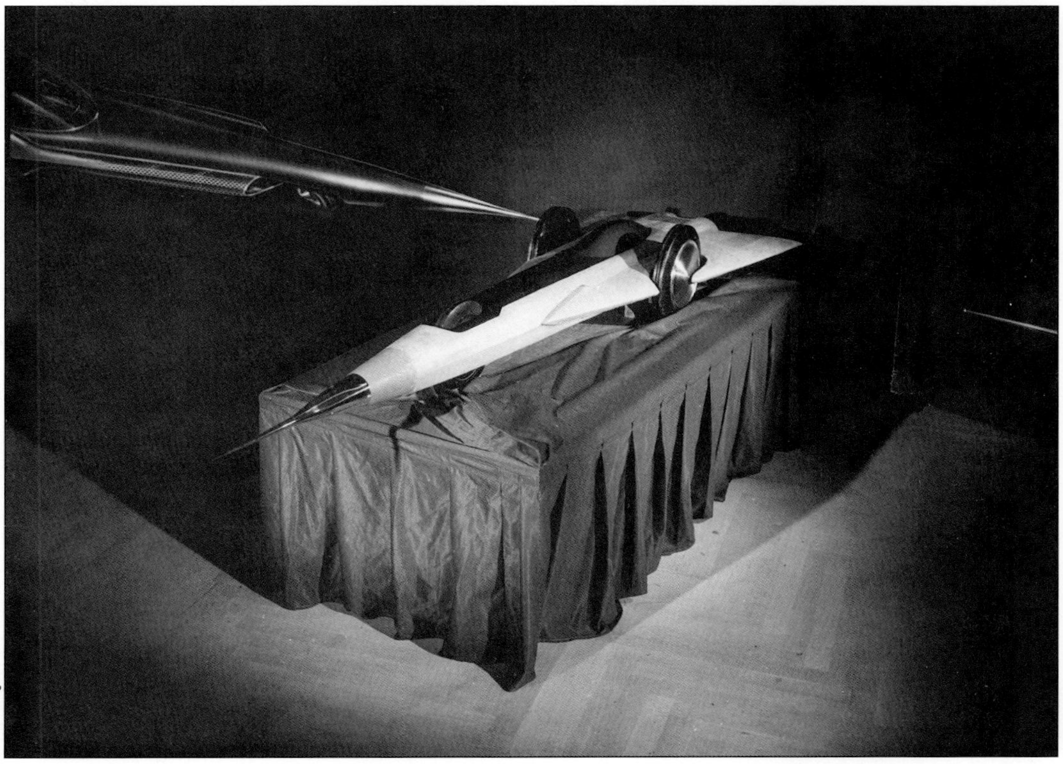

Ford Design Center

The Maxima was Tremulis' first three-wheeled car. It was designed to be jet powered at speeds of up to 500 mph.

Just as the Taj Mahal was Tremulis' proposal for a Mercury sports car, he designed the Madam X as a proposed two-passenger Thunderbird.

Ford Design Center

Madam X

The Madam X was a 3/8-sized model of a proposed Thunderbird two-seater designed by Alex Tremulis in 1955. It had a cantilevered roof and lacked A-pillars. It also had front-fender lines much like those on the Buick LeSabre concept car and rear fenders with blisters or hopups in back of the wheels, a theme taken from Gil Spear's Muroc and D-524. The same front-fender line was later used by Tremulis on the X-1000. Tremulis designed the rear-fender hopups for storage. He claimed to have built them big enough so a lady's 25-inch hat and hatbox could be stored inside. According to Tremulis, the rear-fender storage area was the source of the name Madam X. The name Madam X was first used in 1933 as the name of a full-sized Cadillac concept car, which may also have encouraged Tremulis to chose it.

Although the Madam X is one of Tremulis' lesser known concept cars, next to the LaTosca, it probably had the greatest influence on a production Ford product. When Edsel Division designers saw the Madam X in Tremulis' studio, they told him they wanted to use the front end of the car for the design of the 1958 Edsel. Puzzled, Tremulis asked if they were abandoning their vertical-grille design. The Edsel's designers explained that they wanted the front end of the Madam X, minus the grille, as the design for the rear end of the Edsel.

The Madam X was a serious design proposal Tremulis hoped would become the successor of the two-passenger Thunderbird, which was slated to become a four-passenger car by 1958. Even after the decision was made to turn the Thunderbird into a four-passenger car, both the Engineering department and the Styling Center proposed continued production of the two-passenger car, along with the new four-passenger Thunderbird. By 1955, Robert McNamara replaced Lewis Crusoe as head of Ford Division. McNamara decided there would be only one Thunderbird, and it would be a four-passenger car. That ended Tremulis' hope that the Madam X would become the basis for a production car. ■

Steve Tremulis

(above) The brow over the front fenders of the Madam X became the inspiration for the rear end of the 1958 Edsel.

Steve Tremulis

(right) When designing the Madam X, Tremulis borrowed the humped rear-fender compartments from the D-524.

Although management seldom agreed with him, Alex Tremulis often claimed that much of what others saw as play or a far-fetched design was really in the nature of a scientific experiment or scientific research.

One of Tremulis' scientific experiments involved a cardboard vellum tube with a tight-fitting metal cap. Tremulis would fill it with air from the other end, until the cap blew off. The vellum tube went off with a hollow-sounding "boom," that sent the metal cap flying across the room. Tremulis soon tired of chasing the metal cap shot from his vellum-tube mortar, so he and designer Bob Gurr attached long strings to the caps, so they could retrieve them by reeling in the string. One night, Tremulis fired off his vellum-tube mortar and watched the cap ricochet around the high ceiling of the EEE Building. The metal cap, with the attached string, bounced across the egg-crate ceiling over Bob Maguire's office, eventually falling through. It dangled just above Maguire and a group of designers he was meeting with. Every time Tremulis pulled on the string, the metal cap banged against Maguire's ceiling. Maguire eventually had to stop the meeting, so Tremulis could retrieve the cap to his vellum tube, all the while explaining to Maguire that he was just researching high-speed aerodynamics.

Tremulis really did a lot of research and, based on that research, he began designing ultra high-speed cars and dragsters. Many designers in the mid-1950s believed that future cars would travel on controlled freeways at speeds of up to 500 mph. To Tremulis, that meant there was a real need to start taking aerodynamics into account in designing future cars. Management generally viewed Tremulis' high-speed cars, and especially his proposed dragsters, as having limited production value. As for aerodynamics, management, which was generally much more realistic than Tremulis, felt it had no benefit below 100 mph.

The 999 was a 3/8-sized dragster designed by Alex Tremulis in 1956. He designed it as a tribute to the famous racecar of the same name that Henry Ford built in 1902-03. Features shown on the rendering were toned-down when the clay model was built. About the only similarity between Henry Ford's 999 and Alex Tremulis' concept car is the same sense of brute power conveyed by both cars. Apparently, no design features from the 999 were used on any production Ford product. ■

999

(below) The 999 was designed as a modern dragster. Even though management was not thrilled with the idea of a dragster as a concept car, Tremulis designed it in honor of Henry Ford's famous race car of the same name.

Ford Design Center

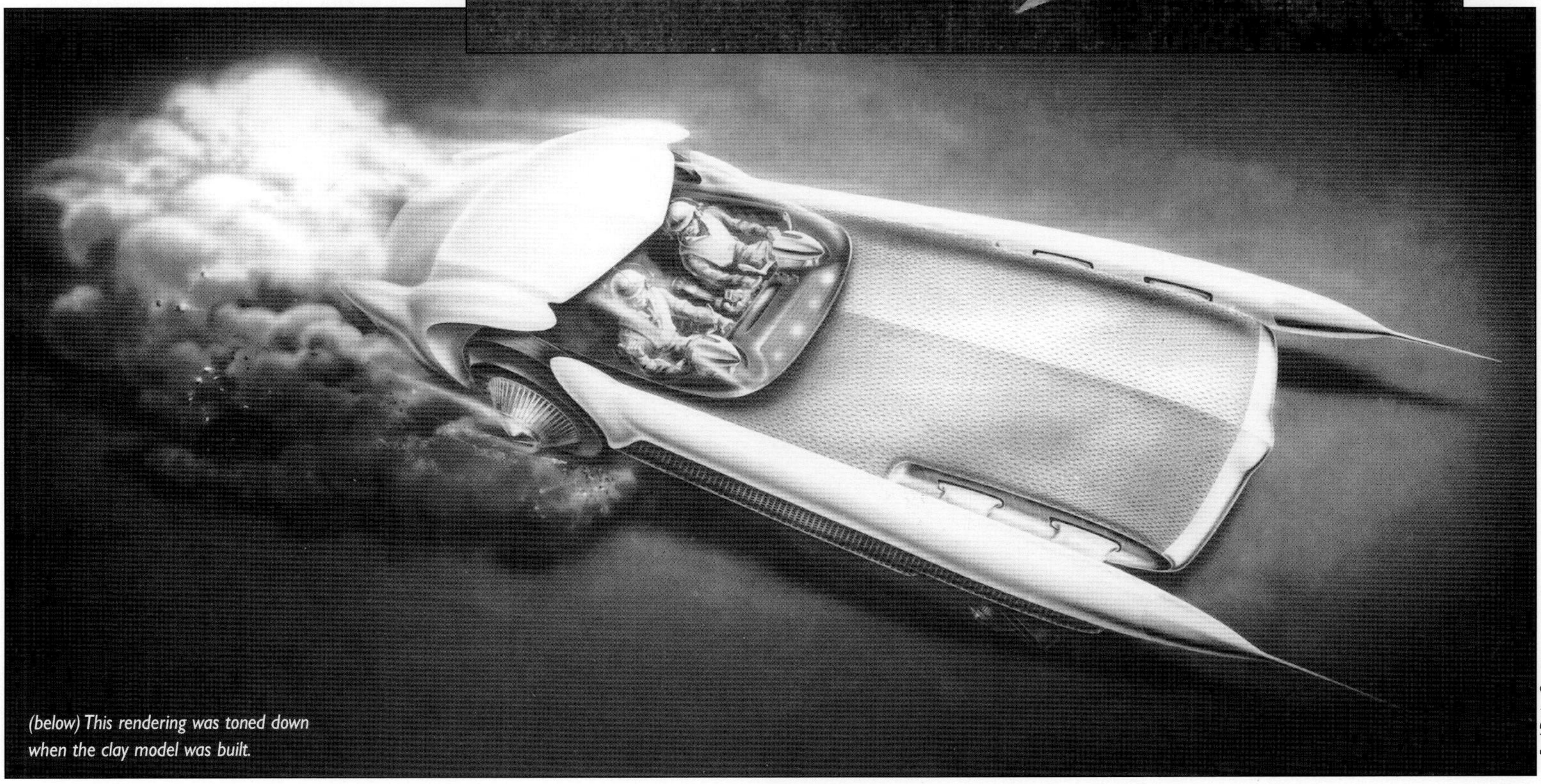

(below) This rendering was toned down when the clay model was built.

Ford Design Center

X-1000

The X-1000 was designed by Alex Tremulis in the Advanced Studio during 1955-56. The X-1000 looks more like a spaceship than a car and, few if any of its features made it into any production car. Tremulis acknowledged that no design features from the X-1000 made it into production except some instrument panel knobs. The most unusual features of the X-1000 were the retractable fins, the taillight pods on the inside of each fin and the two-porthole backlight.

The X-1000 was the first concept car created at the request of George Walker, after he was appointed a Ford vice president and general manager of the Styling Center. Walker decided that the Styling Center was going to have its own display at the 1956 Chicago Auto Show, and he selected Tremulis' X-1000 as the focal point of the display. A finished 3/8-sized model of the X-1000, renderings of the car and a full-sized clay model that was 75% finished (and being worked on) were displayed along with clay modelers and designers, who were rendering and working in clay, so the public could actually see how cars were designed and clay modeled. On the back end of the car, there was even an area where parts of the bumpers were covered with foil, while other parts were still in clay. (During the Walker years, designer George Barbaz remembers being invited to be one of the designers appearing at various car shows Ford participated in, both in the United States and around the world.)

Walker encouraged Tremulis' aerodynamic spaceship designs, which he thought were in keeping with the times. Some designers believe there was a subtle change in policy, and it may have influenced the type of designs Tremulis created. In the mid-1950s, Ford Motor Co. and the Styling Center were surprised to learn that many dealers did not appreciate Ford's showcars on their showroom floors being displayed next to the production vehicles they were trying to sell. At least some of those dealers were telling Ford the buying public preferred to wait to buy their new cars until the features they had seen on the showcars were available — next year. By designing concept cars that looked like the X-1000, that problem was apparently avoided.

Because it was Walker's first opportunity to take part in a major auto show as a Ford executive, he invited his five chief designers to accompany him to Chicago for an advance preview of the auto show, which they and their wives enjoyed the night before it opened. ■

George Walker liked showcars that were on the cutting edge of design. He is shown here with the X-1000, which Alex Tremulis designed at Walker's request for the 1956 Chicago Auto Show, the first major auto show the Styling Center participated in after Walker became a Ford vice president and general manager of the Styling Center.

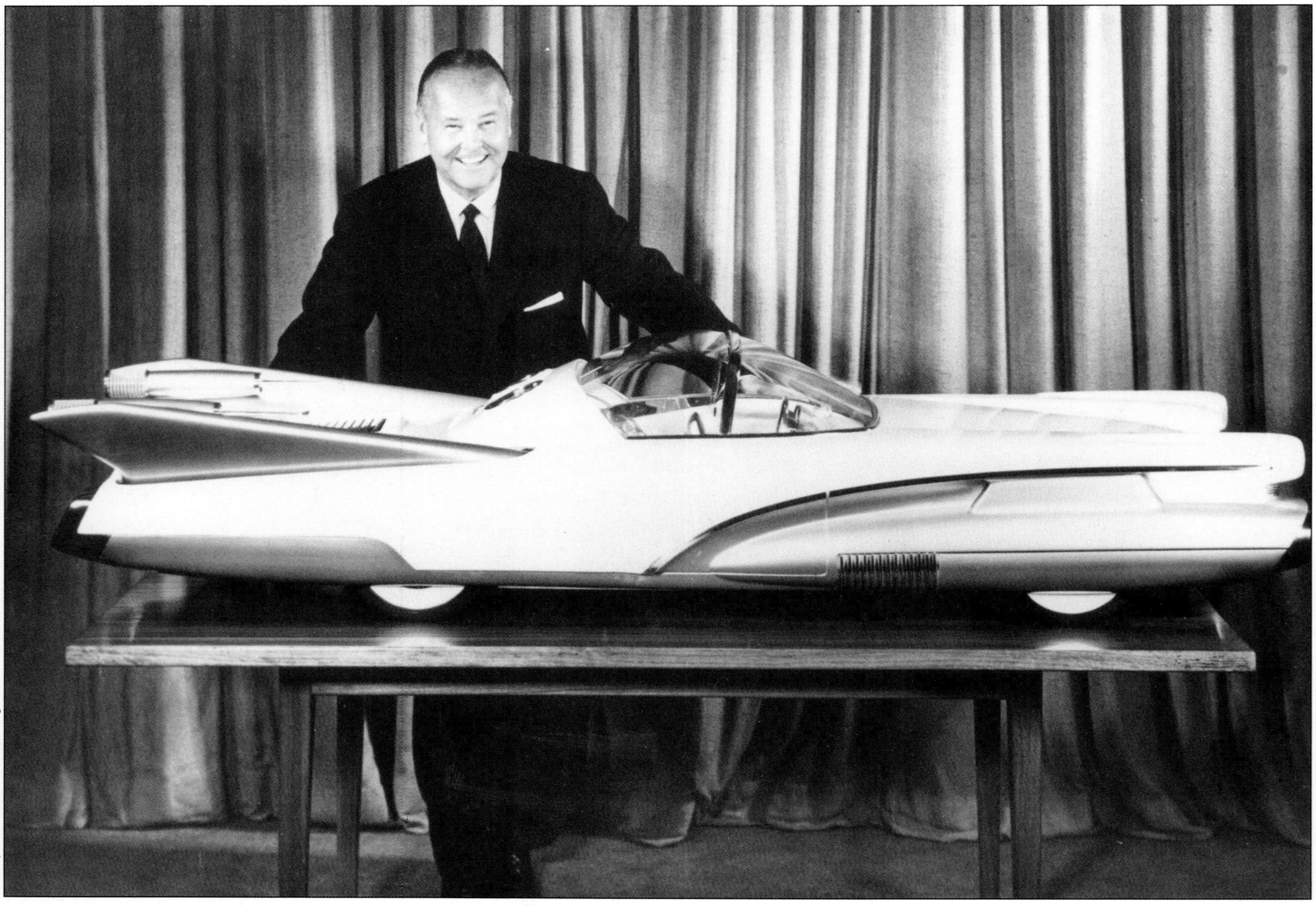

Henry Ford Museum and Greenfield Village

(right) The original colors Tremulis selected for the X-1000 were changed before the Chicago Auto Show.

Steve Tremulis

(below) The X-1000 was designed to include a gas-turbine rear engine. The fins retracted. The dual-porthole backlight may have been influenced by the Briggs Prototype.

Steve Tremulis

John Najjar

John Najjar

(left) A three-eighths sized model of the X-1000 on display at the Chicago Auto Show.

(right) Gil Treweek was one of the Ford clay modelers assigned to clay model the X-1000 at the Chicago Auto Show.

John Ebejer clay modeling the X-1000 during the Chicago Auto Show. Walker wanted Ford's display to show the public how cars were designed and clay modeled.

John Najjar

Ford Photomedia

Henry Ford Museum and Greenfield Village

(above) Larry Wilson (standing at right), clay modeling manager Clyde Trombley (seated at right) and plaster shop technician John Carrol are building a full-sized X-2000.

(left) A three-eighths sized model of the X-2000 after the fins were replaced by bazookas similar to those used on the 1959 Ford.

X-2000

The X-2000 was designed by Alex Tremulis and Bill Balla in Bob Maguire's Advanced Studio in 1956-57. The grille on the X-2000 was meant to forecast, at least in concept, the 1958 Edsel, which was to be introduced in the fall of 1957. The X-2000 was probably the most spaceship-type car Tremulis designed while he was at Ford.

In its original form, the X-2000 had wings or fins that made it look like the modern-day spaceshuttle. On the 3/8-sized model, the fins were later removed and replaced with bazookas, but no other design changes appear to have been made to the car. Both versions of the 3/8-sized X-2000 featured rear engines and gull-wing doors, which were built into the Plexiglas roof. The X-2000 also featured a flow-through ventilation system, and it was designed without windows that opened.

A full-sized clay model of the X-2000 was at least partially completed. According to Gale Halderman, work on the X-2000 was stopped before the full-sized model was completed, because management feared people attending the auto shows where the X-2000 was to be displayed might think future Fords were all going to look like the X-2000. As best Halderman can recall, that decision was made by Ford Division and not at the Styling Center.

The 3/8-sized X-2000 was publicized twice. The first time was in a booklet published by the Styling Center in 1958. It was also shown with the X-1000, the Mexico, the Volante and the Nucleon at the May 1958 opening of the Levacar display at the Ford Rotunda. ■

Ford Design Center

(above)The grille on the X-2000 forecast the 1958 Edsel, which was introduced a year after the X-2000 was built.

Bill Balla

William E. Balla

Bill Balla was born and raised in Cleveland, Ohio. As a child, he loved to draw cars. He attended the Cleveland Institute of Arts, graduating in 1940. His first job was as a designer in George Walker's industrial design firm in Detroit. He was at Walker's firm for only one year before enlisting in the Army at the start of World War II. After the war, Balla took a job in the E.H. Daniels' design firm in California. In 1949, both Daniels and Balla joined the Kaiser-Frazer design department. Balla left Kaiser-Frazer to return to Walker's design firm in 1951. While there, he helped design Walker's entry in the Continental Mark II design contest. During the time Walker was a consultant for Ford, Balla worked on the design of many of Walker's independent Ford, Mercury, Lincoln and Continental proposals.

When George Walker became a Ford vice president and general manager of the Styling Center in May 1955, Walker employees Joe Oros, Elwood Engel and Balla became Ford employees. Balla was assigned to work in Alex Tremulis' Advanced Studio. In 1956, Tremulis and Balla were both transferred to Maguire's Advanced Studio, where they worked on the Stylerama proposals and on several show vehicles, including the X-1000 and the X-2000.

When Maguire was appointed head of the interior studio in 1958, Balla was transferred to the Ford and then the truck studio. He left Ford in 1959 to work at an independent industrial design firm in California with another former Walker employee. After a year in California, Balla returned to Detroit, where he worked for several independent industrial design firms, including Sundberg and Ferrar.

He retired in 1990. ■

Ford Design Center

(above) The X-2000 had gull-wing doors and a simple interior. From left to right are Clyde Trombley, Bill Fox, Larry Wilson and two unknown Styling Center personnel.

Ford Design Center

As originally designed, the X-2000 had huge fins and looked more like a spaceshuttle than a car.

Ford Design Center

The design of the Wind Brake Car was a combination of several of Tremulis' earlier cars.

Wind Brake Car

As Alex Tremulis' interest in aerodynamics evolved, his designs became more experimental and less commercial. Although George Walker demoted him, he liked Tremulis, and he liked his concept cars, because he thought they were on the cutting edge. He continued to encourage Tremulis' futuristic designs, which he also thought were excellent public relations tools for Ford Motor Co. Walker often assigned Tremulis to speak to various groups about the future of car design. Those speeches always included heavy emphasis on aerodynamics, and Tremulis' belief that future cars would travel at very high speeds — up to 500 mph on controlled highways! In his presentations, Tremulis always gave credit to Walker, and the slides he brought with him always emphasized the relationship Tremulis saw between space travel and future car design. Many of the slides Tremulis showed were of his most recent designs, which is probably a good indication they were not considered to have very much commercial value.

One of the aerodynamic cars Tremulis designed was the Wind Brake Car, a 3/8-sized clay model he worked on over the course of several years in the mid and late 1950s. In none of Tremulis' speeches or interviews, however, is there any reference to the Wind Brake Car or the concept. The front fenders and grille of the Wind Brake Car are similar to those on the Madam X and the X-1000. The taillights also have similarities to those found on the X-1000, although there were no wings, retracting or otherwise, on the Wind Brake Car.

Gale Halderman recalls that Tremulis' primary purpose in designing the Wind Brake Car was to explore the use of a retractable panel on the deck lid against which wind passing over the car at high speeds could be channeled to help brake the car. The wind brake was designed to automatically deploy when the brake pedal was depressed. Tremulis theorized that just as the overall shape of a car could help it cut through the air more efficiently, that same air could be used to help stop the car by reversing the aerodynamics. Tremulis was an avid reader of scientific journals and probably got the idea from modern jet aircraft.

According to Halderman, Tremulis asked for support and assistance from Andy Kucher, Ford's vice president of scientific research and the engineers in his department, to fully develop the wind brake concept. Kucher was polite, but his engineers were quick to point out that, for the wind brake to be of any use, a car would have to be going a minimum of 100 mph when it was deployed. That effectively ended work on the wind brake. According to other designers, the overall design of the Wind Brake Car was not well received.

Spoilers used on today's race cars perform approximately the same function as the wind brake by using the air passing over the surface of the car to help keep it on the ground. Aerodynamics has become a valuable design tool, not because it helps control a car at high speeds, but because of fuel efficiency. Thirty and 40 years later, Tremulis, the likable practical joker, is most remembered for his aerodynamic foresight. ■

(left) The Wind Brake Car's taillights resemble rockets or Jato tubes.

Steve Tremulis

(below) When the driver stepped on the brake pedal, the panel on the deck lid raised up to help stop the car.

Steve Tremulis

(below) The deflectors at the rear of the roof were supposed to channel air directly at the wind-brake panel, when it was raised.

Steve Tremulis

(below) From the front, the Wind Brake Car resembles the X-1000.

Ford Design Center

8

Stylerama

In 1949, GM invited the world to come see the latest in automobile technology and the best of its new cars at auto shows in New York and Detroit. Those shows were successful beyond anyone's wildest expectations, with more than 300,000 people accepting the invitation. The extravaganza, renamed the Motorama, was repeated in 1950, and it drew more than 320,000 guests. After missing two years for the Korean Conflict, the Motorama returned in 1953. In addition to New York City, the 1953 Motorama traveled to Miami, Los Angeles, San Francisco, Dallas and Kansas City. Ford executives who attended the '53 Motorama were impressed with the public's response to GM's concept cars, and they couldn't help but notice that 1.5 million people joined them as guests of GM. By 1956, the Motorama shows, displays and concept cars were even more spectacular, and attendance rose to 2,348,241 people. Ford was trying to close in on industry sales leader Chevrolet by competing head-to-head in everything -- and that meant that Ford started looking for an answer to Motorama.

Ford Design Center

Al Mueller

Ford's answer came from John Cuccio. Before he was hired as a Ford designer, Cuccio spent 15 years as a designer for Raymond Loewy and Associates, where he learned something about showmanship. After Loewy lost the Studebaker account, Cuccio was assigned to design and supervise the building of a Loewy showcar for the 1955 Paris Auto Show. While he was in Turin, Italy, supervising the building of the car, he received a telegram from Frank Hershey offering him a job as a designer at Ford, and he was asked to come to Dearborn to meet with Hershey. When Cuccio returned, he called the Design Department to make an appointment, only to find out that Hershey was no longer at Ford, and that George Walker was the new boss. Walker didn't know anything about the telegram but agreed to meet with Cuccio and then hired him as a manager in Buzz Grisinger's Mercury pre-production studio.

After Cuccio settled in, he suggested that Ford design and build a dozen or so full-sized futuristic showcars, put them in specially-designed trailers and truck them across the United States and Canada. Like the one used to display the XM-Turnpike Cruiser, Cuccio suggested that each trailer be designed with sides that opened and with attached decks and awnings. Cuccio gave Walker several renderings that showed the trailers drawn up in a circle, covered wagon style, with people around them looking at Ford's futuristic cars. Walker loved the idea, named it the Stylerama and recommended it to Henry Ford II as Ford's answer to the Motoramas. As soon as he heard the proposal, Henry Ford II became as enthusiastic about it as Walker. He asked for cost estimates, and ordered that the plan be implemented as soon as possible.

Walker asked Bob Maguire to organize the program. In March 1956, Maguire stepped down as head of the Ford studio to organize a new Advanced Studio. Al Mueller was assigned to work with Maguire. Buzz Grisinger was also reassigned from the Mercury pre-production studio to Maguire's studio to help with the Stylerama program. Instead of having two advanced studios, Alex Tremulis' studio was done away with, and he and Bill Balla, a designer in Tremulis' studio, were assigned to Maguire's new Advanced Studio to help with the Stylerama cars.

Robert H. Maguire

Bob Maguire grew up in New York City, where his father was a high school principal. Maguire and his sister were members of the Mensa Society. Maguire studied architecture at Syracuse University but left after one year, when he decided he didn't want to be an architect. He then enrolled at a fine arts school in New York City. After completing the course, he began work as an industrial designer in 1929. In the late 1930s, he worked at Genter Corp. designing buildings for the 1939 New York World's Fair. In 1939, he was hired as a designer by General Motors.

During World War II, Maguire was given a direct commission in the Army as a first lieutenant and was placed in charge of other designers designing camouflage. He then volunteered for and completed Officer Candidate School and ultimately rose to the rank of Major in the Army Air Corps, where he headed a Black American squadron of fighter pilots. After the war, he returned to GM and became head of one of their interior design studios.

In 1947, Maguire left GM to join Ford's Design Department. In 1951, he was appointed head of the Ford studio, but was replaced by Frank Hershey in 1952. Maguire was reappointed head of the Ford Studio when Hershey left in 1955 and was named head of the Advanced Studio in 1956. Maguire later headed Ford's interior studio and was considered for the job as general manager of the Styling Center, when George Walker retired in 1961.

Maguire supervised the design of several concept cars, including the FX-Atmos, the 1955 Thunderbird, all of the Stylerama cars, the X-2000 and the LaGalaxie.

In 1961, when Engel left to become vice president of design at Chrysler, Maguire replaced him as head of the corporate advanced studio. Maguire decided to leave Ford in 1968 because of his wife's failing health. At the Ford family's request, he was given early retirement.

Maguire was an accomplished classical musician, who played the bass saxophone and flute. He taught music appreciation to many Ford designers. ■

(above left) Ford planned to transport its Stylerama showcars around the country in specially-designed tractor-trailers like the one shown above. Plans were to set up the Stylerama display in fields and parking lots, where the public could see the Fords in their future and live entertainment at the same time. Although the Stylerama was canceled before it got underway, it was meant as Ford's answer to General Motor's very successful Motorama.

John Cuccio

John Cuccio

John Cuccio was born in New York City, but grew up in Palermo, Italy, where his father was a large citrus grower and exporter. He moved from Sicily to New York permanently with his family when he was 19. After settling in New York City, Cuccio attended Columbia University and then Pratt Institute. In 1942, he enlisted in the Army Air Corps. After the war, a friend from Pratt, who was working in Loewy's New York office, introduced Cuccio to Raymond Loewy. Cuccio was interviewed by Loewy and hired to work on the Studebaker account in South Bend, Indiana. During the 15 years he worked for Loewy, Cuccio helped design the 1953 Studebaker and the Studebaker Hawk. He also designed products for many of Loewy's commercial accounts, including Frigidaire.

After Loewy lost the Studebaker account, Cuccio designed a custom Jaguar for the Loewy exhibit at the 1955 Paris Auto Show. He was in Italy supervising its construction, when he received a telegram from Frank Hershey offering him an executive position at the Ford Design Department. Because Loewy was cutting back, Cuccio contacted Ford when he returned. Between the time he received the telegram and the interview, Hershey left, and Walker became general manager of the Styling Center.

Cuccio was hired by Walker as a manager in the Mercury pre-production studio working under Buzz Grisinger. While there, he helped design the Blue Job and the L'Avion. Later, he was transferred to the Mercury production studio.

In 1956, Cuccio proposed the Stylerama program to Walker. The proposal was approved by Henry Ford II. It involved the building of showcars to tour the United States and Canada in specially-built trucks. Cuccio was sent to Europe with Walker's administrator, V. Z. Brink, to negotiate with European coachbuilders to build the Stylerama cars. In 1957, after the Stylerama program was canceled, Cuccio left Ford and moved to Connecticut to work for Clare Hodgman and Bob Bourke in their independent design firm. In 1958, he established his own design firm in Westport, Connecticut. ■

The first thing Maguire did was to ask Tremulis and Balla to design several advanced cars. He also asked that the other studios each come up with designs for two futuristic cars for a special show to be held in the Design Rotunda. To encourage the designers' undivided attention to the task at hand, Maguire was given a small room downstairs in the Styling Center, where designers selected from each studio were temporarily assigned until their Stylerama proposals were finished.

On April 3, 1956, a special show was held in the Design Rotunda to pick the cars to be built for the Stylerama program. Fourteen 3/8-sized models were chosen from the 22 designs submitted. Because the XM-Turnpike Cruiser had been built by Ghia from a 3/8-sized model, the plan was to send the selected 3/8-sized models to European coachbuilders, who would then build the full-sized Stylerama cars.

John Cuccio and Walker's administrative manager, V. Z. Brink, left on April 27, 1956, to visit coachbuilders in Italy and Germany. Their purpose was to evaluate facilities, get price and time estimates and assess security at each facility. The Mercury XM-Turnpike Cruiser, which had recently been built by Ghia, was used as the benchmark for cost, time and needed supervision.

During the two weeks they were gone, Cuccio and Brink visited Ghia, Vignale, Boano, Bertone, Farina and Touring in Italy, and Spahn and Drauz in Germany. On May 14, 1956, Cuccio and Brink returned to Dearborn. They submitted a ten-page report to Walker and Maguire. Cost estimates they received ranged from $15,000 to $45,000 for completed full-sized pushmobiles, and it was estimated that it would take four to six months to build each car.

Ford Motor Co. decided to ask each Italian coachbuilder to build two or three inoperable cars -- all they had to do was look real. The next step was to negotiate a fixed price, sign contracts, and then send the 3/8-sized models, full-sized side-view renderings and the needed hardware to Italy. Mario Boano was negotiating with Fiat to become head of their Advanced Studio, but he told Ford he would maintain his separate coachbuilding business, if his firm was awarded a contract to build the showcars.

Instead of shuttling someone from Dearborn to Italy to supervise construction, it was proposed that Cuccio, who had grown up in Sicily and spoke fluent Italian, move to Turin, where he would personally supervise the construction of the Stylerama cars. Cuccio found an apartment in Turin, and he and his family made arrangements to relocate there. Ford agreed to furnish Cuccio with a new Thunderbird and a new Ranch Wagon each year he was in Italy, and he had already ordered cars for his first year.

When the full-sized Stylerama cars were finished and delivered to Dearborn, plans were to put two in each trailer, and send five of the trailers across the United States, while two cars and one trailer were to tour Canada. The Stylerama displays were to be set up on vacant lots, in stadiums, and any other places where the public could see the cars and the accompanying entertainment Ford planned to send with the cars.

Starting with the XM-Turnpike Cruiser, Ford concept cars generally bypassed dealers, who had come to feel that concept cars displayed at their dealerships tended to discourage rather than encourage prospective buyers. Dealers apparently felt that the features on Ford's concept cars only encouraged the buying public to wait until those features were available on the cars they were selling.

The departure date for the first Stylerama caravan was planned for early 1957, but it never happened. Although the program was an open secret at the Styling Center, no public announcement was ever made.

Many designers assume that the Stylerama program was canceled because of the Edsel disaster, but the answer is much more complicated than that. Although some executives at Ford had premonitions of the financial disaster that was about to befall the Edsel, with losses exceeding $235 million, the Stylerama program was canceled because cost estimates for everything but the cars were too high, and some of the cars were judged as a little too far out to be shown in middle America. As Joe Oros recalls it, the program collapsed of its own weight and, by the time the Edsel was introduced, it was too late. Cancellation of the Stylerama program was a great disappointment to John Cuccio, who soon left Ford to open his own industrial design firm ■

Saturn

The Saturn was a Ford studio proposal for the Stylerama program. It was designed by Jim Darden immediately after he was appointed manager of the Ford pre-production studio.

Although the front end had Pontiac overtones, the rear end continued the round taillights that had been a design theme on all Fords since 1952. Joe Oros came up with the round taillight theme for the X-100, and it was used on Fords from 1952 forward. Oros took over as head of the Ford design studio in March 1956, when Maguire was transferred to a new Advanced Studio with directions to organize and spearhead the Stylerama program. Oros chose Darden to run his pre-production studio. The taillights on the Saturn are similar to those used on the 1963 production Ford and, although the front-end theme on the Saturn did not appear on any future Ford, it is similar to the original design proposed for the Quicksilver and the 1961 Thunderbird.

If there was a second proposal for the Stylerama program created in the Ford studio, photographs of it cannot now be located. It's likely that Oros was extremely busy getting his studio organized and, for that reason, he probably submitted only one design proposal for the Stylerama program. In keeping with Oros' preference, the fins on the Saturn were subdued and show some similarity to those on the Quicksilver, which Oros, Darden and designers in the Ford pre-production studio began shortly after the Saturn was finished. The Y-shaped roof support in the center of the backlight was a design feature tried in all studios, but never used in production. The shape of the Saturn's A-pillar was also considered for the 1959 production Ford, but never used. The Saturn's suicide rear doors came from the Continental Division Mark III Berline.

No one recalls what happened to the Saturn after the Stylerama program was canceled, but it is likely it was in Darden's pre-production studio for a while and eventually destroyed. ■

Al Mueller

(above) The Saturn was the Ford studio's proposal for the Stylerama program.

Ford Design Center

(above) The front of the Saturn had seven Pontiac-like chrome trim pieces running from under the grille to the base of the windshield. The quad headlights were in pods placed inboard and along the full length of the sloping hood.

Ford Design Center

(right) The Saturn's taillights were similar to those eventually used on the production 1963 Ford. The Y- shaped roof support in the middle of the backlight eliminated the need for thick C-pillars.

(right) The Scorpion was a future Thunderbird proposal designed for the Stylerama program.

Ford Design Center

(below) The bazookas on the rear fenders of the Scorpion predicted those used on the production 1959 Ford.

Al Mueller

(below) The six headlights and the front-wheel cutouts did not make it into production. The air-conditioning vent on top of the half-glass roof was taken from the Mystere.

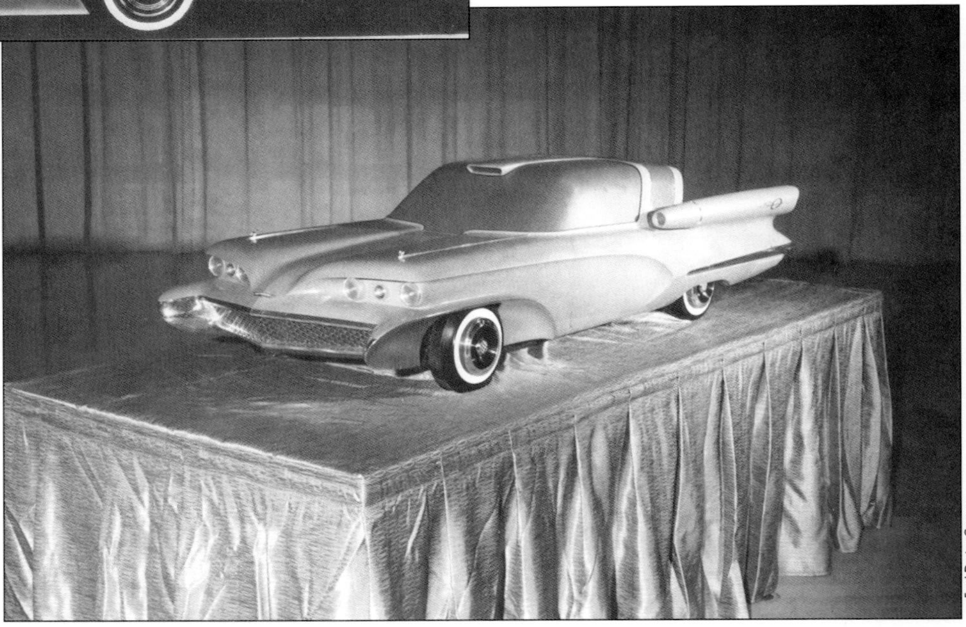

Ford Design Center

Scorpion

The Scorpion was one of Alex Tremulis' contributions to the Stylerama program. It was a proposed four-passenger Thunderbird. By the end of 1955, the decision had been made to convert the Thunderbird from a two-passenger to a four-passenger car starting with the 1958 model year. Although designers recall that Tremulis worked on the Scorpion, several, including Gale Halderman, believe it may include features that were not characteristic of Tremulis' design style. The front-wheel cutouts may have come from another designer. There's another explanation given, and it involves speculation that at least some of the Scorpion's design features, including the front-wheel cutouts and the six headlights, may have been a less than serious effort by Tremulis meant to tweek management.

The production 1959 Ford was the first car designed entirely under the direction of Joe Oros, after he was appointed head of the Ford studio. The more formal roofline and the bazookas on the rear fenders of the Scorpion probably influenced the design of that car. Tony Yuki, one of the designers of the '59 Ford, does not recall the Scorpion, but acknowledges that it may have been an early attempt to develop styling themes that were eventually incorporated on upcoming production cars. ■

Convertible Pickup (Pantera) Light Cab Forward

Traditionally, pickup trucks have utilized about 1/3 of their total length for the cab and engine, and the other 2/3 for payload. In the mid-1950's, designers and engineers at Ford and elsewhere started looking for ways to increase pickup bed space. They determined that the most logical way to do so was by decreasing the space needed for the cab and engine to no more than 1/4 of the pickup's total length. Designers and engineers thought that could best be accomplished by moving the cab forward ahead of the front axle. The engine could then be moved to a position between the passengers or to a transverse position underneath and behind the passengers.

In 1955, GM introduced a concept van called the L'Universelle. Like the big cab-over-engine (COE) trucks, the L'Universelle had no separately-defined hood or engine compartment and, except for the curved-glass windshield and the grille, it resembled a stylized box with a front as blunt as the rear. In 1955, the concept and looks of the L'Universelle were novel, but it did result in making 3/4 of the vehicle available for payload. The L'Universelle was not the first truck to propose moving the engine to a position adjacent to or to the rear of the passengers, but it had a big impact on Ford's truck studio. Soon after they learned about the L'Universelle, Ford designers began designing their own cab-forward van that eventually became the 1962 Econoline.

In early 1956, Bill Wagner's truck studio was asked to submit a proposal for the Stylerama program. The 1957 Ford Ranchero had already been authorized for production, and Stylerama provided an ideal opportunity for the designers in the truck studio to design an even more advanced truck -- an advanced cab-forward pickup that looked similar to the L'Universelle. Instead of having a closed-in storage compartment like the L'Universelle, the Convertible Pickup, as it was called, had an open pickup bed that could be enclosed by an accordion-like canvas top that was stored just behind the cab in a compressed position and, when unfolded, it could be extended all the way to the back of the bed. Like the Ranchero, the Convertible Pickup was designed with many of the styling amenities of a car.

The Convertible Pickup was designed with a built-in accordion-type folding top that could be extended to cover part or all of the pickup bed. The folding top is shown in full-extended position in the rendering on the wall behind the three-eighths sized model.

Ford Design Center

The Convertible Pickup was designed by truck pre-production studio manager Fred Cole and designers George Barbaz, Joe Achor and John Gump and built as a 3/8-sized clay model. According to John Gump, the Convertible Pickup was later renamed the Pantera (Italian for panther), because its designers saw a similarity between the pickup's front end and a crouched panther.

Development of the cab-forward concept did not stop with cancellation of the Stylerama competition. In July 1956, the Convertible Pickup was redesigned and built as the full-sized clay model. When the full-sized clay model was built, the Convertible Pickup was renamed the Light Cab Forward. Although it's design was similar, it was an evolutionary but complete redesign of the Convertible Pickup.

A reverse backlight was incorporated into the design of the Light Cab Forward as a way of increasing the usable cargo area.

Ford Design Center

The Light Cab Forward was redesigned by the same designers who were responsible for the Convertible Pickup, plus McKinley (Mac) Thompson, who was assigned to the truck pre-production studio in June 1956. The Light Cab Forward was a cleaner, less sculptured design than its predecessor, but its designers confirm it came from the 3/8-sized Convertible Pickup.

Development of the Econoline concept, which placed the passengers on either side of and slightly behind the engine, was continuing at the same time. The configuration of the Light Cab Forward went one step further than the Econoline, however, in placing the driver and passenger ahead of the axle and the engine behind them in a transverse position with big grilles on the outside just behind the doors to cool twin radiators. According to Mac Thompson, engine access was through a shallow dog house in the rear wall of the cab. Although a complete engineering study was not done for the Light Cab Forward, its designers tried to incorporate only existing technology when they designed it. The only marking on the clay model was on the trim under the door windows that said "Light Cab Forward."

Unlike the Econoline, the Convertible Pickup and the Light Cab Forward were designed for an upscale audience. The spare tire on the Light Cab Forward was enclosed in a continental kit built into the tailgate, and it had a modern two-tone paint scheme, a

The large vents behind the front wheelwells were designed as air intakes to cool the engine, which was located sideways at the back of the passenger compartment.

Ford Design Center

Mac Thompson

This rendering of the Light Cab Forward was prepared by McKinley Thompson in late 1956.

windshield that arched over and into the roof line, quad headlights, a reverse-slant backlight, and dual exhausts that exited through the bumper ends. Several accessory canopies were designed to enclose the bed on the Light Cab Forward, so that it could be used as a van, station wagon, or camper, but they were all fixed, and none of the proposed variations folded accordion-like as on the proposed Convertible Pickup.

The full-sized clay of the Light Cab Forward and a fiberglass Econoline proposal were shown in the Styling Rotunda in November or December 1956. Development of the Econoline continued, but work on the Light Cab Forward stopped. The designers who worked on it believe the Light Cab Forward lost out because engineering for the Econoline was more conventional, it used more standard parts, and it was cheaper to build.

Although several manufacturers, including Ford, have since built cab-forward pickups, the concept never really caught on, but then again, none of them looked like the Light Cab Forward. ■

To increase usable bed space, the Light Cab Forward's spare tire was placed in an enclosed continental kit on the tailgate. The spare tire was removed from the cargo side of the tailgate.

Ford Design Center

John G. Gump

John Gump

John Gump was born in New York City. As a child, he loved to draw cars. During World War II, Gump served in the Army Air Corps as a crew member on Air Transport Command planes. After the war, he returned home and became a trainee in an industrial design office, and then decided he wanted to become an industrial designer. From 1951-54, he attended Pratt Institute. After graduation, he worked for W. L. Maxon for two years preparing concept drawings and manuals for the military.

In 1956, Gump saw an ad Ford placed in the New York Times seeking car designers. After sending his portfolio, he was interviewed by Bob Maguire and Art Querfeld. Gump was hired and assigned to the truck pre-production studio working for manager Fred Cole. Soon after he got to the truck pre-production studio, Gump helped design the Convertible Pickup (Pantera) for the Stylerama program. After about six months, he was transferred to the Ford interior studio. He was almost immediately reassigned to the Lincoln production studio, where he designed the grilles for the 1962-63 Lincolns. In the mid 1970s, Gump worked on the design of the Bordinat Mark V. He later worked in the Ford studio, where he helped design the grilles for the 1977-78 Fords. He also designed the C-pillar with window used on the 1977 production Thunderbird.

Gump retired in 1980. He did not like retirement so, from 1980-83, he returned to work at GM preparing technical illustrations for their manuals. Beginning in 1984, he was rehired for several years as a free-lance designer in the Ford truck studio. ■

Missile I

When Packard and Studebaker merged in 1955, Raymond Loewy and Associates lost the Studebaker account. Bill Schmidt, who was hired away from Ford to head Studebaker-Packard's design department, asked all of the Loewy designers working on the Studebaker account to stay and work for Studebaker-Packard. Most of Loewy's designers, including Bob Bourke, Bob Koto and John Cuccio, elected to stay with Loewy. After several months at Loewy's New York office designing appliances, Bob Koto noticed that things were slowing down. He wanted to get back to car design, so he suggested that Loewy buy out his contract. Loewy did just that and, in August 1955, Koto was hired as a designer at Ford.

Several months before Koto started at Ford, George Walker was appointed Ford's vice president of design. When Koto started, Walker had not yet completed his reorganization, so his first assignment was in the Edsel studio, marking time. Several months later, he was placed in charge of the Mercury pre-production studio. When Koto was given that assignment, he was asked to come up with six full-sized Mercury idea cars each year. During the first year Koto and the designers and modelers working in his studio built 22 full-sized clay models. Koto later moved to the Mercury production studio but, before he did, he designed the Missile I for the Stylerama competition.

Koto liked the theme of the 1957-58 Mercury, especially the concave ridges along the body sides terminating at the taillights. At the time the Missile I was designed, it was meant as a continuation and refinement of that theme. After the Stylerama program was canceled, the Mercury production studio built a full-sized clay model using several of the design cues from the Missile I and II, but the ideas Koto designed into the Missile I ultimately gave way to the design ideas of Buzz Grisinger, who soon became head of the Mercury production studio and later was placed in charge of all Lincoln-Mercury styling. In addition, McNamara instituted an interchangeability program at about the same time, and it required that the Mercury use sheetmetal from the current year's Ford. ■

Al Mueller

(left) The Missile I was one of the Mercury studio's entries in the Stylerama program. It took many of its design cues from the production 1957 Mercury, but included unique C- pillars far inboard on the rear deck.

(below) The dual grilles on the Missile I contained hidden headlights.

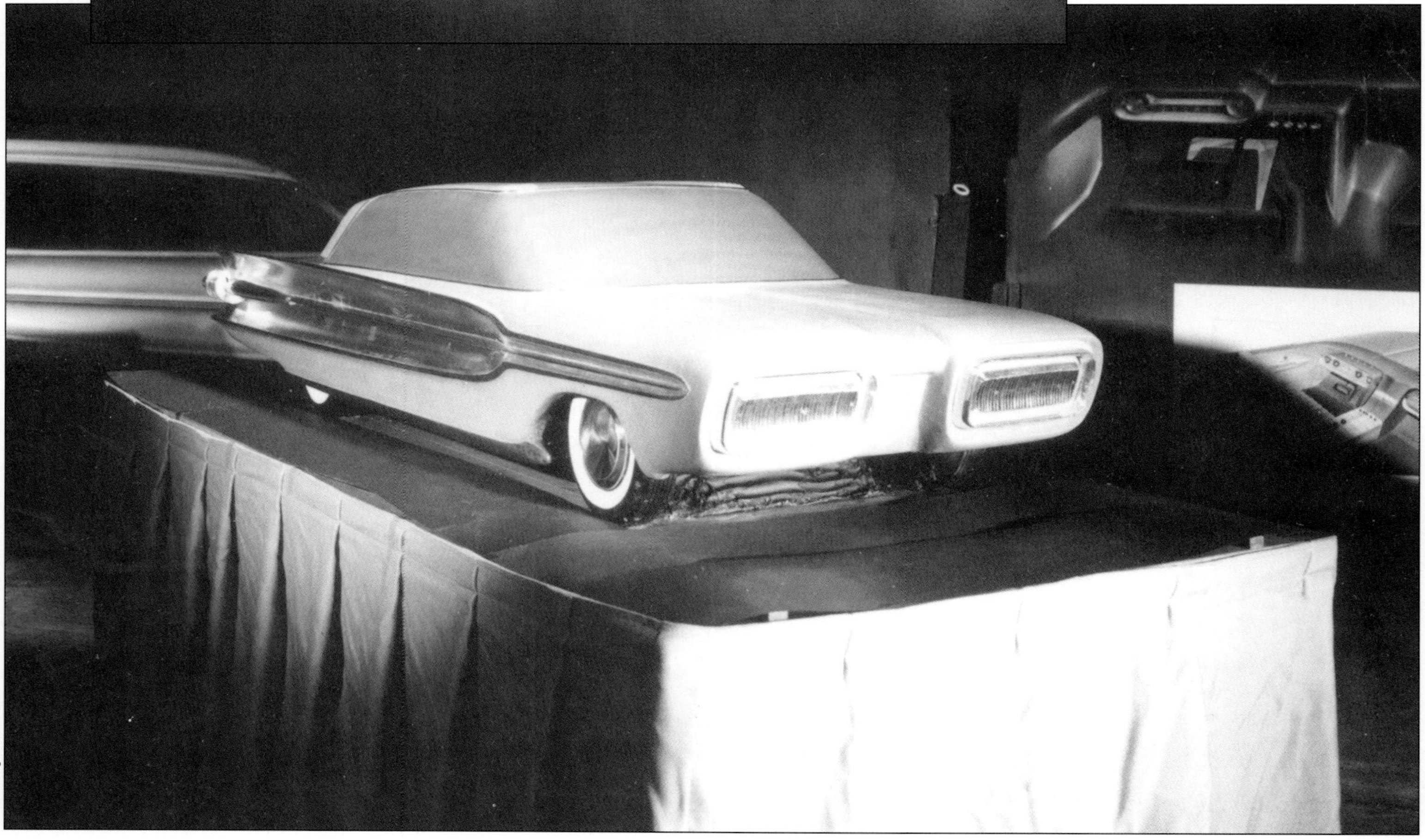

Ford Design Center

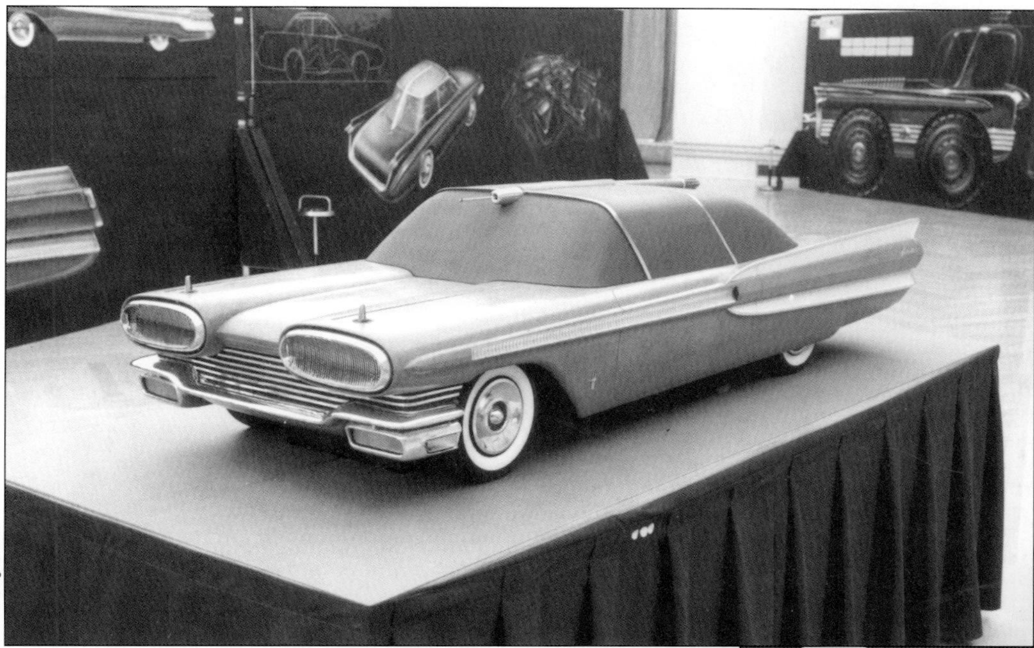

Ford Design Center

(left) The design of the Missile II was similar to that of the Missile I except for the Missile II's complicated Plexiglas-roof structure.

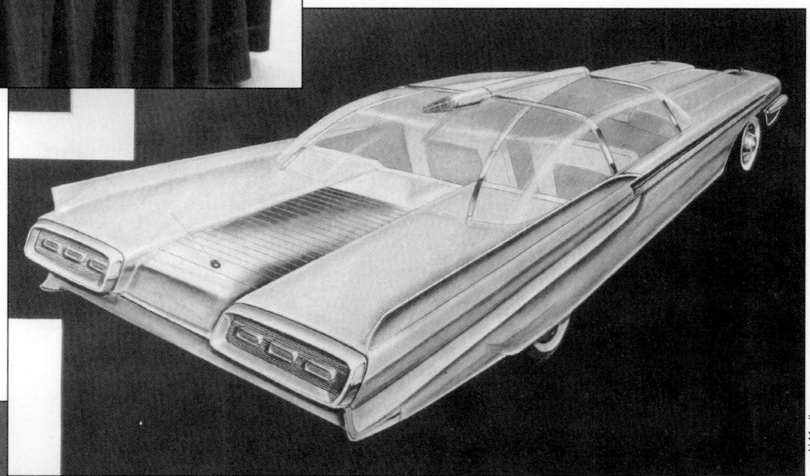

Al Mueller

(above) The Missile II's taillights were eventually used on the production 1964 Mercury.

(below) The Missile II was a combination of design proposals made by two studio designers. For ingress and egress, the front roof section retracted over the rear section.

Ford Design Center

Missile II

The Missile II was the most complicated design selected for the Stylerama program. The section of the roof over the front passengers retracted onto the rear portion of the roof, much like the X-100, except that the spear in the center of the roof acted as a guide, and the portion of the roof that retracted slid over the rear-roof section.

Mercury studio executive designer Bob Weiland believed in design continuity and was insistent on incorporating current Mercury design cues in any new designs coming from his studio. The grilles, inverted fins and taillights were all design cues Mercury studio designers were working with at the time, and it was several years before new ones could be selected. The headlights on the Missile II were designed to retract by rotating in the grille, much like what was later done on the 1967 Cougar.

Design of the Missile II was supervised by Bob Weiland, and it resulted from a combination of designs by two studio designers, Ed Westcott and Hal McCormick. The design of the body from Westcott's car was used with the retracting roof from McCormick's car. The features selected were combined on a new design proposal, which became the Missile II. ■

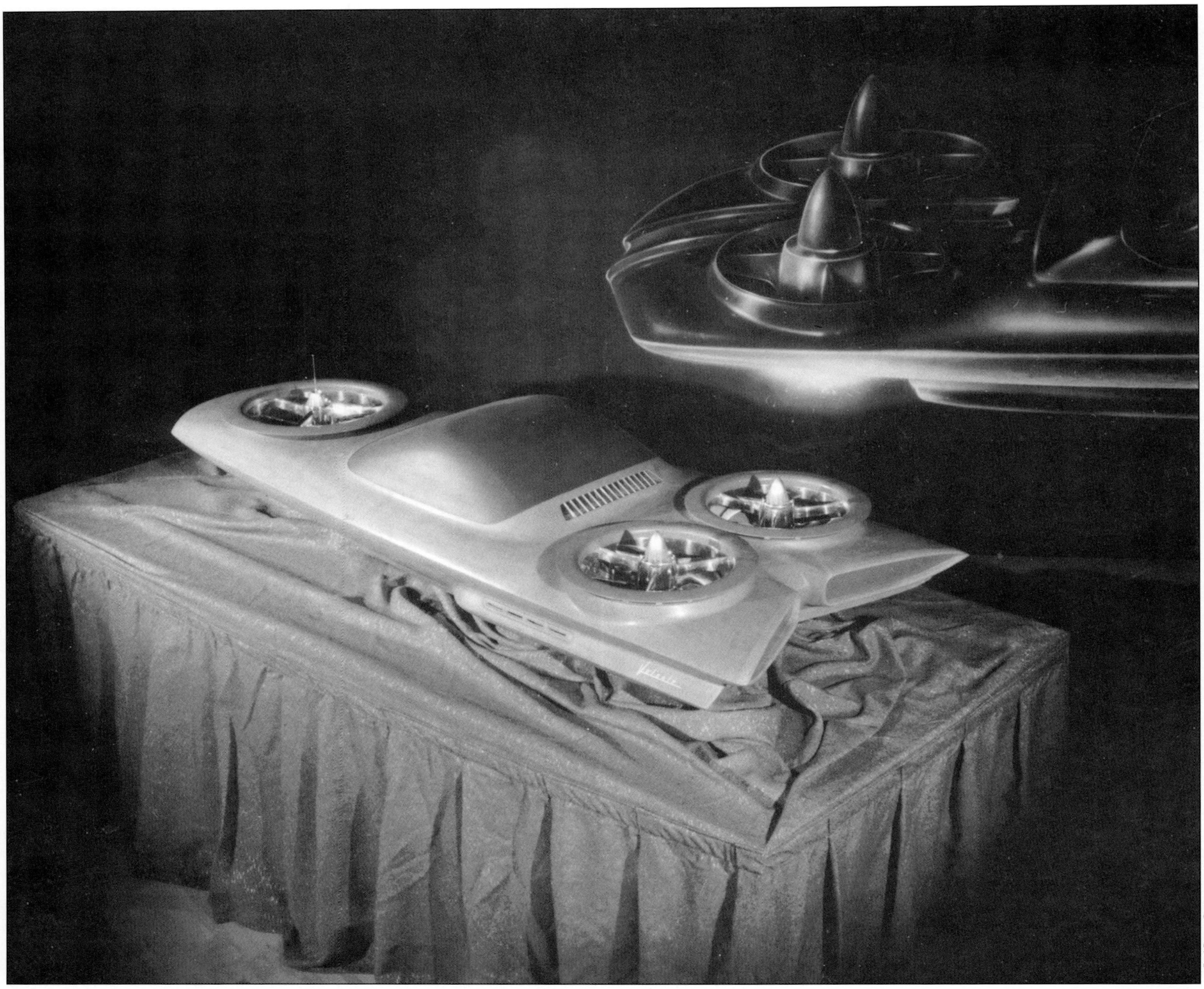

Ford Design Center

The Volante was an Advanced Studio proposal designed for the Stylerama program by Jim Powers. Although the Stylerama program was canceled, the Volante later went on a country-wide tour organized by the Smithsonian Institute.

Volante

The Volante was Jim Powers' idea of what a modern flying car would look like, if Ford had decided to build one. In the teens and 1920s, periodicals were full of articles proposing cars that could easily be converted to planes, fly to their destinations, and then be converted back to cars. Throughout the history of the automobile, there have been a number of flying cars. Some have even been produced and sold, albeit in low numbers. To date, the majority of flying cars have been propeller driven with wings and other parts that disconnected when they traveled on the ground. To save weight, the car part of most flying cars has been rather flimsy, and the airplane part looked more like it was built by Rube Goldberg. The Volante was one of the first proposed flying cars that stayed together in one piece, whether it was on the ground or in the air. Because it used fan jets built into the body, it could theoretically fly like a Harrier jet or be driven on the road. Although the Volante was included in the Stylerama program, it was built several months earlier as a standard Advanced Studio project.

The initial sketches Powers made of the Volante included retractable wings that anticipated longer and faster travel. The final design of the Volante, however, was of a flying car meant for short trips at slower speeds. When on the ground, its small wheels were meant to transport it only over short distances.

The flying platform developed by the United States Marine Corp in 1954, with its single ducted fan, gave Powers the idea for the Volante. The problem with the flying platform was its lack of control. Directional stability was supplied by the operator's body movement. Because it proved too sensitive and too unforgiving, the flying platform proved to be impractical. The Volante was powered in much the same way, but it had three ducted fans. In the front were two sets of counter-rotating blades in a single ducted fan and, in the back were two ducted fans that rotated in opposite directions to balance out the torque. Powers believes the Volante would have been very stable.

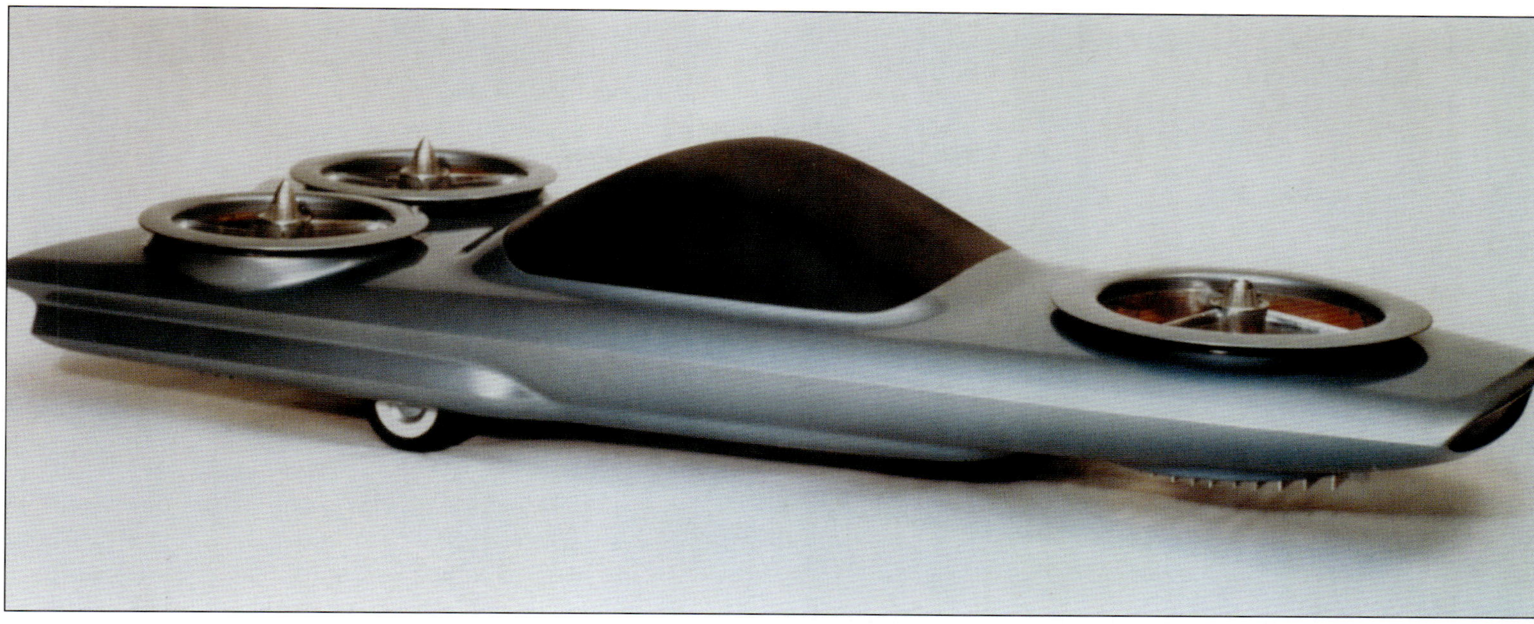

Ford Photomedia

(above) The Volante's fan motors could be powered-up by compressed air. The loud whirring noise they made sounded like the model was about to take off.

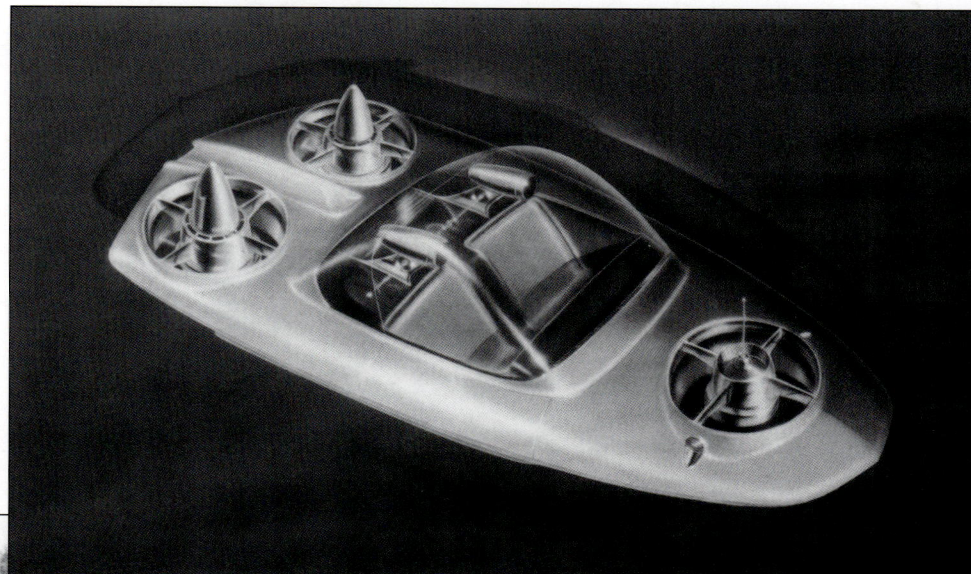

Ford Design Center

(right) According to Powers, this rendering of the Volante was prepared for the Stylerama contest but only after the 3/8- sized model was finished.

Jim Powers

James R. Powers

Jim Powers grew up in Texas. As a teenager, he won the Fisher Body Guild contest for auto design four years in a row at both the state and regional level, and he eventually won the national award, which included a university scholarship. He attended the Art Center College in Los Angeles, California, on additional scholarships from Motor Trend *magazine and Ford Motor Co. Powers completed the four-year course at the Art Center in two years and four months, and then served as an instructor during his last year there. In 1956, he was recruited by Alex Tremulis as a Ford designer.*

Powers' first assignment was in Tremulis' Advanced Studio, where he designed the Nucleon, the Volante, and the original Palomar. In 1957, Powers was assigned to the Thunderbird studio, where he was one of the designers of the 1961 Thunderbird. Powers also designed the interior of the LaGalaxie showcar.

He was later appointed head of the Lincoln interior studio, where he was in charge of the design of the Lido, Lincoln's showcar for 1963, and the interiors for the 1966-69 production Lincolns. At the time he was made head of the Lincoln interior studio, Powers was the youngest design executive at the Styling Center and the youngest designer on Ford's bonus roll.

Powers was a collector of classic cars and drove a Mark II when it was new. By the mid-1960's, he had several collectible cars including his second Mark II and a 1947 Continental Coupe, which he purchased from Paul Wagner, who had taken it in trade when he sold the Mark II Cabriolet.

Powers left the Design Center in 1965 to become a partner in the industrial design firm of Marcks, Hazelquist and Powers and, as a member of that firm, he worked on the Studebaker account. In 1968, he returned to California, where he established his own industrial design firm in Los Angeles. ■

Although Powers' original drawings of the Volante show two and three ducted-fan concepts, three fans were used on the final design, because the Volante was meant to represent a flying car used for short distances, and because it wasn't intended to fly at high speeds, so there was no reason to incorporate retractable wings needed for extra lift. Ford press releases said there was no need for a vertical stabilizer either, because the counter rotation of the ducted fans solved the torque problem. The Volante was designed to be controlled by two sets of adjustable lateral and longitudinal vanes under each ducted fan, which could also be used to maneuver the Volante in all directions. The three wheels the Volante rode on, while on the ground, were about ten inches in diameter. Powers designed the Volante with three wheels, because there was no room for a fourth wheel, given the location of the front fan. According to Powers, it is only coincidental Tremulis was also interested in three-wheeled vehicles at the time.

Henry Ford Museum and Greenfield Village

After the three-eighths sized models were selected for the Stylerama program, Bob Thomas was assigned to Maguire's Advanced Studio to prepare full-sized blackboard drawings of each car to send to the Italian coachbuilders selected to build the full-sized cars. Thomas was only able to finish the drawings of the Volante before the Stylerama program was canceled.

A full-sized Volante would have had a wheelbase of 72.1 inches, a length of 214.3 inches, a height of 43.4 inches and a width of 82.1 inches. The Volante was designed to be powered by either a conventional gas engine, or by battery-powered electric motors located in the middle of each fan. On the lower sides of the Volante there were areas with several small vents, which were meant to exhaust heat from whatever power source was used. The design process did not get far enough along to determine whether batteries were practical, or if the gas engine was to be a turbine or reciprocating. If a gas engine had been used, Powers and the other designers who worked on the Volante thought it would be necessary to use a driveshaft running to the center of the fans.

The Volante was named for a then popular song called *Volare*, which in Italian translates "to fly." The Volante was also referred to as the Volante Tri Anthadyne. The Tri Anthadyne part of the name was thought up by Alex Tremulis and has no meaning other than what Tremulis said it meant. Tremulis claimed that "Tri" referred to the three ducted fans, and "Anthadyne" referred to the turbine fans themselves.

The Volante was painted a silver-blue metallic. There is a well publicized picture of the Volante with a man in the cockpit wearing a spacesuit. The Volante is hovering high above a city with roads and traffic below. As good as it looks, that photograph was made by putting a picture of the landscape underneath the 3/8-sized Volante with a man painted on the windshield of the vehicle.

There were never plans to make an operable full-sized Volante, but the 3/8-sized model was built with fan blades that actually rotated. The shop drawings of the fans and blades were prepared by Gib Giberson, who at the time was Tremulis' assistant manager in the Advanced Studio and, for the Stylerama show, a compressed air tank was hidden under the table and hooked up to the Volante, so that, on command, the fans could be turned on. They whirred very loudly! The roof section on the Volante was designed as a clear Plexiglas canopy hinged from the back and opening as a unit, but budget constraints were such that the canopy on the model had to be fiberglass painted black.

Shortly after it was finished, Army researchers contacted Ford for information on the Volante. Powers was told later that the Army had actually developed military vehicles that flew using three and four ducted fans. Although there is now the technology necessary to build an operable civilian Volante, it is not practical and, unfortunately, never has been. There are fewer than three-quarters of a million civilian pilots in the United States. It is not a growth industry, and there probably isn't a market for a mass-produced flying car.

The Volante and the Nucleon were both loaned to the Smithsonian in 1981 and were on display across the country for several years. After that, they were given to the Henry Ford Museum, where the Volante is now in storage. Although the Volante has been repainted, it does not appear to have been changed in any other way. ■

On one of his trips to the Art Center College, in Los Angeles, California, to recruit students for the Styling Center, Alex Tremulis decided he wanted Jim Powers as a Ford designer, and that he wanted him to start as soon as possible. Since Powers had one term to complete before graduation, Tremulis talked the Art Center into giving him credit for working his last term at Ford. Powers then told Tremulis he wanted $200 more a month than the standard starting salary he was offered. Tremulis smiled, told Powers he drove a hard bargain, and said okay. So, in the fall of 1955, Jim Powers became a designer at Ford Motor Co. making $200 a month more than other beginning designers.

Nucleon

As with other entry-level designers, Powers started in the Advanced Studio. Over the next year, Tremulis and Powers developed what turned into a lifelong friendship and, during his career at the Styling Center, Powers spent many weekends at Tremulis' home in Ann Arbor, Michigan, and came to think of him as a second father.

The first project Powers did in the Advanced Studio was a series of renderings showing futuristic scenes of freeways, cities, watercraft, rockets and cars. The renderings were featured in a 1958 booklet about the Styling Center and, for years, the originals hung on the walls of the main hallway at the Styling Center. During his year-long apprenticeship in the Advanced Studio, Powers also designed three cars that were made into 3/8-sized fiberglass models. They were the Nucleon, the Volante and the Palomar. They were all built at about the same time, and even though they were all completed several months before the Stylerama program got underway, only the Nucleon and the Volante were included in it.

After the Stylerama program was canceled, the Nucleon and the Volante were displayed at the Ford Rotunda. Both cars were also included in a 1958 Design Center booklet and in press releases issued between 1958 and 1961. The Nucleon was always referred to as a "glimpse into the atomic powered future" and as designed "on the assumption that the present bulkiness and weight of nuclear reactors and attendant shielding would someday be reduced."

As hard as it is to believe today, in the early 1950s, nuclear-powered cars were viewed no differently than any other advanced propulsion system proposed for use in an automobile. Just as the dreamcars of the 1930s emphasized streamlined design, many of those in the 1950s featured new power sources as well as advanced-airplane styling, fins and all.

Henry Ford Museum and Greenfield Village

Benson Ford first proposed building a nuclear-powered car in 1951. Although not shown in this photograph, the Nucleon's designer, Jim Powers, recalls that on one of his frequent visits to the Advanced Studio, Elwood Engel commented that the rear grille of the Nucleon would work well on a production car. A similar rear grille was later incorporated on the 1961 Lincoln.

In the April 1951 issue of *Motor Trend,* there was an article called *Tomorrow's Atom Powered Car.* It predicted that advances in technology would make nuclear power safe enough and small enough to be a power source for cars of the future. In an SAE presentation in 1951, Benson Ford proposed building a nuclear-powered car, because he foresaw that the world's supply of oil would someday run out. In 1955, the world's first nuclear submarine went into service, and most people in those pre-Chernobyl days thought it was just a matter of time before small and safe nuclear power plants would someday become available as an economical way to power automobiles. Designers back then saw no reason to fear nuclear power, and believed it would eventually become an acceptable alternate fuel source.

The Nucleon started out as a series of sketches. The designs for the Nucleon, Volante and Palomar all came from projects Powers first developed at the Art Center. When Tremulis saw them, he encouraged Powers to further develop the theme, and later

Ford Design Center

Al Mueller

(top) When the Nucleon was shown in the Styling Rotunda, its nuclear fuel cell was in the process of being changed. The fuel cell was really a small light bulb on a crane made from a modified Dazor lamp.

(bottom) When the clay model of the Nucleon was made, the Plexiglas top was replaced by a cantilevered roof without A-pillars. As built, the Nucleon had no interior. Although this rendering was made after the clay model was completed, the rendering left off the fins.

approved making a 3/8-sized clay model. The Nucleon eventually was built as a 3/8-sized fiberglass model. According to Powers, no changes were made to the Nucleon or the Volante after the clays were started. There was no engineering work done for the Nucleon, and the model was not tested in a windtunnel. The Nucleon was painted a bright candy-apple red and had a silver roof. A portion of the lower body was also painted silver. Because there was not enough money available in the budget to put Plexiglas windows in the Nucleon, it was made of solid fiberglass with the windows painted black. No interior or interior buck was built for the Nucleon.

The twin booms and the cantilevered front end were similar to many of the previous sketches Powers had done at the Art Center College and anticipated the extra weight necessary because of the lead used to shield the power nacelle.

A full-sized Nucleon would have had a 69.4-inch wheelbase, an overall length of 200.3 inches, an extremely low height of 41.4 inches and a width of 77.4 inches. The nuclear reactor or nacelle measured 26 inches across and had a circular opening on top for changing the fuel cell. The opening was surrounded by cooling fins for the reactor. The passenger compartment was cantilevered over the front wheels and included a one-piece windshield without A-pillars and a compound-curved backlight. Ventilation was provided by air intakes at the top of the windshield and at the base of the roof supports. The power unit and drivetrain were designed to separate. It was anticipated that, someday, when nuclear-powered cars were commonplace, power units could be exchanged at charging stations, where drivers would have their choice of power units, depending on how far they wanted to travel and how much power was needed. It was also assumed that electronic torque convertors would take the place of a conventional drivetrain.

For the Stylerama display, the Nucleon was shown with a Dazor articulated lamp that was modified to look like an articulated arm changing the nuclear fuel cell, which was a clear sphere with a pulsating red wheat-grain bulb radiating from the center.

After the first press release, the Argon Nuclear Laboratory in Chicago called and asked for the Nucleon's specifications, but they never said what they wanted them for.

Powers believes that the rear end of the Nucleon may have been inspiration for the rear grille on the 1961 Lincoln. The roof and the cantilevered front end of the Nucleon are also similar to the front-end design of the 1962 Econoline van.

In 1981, historians from the Smithsonian Institute rediscovered the Nucleon, and it was loaned to them for display around the country for about two years. Before the Nucleon was loaned out, it was cleaned up, repainted, and the fins were removed. The white trim on the bottom of the body sides was repainted the same color red as the rest of the body, wide whitewalls were painted on the tires, and a brightmetal-trim piece was added on the body sides at the front of each front fender. After the Nucleon was returned to Ford Motor Co., it was given to the Henry Ford Museum, where it is now on display.

When Powers designed the Nucleon, it was nothing more than blue sky. He personally did not think nuclear-powered cars represented the future. He saw the concept of a nuclear-powered car as strictly fanciful, which it probably still is.

Paint shop employee Ernest Kucksie detailing the Nucleon.

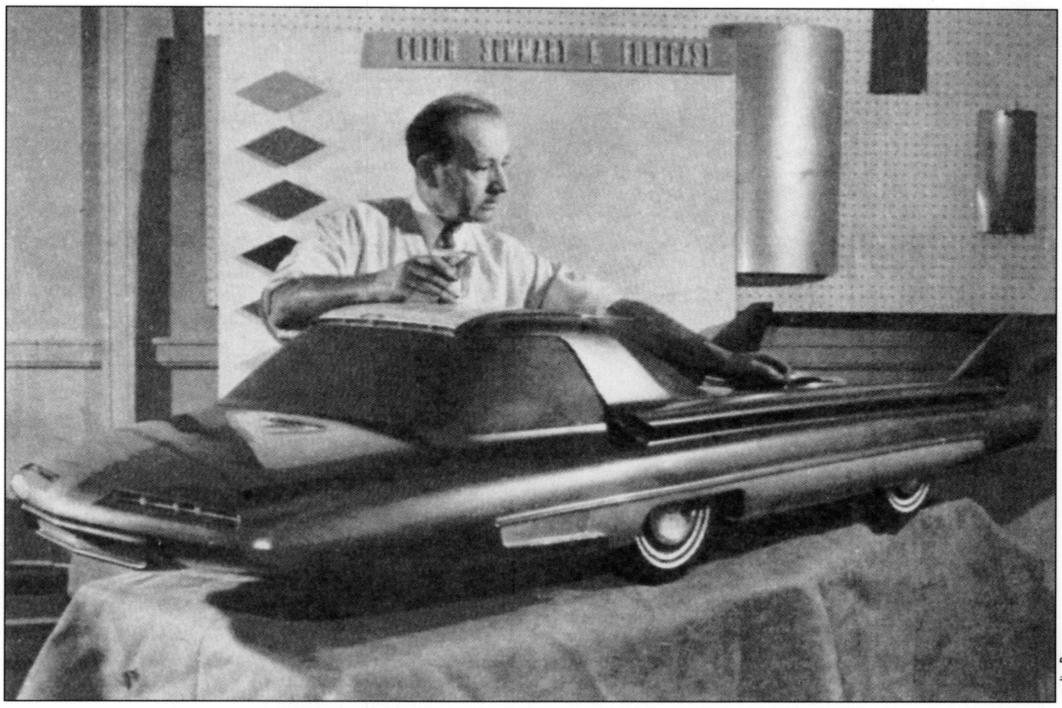

Jim Powers

Currently, it's almost impossible to get accurate information on the present state of nuclear science. Many retired designers believe we may already have the technology necessary to produce a small enough nacelle for use in an automobile, and that it might even be lighter than Tremulis and Powers thought it would be. The biggest obstacle is still the shielding and the ability to safely work on the nacelle. Experts also claim that problem has been solved, but public opinion no longer makes nuclear cars an option. ■

(below) Before it was loaned to the Smithsonian, the Nucleon was repainted red and the fins were removed. It was later returned to the Design Center and eventually donated to the Henry Ford Museum, where it is currently on display.

Henry Ford Museum and Greenfield Village

Buzz Grisinger started at the Ford Styling Center in 1955, after he was recruited by Frank Hershey, but Hershey left just as Grisinger started. After working on the design of the 1956 Thunderbird continental kit for a few weeks, Grisinger was given a small area upstairs, and told he could design anything he wanted, while George Walker, the newly appointed head of styling, decided what studio he would be assigned to. When Walker finished making assignments, Grisinger was named head of the Mercury pre-production studio. About a year later, Grisinger was assigned to work with Bob Maguire, when Maguire was named head of the Advanced Studio. Shortly before the Advanced Studio was moved temporarily to the former Continental Division design studio, the DePaolo was designed in a small upstairs room at the Styling Center.

DePaolo

The DePaolo was a 3/8-sized model designed by Grisinger for the Stylerama program. Part of Grisinger's assignment in the Advanced Studio was to design a small car proposal that became the L'Avion, but he soon became intrigued by a 3/8-sized DePalma Stylerama car Alex Tremulis was designing in that studio. As a result, Grisinger volunteered to design a sports car proposal for the Stylerama program, which he called the DePaolo. The unique head rests on the DePaolo were borrowed from Tremulis' first DePalma design.

Alex Tremulis was assigned to Maguire's Advanced Studio in March 1956 to work on the Stylerama cars. Grisinger and Tremulis had worked together at Kaiser-Fraser, and Tremulis admired Grisinger. They got along well and, even after they retired, Grisinger and Tremulis remained in close contact.

After it was built as a 3/8-sized clay model, the DePaolo was chosen as one of the cars for the Stylerama program. It was then built in fiberglass and painted gray with black body stripes. After the Stylerama program was canceled, the DePaolo was publicized in a Styling Center news release and shown at Styling Center events for several years. It then sat in one of the studios until the mid-1960s, when it was destroyed.

The "high action" styling of the DePaolo was said to represent an "inversion of the form most typical of American automotive styling," and press releases said that the side molding "transmitted the feeling of motion conveyed by the wheels, while the fairings around the wheel housings represent a unique use of sculptured metal around the lower body." Actually, Grisinger got inspiration for the DePaolo from an abstract design he first created in the 1930s while a designer at Chrysler.

The DePaolo was also one of the few Ford concept cars named as a tribute to a real person, Grisinger's friend, racecar driver Peter DePaolo. DePaolo's Ford connection started with the 1934 Elgin Road Race. In 1935, he was team captain of the drivers selected to drive ten Ford V-8 racecars in that year's Indianapolis 500 and, in 1955, he was hired to head Ford's stock car racing efforts.

The DePaolo was designed for the Stylerama program by Buzz Grisinger. It was named in honor of Peter DePaolo, the racecar driver.

Buzz Grisinger

Ford Design Center

(left) The DePaolo at the Stylerama display.

Ford Design Center

After both the L'Avion and the DePaolo were finished, Grisinger began preparing blackboard variations of both, as was standard practice. One of the variations included a parallel-action door system developed by Ray Smith. Grisinger and Smith felt the parallel-action door could be produced at a reasonable cost, and that it was more versatile than the hinged door. Grisinger asked Robert McNamara, who had recently assumed Crusoe's responsibilities, to take a look at their door system. McNamara agreed but, when he came to the studio, he only glanced at the parallel-action study and immediately began reviewing the variations of the L'Avion and the DePaolo on the blackboards around the studio. McNamara told Grisinger that a lot of money could be saved if common body panels for several cars were designed, so they could be produced from the same dies and assembled using separate trim to differentiate them. At almost the same time the Stylerama program was cancelled, McNamara surprised everyone in the studio by proposing that they design small and large Fords, Mercurys and Edsels with interchangeable body panels.

(above) One of the alternate DePaolo proposals had a working parallel-action door.

Ford Design Center

(left) After the design of the DePaolo was completed, Bob Marcks prepared renderings of alternate proposals based on both the L'Avion and the DePaolo.

Grisinger, Advanced Studio manager Al Mueller and designers Bob Marcks and Keith Teter were put to work on a crash program to design and clay model interchangeability cars. After several months of study and lots of charts, about a dozen full-sized clay models were designed and built to demonstrate how Ford products could be made to look different, even though their body panels were interchangeable.

In late August 1957, a presentation was made to McNamara outlining program timing, package dimensions and interchangeability plans for the 1961 Ford, Mercury and Edsel programs. When McNamara showed up to look at what the designers had spent a lot of overtime preparing, he walked around the studio and stopped at a chart showing the number of GM parts that were interchangeable, said that it was just what he thought it was, and left. A memo from McNamara later recommended adoption of the interchangeability program.

When Robert McNamara saw the alternate designs of the DePaolo, he asked for a study on the interchangeability of body panels. That led to use of the same body panels on Fords and Mercurys.

The interchangeability study was then sent to the Ford, Mercury and Edsel studios, and the models were reproduced in plaster for record purposes. According to Grisinger, the end result of McNamara's interchangeability study was to short change Mercury, which was forced to share body panels, including doors, from the full-sized Ford. Grisinger also sees a great deal of irony in the results of the interchangeability studies, because it was his DePaolo that started it and, when he became head of the Lincoln-Mercury studio in 1961, he was the one who had to live with the results every time a new Mercury was designed. ■

Ford Design Center

Bill Ford was a guest at Oklahoma's 50th anniversary celebration, and he brought the three-eighths sized Oklahoma with him.

Oklahoma (DePalma)

(above right) As this rendering shows, the DePalma was not built as originally designed.

(below) The DePalma, before its name was changed to the Oklahoma, is in the Styling Rotunda as a part of the Stylerama display.

The DePalma was a 3/8-sized concept car designed by Alex Tremulis for the Stylerama program. The car was named after Ralph DePalma, the famous racecar driver. It was originally designed as an open racecar.

The headrests on Tremulis' personal Thunderbird were made from 1939 LaSalle headlight buckets he reworked with a Y-shaped base built to support them. He incorporated that design on the DePalma. Although the DePalma was designed as an open car, when the 3/8-sized clay model was built, Tremulis decided to change it to a closed car. The headrests were then used on Buzz Grisinger's DePaolo.

The design of the DePalma was similar to other designs Tremulis developed during this time period. The hooded headlights, the exposed exhaust ports, the large fins, and the squared-off exhaust pods at the back of the car can also be found on the LaTosca, the X-1000, the Wind Brake Car and several others, but do not represent design features that made it into production on any Ford Motor Co. product.

In 1957, Oklahoma celebrated its 50th anniversary of statehood. George Walker's mother was part Cherokee Indian from Oklahoma. Walker was very proud of his Indian heritage and eagerly accepted an invitation from his mother's native state to participate in the celebration, which was held in Oklahoma City during the summer of 1957. The DePalma was dusted off, the name was changed to the Oklahoma, and it was displayed at the anniversary celebration. ■

Attache

Ford Design Center

Al Mueller

(above) The Attache, shown here, the Diplomat and the Envoy were all variations of a common theme.

(right) The Attache was designed by Elwood Engel for the Stylerama program. This rendering was prepared by John Orfe.

The Attache was designed by Elwood Engel with assistance from Lincoln studio designer John Orfe. At the time of the Stylerama contest, Engel was assistant manager of the Styling Center and a consultant in the Lincoln and Mercury studios. He requisitioned Orfe, who prepared the renderings of the Attache according to Engel's directions, although Orfe acknowledges that some of his own ideas went into the design.

The Attache was similar to the Diplomat, although it was a much cleaner design especially in the side and C-pillar areas. Both cars were Lincoln studio entries in the Stylerama contest and, until it was canceled, both cars were on the way to becoming full-sized showcars. The fact that there were several different variations of the 3/8-sized Attache probably indicates there were concerns that it looked too similar to the Diplomat.

Although the Attache and the 1961 Lincoln have little in common, the Attache was Engel's first attempt to produce a clean design considered for the next-generation Lincoln. It's difficult to reconcile the Attache and the '61 Lincoln, until you realize that Engel was looking for a commercially successful design, and it mattered less to him if it was clean or complicated -- as long as it was a commercial success. ■

Ford Design Center

The three-eighths sized Attache before it was displayed at the Stylerama program.

The Diplomat was originally developed in the Lincoln pre-production studio as a Stylerama proposal. After the Stylerama program was canceled, the Diplomat was further developed as a future Lincoln concept car. In final form, the Diplomat was a full-sized fiberglass workout, or pushmobile, without a finished interior. The Diplomat also had different design features on each side. It was never shown publicly, probably because it had too many features that were being considered for the next-generation Lincoln.

The Diplomat was not a basis for the design of the 1958 Lincoln. By the time the Diplomat was started, the design of the '58 Lincoln had already been determined. The LaTosca, with its "X" over "O" design, was the primary design influence for the '58 Lincoln.

Ford Design Center

George Walker with the Diplomat in the Styling Center courtyard.

Diplomat

The design of the 1958 Lincoln and the Diplomat were intended by Lincoln Division to out Cadillac Cadillac. Both cars were developed by John Najjar, who had been appointed head of the Lincoln studio in 1955, after Bill Schmidt left to become head of Studebaker-Packard's design team. When Schmidt left, he recruited several designers from his studio to accompany him to Studebaker-Packard. Although Najjar was asked to accompany Schmidt, he felt there were still opportunities for him at Ford. As a result of Walker's appointment as general manager of the Styling Center, there was a general reorganization and, as a part of that reorganization, Engel recommended Najjar to head the Lincoln studio, and Walker agreed.

When Najjar became head of the Lincoln studio, his first job was to try to tone down the proposed 1957 Lincoln, as it had been designed by Schmidt, before he left. Schmidt's proposed '57 Lincoln took many of its design cues from the Futura, including high fins. Najjar took over two inches in height off the fins and made several changes to the front and sides of the car. When the design of the '57 Lincoln was completed, Najjar and Engel then started on the 1958 Lincoln.

When the 1958 Lincoln was being designed, Najjar and Engel worked as a team, as they had on several past concept cars. Engel was appointed by Walker as an assistant manager of the Styling Center, a position he held in addition to being a Lincoln and Mercury design consultant. Najjar and Engel were given directions by Lincoln Division general manager Ben Mills and other division personnel to come up with a car that fit a specific format; the corners of the car had to be visible from the driver's seat, the interior had to be bigger in all dimensions than the Cadillac, it had to have the quad headlights, etc. To make a long story short, Mills' past experience was mostly in finance, the designers got a mixed message, and time was short. Nobody, including Mills and the car's designers, were completely pleased with the design — even before it went on sale. The end result was that the 1958 Lincoln failed in the marketplace, Najjar was replaced as head of the Lincoln studio by Don DeLaRossa, just as the '58 Lincoln was introduced and, for 1961, the design of future Lincolns went in a different direction. Ironically, the '61 Lincoln was also developed by Najjar and Engel, who, that time, got to design exactly what they wanted.

Although Najjar, as head of the Lincoln studio when the Diplomat was started, supervised its design, the actual design work was done by Bud Kaufman, head of the Lincoln interior studio, with help from Dave Ash, who headed the Lincoln exterior studio. Rulo Conrad also worked on it. Kaufman probably named the Diplomat, but the exact reason the name Diplomat was selected is unknown. According to Najjar, the names Diplomat, Attache and Envoy were used in series because of their diplomatic connotations. Because the Diplomat had no interior, all that could be seen through the windshield, backlight and door windows were the tops of the seats, the top of the steering wheel, and the top of the instrument panel, which sat on a plywood platform painted black. It was located level with the bottom of the door windows.

A finished rendering of the Diplomat prepared for the Stylerama display.

Al Mueller

The Diplomat was painted a metallic powder blue, with a white top. The design of the Diplomat was unconventional and probably came from the same school that produced Dave Ash's LaGalaxie. It was a four-door car without exterior door handles on doors that did not open. The backlight had a reverse slant to it, but it didn't have the LaGalaxie's hooded effect. On both sides, the reverse backlight was also undercut, so it looked separate from the body. The fenders were squared-off and similar to those on the 1958 Lincoln. The sculpturing on the driver's side of the front fender carried all the way back to the middle of the back door. The sculpturing above the passenger-side front fender was less pronounced and in the form of a spear starting above the fender and extending into the middle of the front door. The back end of the Diplomat had its own grille with an integrated rear bumper as its base. The back bumper guards were projectile-style protrusions that started in the middle of the rear grille and had blades extending to the bot-

The full-sized clay model of the Diplomat as shown on the Styling Rotunda turntable with Lincoln studio administrator Bob Pryor.

Ford Design Center

Ford Design Center

Ford Design Center

(above) The Diplomat had different design variations on each side.

Ford Design Center

(above) The Diplomat was clay modeled at the Lincoln interior studio, which was then located across the hall from the Lincoln exterior studio.

(Below) The air-conditioning intakes were designed into the body of the Diplomat and located just behind the reverse backlight.

Ford Design Center

Irving "Bud" Kaufman

Bud Kaufman grew up in Detroit. By age 12, he knew he wanted to be a car designer. At age 15, he enrolled at the Henry Ford Trade School. Because the Trade School did not offer a high school diploma, Kaufman attended night school. He also studied art at Meinzinger's Art Academy on weekends. After graduating from the Henry Ford Trade School in 1938, Kaufman worked in their drafting department but was soon assigned to the Ford tire plant, where he designed the tread pattern used on Ford script tires. He then became an instructor at the Henry Ford Apprentice School, where he established their design program.

During World War II, Kaufman flew B-26s and C-47s. After the war, he returned to the Apprentice School as an instructor. In 1946, classes at the Trade Schools were suspended, so he was loaned to the Design Department, where he was assigned to work with illustrator/designer Ross Cousins. In summer 1946, Cousins left Ford, and Kaufman was asked to finish up his projects. Tom Hibbard later asked Kaufman to stay on rather than return to the Apprentice School.

In August 1947, Kaufman took a higher-paying job working for Virgil Exner designing Studebakers. When Studebaker let Exner go after a dispute with Raymond Loewy, Kaufman was asked to work for Loewy. Kaufman soon became convinced that Studebaker's days were numbered and left in August 1950.

After operating his own industrial design firm for a short time, Kaufman accepted an offer from Gene Bordinat to return to Ford. In February 1955, he became a designer in the Advanced Studio. He then was assigned to the Ford studio, and was later appointed head of Lincoln interior and then the Lincoln pre-production studio. In 1957, Kaufman was placed in charge of the Edsel studio, replacing Roy Brown who was reassigned to Europe. Because the Edsel Division was combined with Mercury and Lincoln to become MEL Division, Kaufman became executive stylist under MEL studio chief designer Gene Bordinat. When the Edsel studio was abolished, he was appointed executive stylist in the Lincoln-Mercury studio. Kaufman later worked in the Thunderbird and Ford studios, and in the truck and tractor studios.

While at Ford, Kaufman helped design several concept cars, including the Lincoln Diplomat, the LaGalaxie and the Edsel 196X.

He retired in 1975. ■

tom of the bumper. The back-bumper guards each had a slightly different shape. The front grille and bumper designs were similar to the 1958 Lincoln and, at the ends, incorporated slanted headlights that were framed by the body sheetmetal. Above the grille was an air intake that extended all the way across the front.

The greatest design differences on the Diplomat were on the rear-quarter panels and door areas. On the driver's side was a small dorsal fin adjacent to the C-panel. The fin was a part of the C-pillar. On the passenger side, there was no fin, but the rear door included a design similar to the looping back-quarter window Ash designed on the LaGalaxie.

(below) The finished plaster and fiberglass Diplomat is in the Styling Center courtyard. It was almost identical to the full-sized clay.

Ford Design Center

During the time the Diplomat was designed, all designers at Ford were encouraged to come up with bigger, ever more distinctive-looking cars with fins, protrusions and lots of chrome. At Ford, GM and Chrysler, unusual shapes and massive cars were the order of the day, because that's what the space-age-inspired public was asking for -- or so manufacturers assumed. Today, most designers and clay modelers who worked on the Diplomat are slightly embarrassed when asked about its design. Some disclaim any recollection or participation in its design, and almost all readily agree it was too big, too awkward, and too excessive.

Ford Design Center

(left) The Diplomat in the Styling Rotunda. The Mark III Berline is behind the Diplomat, and the Lincoln Division proposal for the 1958 Lincoln is in the background at the right .

(below) The driver's side of the completed Diplomat in the Styling Rotunda.

Ford Design Center

The design of the Diplomat and similar cars sparked a fundamental split between designers, who thought it forecast the future, those who believed Lincoln needed the continuity the Diplomat represented and a third group, made up of mostly young designers who desperately wanted a new design direction for Lincoln.

The Diplomat also became a benchmark of sorts. For approximately the five-year period it remained in the Lincoln studio, it was often used for comparison with newer proposals. For awhile, it also influenced designs in other studios, but that began to change about the time the Lincoln Marlin and the Quicksilver were created. The Diplomat probably had a greater influence on the 1961 Lincoln than is generally realized, because designers who disliked the design direction it represented won converts each time the Diplomat was shown and with each new design that built on its features -- including that of the Envoy.

Najjar is proud of what he and his designers came up with under most difficult circumstances, but he is even prouder of the 1961 Lincoln.

What happened to the Diplomat? Najjar remembers seeing it around the Design Center for years, and then it disappeared. He assumes it was crushed. There was a scrap pile next to the Styling Center building, where parts of crushed concept cars were piled and picked up about once a month with the rest of the scrap. ■

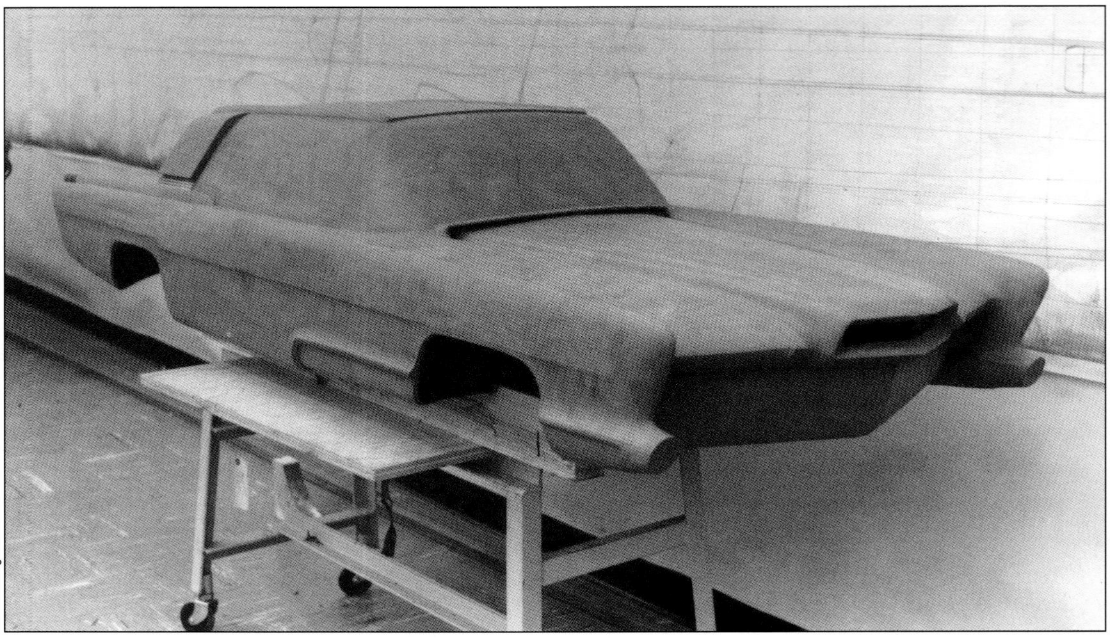

Ford Design Center

Continental Mark X

The Continental Mark X was designed by John Reinhart, former head of the Continental Division design studio. Continental Division was in the process of closing down at the time the Mark X was designed, and the Mark X may even have been designed partially at the Continental Division design studio and partially at the Styling Center during the time Continental Division's designers and clay modelers were being transferred to the Styling Center or let go.

The Mark X was not shown in the Styling Rotunda with the other 3/8-sized models, and it was not selected for the Stylerama program. Although it was built for the Stylerama contest, it is unlikely the Mark X was finished beyond the stage shown in the accompanying photographs.

The Mark X was probably the last Continental Reinhart designed while at Ford. It was a radical departure from the classic design theme of the Mark II and the Mark III Berline. At the time the Mark X was designed, the Mark II had been branded a failure, and the Continental Division Mark III Berline had been rejected. It would take several more years before the Mark II and the Mark III Berline would gain general recognition as design classics. The 1961 Lincoln took its design from the Continental Mark II and III, but at the time the Mark X was done, it is probably not surprising that it looks totally different from them.

The long snout-like grille in the rendering was toned down, when the 3/8-sized clay model of the Mark X was started as were many of the other exaggerated features shown in the renderings. Unfortunately, the roll-top roof feature was also eliminated. ■

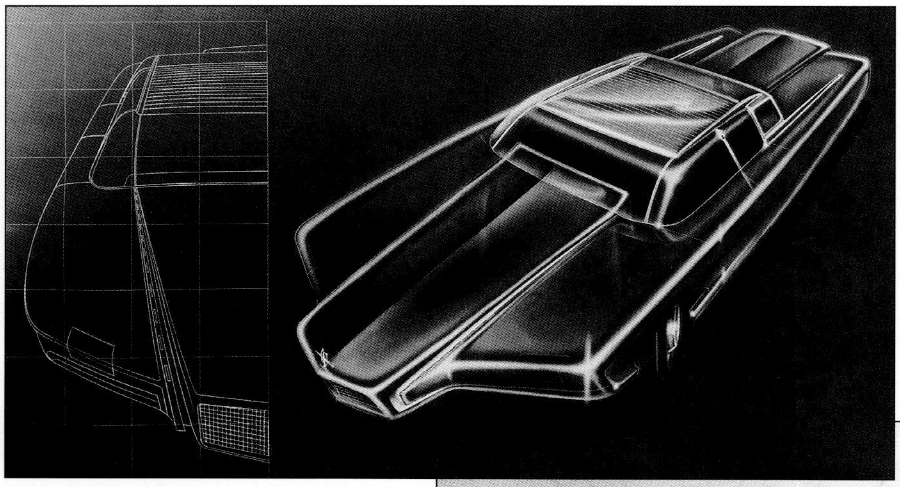

Al Mueller

(top) The Continental Mark X proposal was designed by John Reinhart for the Stylerama program, but it was never finished or shown.

(middle) The front-end design of the Mark X was changed when the three-eighths sized clay model was built as was the roof, which was originally designed to slide open and close like a roll-top desk.

(bottom) The rear end of the Continental Mark X as built was more conservative than the rendering.

Ford Design Center

Shrike

The Shrike was designed by Edsel studio designers Ken Spencer and Romeyn Hammond for the Stylerama program. At the time, Spencer was executive stylist in the Edsel exterior studio and Hammond was Spencer's design manager.

The name Shrike came from a little known and relatively unsuccessful 1930s Curtis-Wright pursuit plane of the same name. The name derives from a vicious bird, but that was not the connotation meant, when Spencer, an avid private pilot, named the car.

Before they started the 3/8-sized clay model, Spencer and Hammond prepared a full set of renderings, which included an interior that featured a rear-seat cowl and windscreen. The primary design feature of the Shrike was the triangle over rectangle shape that started at the grille, proceeded through the passenger compartment and continued onto the deck lid, where it terminated in horizontal fins that were as wide as the body. Although the rear end of the clay model was never finished, the design of the Shrike influenced Bud Kaufman's design of the Edsel 196X, a 3/8-sized model that was publicized in June 1958, as a proposed next-generation Edsel. ■

Ford Design Center

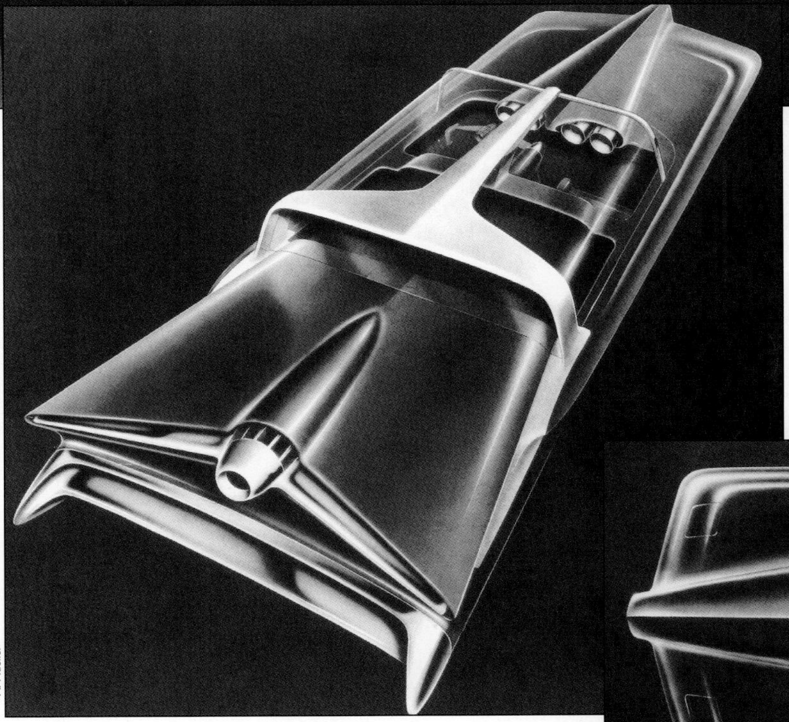

Al Mueller

(above) The rear end of the clay model of the Shrike was never completed because its designers ran out of time.

(left) If the clay model had been completed, the rear end of the Shrike would most likely have been changed from this rendering.

(below) The Shrike was designed as a triangle placed over a rectangle. The design theme continued through the interior of the car.

Al Mueller

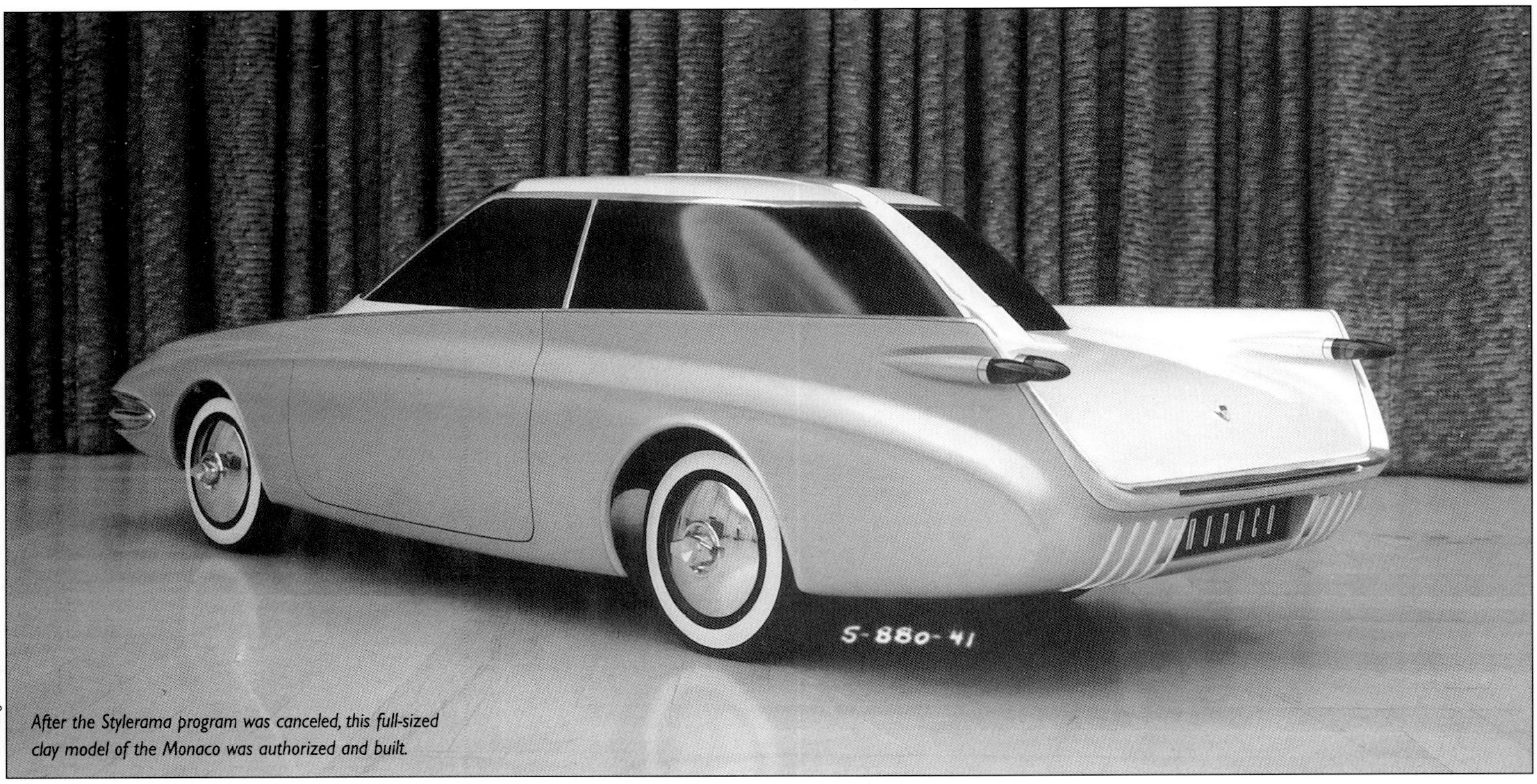

Ford Design Center

After the Stylerama program was canceled, this full-sized clay model of the Monaco was authorized and built.

Monaco

The Monaco was designed by Wes Dahlberg for the Stylerama program. It was the entry from the international and Canadian studio. It started out as a 3/8-sized model called the Freya but, after the Stylerama program was canceled, it was renamed the Monaco and made into a full-sized clay model. According to Fred Hoadley, head clay modeler in the international studio at the time, production of the Monaco was briefly considered by one of Ford's international car companies (He doesn't recall which one), but it was not developed beyond the full-sized clay model. No one recalls whose idea it was to change the name to Monaco. Freya in Norse mythology is the goddess of love and beauty. Given Dahlberg's Scandinavian background, he likely chose that name. The Monaco was probably named after the Principality of Monaco, which was in the news in 1956, because of the wedding of movie star Grace Kelly to Prince Rainier of Monaco. Whatever the reason, according to Dahlberg, the name Monaco came from the list of names Gil Spear began in the late 1940s for proposed new models and concept cars.

Dahlberg started the design of the Monaco by making multiple renderings of the car's exterior. He does not believe interior renderings were made, nor was an interior buck built. Dahlberg also recalls that at the time the Monaco was built as a full-sized clay model, consideration was given to making it into a full-sized operable showcar. When he left for Germany in 1958, those plans went by the wayside.

Ford Design Center

The Monaco had hidden headlights and featured a unique two-toned paint scheme.

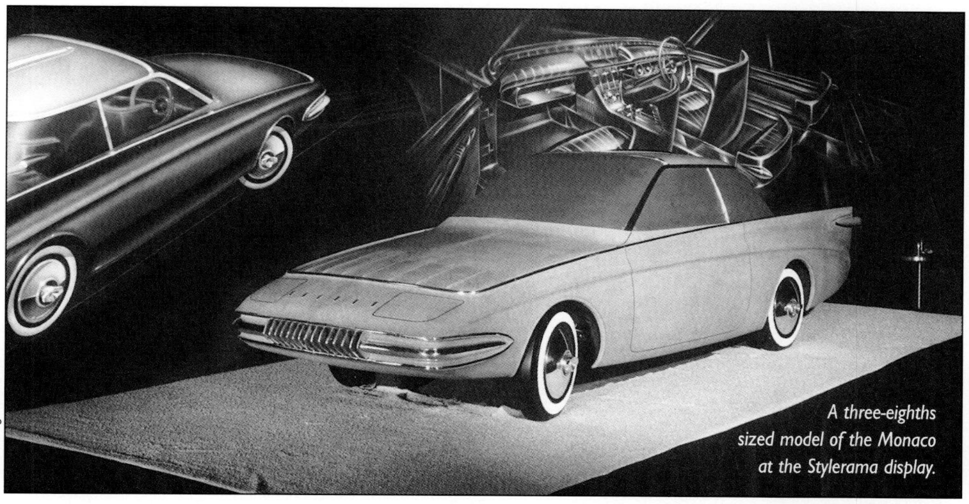

A three-eighths sized model of the Monaco at the Stylerama display.

Ford Design Center

Because Dahlberg wasn't fond of the architectural and finned excesses of car designs of the late 1950s, the fins molded into the body of the Monaco were as modest a concession to convention as he felt he could get by with. The pillar in the middle of a backlight was an idea Dahlberg tried to incorporate into several designs, but it was never used on a production Ford product. Dahlberg incorporated what he calls a flying buttress in the middle of the backlight to support the cantilevered roof of the car in a way that emphasized the split backlight. The mild fins, which Dahlberg incorporated into the character lines on the sides of the Monaco, were terminated in a way that, from the rear and with the split-window, the Monaco resembles the Bertone-built, Franco Scaglione designed, Alfa Romeo Bat 5 automobile.

The subtle elliptical form used on the sides of the Monaco was copied on the 1961 German Taunus, which was the first car Dahlberg designed after he became chief designer at Ford of Germany in 1958. Dahlberg tried to use the sloping front end from the Monaco on the '61 Taunus, but cost constraints required that he design a squared-off front end without hidden headlights.

The general design of the Monaco is consistent with design trends used on production cars of 30 years later. Today, Dahlberg believes that because of the higher cost of gasoline for cars Ford designed for the European market, there was emphasis on aerodynamics and economy of operation much earlier than on cars designed for the U. S. market. Dahlberg believes he was one of the only designers at Ford to pay serious attention to European design. Although it was not popular then, by the mid-1980s, all American car makers were heavily influenced by the Euro-design look.

Many of the designs Dahlberg later did in Germany were sent back to the States, some of them as full-sized clay models. Dahlberg believes features from several of his cars were incorporated on cars produced for the American market. ■

Gil Treweek

Wesley P. Dahlberg

Wes Dahlberg grew up in Jamestown, New York. His father was a woodcarver, and his mother was an artist. Dahlberg graduated from the Carnegie Institute of Technology with a Bachelors Degree in fine arts. In 1940, his first job was as a teacher at the St. Louis School of Fine Arts. At the start of World War II, he was drafted, allowed to finish out the school year and then entered the U.S. Army Air Corps, where he served as an operations and intelligence officer.

After the war, Dahlberg took jobs as an industrial designer in Minneapolis and then Seattle. When the job in Seattle ended, his college roommate, who was employed at the Chrysler design department, suggested he apply for a job as a designer in the more stable automobile industry. Dahlberg was hired at Ford in April 1947, and started in the plaster model shop. His first job was working on the plaster models of the 1949 Ford. As promised, he soon became a designer in Bill Wagner's truck and international studio, where he worked on redesigns of the Anglia, Taunus and Vedette. He later designed the Monaco concept car and Debonaire (Falcon) proposal.

In the mid-1950s, Dahlberg was one of a small group of designers at Ford who openly opposed design fads, including bombs and fins.

In 1958, Dahlberg was appointed chief of design at Ford of Germany. He was in Germany until 1967, at which time he returned to Dearborn, where he was assigned to Don DeLaRossa's corporate projects studio. Dahlberg worked on the Pinto and was also responsible for the exterior design of the Lincoln Continental Mark IV. He was then reassigned to the Lincoln-Mercury interior studio and was responsible for the design of the interior of the Mark IV. He was later transferred to the truck and international studio. Dahlberg took early retirement in 1973 to pursue a successful second career as an artist. ■

Ford Design Center

Ford Design Center

(above) The roof support in the middle of the backlight made the Monaco into a splitwindow.

(left) Designer Wes Dahlberg incorporated the mandatory fins into the design of the Monaco in a most unique way.

The Goliath, originally designed for the Stylerama program, was resurrected in 1960, and its name was changed to the Typhoon II for an agricultural convention.

John Najjar

Ford Design Center

The Goliath at the time of the Stylerama program.

Goliath (Typhoon II)

The Goliath was one of several tractor studio entries in the Stylerama contest. It was designed by Wes Dahlberg and built as a 3/8-sized clay model. Bill Wheeler assisted Dahlberg on the design of the Goliath. He recalls designing the exterior trim. The Goliath was designed before the tractor studio was separated from the truck and international studios. Except for a few designers who really liked designing tractors, that studio was not considered a plum assignment. Designers were generally rotated into whatever studio was in charge of tractors -- and usually got out of there as fast as they could arrange a transfer. Once the studios were separated, Dahlberg remained as a designer in the international studio and, in 1958, he was transferred to Germany as head of Ford's design department there. At that time, he lost track of the Goliath. The name Goliath was taken from a list of names kept for new vehicles, because Dahlberg thought the name fit the tractor's bulldog looks.

Three years after the Stylerama program was canceled, the Styling Center received an invitation to send a designer to talk about the future of farm equipment at the American Society of Agricultural Engineers' convention, December 4-7, 1960, in Memphis, Tennessee. George Walker appointed John Najjar, who was then executive stylist in the corporate advanced studio, to make the presentation, which, of course, had to include a farm implement of the future. For the convention, the exterior of the Goliath was mildly redesigned by corporate advanced studio designer John Orfe, who also designed a full interior for it. The Goliath was also renamed the Typhoon II. Ford's Implement and Tractor Division had recently come out with a new full-sized experimental tractor called the Typhoon. A 3/8-sized see-through fiberglass model was made of the Typhoon II, and it was sent with Najjar to the agricultural engineers' convention.

Najjar told the agricultural engineers that the proposed Typhoon II tractor, represented by the 3/8-sized model he had with him, was designed to give future farmers all the conveniences of their living rooms. Passenger cars were going to be smaller in the future, but Najjar said tractors would be bigger with more features and creature comforts. The features on the Typhoon II included: (1) power steering; (2) automatic transmission; (3) an automatic guidance system; (4) a rear engine and independently-operated rear wheels; (5) power takeoffs on the front, back and both sides; (6) a complete repair and service kit in a side-storage compartment; (7) electrically-operated magnetic couplings in the front and rear; (8) front and rear-drive wheels that steered independently and simultaneously; (9) gull-wing windows; (10) hydraulic controls to increase or decrease the width of the track and change ground clearance; (11) a two-way phone; (12) heating, air conditioning and radio speakers provided in ductwork on three sides of the cab's interior; (13) heater coils built into the stainless steel floor; (14) a TV screen to watch rear-mounted implements; (15) weather-forecasting equipment; (16) foot-pedal steering; (17) a warning-light system to replace the gauges; (18) a food warmer, coffee maker and a refrigerator; and (19) a contoured swivel seat with free-flow air inserts that could be moved in any direction and could be tilted to let the driver sit upright even when the Typhoon was on a slope. Although Najjar was careful to explain that no engineering had yet been done on the Typhoon II, he told his audience that Ford could be making similar tractors within the decade, which was a little off the mark. The Typhoon II was never seriously considered for production.

No one knows what happened to the 3/8-sized model of the Typhoon II after the convention, although Najjar remembers bringing it back to the Styling Center. ■

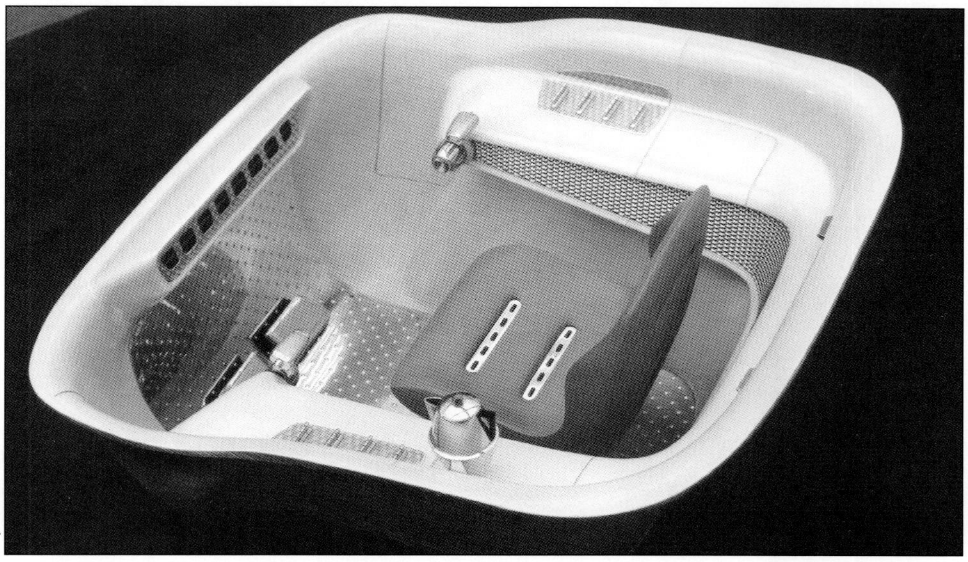

John Orfe

(above) After the Goliath was resurrected as the Typhoon II, John Orfe designed its interior, which was completed and installed when the model was shown in 1960.

Ford Design Center

The Goliath being clay modeled.

Pete Renz

Willard L. "Bill" Wheeler

Bill Wheeler grew up in New York. When World War II began, he joined the U.S. Navy and spent 42 months on active duty before being discharged in 1946.

Because he loved cars and art, he applied at Pratt Institute in Brooklyn, New York. Pratt's 1946 class was full, so Wheeler started in 1947. While he was waiting to start at Pratt, he took a job as a type-face designer for IBM. When he started at Pratt, he took an educational leave of absence from IBM, because he intended to return.

After he graduated from Pratt in June 1950, Wheeler and two classmates, O.M. Renz and John Weisbrod, were interviewed at Ford's Design Department by Gene Bordinat. All three were hired as Ford designers. Wheeler planned to work at Ford until IBM got its design department organized, but changed his mind and stayed at Ford.

Wheeler's first assignment was in the Advanced Studio. Six months later, he was transferred to the truck, tractor and international studio, working for Bill Wagner. During the time he was in that studio, he and Wes Dahlberg designed the Goliath advanced tractor.

In late 1958, Wheeler was assigned to Germany for several months to work with Dahlberg on the Taunus 17M. After he returned to Dearborn, he was assigned to the MEL studio working for Bud Kaufman. After the MEL studio was dissolved, Wheeler was assigned to the Lincoln-Mercury studio. He later returned to the truck studio.

Before retiring in 1986, Wheeler spent several years in the graphics studio. ■

9

Special Vehicles Department

D-523 (Cougar)

Gil Spear's Advanced Studio was a place where ideas for the Fords, Lincolns and Mercurys of five to ten years down the road were developed. During the almost six years he was in charge, Spear issued an open invitation to designers in the production studios to visit the Advanced Studio regularly. If they saw an idea they wanted to incorporate on a production car, it was theirs to use. The Advanced Studio was also a place where new designers could be evaluated and, more importantly, taught how to design production cars. After they learned the ropes, studio heads, who had watched and evaluated their progress during the months they were in the Advanced Studio, could then invite them to become designers in the production studios.

Ford's management, from Henry Ford II on down, also visited Spear's Advanced Studio regularly. Ernest Breech, executive vice president, Lewis Crusoe, head of Ford Division and Benson Ford, head of Lincoln-Mercury Division, regularly discussed with Spear what they wanted to see in future car designs, and Spear tried to incorporate their ideas into the cars designed in his studio.

Ford Photomedia

After several years in storage, the D-523 was dusted-off, repainted, reupholstered, had a new engine installed and was renamed the Cougar for the 1962 New York Auto Show.

Ideas from the designers themselves were also harvested for use on future concept and production cars. Spear kept a book on his desk, where designers were encouraged to write down their suggestions for new features, or improvements they thought should be incorporated on future car designs. The loose-leaf book was titled *Inventive Styling*, and if Spear thought an idea placed in the book was feasible, it would be incorporated into the design of one of the next 1/8 or 3/8-sized concept cars built in the Advanced Studio. Once successfully incorporated in the design of a concept car, designers from the production studios would hopefully incorporate the idea in one of the production cars they were designing.

As time passed, Spear began to believe that many of the best ideas his Advanced Studio designers came up with were not being utilized. He also felt there were some serious design problems that needed solving, including the blind spot created by the A-pillars and other safety features. That led him to propose creation of a new studio at the Design Department, where the backlog of new ideas could be incorporated into full-sized cars he helped design.

By September 1953, Design Department general manager Charlie Waterhouse and his assistant managers, Gene Bordinat and Frank Hershey, also concluded that the Design Department needed a new studio they referred to as a special projects studio. At the time, the Design Department was still a part of Engineering and under the direct supervision of Henry C. Grebe, Ford's chief body engineer. On September 3, 1953, Waterhouse, Bordinat and Hershey submitted a written proposal to Grebe. As the proposal explained, there was a "missing link" in the studio structure at the Design Department, and they recommended it be fixed.

Ford Photomedia

Earle S. MacPherson

In 1915, Earle MacPherson graduated from the University of Illinois with a degree in engineering. During World War I, he worked on aircraft engines for the U.S. Army. In the early part of his career, he also worked for Chalmers, Liberty, Harley-Davidson, Pierce-Arrow and Studebaker. In 1934, he was hired by GM as an engineer and, in 1935, he became chief engineer for Chevrolet Division. MacPherson was an innovator, who stressed lightness and simplicity in the cars he engineered. When GM terminated MacPherson's Cadet small Chevrolet program in September 1947, he left to accept a position at Ford offered him by Harold Youngren, a former GM engineer recently hired as Ford's chief engineer.

MacPherson was first assigned to Ford of England, where he helped develop Ford small cars, using the engineering principles he had developed during the GM small-car program including the MacPherson strut suspension. After he returned to Dearborn, MacPherson oversaw development of Ford's overhead-valve engine and new suspension and chassis designs first used on the 1952 Lincoln. When Youngren retired in 1952, MacPherson succeeded him as Ford's chief engineer.

At the time of George Walker's appointment as a Ford vice president and general manager of the Styling Center, in May 1955, MacPherson established a separate design studio (Special Vehicles Department) in the Engineering Department with Gil Spear as head designer. The D-523, D-524 and Bimini concept cars were completed in the Special Vehicles Department as engineering test beds. After the Special Vehicles Department was dissolved, several projects started in that department were transferred to the Experimental Engineering Department.

In 1957, MacPherson suffered a heart attack and retired as Ford's chief engineer. The Edsel crisis, a new set of priorities, and George Walker's appointment to Ford's finance committee assured the end of his Special Vehicles Department. ■

Their proposal started out by identifying the purpose of the existing studios. It said the two-fold purpose of the Advanced Studio was to discover future design trends and to teach new designers the "Ford way of design," and that the Ford and Lincoln-Mercury body development studios were not only liaisons with engineering and manufacturing concerning the surface layout of future production vehicles, they were also responsible for interchangeability of parts and development of the packages the production studios worked within. The report ended by pointing out that the Ford and Lincoln-Mercury production studios took what the body development studios produced and refined them into production vehicles.

The next section of the report went on to identify the proposed special projects studio as a place "where full-sized futuristic cars could be developed without interfering with the scheduled work load of the body development or production studios," and where "different approaches could be explored by following present styling trends to their logical conclusion." The proposal identified, in a way somewhat critical of the Engineering Department, why there was a problem:

J.R. Samsen

"An example of this type of exploration can be drawn from our recent past experience relative to windshields. If Engineering Staff had been able to completely evaluate full-wrap windshields several years prior to their adaption to certain 1955 body styles, considerable last minute improvisation could have been avoided and a more confident approach could have been taken based on our own experience. The windshield is, of course, only one area of the car. The future imposes even more complex problems. If the trend towards lower cars continues, the problem of entrance room must be explored. This could well change our current concepts of door opening, drip rail lines and basic body shapes. If increased visibility continues as a trend, certain styling, engineering and manufacturing problems will be created that have not yet been considered. The list is endless and one new problem breeds another."

The report next compared Ford's Design Department to its unnamed chief competitor, pointing out that the Y-job, the LaSabre and the XP-300 (all running models) had been developed in that competitor's special projects studios, which had been in existence for over 20 years. It was also pointed out that that same competitor's special projects studio was planning to turn out 14 new concept cars for the upcoming 1954 Motorama! In other words, if Ford was to stay competitive with GM, the Design Department needed a special projects studio to do so. In addition, the report claimed that new ideas required for the next generation of Fords, Lincolns and Mercurys could be better developed in separate special projects studios. The report ended by actually proposing two new studios; a Ford Special Projects Studio under the supervision of Frank Hershey and a Lincoln-Mercury Special Projects Studio under the supervision of Gene Bordinat.

J.R. Samsen

(above right) Designer Dick Samsen's original rendering of the D-523 was completed in January 1954.

(right) An alternate rendering of the D-523 prepared by designer Dick Samsen in April 1954.

The response from chief engineer Earle S. MacPherson and Grebe must have shocked Waterhouse, Bordinat and Hershey and surprised everyone else at the Design Department because, in late January 1954, Grebe announced an organizational realignment of the Design Department. Instead of just creating the new studios as requested, the Design Department was split up into four separate departments. Waterhouse was no longer general manager but, instead, was named manager of the new Special Projects and Training Department, which had responsibility for styling of vehicles for Ford of Canada and the international division, tractors and full-sized advanced vehicles. (According to Spear, during the six years he ran the Advanced Studio, he was also responsible for supervision of design done for Ford's international and Canadian cars and tractors.) Hershey was promoted from assistant manager to general manager of a separate Ford Design Studio, and Bordinat was promoted from assistant manager to general manager of a separate Lincoln-Mercury Design Studio. Also created, at the same time, were the Styling Fabrication Department with Charlie Stobar as general manager, and a Scheduling and Records Section with R. A. Ketchman as general manager.

The D-523's fuel-injection system as seen from the right side. The D-523 was Ford's first fuel-injected car.

Gil Spear

When the Special Projects Studio was formed with Waterhouse as its manager, Spear was named its chief designer. The first order of business was to staff the new studio. About ten designers, six clay modelers and two studio engineers were selected from the other studios and assigned to the Special Projects Studio. Because the new Special Projects Studio also had responsibility for Canadian, international and tractor design, Wes Dahlberg was assigned to supervise that part of the studio's work. Spear and the designers assigned to him then set out to design full-sized, fully-operable cars incorporating what they and the Engineering Department saw as future styling and engineering trends.

(below) The D-523's fuel-injection system as seen from the left side with the air cleaner installed.

Gil Spear

The first full-sized car started in the Special Projects Studio was the D-523. (Some designers remember it as the D-526, but it is referred to here as the D-523, the designation most designers recall it as having.) There were three problem areas Spear was trying to address in designing the D-523: (1) It was designed to test gull-wing doors, first used on the then new Mercedes-Benz 300 SL Gullwing; (2) it was to be Ford's first fuel-injected car and; (3) it was decided that if the Thunderbird sold well, the D-523 would be developed as a Mercury sports car.

Studio designers were asked to submit design proposals for the D-523. The package given them included the gull-wing doors and a 102-inch wheelbase, the same as the Thunderbird. One of the designs sketched by J. R. "Dick" Samsen was selected, and he was asked to prepare a full-sized side-view rendering of his proposal. After the full-sized side-view rendering was completed and approved, Samsen completed a full-sized airbrush rendering of the car on black paper. After more approvals and some refinements, Samsen was asked to oversee completion of a full-sized clay model of the car. A separate wooden buck was also built of the doors, so the gull-wing mechanism could be worked out before the car was actually built. As the full-sized clay model was nearing completion, MacPherson visited the studio. He liked the car, but insisted that 10-12 inches be taken off the front of it. Samsen did so, but he thought it made the front end too blunt.

(below) The left side of D-523's fuel-injection system with the air cleaner removed. Ford engineers weren't happy with the reliability of their first try at fuel injection.

Gil Spear

Spear's original idea for the gull-wing doors was to make them as light-weight as possible. He designed very thin lower sections of the doors, so there would be as much headroom as possible when getting in and out of the car. During the clay modeling process, a disagreement developed with Engineering. To make the doors thinner, Spear recommended jalousie-type side windows, but he was overruled by chief body engineer Henry Grebe, who insisted on electric windows and padded wrap-around armrests, that started on the doors and wrapped over the instrument panel creating coves of protective padding over the steering wheel and the glove-box. As built, the doors were opened and closed by hydraulic cylinders actuated by electric motors.

Gil Spear

(above) Access to the Cougar was by means of electrically-operated gull-wing doors.

As he had first proposed for the Thunderbird, Samsen designed the D-523 with a shaker-style hood and a designer air cleaner sticking up through the hood. The fuel-injection system eliminated the need for additional hood height, so the shaker hood was eliminated. Fuel injection was one of chief engineer Earle S. MacPherson's pet projects, but Spear recalls that MacPherson's engineers were not pleased with the way the fuel-injected D-523 performed. The D-523 was also the first full-sized fiberglass car completely built in-house at Ford. Before the D-523, Ford's full-sized fiberglass showcars were usually built at Creative Industries in Detroit.

During the 17 months that Hershey, Bordinat and Spear were, more or less, co-equals, relations between the Design Department, design consultant George Walker and Engineering became more strained. It was the impression of designers there at the time that conflicts arose because of a lack of cooperation between the managers of the design studios and Walker. Designers report that, at best, Walker was ignored and, at worst, Walker and his employee, Elwood Engel, were not welcome in the Lincoln-Mercury studio. Hershey apparently got along somewhat better with Joe Oros. Engineering sided with Hershey and Bordinat against Walker, which made matters even worse and, during that time, designers who cooperated with Walker, Oros or Engel were often criticized for doing so by Henry Grebe. The solution to the problem was again different than expected.

(right) The D-523 was considered for production as a Mercury sports car.

Gil Spear

On May 22, 1955, George Walker was appointed general manager of the Design Department and a Ford vice president. His former employees, Joe Oros and Elwood Engel, became Ford employees, and were named assistant managers and advisors to the Ford and Lincoln-Mercury studios respectively. Bordinat became head of the Mercury Studio, and John Najjar was appointed head of the Lincoln Studio. Hershey would probably have become head of the Ford studio, but he chose to resign rather than work under George Walker. Hershey left Ford for a job at Kaiser Aluminum, and Bob Maguire was appointed head of the Ford studio to replace Hershey. (Joe Oros became head of the Ford studio in March 1956.)

(below) Chief engineer Earle MacPherson did not like the extended front end originally designed by Dick Samsen and requested that it be made shorter. The redesigned front end had a more rounded look that matched the back end.

Gil Spear

At the time of Walker's appointment as general manager of the Design Department, it became a separate, autonomous, co-equal department with Engineering. Chief engineer Earle S. MacPherson wasn't happy about losing control over the Design Department, but he made a compelling argument that Engineering needed its own design function, especially if it still had responsibility for engineering the design features anticipated on future Ford products. As a result, the Design Department's Special Projects Studio became the Engineering Department's Special Vehicle Department and was moved across the street to the EEE building annex.

MacPherson asked Spear to become chief designer in the Special Vehicles Department. Walker also asked Spear to remain at the Design Department but, after several meetings with both MacPherson and Walker, Spear chose to move to Engineering, and he was followed by many of the designers and modelers in the Special Projects Studio. Dick Samsen was one of the designers who chose to remain at the Design Department, and he was reassigned to the Ford studio to work on the 1957 Ford passenger car then being designed. Samsen soon left for a job at Chrysler. According to Samsen, when he was reassigned to the Ford studio, the full-sized clay models of both the D-523 and D-524 had been completed, and they were ready to make the plaster molds from which fiberglass bodies would be made.

Once the move across the street had been completed, MacPherson gave Spear a budget of $185,000 to complete and test the D-523, the D-524, to design and build the Bimini, and for a couple of other projects. When the fiberglass body of the D-523 was completed, it was placed on a slightly altered 1955 Thunderbird chassis, and a Thunderbird 312 CID fuel-injected engine was installed. The Engineering Department wasn't pleased with the way the fuel-injected engine ran. Performance and reliability problems developed, and the D-523 did not lead to a production fuel-injected Ford product. After it was moved across the street, what design work was left to do on the D-523 was supervised by Jack Mills, who acted as Spear's executive designer.

The car was not known as the D-523 while it was at the Design Department, but was first assigned that name after the move across the street. During the time it was at the Design Department, the car was unnamed but was informally referred to by Spear as the Ornithopter, because the car's gull-wing doors, when opened, reminded him of an unsuccessful early-day wing-flapping airplane. After it was moved to the Special Vehicles Department, MacPherson decided that the car should be known only as the D-523. MacPherson didn't believe test bed vehicles should be named and didn't want either the D-523 or the D-524 to become showcars. (The D-523 was named the Cougar when it was placed on the show circuit in 1962.) According to Spear, all went fairly well at the Special Vehicles Department until late in 1957, when MacPherson suffered a heart attack that eventually forced his retirement. After MacPherson's heart attack and, even before he officially retired, Andy Kucher took over as chief engineer and was eventually appointed to that job. Kucher did not share MacPherson's enthusiasm for a separate design studio in the Engineering Department.

Ford Photomedia

(above) The interior of the Cougar had bucket seats, a padded instrument panel that wrapped around and into the armrests, a center console, telescoping steering wheel, floor shift and roll-up windows.

At about the same time MacPherson suffered his heart attack, George Walker was appointed to Ford Motor Co.'s finance committee and was instrumental in cutting off funding for the Special Vehicles Department, which effectively put it out of business. When the Special Vehicles Department was shut down in early 1958, Spear found jobs for most of his people, and then he and Jack Mills returned to the Styling Center. Designer Doug Miller and clay modeler Walt Jennfeldt remained with Experimental Engineering, although Miller eventually returned to the Styling Center, and Jennfeldt was eventually reassigned to the Scientific Laboratory.

When the Special Vehicles Department was disbanded, the D-523 was finished and fully operable, although Spear never had a chance to drive it. When Spear returned to the Styling Center, the D-523 and the D-524 were placed in storage in the basement of the EEE building annex under cover. From time to time after that, the D-523 was taken out of storage and brought to the Experimental Engineering garage to test some new idea and, from time to time, it was driven on the street.

Walt Jennfeldt recalls that Kucher, who was also head of the Scientific Laboratory, had particularly bad luck every time he tried to drive the D-523. On one occasion, he cut his head getting out of the car and needed first aid. On another occasion, the battery went dead with Kucher inside the car. Since the doors could only be opened electrically,

(below) Doug Miller designed aerials that were attached to the rear-door hinges on the roof.

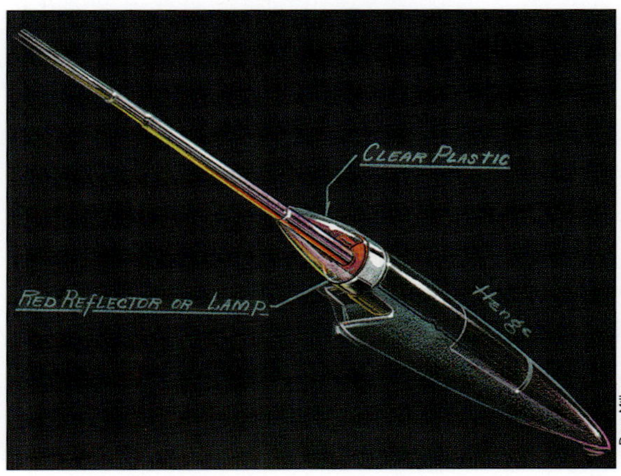

Doug Miller

Ford Photomedia

After the D-523 was rediscovered in 1962, it was removed from storage and turned into a showcar. It was named the Cougar, a 406 CID Ford engine was installed, the fender skirts and roof aerials were removed, Kelsey-Hayes wire wheels from a 1962 Thunderbird Sports Roadster were installed, the car was reupholstered in white leather and repainted metallic blue.

Kucher was trapped inside until a new battery could be installed in the car. Jennfeldt recalls that almost the entire engineering staff came to watch their compatriots trying to get Kucher out of the car. As Jennfeldt remembers it, Kucher took it all in good humor.

In 1959, the D-523 was displayed on a Ford Motor Co. float for the Veterans' Day parade in Detroit and, in late 1961, the car was also displayed at the Ford Rotunda. When Spear learned that the D-523 was at the Ford Rotunda, he was surprised that it still existed, and that it had not been destroyed as he had assumed. He contacted the Ford Public Relations Department and suggested that the D-523 be used on the show circuit. A Ford public relations employee looked at the car and quickly agreed.

Fire destroyed the Ford Rotunda soon after the D-523 was removed and taken to Dearborn Tube and Steel, a local job shop, where the rear fender skirts and the horizontal roof aerials were removed. It was repainted a metallic blue, reupholstered, the original fuel-injected 312 CID engine was replaced by a newer 406 CID engine, and it was fitted with a set of Kelsey-Hayes wire wheels from a Thunderbird Sports Roadster. Starting in 1962, the repainted and cleaned up D-523 became the Cougar showcar. Spear has no idea why it was given the name Cougar. It was re-introduced at the 1962 New York Auto Show, and was also shown at the 1962 Chicago Auto Show.

Someone from Hollywood, who was connected with a movie about to go into production, saw the Cougar at the New York Auto Show and proposed to the Ford public relations staff that the car be used in the movies. One thing led to another, and the Cougar was sent to California. George Barris repainted it red, and it was featured in the 1964 Jerry Lewis movie, *Under The Yum Yum Tree*.

This photograph was made from an 8- mm home movie taken of the D-523 at a 1959 Veterans' Day parade in Detroit.

Joe Mooradian

After the movie was completed and in about 1965, the car was sold to Joe Mooradian who, at the time, was a partner in Autocrafters Inc., a Detroit area restoration shop. Mooradian had first seen the car at Experimental Engineering soon after it was finished and fell in love with it. He took home movies of the car during the 1959 Veterans' Day parade and stopped to see it several times a week, when it was on display at the Rotunda. After he purchased it, Mooradian brought the car home to the Dearborn area, where he drove it during good weather and on special occasions. The car is not for sale and is currently undergoing a ground-up restoration. Mooradian reports that the original color of the car was pearlescent light lime-gold. It was subsequently repainted gold, then metallic blue, and finally red. He plans to return the D-523 to the same condition it was in when shown in 1962. ■

Bob Puffer

The restored D-524 at the 1997 Meadow Brook Concours.

D-524

Design studies for the D-524 began in early 1954, immediately after Spear became chief designer in the new Special Projects Studio. It was designed and built to test advance concepts in air conditioning, seating, ingress and egress, lighting, front frame crash absorption and as Ford's first car without A-pillars. Even though the D-524 now has only two headlights, it was also designed as Ford's first car with four headlights. (Some designers remember the D-524 as the D-525. Because most designers refer to it as the D-524, that designation is used here.) The D-524 was never intended as a showcar but, rather, was designed and built as a "research project on wheels." The exterior was designed by Gil Spear and Wes Dahlberg and the interior by J. R. "Dick" Samsen.

In 1952, Spear first developed a working relationship with Ford's chief engineer Earl S. MacPherson, who appreciated the fact that he designed into his cars advanced engineering concepts they were both interested in. During development of the Syrtis, MacPherson told Spear that he thought like an engineer — unlike the rest of the designers. That compliment didn't endear Spear to his superiors in the Design Department, and it embarrassed him. But MacPherson had a point — Spear understood that sound engineering was a necessary component in the cars he designed.

When designed, the exterior of the D-524 was meant to be more functional than beautiful. After the exterior design was established, alternate renderings were prepared by designer Barney Harb. After the final designs were selected, a 3/8-sized clay model was built. Then a full-sized clay model was built and, from that, plaster molds were cast, from which a fiberglass body was made. The surface development on both the 3/8-sized clay model and the full-sized clay model was supervised by Wes Dahlberg, who was assistant manager of the studio. After the move across the street, Dahlberg, who stayed at the Styling Center, became manager of the international studio, and Jack Mills became the new assistant manager of the renamed Special Vehicles Department. The spare tire humps at the rear of the back fenders were givens and part of the package set by Spear, but it is because of Dahlberg that the rest of surface of the car looks more rounded and integrated than it did in the renderings.

The interior design of the D-524 was selected by Spear from several alternative designs Dick Samsen prepared. In 1952, when Samsen was hired as a Ford designer, Frank Hershey and Damon Woods accompanied him on a get-acquainted tour of the Design Department. During that tour, Samsen got his first close-up look at the XL-500. He was especially impressed with its split-bumper front end, which he remembered and later used as inspiration for the instrument panel design selected for the D-524.

At the time the D-524 was designed, it was anticipated that air-conditioning would, in future years, become an accessory purchased by more new-car buyers. Because air-conditioning evaporators were so large they had to be located in the trunk, Spear was con-

Doug Miller

This full-sized rendering of the D-524 was made before the clay model was started. During the clay modeling process, many changes were made in the car's design. The final design of the D-524 was refined and completed by Wes Dahlberg during the clay modeling stage.

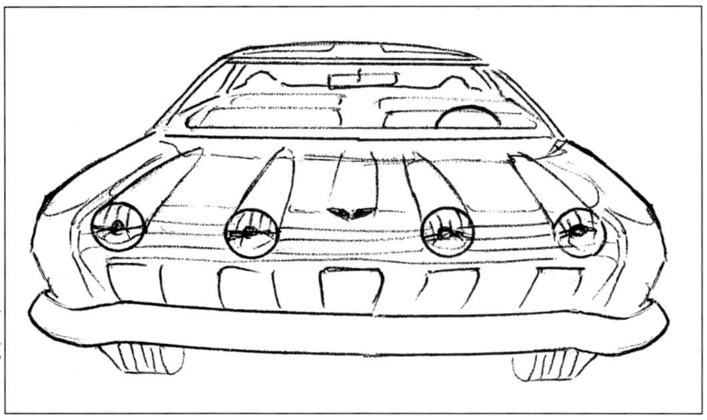
drawing by Gil Spear

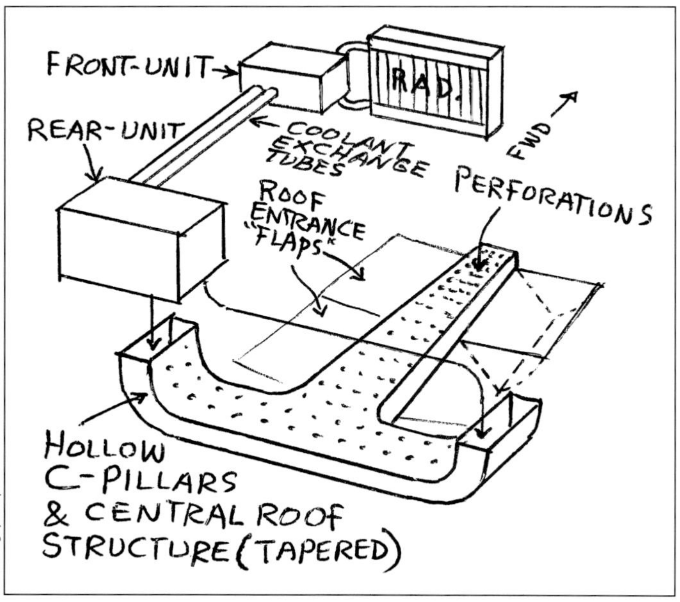

drawing by Gil Spear

Jack E. Mills

Gil Treweek

Jack Mills grew up in the Detroit area. His father and grandfather were draftsmen. Mill's grandfather helped build Fairlane, Henry Ford's home.

Except for art courses at Cass Technical High School, Mills had no formal training as a designer. His first job was working in the Ross-Roy advertising firm. When World War II started, he enlisted in the Army Air Corps but was soon discharged, because he was color blind. He returned to Detroit and started work at Ford's Rouge plant. He spent several months preparing a portfolio and, in 1942, was hired to work with Bill Schmidt at the Willow Run bomber plant illustrating plans used in building the B-24. In 1945, Mills and Schmidt were transferred to the Design Department to begin their design careers.

During his career at the Design Department, Mills worked in the Ford, Lincoln-Mercury, Special Vehicles and truck studios. In the Lincoln-Mercury studio, Mills worked on the exterior design of the Futura. Mills was Gil Spear's executive designer in the Special Vehicles Department, where he worked on the D-523, the D-524, the Bimini, the Midshipman and the Personal Safety Vehicles before returning to the Styling Center, where he was assigned to the Ford studio.

Mills was an avid pilot, who designed and built six of his own planes. His wife Imogene was also an employee at the Design Department, and three of their sons currently work at the Ford, Chrysler and GM design centers. Mills retired from the Ford Design Center in 1988. ■

(above left) According to Gil Spear, the grille and headlights now on the D-524 are probably different than they were when the car was first completed. Earle MacPherson wanted the D-524 to be the first Ford product with four headlights. Spear was successful in talking MacPherson out of the configuration shown here.

(middle left) As originally built, air-conditioned cold air passed through the hollow C-pillars of the D-524 and was discharged into the passenger compartment through perforations in the headliner. The perforations were larger towards the front of the car to help assure an even discharge of cold air.

(below) As originally built, the D-524 had Plexiglas-roof flaps that automatically raised and lowered when the doors were opened and closed. The same concept was later used on the Mercury XM-Turnpike Cruiser.

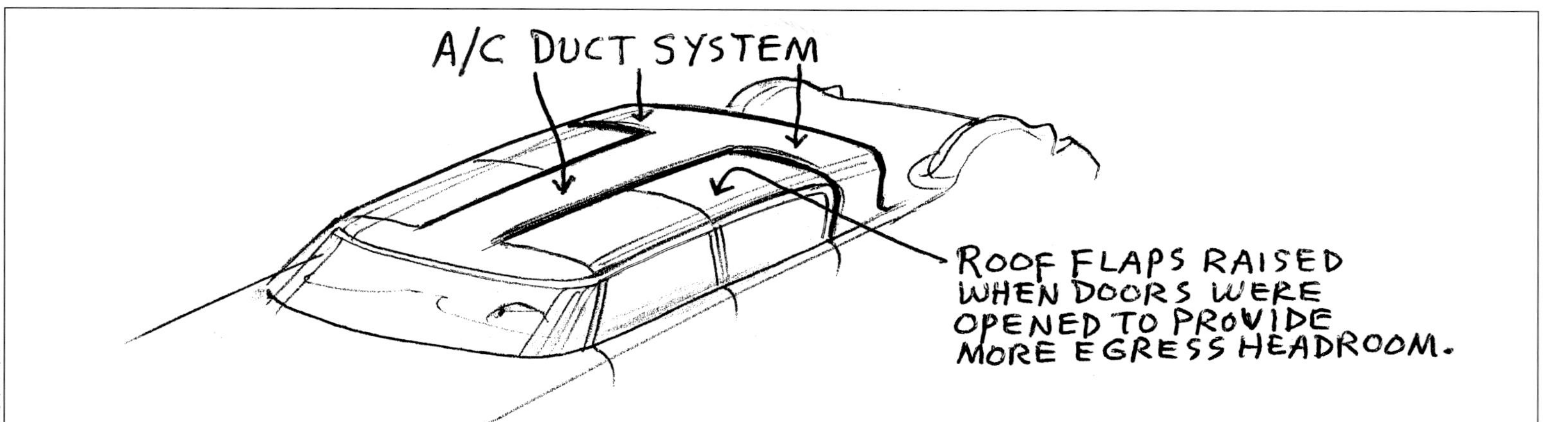

drawing by Gil Spear

cerned they restricted available luggage room. Spear's answer to the problem was to relocate the spare tire to the back of one fender and the gas tank to the back of the other, with flip-top lids providing access to each. (Several years later, technical advances reduced the size of the evaporator, and it was relocated to the engine compartment.)

MacPherson originally wanted four headlights spaced equidistant across the front of the D-524 but, after making several alternate front-end proposals, each with a different variation of the four-headlight theme, Spear prevailed on MacPherson and was able to group the headlights in pairs. No photographs have been located of the D-524 with four headlights, and it may be that studies of the car with four-headlights were done on clay models only. Spear, however, believes that, when first completed, the D-524 had stacked-dual headlights, much like the 1957 Lincoln. The D-524 now has one headlight on each side. Spear also believes the grille originally on the D-524 had a different configuration than it does now.

Gil Spear

Gil Spear

As originally designed, the roof of the D-524 was shaped like a "T" with the top of the "T" folding downward to form the C-pillars. The roof was hollow and acted as a conduit for cold air coming from the air-conditioning evaporator located in the trunk. Cold air from the evaporator flowed up and through the hollow C-pillars and into the hollow roof section exiting through perforations in the headliner. To assure an even distribution of cold air, Spear designed the perforations in increasingly larger diameter as they got closer to the front of the car. Over the doors, and extending into the roof by about 18 inches, Spear designed Plexiglas panels that were hinged towards the middle of the car's roof. For ease of entry or exit, each panel was automatically raised or lowered by an electric motor when a door underneath it was opened or closed.

(top) Dick Samsen designed the instrument panel after the split front end of the XL-500 showcar.

(above) The D-524 was built without A-pillars. Because the windshield cracked, supports were installed between the half B-pillars and the roof.

(below) The spare tire is housed in the right rear fender and the gas tank is located in the left rear fender.

The hooded reverse backlight on the D-524, as Spear designed it, rolled up and down manually. Reverse backlights were later used on the production 1958-60 Lincoln Continentals, the LaGalaxie showcar and on some production Mercurys of the 1960s. Several years earlier, Spear had designed and built a 1/8-sized concept car with a reverse backlight. MacPherson liked it and incorporated the idea on his personal Lincoln. From then on, he was an advocate of the reverse backlight. Spear's original idea for the reverse backlight included a thick piece of clear Plexiglas with a prismatic surface on the inside of the backlight, so that the driver's rear-vision lines would be deflected on a more upward angle. Spear designed his prismatic Plexiglas backlight so the rearview mirror could be placed higher on the windshield, which Spear thought interfered less with forward vision. Unfortunately, the cost to manufacture the prismatic-Plexiglas backlight was too high, so that feature had to be eliminated when the D-524 was built. Another idea Spear proposed, but could not get MacPherson to authorize on the D-524, was an elevated-brake light above the backlight.

Gil Spear

Wes Dahlberg, a designer in the international studio, supervised the clay modeling of the D-524. He gave it a more rounded look and eliminated the squared-off look present in the renderings, but his participation ended when Spear and the car left the Styling Center. While the D-524 was still at the Design Department, much of the clay modeling work on it was done by Fred

Gil Treweek

Douglas A. Miller

After serving as a tail gunner on a B-17 in the European Theater during World War II, Doug Miller returned home to Detroit and began studying art at Meinzinger's Academy. He loved cars and wanted to design them. While attending Meinzinger's Academy, he worked on the spot-weld line at Briggs Manufacturing and did contract airbrush work on the side. He was hired by Herb Weisinger as a designer at Kaiser-Frazer in 1949. While at Kaiser-Frazer, he worked with Buzz Grisinger, Rhys Miller and Alex Tremulis designing the Kaiser, Frazer and the Henry J. In 1951, after he was laid off at Kaiser-Frazer, Miller worked at several Detroit area job shops including Creative Industries, until he was hired as a designer at Ford in 1952.

Miller's first assignment at Ford was in Gil Spear's Advanced Studio, where he worked on many of Spear's 3/8-sized concept cars including the Syrtis. After six months in the Advanced Studio, Miller was assigned to the Lincoln production studio but, when the Special Projects Studio was formed in early 1954, he was reassigned to work in that studio. While in the Special Projects Studio, Miller worked on the design of the D-523 and the D-524. After the Styling Center was separated from Engineering, Miller moved to the Special Vehicles Department and helped complete the design of the Bimini, the Midshipman and the Personal Safety Cars. Miller and Norm Nebozenko also designed the Glideair.

After the Special Vehicles Department was disbanded, Miller and Walt Jennfeldt were transferred to Ford's Experimental Engineering Department, where Miller designed the Aeolis (Levatrain) and worked on Big Red and the Bush Buggy.

In 1961, he was transferred back to the Styling Center and was assigned to the Lincoln exterior, and then to the Lincoln interior production studios. Before he retired in 1982, Miller also worked as a designer in the truck studio and finished his design career as a designer in the tractor studio. ■

Bob Puffer

(above) The D-524 originally had four headlights, but was later rebuilt with single headlights on each side.

Bob Puffer

(above) From the rear, the D-524 looks very similar to Alex Tremulis' Madam X.

Gil Spear

The D-524 was the first full-sized car built at Ford with a reverse-slant backlight.

Hoadley, who also chose to stay at the Design Department rather than move to the Special Vehicles Department. Clay modeler Bill Leverenz left with Spear and became modeling manager in the Special Vehicles Department. He supervised completion of the full-sized D-524. When they left the Design Department, the D-523 was closer to completion than the D-524. The D-524 was finished at the Special Vehicles Department, but it was also revised several times while there, and was not in completed form when the department was disbanded.

When Spear returned to the Styling Center, the D-523 and the D-524 were placed in storage in the EEE building annex basement. According to Walt Jennfeldt, the D-524 was moved from place to place for several years and, every once in awhile, it was used to test some new engineering concept. Because the D-524 was originally built without A-pillars and relied on the windshield to support the front part of the roof, it had a habit of cracking or popping out windshields when driven. As Jennfeldt recalls it, half B-pillars were later installed to provide support for the roof, so that the D-524 could continue to be used to test new ideas. The only other use Jennfeldt recalls the D-524 was put to included something to do with an interchangeable seat idea.

Although the D-523 was renamed the Cougar and was used as a showcar by Ford in 1962, it was decided not to attempt to make a showcar out of the D-524. In late 1962 or early 1963, however, the D-524 followed the D-523 to Hollywood for use in the movies. After the D-524 arrived in California, George Barris, who later turned the Futura into the Batmobile, repainted it, removed the name D-524 from the front fenders, and put block chrome letters on the front fenders and doors that spelled out "Beldone," a stage name for the car selected by Paramount Pictures. Barris had movie industry contacts, and convinced Ford he could get movie and TV coverage for concept cars Ford no longer had a use for.

The Beldone first appeared in a 1964 Jerry Lewis movie called *The Patsy*. For the movie, the Beldone was equipped with radio remote-control devices that opened and closed the doors, trunk and hood for one of the comedy scenes in the movie. After the studio was through filming *The Patsy*, the car was stored on the Paramount backlot and infrequently used in other movies and TV programs, including a 1964 TV series called *The Duplicate Man*. It also had a bit part in the movie *Back To The Future III*.

In 1984, Bob Butts, the owner of a San Diego-based business that provided cars for the movie industry, purchased the D-524 from George Barris, who had recently acquired it from Paramount Pictures and repainted it again. The VIN number on the car described it as an experimental 1953 Mercury Monterey, and Butts was eventually able to confirm the car's Ford heritage.

In 1995, Butts sold the D-524 to The Guild Of Automotive Restorers, a group of Canadian investors, and the car is now located in Ontario, Canada. After they bought the D-524, the Guild had trouble finding a carburetor with a low enough rise to permit the hood to close. The D-524 now has a suitable carburetor and is operable again — and the hood closes. It has been repainted its original copper and black two-tone, and was most recently shown at the 1997 Meadow Brook Concours. ■

Gil Treweek

Walter Jennfeldt

Walt Jennfeldt grew up in the Detroit area. After graduating from high school, he went to work for Ford building car frames. In 1950, after less than a year building frames, he prepared a drawing of a ship, applied and was accepted in Bill Leverenz' first clay modeling class. After completing the class, he was hired as a clay modeler at Ford's Design Department, where he was assigned to the International and truck studio. In 1955, when Gil Spear left the Design Department to become chief designer in the Special Vehicles Department, Jennfeldt was among those who went with him.

While at the Design Department, Jennfeldt was one of the clay modelers who worked on the D-523 and the D-524. At the Special Vehicles Department, he and Bill Leverenz finished up those two cars and built clay models of several of the Personal Safety Vehicles, the Bimini and the Midshipman.

After the Special Vehicles Department was dissolved, Jennfeldt was assigned to Experimental Engineering and then to the Scientific Laboratory, where he clay modeled the Levatrain (Aeolis) and the Bush Buggy. He also assisted in assembling the cab of the gas-turbine truck (Big Red), and used blueprints to clay model its full-sized gas-turbine engine before it was built.

While at Experimental Engineering, Jennfeldt spent a substantial portion of his time building and testing models or portions of models for use in the University of Michigan windtunnel.

He was offered and took early retirement in September 1977. ■

Bob Puffer

The D-524 was repainted its its original colors before it was shown at the 1997 Meadow Brook Concours.

Bimini

After World War II, designers and engineers at Ford foresaw the trend towards ever-lower cars, and believed that, eventually, that trend would require changes in the way people exited and entered their cars. Traditional doors were seen as an impediment to making lower cars. Gull-wing doors, such as on the D-523, weren't seen as a complete answer nor were doors with indented roof panels over each door, such as were originally included on the D-524. Doors were also the primary reason structural A-pillars were needed, and Ford's designers and engineers wanted to try to eliminate the blind spot created by the car's A-pillars. In the quest to find a new entry system, Ford engineers and designers were searching for a break-through that would allow them to build lower cars and also eliminate the blind spot the A-pillars created.

With standard doors placed, as they were, behind the A-pillar, there was very little other structure available to support the roof if the A-pillars were removed. Given the technology then available, attempts to use the windshield itself as a structural element to support the roof failed. It was tried on the D-524, but eventually the car had to be retrofitted with B-pillars to help support the roof.

One of the projects Gil Spear and his Special Vehicles Department undertook was to come up with a radical new entry system that, if it could be made to work and wasn't too expensive, solved both problems. Spear had been working on a way to eliminate the A-pillars since at least 1948, when he first became head of the Advanced Studio. Of the various proposals tried, the most radical one looked the most promising. That proposal was to make the doors, the front part of the roof and the windshield raise-up and tilt forward as a unit, so that passengers could enter and exit the car standing up and by walking in or out of it. The system was named the strato-door by Spear.

As the D-523 and the D-524 were being built, Spear began to design a third car. That third car was a front-wheel-drive, strato-doored, two-door station wagon he called the Bimini. One of Spear's hobbies was skin diving, and the Bimini was probably named after the Caribbean location of several of his skin diving vacations. At first, the car also had a number designation as was chief engineer Earle MacPherson's practice, but Spear does not recall what D-number was assigned to it.

The Bimini was designed to seat four passengers across its single front seat. Several auxiliary rear-seat proposals were also prepared for children. Although eliminated when the Bimini was built, it was originally designed with a removable built-in shopping cart on wheels that could be collapsed and stored in the rear of the car. It was similar to an ambulance gurney and could be removed for shopping and later wheeled back to the car and collapsed into the back. At home, the shopping cart could be removed, taken in the

Designer Bud Ward is shown demonstrating the Bimini's proposed entry system on a rendering air brushed on black paper, tacked to plywood and cut out. A seat and several appendages were placed behind the rendering for a realistic view of what the car would look like. The door was a separate piece of plywood hinged to the cowl area.

Doug Miller

house, unloaded and then returned to the car. Another safety feature Spear incorporated into the Bimini was a high brake light on each side at the rear of the station wagon's roof.

A full set of package drawings were made for the Bimini, as were a 3/8-sized clay model and a full-sized engineering buck used to demonstrate and test the strato-door mechanism. The full-sized buck of the strato-door system used on the Bimini was built and refined until engineers were satisfied they had worked all the bugs out. As Spear remembers it, the full-sized clay models of the Bimini were not built until after the move across the street to the EEE building annex. Several different door configurations were also considered, and a 1/8-sized engineering model of a sedan with the strato-door system was also built. By 1957, the station wagon market had grown to 15% of the new-car market, and designers and engineers all seemed to agree that more was to be gained by building the Bimini as a station wagon than as a sedan.

(above) Before the Bimini was built, several different entry systems were considered including one with an articulated door that opened upward.

(left) This full-sized clay model of the Bimini, which provided for small roll-up windows in the doors and eliminated the A-pillars, was very close to the finished car.

According to Spear, the Bimini was the pet project of Victor G. Raviolo, who was responsible for development of the Y-block engine, and later became Ford's chief engineer. Spear recalls that Raviolo was a private pilot and was particularly inclined to incorporate airplane technology into his ideas.

Engineering was not overly interested in the way its test models looked, primarily because it was not intended that the public ever see them — they were considered engineering tests beds and not much more. The design work on the Bimini was completed by studio designers, including Jack Mills, Bud Ward and George Gobulics with assistance from Gil Spear and Doug Miller.

(below) This alternate clay model was considered, but one with a curved windshield with A-pillars in the middle of the doors was selected instead.

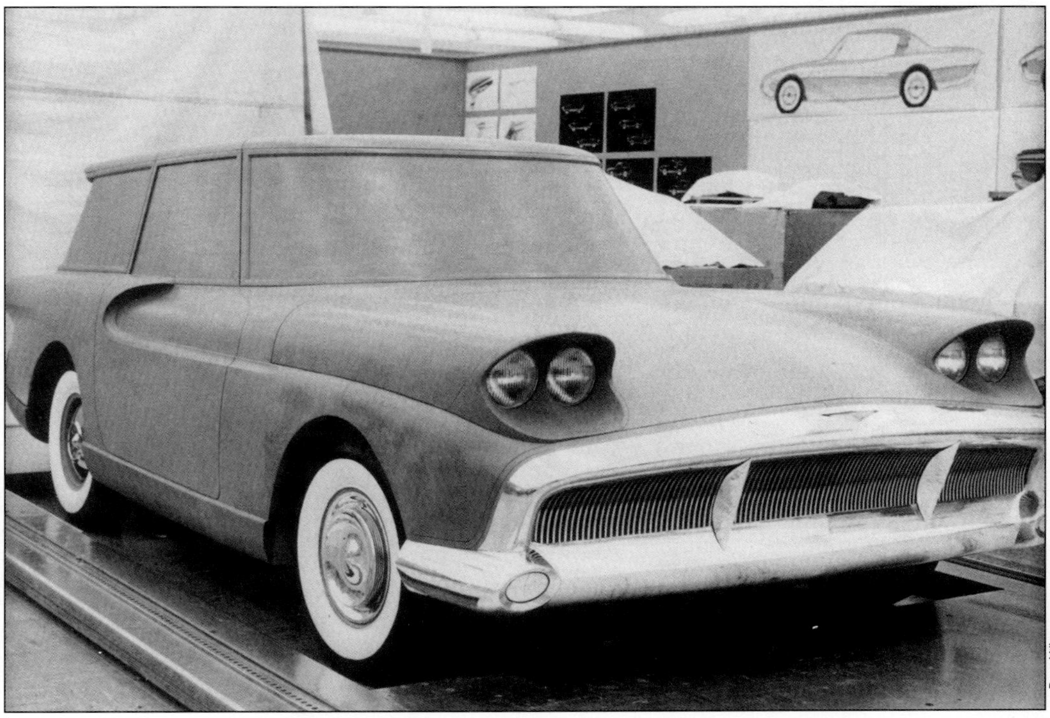

According to Doug Miller, Engineering arranged for the car to be built by Ghia in Turin, Italy, because the Styling Center told them they did not have the time to build the body of the car in fiberglass. After a full-sized clay of the Bimini was completed, the hardware needed to complete the car, a plaster model of the car, a set of engineering drawings and, possibly, the buck of the strato-door system were sent to Ghia, where the Bimini's body was built. No chassis or running gear were sent, because it was anticipated the body would be mounted on a running chassis when it was returned. Ghia completed the car in 1956, and it was returned to Dearborn by ship. The Bimini cost approximately $80,000 to build, the same amount Ghia charged for the Turnpike Cruiser, which was also built at Ghia's Turin facilities in late 1954. Although records can-

(right) On the competed Bimini, the doors, windshield and the front-half of the roof lifted as a unit, via electric motors, to permit entry and exit from the car. This photograph was taken at Ghia in Turin, Italy, before the body was shipped to Ford, where it was mated with a running chassis.

Ghia s.P.a.

(below) The interior of the Bimini with the strato-door in the raised position and the front-passenger seat in the down position. Because the occupants of the Bimini stood up before they stepped out of the car, the steering wheel was designed to swing forward, when the roof section was raised. The speedometer was located in the center of the steering wheel like the Futura.

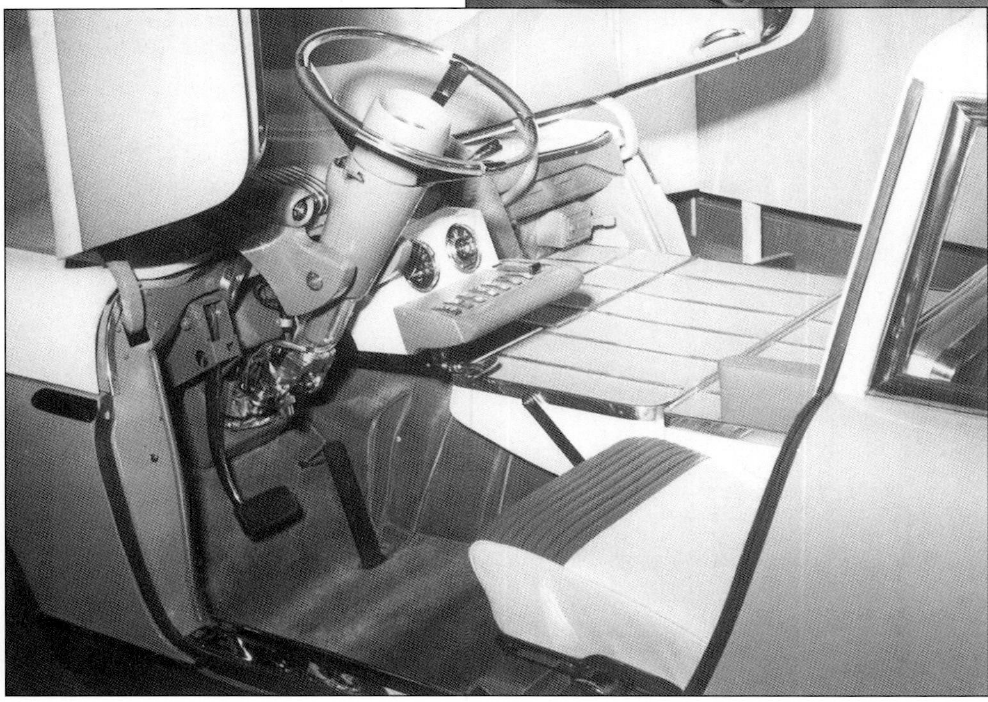

Ghia s.P.a.

not now be located indicating the type of engine used, it was probably a 312 CID Thunderbird V-8. During the time it was being tested by Ford engineers, Joe Mooradian, who later acquired the D-523, had a chance to drive the car. He remembers it as being very fast, but a little bit "squirrely," probably because of the short wheelbase.

When Spear returned to the Styling Center, after the Special Vehicles Department had been dissolved, designer Doug Miller and clay modeler Walt Jennfeldt remained as employees of the Engineering Department and were assigned to complete various projects of interest to that department. Their recollection is that the Bimini had been completed by that time and had already gone through most of its testing.

The Bimini has since disappeared, although there are persistent rumors it still exists. It is unknown whether the Bimini, as built, actually had front-wheel-drive. It is unlikely it did because of the center hump shown in the photos. The designers who worked on it believe the Bimini contributed to the smaller, compact Fords and Mercurys built during the 1960s. The strato-door system used on the Bimini was never used in production probably because of body stresses and cost. ■

(below) The Bimini was originally conceived by Gil Spear as a second car that could be used for shopping and other around-town errands. Although it was designed with only a front seat, optional rear seating for children was considered.

Gil Spear

Ghia s.P.a.

(left) The full-sized metal body of the Bimini as built by Ghia. A 312 CID engine was reportedly installed in the competed car.

(below) One person who drove the Bimini recalls that its V-8 engine made it go "like a scalded cat," but the short wheelbase made it "a little squirrely."

(bottom) This photograph was taken before the Bimini left Ghia. The interior was designed by Doug Miller. The roof section over the rear-cargo area was also removable.

Ghia s.P.a.

Gil Treweek

William M. Leverenz

Bill Leverenz was born in Alsace-Loraine and trained in Berlin, Germany as a sculptor and decorative plasterer. After he immigrated to the United States, his first job was as a lamplighter in Chicago. Before becoming a clay modeler, Leverenz worked as a decorative plasterer and helped finish many of the theaters in Chicago and Detroit. In Detroit, he worked on the decorative-plaster ceilings of the Fox Theater and the Fisher Building. Leverenz' first clay modeling job was at GM's Fisher Body Division. In 1939, he was hired by Bob Gregorie to work in Ford's Design Department. During World War II, Leverenz worked at the Willow Run bomber plant with Bill Schmidt. After the war, he became a modeling manager in Schmidt's Lincoln studio.

Because Ford was constantly short of qualified clay modelers, in January 1950, Leverenz started the Detroit area's first automotive clay modeling school at Ford's Rouge plant. The six-month long course was offered to all talented hourly and salaried employees. Leverenz' first class had 93 students, who were trained in drawing as well as clay modeling.

In 1955, Leverenz followed Gil Spear to the Special Vehicles Department as his clay modeling manager. He remained there as head clay modeler, until funds ran out and the studio closed in 1958. Leverenz volunteered to retire, so that the younger clay modelers could take the available positions at the Styling Center and elsewhere. Leverenz was clay modeling manager at the time the D-523, the D-524, the Midshipman and the Bimini were designed and built.

Leverenz was also an expert fencer and archer, who taught fencing courses to many Ford employees through the Ford Sports Club. ■

Ghia s.P.a.

Gil Spear

Although the Midshipman was designed with many innovative features, its looks were unorthodox. In addition to renderings and plywood cutouts, this one-eighth sized clay model was as far as the Midshipman was developed.

Midshipman

In 1949, two Volkswagens were sold in the United States and, in 1950, Max Hoffman was appointed the exclusive U.S. east coast importer of Volkswagens. By 1955, there were over 250 Volkswagen dealers in the U.S. They sold 29,000 cars that year and, by 1958, their sales exceeded 100,000 cars. Volkswagen obviously capitalized on an unrecognized need. Its quality was also seen as better than what was being produced by Detroit, and it quickly became a popular second car -- and a reverse status symbol.

Whether they publicly admitted it or not, American automobile manufacturers initially underestimated the Volkswagen's appeal and were taken by surprise, when the small, inexpensive, no-frills car that capitalized on its lack of design changes became so popular. The growing sales success of the Volkswagen was not lost on Ford's engineers and designers, some of whom had foreseen the reason for the Volkswagen's popularity, even before Ford's market research people did. Robert McNamara liked the Volkswagen, and encouraged efforts by Ford's engineers and designers to develop a similar car.

By 1957, the Special Vehicles Department proposed a rear-engine, blunt-nosed, front-entry car called the Midshipman. The Midshipman was designed by Doug Miller, Bud Ward, Norm Nebozenko and Jack Mills. It was not a Volkswagen, but it was as close to it as Ford's engineers and designers were allowed to come at the time. The Midshipman was more complicated than the Volkswagen, had more interior room, was easy to get in and out of, and was more economical than most cars Ford produced. Instructions given to the designers of the Midshipman were that its looks were less important than the engineering concepts included in the design. The designers at Special Vehicles Department all wish they had been given more time to refine its design.

The Midshipman was designed as a front-entry, front or rear-engined, front or rear-wheel-drive car, with an engine, transmission and drive unit that could be easily removed and replaced with a loaner, while the dealer made repairs to the driver's unit. As designer Doug Miller remembers it, the Midshipman was so versatile that, in addition to being set up for both front and rear-wheel-drive, it was planned for either reciprocating or gas-turbine engines and for air and water-cooled engines.

The entry system on the Midshipman was similar to that of the Bimini, except that, on the Midshipman, the whole front end of the car opened clam-shell style, much like on the Isetta or Messerschmit. The Midshipman got as far as several full-scale side-view renderings, a full-scale seating buck and a 1/8-sized clay model. Renderings were made of several different models including a convertible. Development of the Midshipman was stopped in early 1958, when the Special Vehicles Department was disbanded.

In late 1959, Ford introduced its own small car called the Falcon, also a favorite of Robert McNamara. The 1962 Econoline van is closer in concept to the Midshipman, but neither was influenced much by it. The Midshipman was ahead of its time — and it still may be. ■

Norman A. Nebozenko

After he graduated from high school, Norm Nebozenko took a job at the J. L. Hudson department store and also started art classes at night at both the Center for Creative Studies and Meinzinger's Academy. In 1951, he was hired by Ford Division to prepare charts and graphs. After reading an article in the Rouge News, Nebozenko enrolled in Bill Leverenz' clay modeling class and, in 1951, he was hired as a clay modeler in the Mercury studio. Because of a shortage of clay modelers in the Lincoln studio, he and several other Mercury studio clay modelers were temporarily assigned to work on the Futura. After the Futura was completed, he worked on the 3/8-sized model of the Turnpike Cruiser that was eventually sent to Ghia.

While working on the full-sized clay model of the production 1957 Mercury Turnpike Cruiser, Elwood Engel promised him "anything he wanted" if he could straighten out the vents and aerials at the top of the windshield. Lewis Crusoe was unhappy with the production car, and DeLaRossa and Engel disagreed about its design. When the car was finished and, after Crusoe approved it, Nebozenko reminded Engel of his promise and asked for his help in becoming a designer. Engel secured Walker's agreement and, after six months in the Advanced Studio, Nebozenko became a designer in the Edsel studio. He is responsible for the production Edsel hood ornament.

In 1957, after two years in the Edsel studio, Nebozenko was laid-off. Gene Bordinat helped him secure a job at the Special Vehicles Department, where he and Doug Miller designed the Glideair. Nebozenko also worked on alternate designs for the Bimini and the design of the Midshipman. When the Special Vehicles Department was dissolved, Nebozenko took a job with Trombley and Passini, an independent design firm and helped design their independent proposal for the Ford Falcon.

After several months at Trombley and Passini, Nebozenko secured a job at Chrysler as a designer in their ornamentation studio. He was later moved to the interior studio. He retired from Chrysler in 1989. ■

Doug Miller

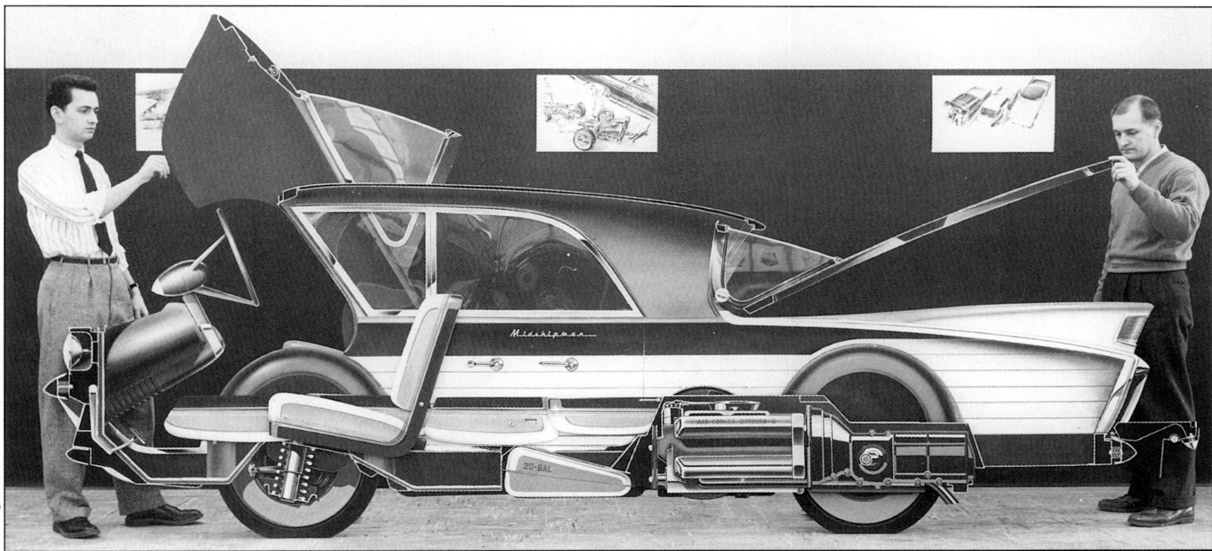

Doug Miller

(left) Designers Norm Nebozenko (left) and Bud Ward (right) demonstrate some of the features of the Midshipman on a plywood cutout model of the car.

(below) The Midshipman was designed to use either an air or a water-cooled engine. This drawing has the engine located near the center of the car.

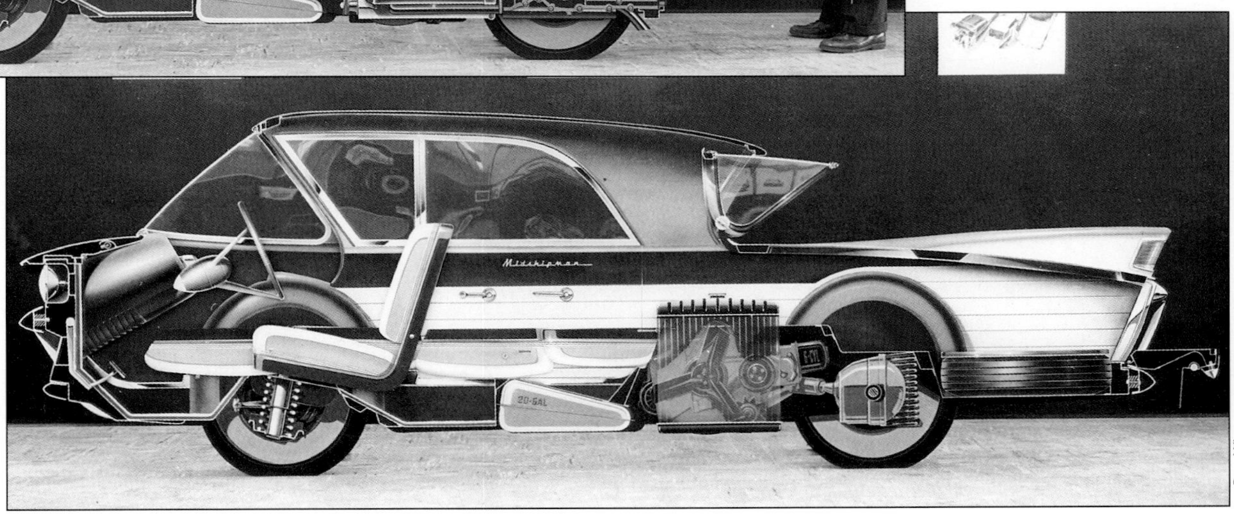

Doug Miller

(below) Originally, plans called for several different models of the Midshipman. All models, including a convertible, had forward-opening doors.

Doug Miller

In 1950 or '51, Gil Spear was in a head-on collision. He was driving a new Ford station wagon and, although his car was destroyed, neither he, his wife nor their three children were seriously injured. Several months before the accident, Spear had reversed the back seat in his station wagon, so his children rode facing towards the rear. His children liked riding backwards, and he also built a table to the rear of the rear-facing back seat. When he surveyed the damage done by the collision, Spear felt certain that, but for the rear-facing back seat his children were in, one or more of them would have been seriously injured or worse. From then on, Spear took a personal interest in safer cars.

The sliding door designed for the Personal Safety Car was a forerunner of the sliding doors on today's vans.

Gil Spear

Personal Safety Vehicle

After he moved to the Special Vehicles Department, Spear's quest for safer car designs received added support from Robert McNamara, who later became president of Ford and then President Kennedy's secretary of defense. In 1956, at McNamara's urging, Ford Motor Co. made safety a selling feature on all of its cars. Ford called the safety features built into its cars *Lifeguard Design*. Included as standard equipment on all Fords, Lincolns and Mercurys were energy-absorbing, deep-dish steering wheels and safety door latches. Also offered were optional front seat belts, padded instrument panels and sunvisors, and improved safety glass that reduced injuries when shattered.

Until halfway through the 1956 model year, the safety features Ford Motor Co. offered on its cars sold well but, after that, it was generally felt that selling safety scared prospective buyers away. At the end of the model year, when Chevrolet outsold Ford by 123,000 more cars than it did in 1955, Henry Ford II blamed it partially on McNamara's safety campaign. The safety features Ford offered in 1956 weren't considered one of its better ideas, but they paved the way for government-mandated safety features in the late 1960s. What Ford started in 1956 was also quietly adopted by other manufacturers.

No matter what Mr. Ford thought, Spear became even more convinced of the need to design and engineer better safety features into future cars. So, beginning in 1956, Spear's Special Vehicles Department designers developed several Personal Safety Vehicles or PSVs, as they were referred to. Windshields that popped-out in accidents, recessed knobs and controls, sliding doors, padded interiors, structures that absorbed energy in crashes, better lighting and safer seating were just some of the ideas Spear's group investigated.

At the time the Special Vehicles Department was disbanded, there were plans to build at least one full-sized PSV. Several proposals were rendered full-sized, and at least two of them got as far as 3/8-sized clay models. The primary safety features incorporated into the design of one or the other clay models included rear seats that faced backwards, sliding doors and safety instrument panels. Designers who worked on the PSVs included Gil Spear, Jack Mills, Norm Nebozenko, George Golubics, Barney Harb and Doug Miller.

After the Special Vehicles Department was dissolved, Doug Miller and Walt Jennfeldt designed and clay modeled an instrument panel buck for a PSV inside a production Edsel. They designed all instruments and knobs behind the steering wheel, and the instrument panel itself was constructed so that, in an accident, when seat-belted occupants jackknifed forward, they avoided contact with any part of the car.

Spear believes the sliding-door mechanism his designers developed for the Personal Safety Vehicle became the model for those now used on most vans. ■

(right) *The sliding door on the Personal Safety Vehicle was wide enough to permit ingress and egress to the front and back seats at the same time.*

Gil Spear

Walt Jennfeldt

(left) *The shape of the roof, with the reverse backlight and the finned rear fenders, was ideal for sliding doors.*

(below) *Several three-eighths sized clay models were made of Ford's proposed Personal Safety Vehicle. They incorporated the same safety features but were of slightly different exterior designs.*

(below) *George Golubics is at the wheel of this full-sized seating buck made for the Personal Safety Vehicle. Doug Miller is in the passenger seat, and Barney Harb is seated behind Golubics. One of the alternate designs had a rearward-facing back seat as in this seating buck. The next step, absent budgetary constraints, would have been to build the real thing.*

Walt Jennfeldt

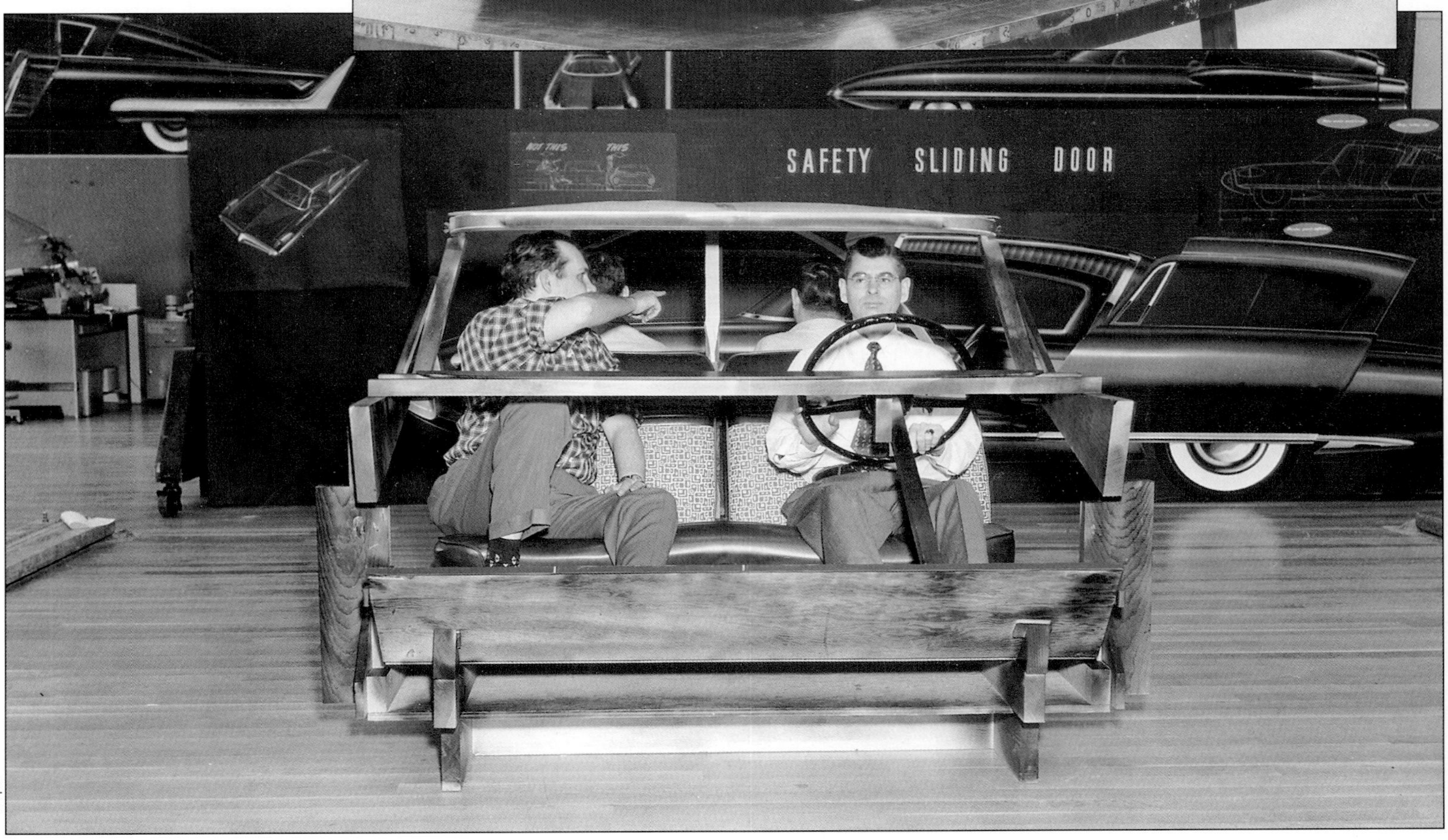

Gil Spear

10

The Rise and Fall of George Walker

XM-800 (Javelin)

XM-Turnpike Cruiser

Blue Job

L'Avion

LaGalaxie

Edsel 196X

Quicksilver

Levacar

Gyron

Envoy

Lincoln Marlin

1961 Continental

Continental Mark IX

X sixty five

Mercury Palomar

Unitron

Firefly/Astrion

George Walker was born in Chicago. He attended school in California and the Cleveland Art Institute, where he studied illustration. In 1923, he began his design career illustrating magazine ads for women's clothing. In his fashion illustrations, Walker customarily included a car of his own design in the background. One of his illustrations came to the attention of the president of the Peerless Motor Car Co., who hired him to prepare Peerless advertising illustrations. Walker later joined Baker-Raugh-Lang designing car bodies. He also played semi-professional football in his spare time. In 1929, Walker moved to Detroit, where he opened his own industrial design firm. By the 1940s, he was a respected designer of hardware for many automobile suppliers and had done automobile design work for Pierce-Arrow, Packard, Willys, Graham-Paige, Nash and International Harvester.

Walker's ties with Ford began in July 1946, when Ford's new chief engineer, Harold Youngren and Ernest Breech, Ford's new executive vice president, decided that Bob Gregorie's design for the 1949 Ford was too big, and that it would likely be unprofitable. Breech asked his friend, independent designer George Walker, for his professional opinion. Walker concurred. Breech then made him a consultant at Ford, and proposed a design contest between Gregorie and Walker's firm for a new, smaller '49 Ford. Walker's proposal won, and Gregorie resigned.

After Walker and his employees completed design work on the 1949 Ford, their contract was terminated because Breech decided he wanted to try to build up Ford's own design department. Less than a year later, Walker and his employees, Joe Oros and Elwood Engel, were asked to return, this time as long-term design consultants. Walker was given the right: (1) to control all publicity concerning Ford design, (2) to retain his own industrial design firm and his ownership interest in Trim Trends, a business that sold automobile trim parts to Ford and other manufacturers, and (3) to design competing proposals, if Oros or Engel did not like what was being proposed by the Ford designers. Although he did not spend a lot of time at the Design Department as a consultant, Walker was the guiding force behind the use of Plexiglas roofs on Ford Motor Co. concept, prototype and production cars built in the early 1950s.

Walker was personable and knew everyone in the automobile business. By the time he was appointed vice president of design and general manager of the Styling Center, on May 22, 1955, Walker had designed 3000 products and was doing nearly a million dollars of business annually. After he became head of styling at Ford, Walker spent long hours at the Styling Center and was actively involved in the selling of car designs to Ford management. He was also responsible for raising the status and salary of Ford designers, he insisted that they regularly attend auto shows and arranged for them to see and drive the best competitive and exotic cars available. On the otherhand, Walker instigated conflict and competition between designers and modelers. Although they admired his skills and his ability to get things done, many Ford designers and modelers believe Walker was too political, too much of an opportunist, and that he played favorites -- with designers and designs.

Ford Photomedia

George W. Walker

Walker was required to retire when he reached Ford's mandatory retirement age in 1961. During the time he was connected with Ford design, he was a showman who recognized the power of publicity, and the need for designers to have creative alternatives to the usual production work. He encouraged modernistic and sometimes fanciful concept cars. He deserves much of the credit for Ford's increased market share during the 1950s and early 1960s. After he left Ford, Walker moved to Florida and was elected mayor of Delray Beach. ■

Benson Ford proposed building the Mercury Monterey XM-800 as a second Mercury car line beginning with the 1956 model year.

XM-800 (Javelin)

By the early 1950s, Ford Motor Co. discovered it had a problem. People who bought Plymouths moved up to DeSotos. Chevrolet owners moved up to Buicks, Oldsmobiles or Pontiacs. People who bought Fords — they moved up to DeSotos, Buicks, Oldsmobiles, or Pontiacs, but very few of them bought Mercurys. In June 1952, the Davis Committee made recommendations to Henry Ford II as to how Ford Motor Co. could better compete with General Motors. Those recommendations included producing at least two more cars, a mid-sized car to compete with Buick, Oldsmobile and Pontiac and a top of the line luxury car. The top of the line car became the Continental Mark II. The mid-sized car was the Mercury Monterey, which was to share a platform with the production Mercury and sell for more than the Mercury but less than Lincoln.

The whole idea of the Monterey was to capitalize on the interchangeability of parts, like GM was doing, and to market an upwardly-mobile car Ford owners would buy rather than the non-Ford products they had been moving up to.

The car that became the XM-800 was designed in 1952 as a showcar, but it did not start out to be the car proposed for production as the Mercury Monterey. The name was changed to the Monterey only after the full-sized clay model was finished, and when it was decided to produce the car as the Mercury Monterey. It was designed in the Mercury pre-production studio by John Najjar, manager of the studio, and Elwood Engel, a Walker employee and design consultant.

At the same time Engel and Najjar started on the design of the Javelin show car, as the XM-800 was first named, the Mercury production studio was also asked to prepare a competitive design. Accordingly, Mercury staff designer Bob Jones was assigned by studio head Don DeLaRossa to prepare their competitive proposal for a new concept car. Jones' proposal was approved, and he was authorized to prepare a full-sized rendering of the car, but he was unhappy with the roof design on his proposal, because he felt it was not very fresh. After preparing several perspective drawings with revised roof lines, Jones came up with one he really liked. He made a small rendering of his new roof design on black paper and stapled it to the left side of the board on which he had made his original full-sized rendering. He left that night intending to next prepare a full-sized roof overlay and place it on the rendering. The following day, Jones came into the studio only to find the rendering of his new roof was gone. Later that day, Walker asked both DeLaRossa and Jones to stop work on their design for the new Mercury showcar. Walker told DeLaRossa and Jones that he wanted to get Engel as much visibility as possible, because he wanted him to eventually succeed him at Ford, and one way of getting Engel exposure was for his proposal to become the new Mercury showcar. Neither DeLaRossa nor Jones thought it worth a fight, so they acquiesced. Several days later, Jones found the door between the two studios open, so he looked in on Engel and Najjar's proposed showcar. Jones very diplomatically says that he was pleased to see that it incorporated his new roof design.

After Engel and Najjar selected various features from renderings made by studio designers, a 3/8-sized clay model, and then a full-sized clay model were made. The decision to replace the roof on the Engel-Najjar proposal was made after their full-sized clay model was under way. Jones' roof design was mated to the full-sized clay model under the supervision of Joe Achor, the design engineer assigned to the project.

When the production Monterey was authorized, Benson Ford wanted it to include a retractable model, and he asked Gil Spear to incorporate his retractable concept onto the XM-800. Spear prepared a layout drawing of the XM-800 with a retractable roof. Unfortunately, he determined that the windshield had to be moved seven inches forward and the backlight (Jones' new roof design) had to be radically redesigned to accommodate the retractable mechanism. In addition to layout drawings, Spear also prepared at least one rendering of a proposed XM-800 retractable.

There was another reason why the XM-800 was not built as a retractable. It was because Bill Ford wanted exclusive rights to a retractable model for his Continental Mark II.

The final XM-800 showcar, without a retractable top, was built at Creative Industries from fiberglass applied over a full-sized plaster plug, which was made from the full-sized clay model. Even the bumpers and most of the chrome trim on the exterior of the car were chrome-plated fiberglass. The curved windshield and the backlight were Plexiglas. The XM-800 had a 312 CID Mercury engine and a Merc-O-Matic transmission. The 312 engine would not appear on production Ford and Mercurys until 1956, the year the Monterey was slated to go into production. Although the engine looked complete, it was not operable. It was painted with a thick rubberized paint that covered up all the casting imperfections. The engine had all the correct internal parts in it, but the Merc-O-Matic transmission case had nothing inside. Like with most concept cars, shortcuts were taken to cut costs and to save time.

The XM-800 was built on a modified Mercury frame, had a 119-inch wheelbase, one inch longer than the standard 1954 Mercury, and was 55.6 inches high, which was over ten inches lower than the standard '54 Mercury. At the time, it was the lowest two-door hardtop Ford Motor Co. had ever built. It was painted a pearlescent white, with a pearlescent copper-colored roof. The interior was two-toned in matching colors. The front-end

(top) The full-sized clay model of the XM-800.

(above) Benson Ford asked Gil Spear to design a retractable top for the XM-800. As this rendering indicates, building the XM-800 as a retractable would have required major changes in its design.

(below) Gil Spear's drawing of the XM-800 as a retractable.

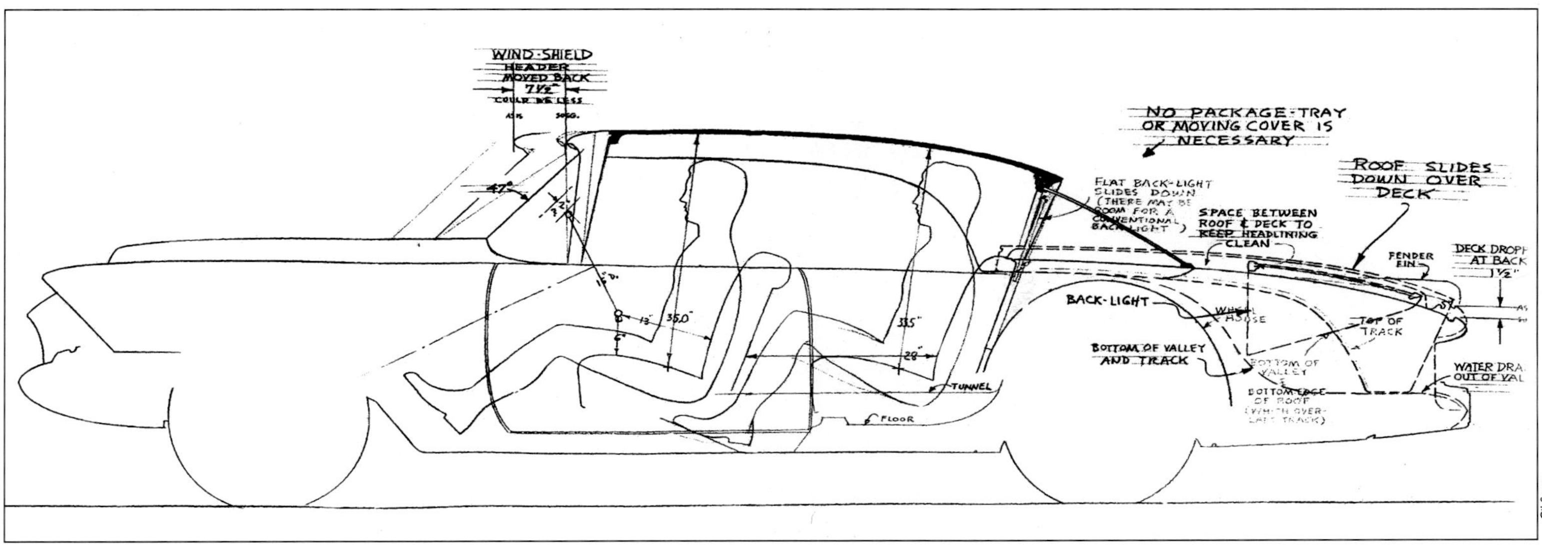

styling of the car included an integrated bumper and a concave grille that press releases said cooled the engine better. The XM-800 was also the first Ford Motor Co. car to use fins as a design feature, although they were fat fins.

The instrument panel was meant to look like it was fully functional, but none of the gauges or instruments worked. Press releases said there were individual gauges for water temperature, oil pressure, battery and fuel. Each gauge was supposed to have a red warning light built in. Instruments included a speedometer, tachometer, a clock with a sweep-second hand and a radio in the middle of the dash. Other innovative features on the XM-800 included inside door pulls and door handles combined in a single built-in unit and something called a fluted-type steering column. The front and back seats were individual buckets with a console between them running front to back. (It would be six years before an American production car would be built with bucket seats.) In the front and the back-seat areas, the console had control boxes containing the power-window and power-seat switches. The front seats also fully reclined. As on the X-100, the sunvisors were tinted Plexiglas. The car was advertised as having power steering, brakes and windows, but it was really a pushmobile.

The Thunderbird and the Mercury Monterey XM-800 were introduced at the revived Detroit Auto Show on February 24, 1954. On March 13, 1954, the radical FX-Atmos car of the future was introduced at the Chicago Auto Show, where it was shown with the Thunderbird and the XM-800. Several weeks later, all three concept cars were at the New York Auto Show. For the show circuit, a big timer and an electrical cord were installed in the trunk of the XM-800. When plugged in, the timer could be set to open and close the windows, doors, hood and trunk on command. There was good press coverage for all three of Ford's 1954 concept cars, although the Thunderbird was clearly the star.

At the time the Thunderbird was introduced, it was announced that it would be produced the following year. The FX-Atmos was not proposed for future production, and the press releases said so. The press releases on the Mercury Monterey XM-800, however, said it would be

(top) The full-sized clay model of the XM-800 between the rails of a styling bridge, after the bridge itself had been removed.

(bottom) Subtle changes were made to the XM-800 after the full-sized clay model was completed.

John Najjar

John Najjar

From left to right are clay modelers John Morris and Marty Hrbal working on the XM-800.

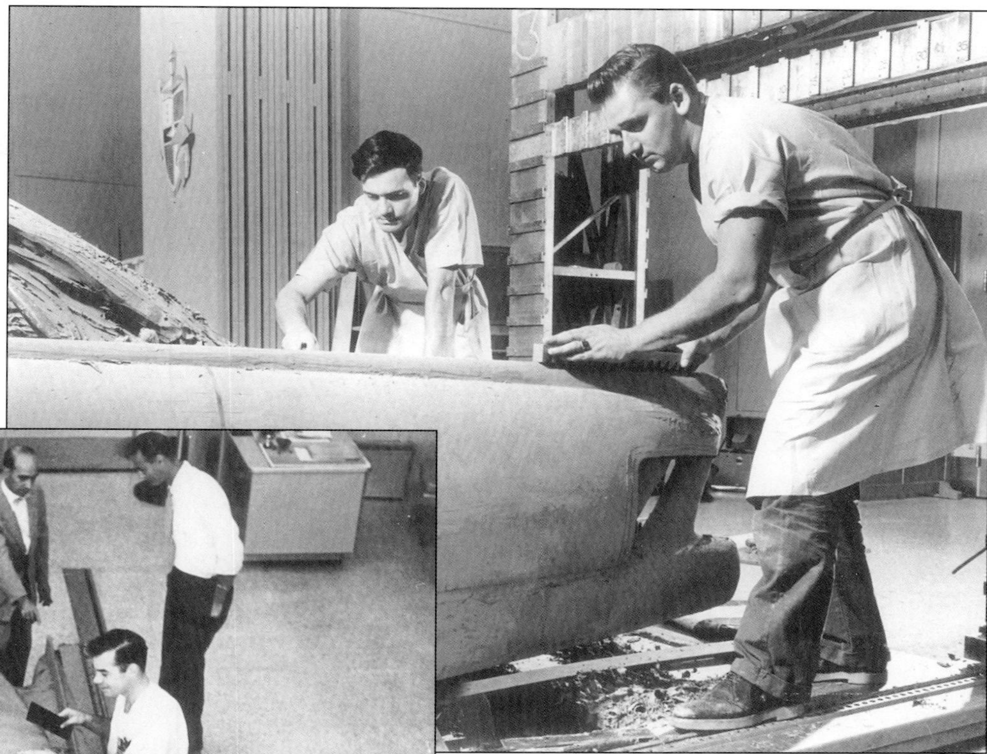

John Najjar

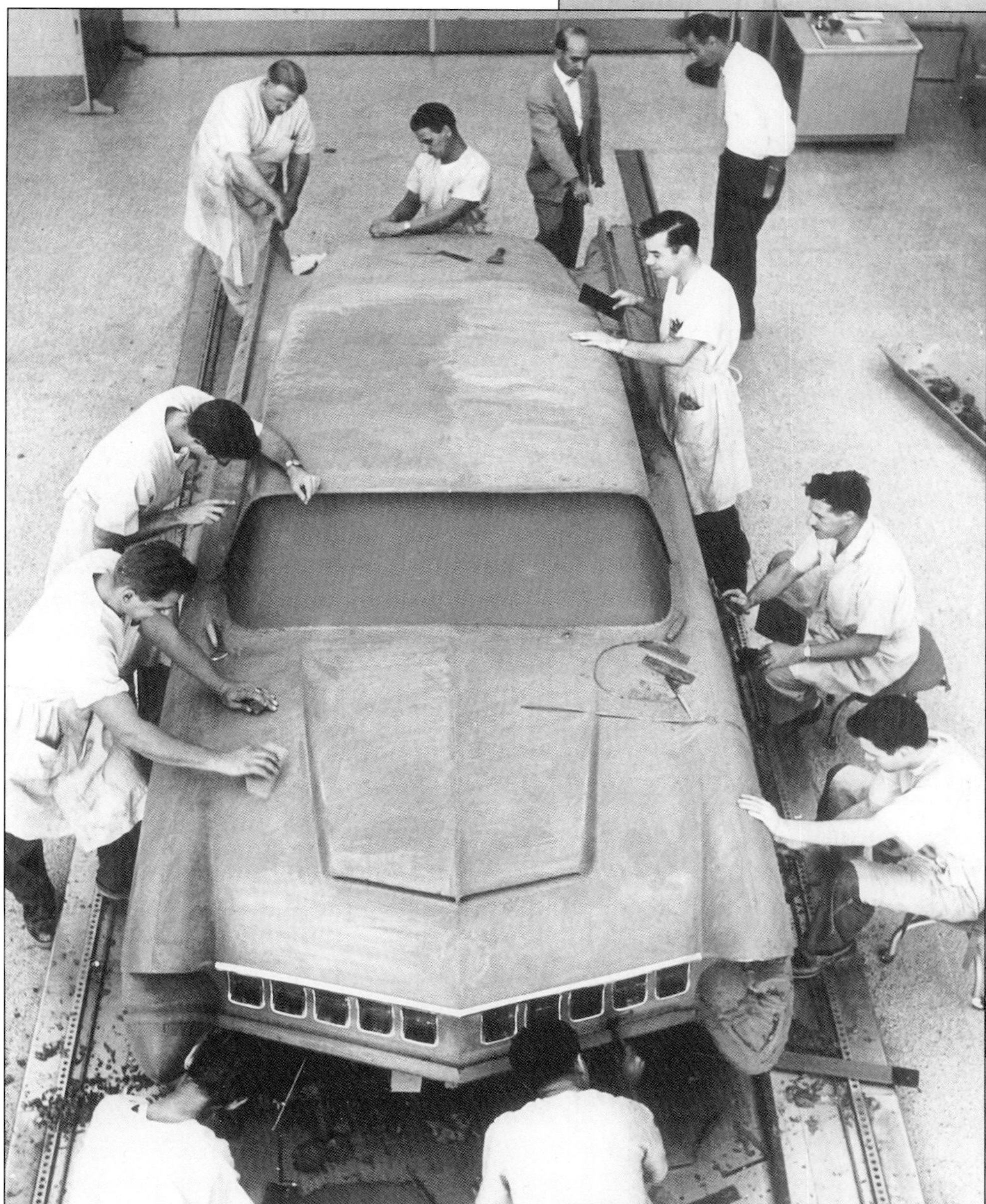

John Najjar

(left) The full-sized clay model of the XM-800 was nearing completion when this photograph was taken. John Najjar is in the suit at the back of the car. Design engineer Lloyd Kamm is to Najjar's left. Clay modelers in the front of the car are Delmar Dustman, and to his right is Joe Apap. From left to right, around the car, are clay modelers Marty Hrbal, Gil Treweek, Mike Nowicki, Paul Baud, Bill Neu, Romeo D'Angelo and John Morris.

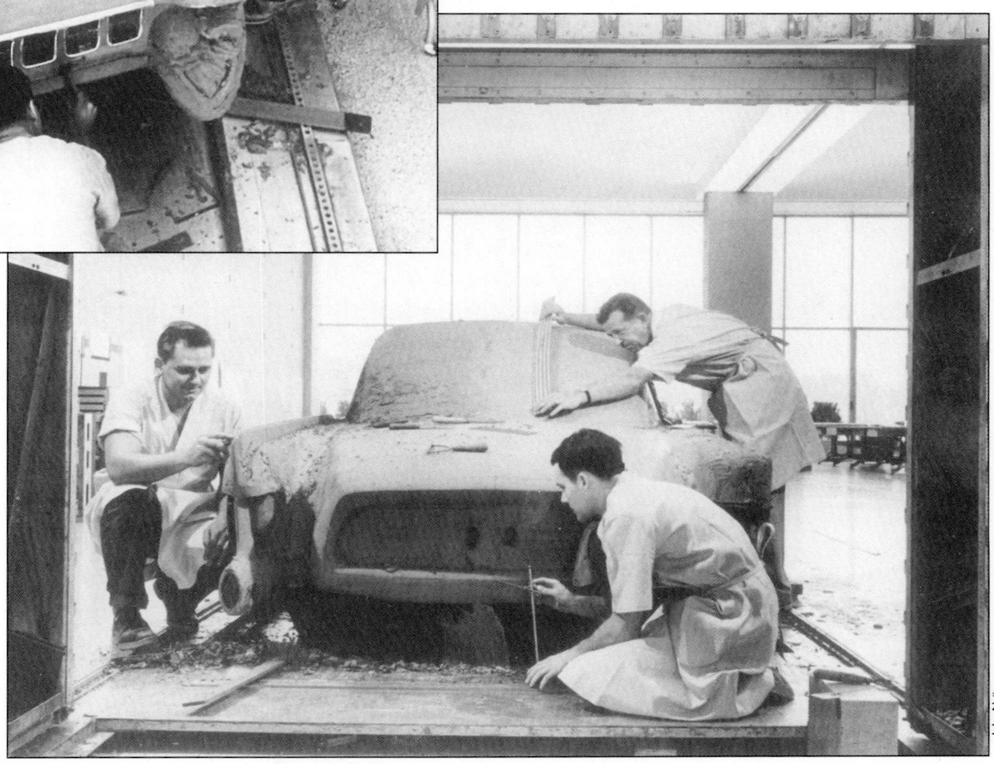

John Najjar

(right) From left to right, clay modelers Marty Hrbal, Al Kellum and John Morris are working on the XM-800.

Ford Design Center

Elwood P. Engel

Elwood Engel was born in New Jersey and raised in New York. His father was a silversmith, who died when Engel was seven. Engel studied industrial design at Pratt Institute, graduating in 1939. He designed locomotives, women's wear and several 1939 New York Worlds Fair displays before starting his automotive design career at GM in 1939. At GM, he met Joe Oros who became his life-long friend. After he finished his schooling, he worked in GM's advanced studio under Jules Agromonti and Tom Hibbard and, later, in the Pontiac studio under Frank Hershey.

At the beginning of World War II, Engel volunteered for the horse cavalry. When it was disbanded, he joined the tank corps and became a cartographer. Near the end of the war, Oros wrote Engel and suggested he apply for a job at George Walker's industrial design studio, where Oros was working. Engel was hired, and Oros and Engel became two of Walker's top designers. In 1946, Walker was asked to design an all-new 1949 Ford in competition with Bob Gregorie. Walker's proposed '49 Ford, which was completed by Oros and Engel, won the contest but, once that job was done, Walker, Oros and Engel left Ford.

Less than a year later, Ernest Breech decided the talent of Walker's firm, including Oros and Engel, was still needed at Ford. With Walker's return as a long-term design consultant, Oros and Engel, as Walker employees, were assigned to the Ford Design Department, where Engel oversaw the design of Lincolns and Mercurys. When Walker became Ford's vice president of design in May 1955, Engel became a Ford employee and Walker's assistant manager. In 1958, he also became head of the corporate advanced studio.

In addition to many production vehicles, Engel worked on the design of the XM-800 and the XM Turnpike Cruiser. In the corporate advanced studio, Engel and his designers designed the 1961 Thunderbird/Lincoln, the Mark IX, the Gyron, the X sixty five, the Palomar, and the Unitron.

When Walker retired in 1961, he wanted Engel or Oros to succeed him, but Gene Bordinat was appointed vice president of design instead. In November 1961, Engel, who had been contacted by an executive search company, was offered the job of vice president of design at Chrysler. He held that position until he retired in 1984. ■

produced "if there was enough public demand." The TV script prepared to go with a one-minute film clip introducing the Mercury Monterey XM-800 showcar said that "A step toward easing unemployment in Detroit's huge automotive industry will depend upon public demand for an entirely new designed car unveiled today by Benson Ford." Other Lincoln-Mercury Division press releases claimed that engineering on the XM-800 had been completed, and "with completion of normal tooling it could be put into production." Although the XM-800 showcar was made out of fiberglass, Lincoln-Mercury Division said production models would have steel bodies. Because the new Monterey was to be offered in several different models (e.g. convertible, four-door, etc.) the Design Department also built two full-sized clays of different models of the car and six 3/8-sized clay models, representing the full range of Monterey models to be offered, but they were never publicly displayed.

To build interest, Ford Motor Co. made a deal with Post Grape Nut Flakes. Kids and their parents could order small plastic models of the Thunderbird, 1954 Mercurys, 1954 Fords and the XM-800. For 25 cents and a boxtop they could send for either set A or set B, each of which consisted of five cars. The XM-800 was in set A, and the Thunderbird was in set B. Ford Motor Co. saw this as a good way to advertise its current and future production cars. Unfortunately, the offer expired December 31, 1956.

Before the Mercury Monterey could be put into production, one of the original "Whiz Kids," Francis C. "Jack" Reith, returned from Europe. He had turned around and then unloaded Ford's French subsidiary to Simca at a handsome profit and returned a hero. On February 7, 1955, Reith made a surprise proposal to the Product Planning Committee. He proposed scrapping the Monterey planned for production that fall and, instead, proposed to make the Mercury into a distinct mid-sized car produced by a separate Mercury Division. Then to do the Davis Committee recommendation one better, he proposed another new division producing two separate models of what became the Edsel. Reith proposed that the $100 million authorized for development of the Monterey be increased to $250 million! At that same meeting, Reith displayed one of the models of the Monterey, and a model of his XM-Turnpike Cruiser as the all-new Mercury he proposed to market in 1957. Reith's argument was that Ford Motor Co. could better compete with GM's mid-sized cars with an all-new Mercury like the XM-Turnpike Cruiser. Over Benson Ford's objection and with Henry Ford II's support, Reith's proposal was accepted.

The Monterey was out, and the XM-Turnpike Cruiser was in. A couple of months later, the second part of Reith's plan, the Edsel, was also approved. Less than a week after that, a new Special Products Division was organized to build the Edsel. At the same time the old Special Products Division was renamed the Continental Division. Lincoln-Mercury Division was split, with Reith as general manager of Mercury Division, and Ben Mills as general manager of Lincoln Division. Benson Ford, the head of Lincoln-Mercury Division for seven years, was put in charge of dealer relations -- a dead end job.

As part of the fallout from cancellation of the XM-800, some executives in the Engineering Department chose the occasion to rebuke Najjar because they thought he had shown a lack of loyalty by working so closely with Walker and Engel. Although he wasn't particularly surprised, his answer was that he thought he was working in Ford's best interest for a common goal —- designing good-looking cars.

The decision not to build the Monterey was also the death blow to the XM-800 showcar. It was soon removed from the show circuit and, in late 1955 or early 1956, padding was added to the dash, and the radio knobs were made flatter apparently in an effort to turn the XM-800 into a safety car. In addition, the car's flat steering wheel was replaced by a safer deep-dish one.

In December 1956, because there was no further use for it, the XM-800 was given to the University of Michigan as part of a package that included Henry Ford's home, Fairlane. For about five or six years, the XM-800 was used as a teaching aid in the University's automotive design courses to provide "incentive for future automobile engineers." As time passed, styling trends changed and, in the early 1960s, the XM-800 was sold at a surplus property disposal sale to a now unknown buyer, who promptly took it to a friend's farm in Manchester, Michigan, where he paid for a year's storage in the farmer's barn. The new owner was never heard from again. A year or two later, the farmer pulled the XM-800 outside and parked it in the open, right up against the side of the barn, but it could not be easily seen by people driving by.

In the early 1980's, 17-year old Dan Brooks and his friend, both budding car collectors, often spent weekends prowling Michigan's back roads looking for old cars. One weekend, while out prowling the countryside on his own, Dan's friend caught a glimpse

John Najjar

(above) This full-sized clay model of the XM-800 was called the Javelin. Studio manager John Najjar was not aware, until years later, that the roof design used on the XM-800 was developed by Bob Jones in another studio.

Ford Photomedia

(below) The interior of the XM-800 was completely finished, although only the lights worked.

(below) The inside door handles on the XM-800 were built into the armrests.

Ford Photomedia

Bob Jones

Robert B. Jones

After high school, Bob Jones started his design career as a design and color engineer for a North Carolina textile mill, where he worked from1937-45. He also attended Davenport College and later Meinzinger's Art Academy and the University of Michigan, where he studied design, color theory and sculpture.

After a year as a designer at Briggs Manufacturing, Jones started at the Ford Design Department in April 1947. His first assignment was in the truck and international studio. In 1948, he was transferred to the Ford studio, where he worked on the redesign of the 1950 and '51 Fords. In 1949, Jones left Ford to accept a position as a senior designer at Packard, where he designed the vertical grille for the Packard Request and Predictor.

In mid-1952, Jones was rehired by Ford as a designer in the Mercury studio. Although Jones worked mostly on production Mercurys, his unique roof design was incorporated into the design of the XM-800.

In 1954, Jones was assigned to Roy Brown's Edsel studio, where he was assigned to work on the car's exterior design. While in the Edsel studio, Jones helped design a proposed small Edsel that eventually became the Comet. When the Edsel program was canceled, Jones returned to the Mercury studio.

In 1962, Jones was made a design manager in the truck studio and placed in charge of exterior design. In 1968, he became a manager in the Mustang, Fairlane, Falcon and Maverick interior studio. Jones was later transferred to the truck and international interior studio as a manager. During his last ten years at Ford, he was manager of the trim and color studio. ■

Ford Design Center

Dan Brooks

John Najjar

In high school, John Najjar developed an interest in automobile design. He studied drawing and, in his senior year, a Ford recruiter visited one of his classes and urged Najjar to apply for a design job at Ford. When he started at Ford, Najjar was placed in Ford's Apprentice School learning how to operate milling machines and lathes as a prelude to a job in the Rouge plant. While working on a lathe, Najjar felt a tap on his shoulder. When he turned around, Henry Ford asked him if he was happy doing what he was doing. Najjar said "no," and that he would rather be drawing cars. Mr. Ford asked Najjar to bring some of his drawings to work the next day. The next day, when Mr. Ford returned, he looked at Najjar's drawings, and asked if he could keep them for awhile. Najjar said he could, and Mr. Ford left with them. Later that day, someone came to tell Najjar to report to the Engineering Department as a drafting trainee. In September 1937, Najjar started a 42-year design career at the Design Department.

While at Ford, Najjar developed organizational skills that were valued by those in charge of the different studios he worked in. Najjar was an excellent administrator, and worked well with almost everyone. He also participated in the design of many concept, show and prototype cars, including the original Continental, the 05L, the 1941 Plastic Car, the X-100, the Futura, the XM-800, the XM-Turnpike Cruiser, the Levacar, the Diplomat, the Gyron, the 1961 Continental, the Mark IX, the Mustang I and the Mustang II.

From 1955 to late1957, as head of the Lincoln studio, Najjar was responsible for the design of the 1957 and 1958 Lincolns. In 1957, he was transferred to the truck studio as executive stylist. Several months later, he was reassigned to Elwood Engel's corporate advanced studio as executive stylist. In 1963, Najjar was appointed head of the international and truck studio. In 1969, he was made the head of Ford's Industrial Design Office, where he coordinated the design of the Renaissance Center, the Fairlane Housing Project, the Fairlane Shopping Center, the Fairlane Town Center and Mall and various Ford trade-show exhibits. As head of the Industrial Design Studio, Najjar was responsible for the design of Ford's Guest Center, the People Mover and the interior of Ford's Boeing 727. In 1974, Najjar returned to the truck studio. He retired in 1979. ■

of the taillight of the XM-800 as he drove by the farmer's barn. When he investigated, he was told by the farmer that the car was about to go to the dump, which was right next door. The farmer told Dan's friend that if he hauled the car away he could have it for free. Dan's friend then contacted Dan and asked him what an XM-800 was. When they figured out that the XM-800 was a one-off showcar, they both headed back out to haul it home before the farmer changed his mind. Later, Dan's friend needed money to pay for new upholstery for his Mustang and sold his interest in the XM-800 to Dan for $100.

When Dan Brooks bought the XM-800, it was in pretty sad shape. The roof had a droop in it, the windshield pillars had begun to collapse, and everything else was badly deteriorated. Dan partially disassembled the car. He also contacted John Najjar, who was able to reproduce the design of the original instrument panel, so the XM-800 could be restored to its original condition. After several years, Dan realized that restoration of the XM-800 was going to be very expensive, and he decided to sell it. The estimate he got just for rechroming the fiberglass bumpers and plated parts was $10,000.

He eventually advertised the car for sale in *Hemmings Motor News* but, before the offers started coming in, he changed his mind. One of the callers responding to the ad was Chicago nightclub owner and collector of concept cars, Joe Bortz. Bortz was very nice but also very persistent. He called Dan every six months for the next six years until, in 1986, he was able to buy the car. Dan says the price Bortz paid, reportedly $20,000, was fair. Bortz has not yet completed restoration of the XM-800, but says it will be done right.

Because future Mercurys went in the styling direction set by the XM-Turnpike Cruiser, not many features of the XM-800 can be found on later Mercurys. The XM-800 was predictive, however, of many of the styling trends seen on future Lincolns. The taillights are predictive of the 1955 Lincoln; the fins showed up on the 1957 Lincoln; the hooded headlights appeared on the 1956 and '57 Lincolns; the band over the top and the back-quarter window showed up on the 1956 Lincoln Landau; the front wheel-well sculpturing showed up on the '56 Lincoln, and the dog leg on the A-pillar of the XM-800 was copied on many Ford products. ■

Dan Brooks

(top) The XM-800 in the condition it was when found in the early 1980s. It was just about to be hauled off to the junk yard.

(bottom) The XM-800 sat outside for many years. The trim panel on the rear deck rusted. Most of the car, including the bumpers, was fiberglass. Although the roof sags from years of neglect, the car is restorable.

XM-Turnpike Cruiser

The idea for the XM-Turnpike Cruiser came from John Najjar, who made the proposal to build a new showcar in April 1954. Najjar's proposal included five full-sized renderings of alternate concept cars and a booklet describing the features proposed for each car. The overall dimensions of the alternative showcars Najjar proposed were: a 119-inch wheelbase, 212-inch overall length and a height of 49 to 51 inches. Najjar's five proposals included the following:

1) Malibu: A four-passenger car with twin bombs high on the rear-quarter panels, which were the C-pillar supports for a shelf-type roof. The Malibu was proposed with an offset front, a wider rear tread and a wild two-tone paint scheme.
2) Monte Carlo: A four-passenger, two-door car with sliding doors and fluorescent lights running the length of the sides of the car.
3) Mystair: A four-passenger car with a floating roof. The roof was designed with a tubular section that utilized water evaporation for air conditioning and ventilation. Sports car-type flares over the front wheels served as exhaust ports.

Ford Photomedia

The XM-Turnpike Cruiser was built in metal by Ghia in Italy at a cost of $80,000. It was a completely-operable car that, with some design modifications, became the 1957 Mercury.

4) Mandalay: A four-passenger cross-country turnpike cruiser with exhaust ports exiting behind the front wheels. An air-conditioning saddle covered the upper area of the body, and tailfins held aircraft-type taillights. Chromed rear tubes on each side of the car carried pint-sized Jato (jet assist take-off) bottles for extra power in case of emergency.
5) Montauk: A four-passenger convertible with clear or chromed mylar top material. The Montauk had the distinctive V trim on the side that appeared on the sides of the 1955 Mystere and on the production 1957 Ford Fairlane. The Montauk also had fresh-air intakes just behind the front-wheel openings.

At the time Najjar's proposals were made, Elwood Engel was a George Walker employee assigned to the Lincoln-Mercury studio. Gene Bordinat had recently been appointed the semi-autonomous manager of the Lincoln-Mercury studios. Under him were Bill Schmidt, chief designer in the Lincoln Studio and Don DeLaRossa, chief designer in the Mercury Studio. Najjar was in charge of the Lincoln pre-production studio.

When Engel saw Najjar's proposals, he fell in love with the Jato bottles and the name Turnpike Cruiser associated with the Mandalay proposal. He changed the name of the car to the XM-Turnpike Cruiser, and its design eventually evolved into a much different car than any of those Najjar proposed.

As a result, in mid-1954 two teams of designers were assigned to design and build competing full-sized clay models of the new concept car. Gene Bordinat and Don DeLaRossa started on one full-sized clay, and Elwood Engel and John Najjar started on another. As with most of the concept cars designed at the Design Department in the 1950's, there were two design teams working in competition with each other. The proposal Engel and Najjar were building was placed in a stall next to the Futura and, for awhile, both were under way at the same time.

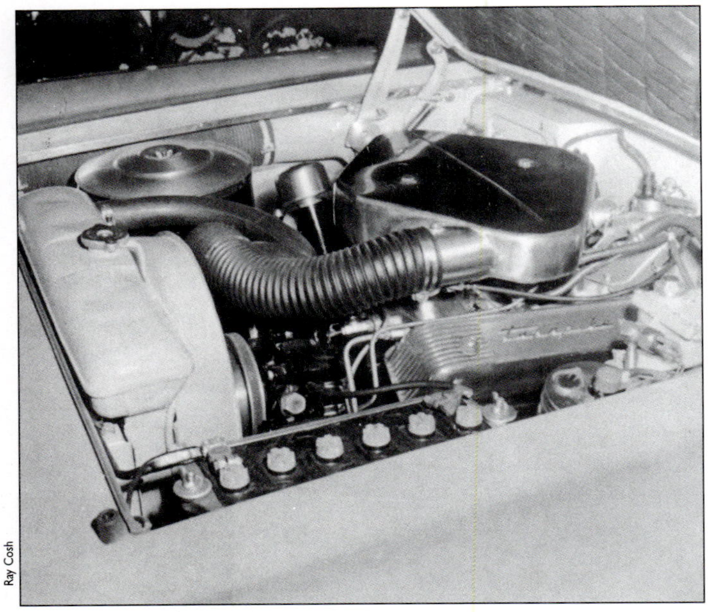

Ray Cosh

The 312 CID engine in the Mercury XM-Turnpike Cruiser had two four-barrel carburetors and Thunderbird dress-up kit valve covers that were modified by placing Turnpike script on them.

The XM-Turnpike Cruiser project began in earnest with several studio designers filling two blackboards with small sketches of themes and ideas for Engel and Najjar to select from. As they selected the best of the ideas, the clay modelers began work on a 3/8-sized clay model. Walker would often come by and make suggestions.

Engel and Najjar worked well together, and Najjar came to look upon Engel as his mentor. Najjar's personality leaned more towards the serious, while Engel was the nervous, high-strung practical joker, who liked to pitch pennies with the younger designers in the studio. Engel was not much of an administrator, but Najjar was. Especially during the long hours they worked on the Turnpike Cruiser, they became even closer friends.

In the fall of 1954, Francis C. "Jack" Reith returned from an assignment in Europe, where he had turned around and then sold Ford's French subsidiary to Simca at a handsome profit. Reith was the only one of the original "Whiz Kids" who was a true car guy and, after he returned from Europe, he started dropping by the Design Department after his duties at the office were finished. Soon, he was helping design the XM-Turnpike Cruiser with Engel and Najjar, even though he had no formal training as a designer. Since those were days of lots of overtime, Engel and Najjar would often walk down the street to the Dearborn Inn for dinner, and then return to the studio to meet Reith and work late into the night. Reith would take off his jacket, pull up a stool and trade ideas with Engel and Najjar about the clay model in front of them. As Engel was explaining the features being sculpted into the clay, Reith would suggest changes or features he thought would make the design better, and Engel and Najjar incorporated many of those ideas into the clay. After Reith would leave each evening, Engel would outline the work to be done as a result of his suggestions and before his next visit. As Najjar remembers it, Reith was enthusiastic about the design of the car and thought the name Turnpike Cruiser was a natural.

For his first assignment as a new designer at Ford, Larry Shinoda was assigned to work on the design of the XM-Turnpike Cruiser. Although most of the design of the car is attributed to Engel and Najjar, Shinoda designed what he calls the "bookends" on the rear fenders. Shinoda's inspiration was simple: if two fins were enough on other cars, four fins on the XM-Turnpike Cruiser should be better. Shinoda later suggested taking the jet pods from the front of the Mystere and putting them at the upper-outside edges of the front windshield on the production Turnpike Cruiser.

When the competing models were finished, the Engel/Najjar and the Bordinat/DeLaRossa full-sized clay models were shown in competition. The Engel/Najjar clay model won. By that time, Reith had also come to believe that the XM-Turnpike Cruiser would make a better production Mercury than the Mercury Monterey XM-800, which had already been approved for production as a 1956 model.

In late 1954, approval was given to build the XM-Turnpike Cruiser as a fully-operable showcar, and it was to be built by Ghia in Turin, Italy. Ghia worked in metal and could produce fully-operable vehicles. A 3/8-sized plaster model cast from a 3/8-sized clay

The Mercury XM-Turnpike Cruiser toured the country with its own tractor-trailer. The trailer had built-in stairs, and it swiveled and expanded, so that visitors could walk up and around the car.

Ford Photomedia

model, and detailed full-sized layouts drawings were made as well as full-sized drawings from the full-sized body surface. Designers Ken Spencer, Jim Arnold and clay modeler Norm Nebozenko built the 3/8-sized clay of the XM-Turnpike Cruiser, from which the plaster model was made to send to Ghia. They were sent with a chassis, engine, drivetrain and body hardware to Ghia. As with the Futura, Don DeLaRossa travelled to Italy at least once to make sure the car was built as the designers wanted it. When it was finished, Ford paid Ghia $80,000 to build it.

There were several subtle changes made by Ghia when building the full-sized car. Those changes included a different shaped windshield and the lack of vent windows on the finished car. As noted in a report prepared several years later by Styling Center Administrator V. Z. Brink, several problems arose at the time the XM-Turnpike Cruiser was built, and it was thought best that, when future cars were built at Ghia, two and possibly three visits be made to make sure what Ghia was building was being done the way the Styling Center wanted.

In February 1955, Reith proposed that a modified XM-Turnpike Cruiser be produced as the 1957 Mercury. Reith's proposal was to make the XM-Turnpike Cruiser a distinct mid-sized Mercury — and to forget the Monterey. Reith convinced Ford's executive committee that his proposal would make a bigger, quicker dent in the mid-priced car market. He first proposed to replace the Monterey with the XM-Turnpike Cruiser. Once he had won that point, he displaced Benson Ford as general manager of a new, separate Mercury Division.

(below) The roof panels on the XM-Turnpike Cruiser opened and closed automatically when the doors were opened and closed.

(bottom) The instrument panel in the XM-Turnpike Cruiser was dramatic, but much too expensive to be considered for the production 1957 Mercury.

Ford Photomedia

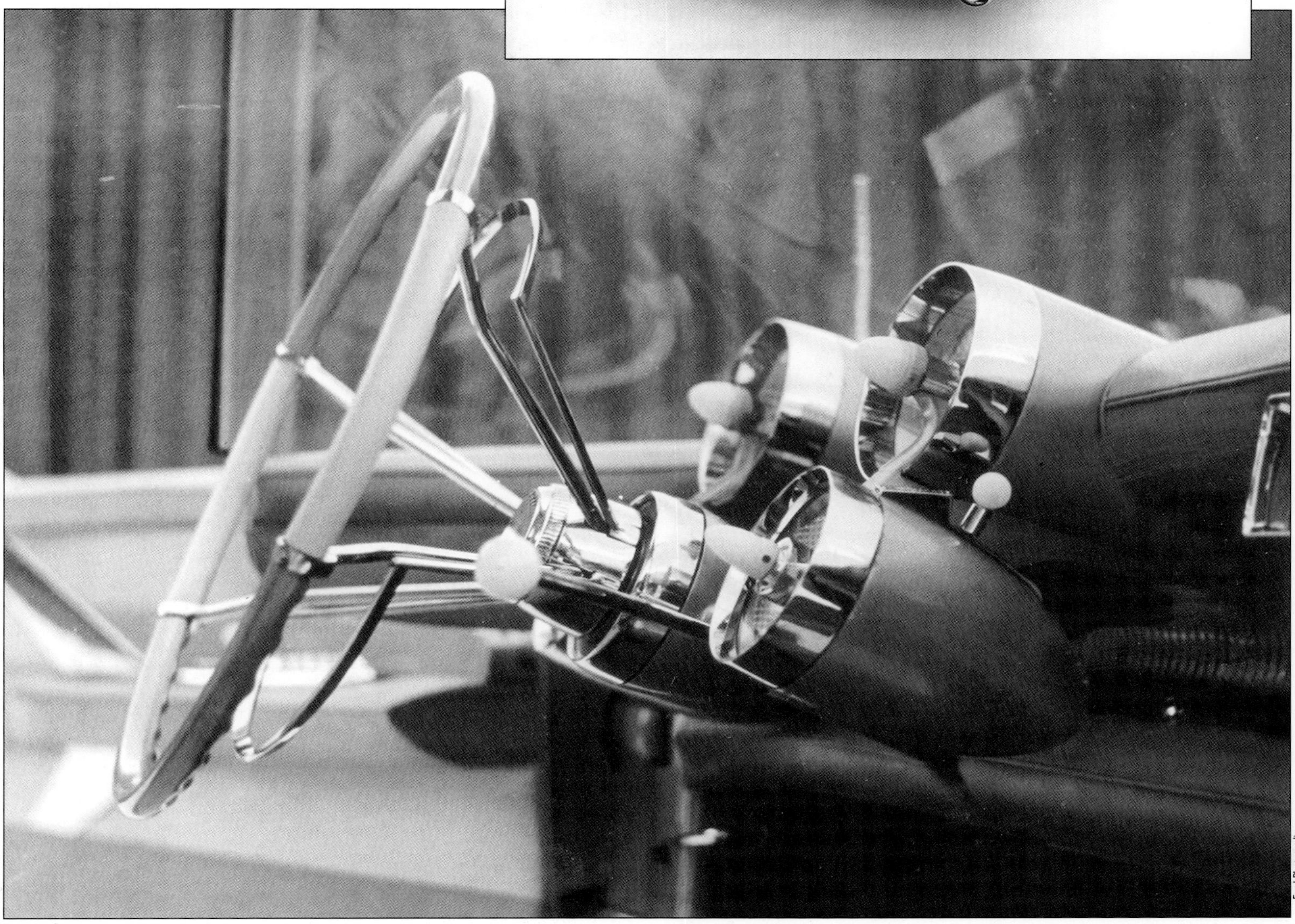

Ford Photomedia

While Ghia was building the XM-Turnpike Cruiser showcar, Reith as the new general manager of Mercury Division, gave the go-ahead to move the original full-sized clay model of the XM-Turnpike Cruiser into the Mercury studio, so it could be reworked into the production 1957 Mercury, and so that body series types, hardware and colors could be determined. After the car was moved into the production studio, control of it passed to Bordinat and DeLaRossa, although changes necessary to turn the XM-Turnpike Cruiser into the production '57 Mercury were approved by Engel and Walker. After the normal design and engineering process and approvals, the production '57 Mercury was introduced in September 1956.

Ray Cosh

The XM-Turnpike Cruiser had armrests in between the front and rear seats.

Meanwhile, in December 1955, Ghia delivered the fully-operable XM-Turnpike Cruiser to the Styling Center. It was 52.4 inches high, 18.4 feet long and 77.1 inches wide. After it was returned to the Styling Center, "Mercury," in block letters was placed on the front of the car below the hood, bigger block letters spelling out "Mercury" were placed just ahead of the taillights in the side channels, "XM-Turnpike Cruiser" script was placed on the sides of the front fenders, and a big "M" was installed as a hood ornament. Molded-stainless steel panels were added to the side channels and under the Mercury script before the car went on the show circuit. Photographs show that the side channels were painted a darker color before the stainless steel inserts were installed. After the XM-Turnpike Cruiser was returned from Ghia, blackwall tires were replaced by 15-inch Firestone Deluxe Champion whitewalls. The interior rearview mirror was also changed. The original rearview mirror was fastened through the windshield to a short spear that stuck straight out from the windshield parallel with the ground. Although it looked like a radio aerial, it was not. The radio aerial was located on top of the left front fender. The replacement rearview mirror was hung by an arm from the upper-interior windshield trim.

Press releases said the XM-Turnpike Cruiser was "a fully-operative automotive styling laboratory dedicated to the specific task of pioneering, testing, and perfecting new ideas in design, new features, new safety and new power application." When the doors were opened, T-top panels called butterfly top panels raised up, gull-wing style, to allow easier exit and entry. The top panels were made from Plexiglas with chrome surrounds and were hinged towards the middle of the roof. They operated automatically by electric cylinders with U-shaped pistons made from aluminum, which were hidden under the headliner in the center section of the roof. There was also an override feature that allowed the roof panels to open with the doors closed. The outside door handles were concealed at the top of the side channels by spring loaded flaps similar to those used on the 1968 Corvette.

Ghia built the XM-Turnpike Cruiser from this 3/8-sized plaster model sent to Italy along with a running chassis and other standard Mercury trim parts.

Ford Photomedia

At the front of the car, as part of the split bumpers, were twin jet pods that contained parking and turn-signal lights. The XM-Turnpike Cruiser was the first Ford product with delayed-action headlights that stayed on for thirty-to-forty seconds after being turned off.

The car was painted a pearlescent-persimmon color and had matching glove-soft leather upholstery. There were four individual bucket seats separated by a center console that extended from front to rear. There were raised center armrests between the seats that contained ashtrays and cigar lighters and an oversized light between the rear-seat backs. The rugs were persimmon-colored, and the radio speakers were in the headliner. The glovebox was even lined in persimmon-colored leather and was located in the instrument panel in front of the front-passenger seat and had a chrome plated, concave-shaped door. The shift lever and the steering column were similar to other mid-1950s Ford products, except that the outer-steering column had an XF part number on it. The shifting quadrant was standard Ford, but the plastic PRNDL plate was unlike any other Ford product. The shift and turn-signal levers had white balls on the ends as did the inside-door handles. The instruments were contained in four large chrome-edged cylinders clustered in front of the driver. They housed a speedometer, tachometer, fuel and temperature gauges and oil and battery gauges. The car had power steering, brakes and windows. The windows were standard Italian-style power windows operated by intricate cable systems hidden in the doors. The ignition switch and key were unlike any produced by Ford and were of Italian design.

Ford Photomedia

(above) Because the Mercury XM-Turnpike Cruiser was planned as a fully-operable car that could be driven on the street, it was designed with only two headlights. When it was built, quad headlights were not legal in all states.

(left) The XM-Turnpike Cruiser had full instrumentation, including a 150 mph speedometer and a tachometer.

Ford Photomedia

The ignition switch has a heavy chromed bezel located on the instrument panel to the right of the steering column. The hood release was a chromed lever located next to the right-hand kick panel, and the hood was hinged at the front.

The engine on the XM-Turnpike Cruiser was a 312 CID Mercury engine with a low-rise manifold, two four-barrel carburetors and a special air cleaner similar to what later appeared on stock Turnpike Cruisers and Lincolns with two four-barrel carburetors. The exhaust manifolds were black porcelain and exited the engine compartment to the rear through holes in the frame and into two large mufflers. The tailpipes went over the rear axle to exhaust-splitters similar to those on the 1964 Pontiac GTO, and then through chrome-plated brass ports on the rear-quarter panels. The automatic transmission was a standard Merc-O-Matic.

John Najjar

John Najjar and Elwood Engel worked lots of overtime designing the XM-800. When they worked late, they usually took a break for dinner. While at the Dearborn Inn for dinner one night, Najjar drew this caricature of Engel on his place mat.

Lawrence K. Shinoda

Larry Shinoda

Larry Shinoda grew up in Southern California and was a part of its hot rod car culture. He not only liked to draw cars, he also built and raced them. After being interned during World War II, Shinoda spent two years at Pasadena City College before entering the U.S. Air Force during the Korean War. In 1951, after he was discharged from the service, he enrolled at the Art Center College while continuing his racing career. After two years at the Art Center he left but, a year later, he was invited to interview with Gene Bordinat, who was at the Art Center recruiting for the Ford Design Department. Shinoda was hired and started at Ford in 1954.

During the year he was at Ford, he helped design the XM-Turnpike Cruiser and the production 1957 Mercury. In 1955, Shinoda left Ford to accept a job in Packard's design department. He was at Packard for three months but left to design Indianapolis race cars. In 1959, he was hired by GM's Styling Section. While at GM, he was assigned to the Chevrolet and Pontiac studios and was chief designer in their special vehicle department. In addition to several production cars, Shinoda worked on the designs of the 1963 Corvette, the Mako Shark, the Corvair Super Spyder and the Monza GT.

In 1968, GM vice president Semon E. "Bunkie" Knudsen was hired by Henry Ford II as president of Ford. Knudsen brought Shinoda with him to Ford. In 1970, Knudsen was let go by Henry Ford II, and Shinoda left with him. While at Ford the second time, Shinoda helped design the Mustang Boss 302, the King Cobra and the Talladega.

Form 1970-75, Knudsen and Shinoda were in the recreational vehicle business. From 1975 until 1997, Shinoda operated his own independent design firm in the Detroit area. ■

The engine compartment, including the inner-fender panels, the power-brake unit and the fan shroud were painted persimmon. The air cleaner, pulleys, generator and many other engine parts were chromed and had a brushed appearance. The valve covers were modified 1955 Thunderbird dress-up kit with Turnpike Cruiser inserts in place of the raised Thunderbird insignia. The radiator was thicker, wider and taller than those used on stock Ford products of the day and was probably made in Italy. The car had a 12-volt electrical system.

The frame of the XM-Turnpike Cruiser was from an early 1954 Mercury convertible with an X-member. It was boxed and reinforced with X-shaped extensions of approximately two-feet at the rear. The rear-leaf springs had extra leaves, and metal flaps concealed the springs from sight.

The trunk lock was standard 1953 Ford and had a spring-loaded cover. The trunk was hinged and opened the same as other Ford products of the era. The trunk was full-sized and was covered with persimmon-colored carpet. The gas filler was in the trunk, and there was a funnel-shaped fill pipe sealed by a rubber disc on the trunk lid.

The XM-Turnpike Cruiser had its own custom enclosed trailer pulled by a Ford COE tractor with big wide whitewall tires. The trailer had Plexiglas panels on the sides that were as big as the car itself. When the XM-Turnpike Cruiser was displayed, the right side of the trailer opened up and extended, so the car could be turned at a 90-degree angle. The trailer platform the car was on had steps and handrails on either side, so people could walk up and around the car and down the other side. The truck and trailer were painted in the same colors as the XM-Turnpike Cruiser and lettered to identify the car inside the trailer as the XM-Turnpike Cruiser.

The XM-Turnpike Cruiser was first shown at the Cleveland Auto Show in January 1956. It was displayed in the parking lot in its special trailer. From there, it went to the Detroit, Chicago and New York Auto Shows. After that, it was sent around the country to other auto shows and was also displayed at selected Lincoln-Mercury dealers. Because it was in its own display trailer, it could be shown almost anywhere. An advance man usually placed ads in the local newspapers, telling people the places and dates the car would be shown in their area.

In January 1957, Reith arranged for a production 1957 Mercury Turnpike Cruiser to be the Pace Car in the 1957 Indianapolis 500 race. Even the publicity from the Indianapolis 500 that year could not forestall what lay ahead for Reith. Although the 1957 production Mercury increased market share from 17% to 21% through the first half of the 1957 model year, it was not enough, and it didn't come anywhere close to the expectations Reith had led management to believe they could expect. He was removed as head of Mercury Division and offered a face-saving job as head of Ford of Canada but was too proud to take it. On August 30, 1957, he resigned. Another one of the original "Whiz Kids," Ben Mills, was appointed to take Reith's place and, several months later, Mercury was combined in a single new Division with Edsel and Lincoln.

As Reith's car, the XM-Turnpike Cruiser became an outcast, the car and its hauler were parked behind Lincoln-Mercury Division offices in a parking lot visible from the intersection of Schaefer Road and an I-94 off-ramp. Sometime in 1958, Joe Bayne, Lincoln-Mercury sales manager and a friend of Benson Ford from the days he headed Lincoln-Mercury Division, came into the office, and said he was tired of looking at Reith's car and wanted it disposed of. That job was assigned to Tom Berry. For $300, Berry sold the XM-Turnpike Cruiser to Jim White, vice president of Dearborn Tube and Steel, one of the contractors who built some of Ford's pre-production vehicles.

The special trailer the XM-Turnpike Cruiser was transported in was sold separately, but here the story gets murky — because there are three different stories. The most probable version of the story is that it was sold to someone in Ohio, who used it as a mobile radio-broadcasting facility. A second version of the story has it that it was sold to Holman and Moody to haul the 100,000 mile Comets, that were a part of Mercury's publicity campaign that year. There are also some Lincoln-Mercury employees who believe the hauler was sold to a Michigan farmer and cut up for scrap. Whether the trailer still exists is

Although the trunk of the XM-Turnpike Cruiser was completely finished and contained the spare tire, Ghia formed all the metal by hand, and no pressed stampings were used.

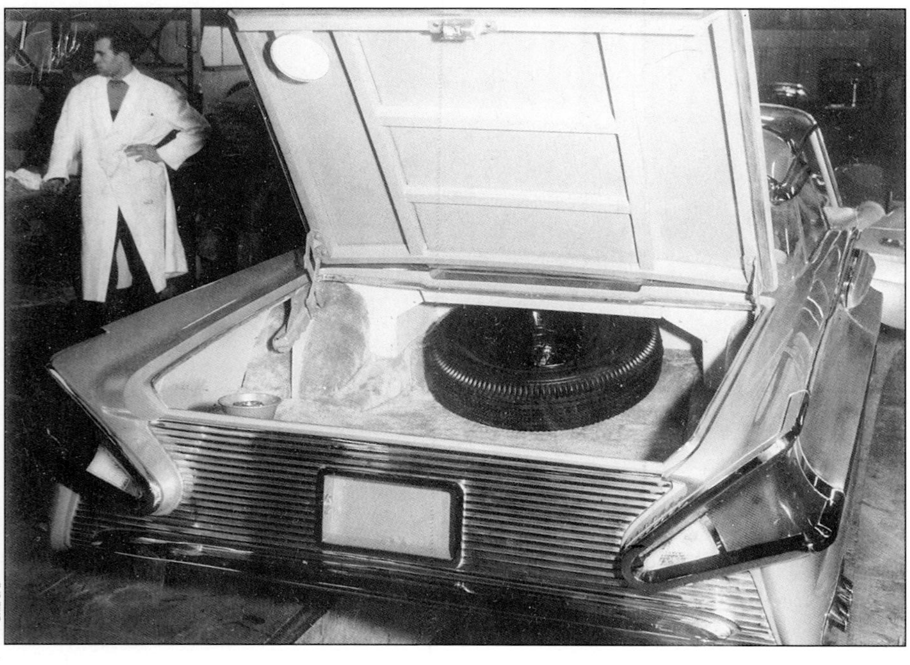
Ghia SpA

J.R. Samsen

Gale Halderman

James A. Arnold

Jim Arnold was born and raised in Gadsden, Alabama. He loved to draw as a child, but didn't develop an interest in designing cars until after college. He attended Auburn University, graduating in 1952 with a degree in Industrial Design. In 1952, he won a national design award sponsored by the Wood Office Furniture Institute. As an ROTC graduate, he then spent two years in the Air Force as a photographic officer stationed at Selfridge Field near Detroit. Because his wife was still attending Wayne State University when his tour of duty ended, Arnold applied for a job as a car designer and was hired by Roy Brown to work in the Edsel studio. While waiting for the Edsel studio to be formed, his first six weeks at Ford were spent working with Ken Spencer building and refining the design of the 3/8-sized model of the XM-Turnpike Cruiser sent to Ghia. Arnold then spent the next three years mostly in the Edsel interior studio and was then assigned to the MEL studio, and then to Colin Neale's advanced interior studio.

Later, he worked as senior designer in the Corporate Projects Studio on the Continental Mark IV. During tours in the Ford interior and exterior studios, he became head stylist. After tours at Ford of Australia and Ford of England, Arnold returned to Dearborn and worked in the luxury car studio on the design of the Mark VII. He retired in 1990 as an executive designer. ■

unknown. Whatever happened to it, when the Stylerama program was proposed, the truck and trailer became the model for six tractor-trailers, which were planned to haul Ford concept cars around the country much as was done with the XM-Turnpike Cruiser.

Shortly after Jim White bought the XM-Turnpike Cruiser, he took it to his parents' property and stored it on their acreage. White's parents operated an auto parts store in Flatrock, Michigan. For many years, it sat in a fenced field behind the auto parts store slowly deteriorating. The hood, trunk, and top rusted. A man by the name of Baker, who owned a body shop in Lima, Ohio, often saw the XM-Turnpike Cruiser, when he was in Flatrock and tried on several occasions to buy it. For several years, he was told it was not for sale. In 1971, Baker inquired again, and this time he was directed to Jim White. He was finally able to purchase the XM-Turnpike Cruiser for $500. By the time Baker purchased the car, it had been vandalized several times and was missing the intake manifold, air cleaner and both four-barrel carburetors. One of the Plexiglas T tops was also broken. The hood was very rusty and was later reskinned in Baker's body shop with a similar, but not identical, hood skin. The trunk lid was also rusty around the edges, but the chassis and the floor pans were solid. The upholstery was all there but, because the windows had been left opened, it was sadly in need of replacement.

In July 1979, Ray Sabo and his son Al were selling parts at a swap meet in Ohio. The Sabos owned a 1957 Mercury Turnpike Cruiser, which must have been obvious from the parts they had for sale. Baker was at that same swap meet and stopped by the Sabos' space. He asked them if they were interested in buying a prototype 1957 Mercury. The Sabos made an appointment to see the car but were skeptical, because they didn't believe such a car could be for sale. They did a little investigating, figured out what the XM-Turnpike Cruiser was and, after they saw it, bought it for $3000. The Sabos trailered the car back to their home in Michigan, where it was placed in storage.

In September 1982, the Sabos decided to sell the XM-Turnpike Cruiser. They placed an ad in *Hemmings Motor News*. Ray Cosh, a former Michigan resident who had moved to Northridge, California, got a call from a friend in Michigan, who asked if he had read the latest copy of *Hemmings*. Cosh hadn't but quickly got out his copy. He was the first to call in response to the Sabos' ad. A price of $10,000 was agreed upon, and the Sabos held the car until Cosh got there. Cosh flew to Michigan, looked the car over, paid the agreed price, and had the car transported to California. Cosh now lives in Ojai, California. He has not yet found time to restore the XM-Turnpike Cruiser, although he still plans to do so. ■

(upper left) The XM-Turnpike Cruiser at the 1956 Chicago Auto Show.

(below) The interior of the XM-Turnpike Cruiser was upholstered and painted to match its persimmon-colored exterior, which made for a very eye-catching car on the show circuit.

J.R. Samsen

Buzz Grisinger

The Blue Job, as it came to be known, was designed in Buzz Grisinger's Mercury pre-production studio, soon after he was hired by Ford.

Blue Job

Buzz Grisinger left Kaiser-Frazer in 1952 to form an independent industrial design partnership with Rhys Miller, a good friend from their Chrysler days. Grisinger and Miller were encouraged to go into business for themselves, so they could participate in the Mark II design contest. They received a $10,000 payment for entering that competition, and were told they would be paid more if they won. Their Mark II design wasn't selected, but it did enhance their reputation especially with Ford Motor Co.

After the Mark II design contest, Miller got a call from Frank Hershey, who offered both of them jobs. Grisinger suggested they accept the offer, and Miller agreed. They decided that Miller should start at Ford first, while Grisinger closed down their business. Hershey assigned Miller to the Thunderbird studio and, when Grisinger came on board several months later, he was temporarily assigned there also. Grisinger designed the continental kit used on the 1956 Thunderbird.

Almost immediately after Miller and Grisinger came to Ford, George Walker was appointed general manager and vice president of styling. Hershey and Walker never got along and, most recently, had argued about using the swoosh trim on the Thunderbird. Hershey won that argument but, to make a long story short, when Walker took over, Hershey left rather than have Walker as his boss. It took Walker a month or two after he took over to make all of his studio assignments but, when he did, Miller and Grisinger were assigned to the Ford pre-production and the Mercury pre-production studios respectively.

Miller was not particularly happy with the pace demanded by Bob Maguire, who had been appointed by Walker to replace Hershey as head of the Ford studio. Miller soon left and, after a few months at Chrysler, he accepted a position as Hershey's assistant at Kaiser Aluminum. Grisinger, on the other hand, "started cranking out cars" in the Mercury pre-production studio "like mad." Grisinger's immediate superior was Gene Bordinat, who was head of the Lincoln-Mercury studio. Although he was aware of the political tug-of-war between Engel and Bordinat to succeed Walker, Grisinger got along well with Bordinat and, in 1961, when Bordinat was appointed general manager of the Design Center, Grisinger succeeded him as head of the Lincoln-Mercury studio.

The Blue job, as it came to be called, was the first car Grisinger designed, after he was appointed head of the Mercury pre-production studio. As Grisinger recalls it, some of the initial design work on the Blue Job was done upstairs around a conference table that acted as his makeshift studio, while he and the other designers assigned to his studio were waiting for their quarters to be finished downstairs. Although the new building was occupied in early summer 1953, the basement wasn't finished until later.

When Walker put Grisinger in charge of the Mercury pre-production studio, he told him that he was free to design whatever cars he wished, but he also told him it was unlikely they would be considered as future Mercurys, because the design of the next Mercury was already set. The Blue Job was Grisinger's idea of an evolutionary Mercury design. It was started in 1955, after the XM-Turnpike Cruiser was designed, and probably just after the full-sized clay model of the XM-Turnpike Cruiser had been moved into the Mercury production studio, so it could be turned into the 1957 production Mercury.

Walker told Grisinger to show him what he could do and, according to Grisinger, the Blue Job was designed primarily to show Walker and Ford management what he was capable of. Grisinger was well aware that what he was doing was a test and, for that reason, he did most of the design work on it himself. He also actively participated in clay modeling the car. The other designers in the studio remember their input on the Blue Job as very limited. Other designers who worked on the Blue Job include Ed Westcott, Dick Kessler, Bob Chieda, Jerry Morrison, Ed Iser, Herb Tod and Chuck Berry. Doug McCombs was the lead clay modeler.

The design of the Blue Job took several months to complete. After it was clay modeled, it was cast in fiberglass, but it had no finished interior. It had a Plexiglas windshield and backlight, but the top of the instrument panel, the top of a steering wheel and the top of seats were built on a plywood panel that was level with the beltline. When finished, the car was painted blue and trimmed with chromed fiberglass bumpers, but it did not have opening doors, hood or trunk, nor did it have any running gear.

The most unique feature of the Blue Job was a proposed retractable top developed by Dick Noe, a graduate of MIT, and an innovative design engineer Grisinger had recruited from Chrysler. Noe's proposed retractable mechanism was less costly than the one developed by Continental Division. Noe proposed the retractable top after the car was built, and because there was no money left in the budget to actually build the Blue Job as a retractable, only renderings and a full set of engineering drawings were prepared. Noe's retractable mechanism was also lightweight, multi-purpose and easy to service, and it included the following special features:

- A light-weight Plexiglas sliding-top panel over the front-seat area that retracted back into the rear section of the roof, much like the roof on the X-100.
- An anodized-aluminum sunshade that closed off the Plexiglas panel from the elements. The aluminum sunshade was manually operated, and it could be textured and/or painted in matching or contrasting colors.
- With removal of roof-trim pieces on the inside of the car, the top-channel guides could be easily adjusted.
- The entire retracting-top mechanism slid on inexpensive nylon bushings and also utilized nylon weatherseals and header bumpers.
- The header was attached with eight sheetmetal screws, and included a built-in drain system that channeled water down through the A-pillars.

Esther Kessler

Richard L. Kessler

Dick Kessler grew up near Cleveland, Ohio. He attended the Cleveland Institute of Arts, majoring in fine arts. He graduated in 1954. Because he wanted to design cars, he applied at the Ford Design Department. Buzz Grisinger suggested that he acquire more industrial design experience, so he returned to the Cleveland Institute of Arts for an additional year of schooling. In 1955, he and Ed Westcott applied and were hired at Ford.

Kessler's first assignment was in Grisinger's Mercury pre-production studio, where he worked on the Blue Job and the L'Avion. He was later transferred to the Ford studio.

He left Ford in 1961 to work for former Ford designer John Cuccio in his independent design firm. His last project at Ford was the Mustang, which was in the beginning stages. While employed by Cuccio, Kessler worked on the Volkswagen and Budd accounts and designed many other commercial products.

In 1971, Kessler joined Ed Westcott's industrial design firm. He retired in 1991. ■

Each side of the Blue Job had a different design. As built, the car had no interior.

Buzz Grisinger

Douglas R. McCombs

Doug McCombs grew up in Dearborn, Michigan. Beginning in 1941, he attended the Henry Ford Trade School. After he graduated in 1945, he served in the Army for two years and then returned to Dearborn to work in the Rouge plant as an apprentice tool and diemaker. In 1950, McCombs enrolled in Bill Leverenz' first clay modeling class. After he completed the class, McCombs returned to the Rouge plant for awhile. He started as a clay modeler at the Design Department in 1951.

Gil Treweek

McCombs worked on the 1955 Thunderbird and, later, on most of Buzz Grisinger's cars including the L'Avion and the Blue Job. He also worked on the Envoy and was the modeler who did the winged roof used on one of the Envoy's alternate designs.

After his retirement in 1984, McCombs taught clay modeling at the Center for Creative Studies until 1991. He also worked at Ford's Concept Center in California. During a three months stay at the Concept Center, McCombs recruited his friend Buzz Grisinger to work with him as a clay modeler, a job Grisinger says he enjoyed immensely. ■

- The headliner was made of molded plastic that was cemented to the roof panel.
- Flexible driveshafts were used that simplified the installation and removal of the reversible electric-drive motors, which were located at the centerline of the roof and operated the top at nine inches per second.
- The cantilevered support for the Plexiglas top was designed with built-in moisture drains.
  - The retractable backlight, trim, molding and glass run folded into the C-pillars and roof-panel structure and left a simple and light-weight roof section, that could be retracted into the trunk.
  - The folded and stamped-roof structure had interchangeable right and left side panels, which saved on stamping costs.

The central premise of Noe's retracting roof was its simplicity and low cost. The front part of the roof retracted into the back part, and then the remaining structure slid on channels down into the trunk. The deck lid was hinged and opened from the back. The lightweight anodized-aluminum panel that closed-off the Plexiglas panel was designed to solve the heat-buildup problem on the production Sun Valleys and Skyliners. The system Noe designed closely followed one of the alternate designs Gil Spear had earlier patented for the Syrtis and emphasized the simplicity Spear advocated, but there was apparently no direct contact between Spear and Noe.

Each side of the Blue Job, including the C-pillars and rear-quarter panels, had a different design and represented variations Grisinger thought should be considered on future Mercurys. The sculpted rear-quarter panel on the passenger side made use of an anodized-aluminum insert, probably a first at Ford. The Blue Job was never displayed publicly, but

Ford Design Center

(left) The Blue Job was sometimes referred to as the Strato-Top but, according to its designers, that name referred primarily to its proposed retractable-top system.

The Blue Job was designed with a retractable roof that was less complicated and cheaper to build than the system developed by Continental Division.

Buzz Grisinger

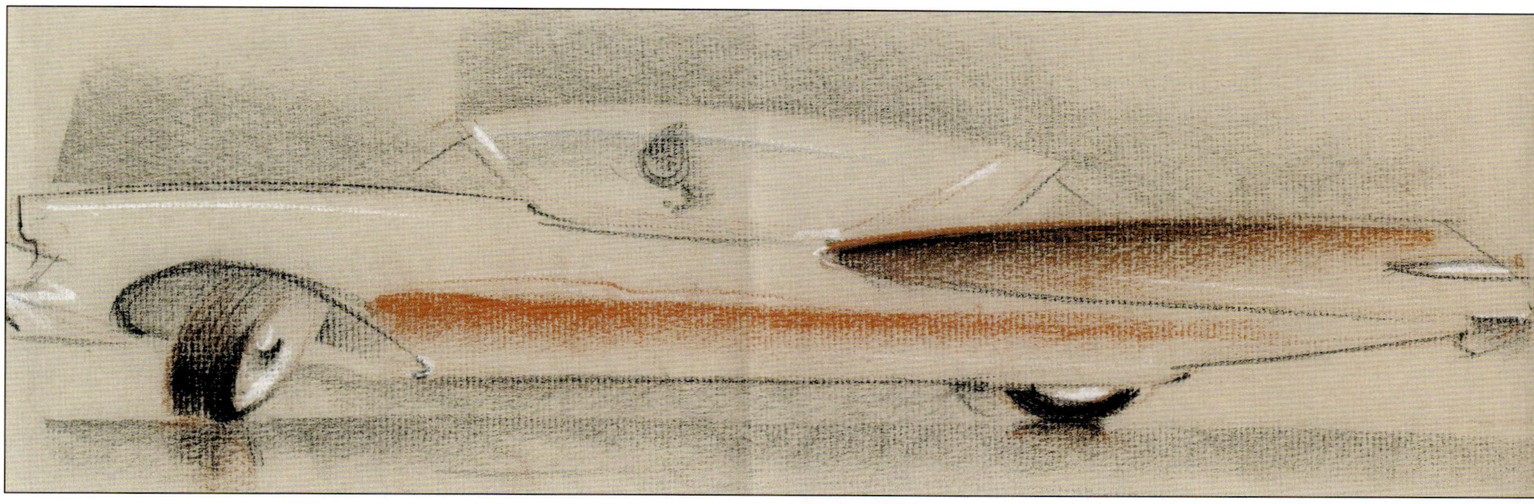
Buzz Grisinger

One of Buzz Grisinger's early sketches of the Blue Job, which was obviously not called the Blue Job at this stage of its development.

it was included as part of the background in a photograph of Walker's design staff in a 1956 booklet about the Styling Center.

Lewis Crusoe, who by that time had been placed in charge of all car and truck production at Ford, closely watched the development of the Blue Job. He was impressed with what he saw and told Grisinger so. Crusoe asked Grisinger to give up the Mercury pre-production studio to design a new small car he was considering for production. Grisinger readily agreed.

In March 1956, at about the same time Grisinger was asked to design Crusoe's small car, Walker appointed Joe Oros to replace Bob Maguire as head of the Ford studio. Maguire was named head of a new Advanced Studio and asked to oversee the design of the Stylerama cars. Grisinger started work on the design of Crusoe's small car even before he left the Mercury pre-production studio but, when space became available, he moved to Maguire's Advanced Studio and began designing Crusoe's small Mercury, which he named the L'Avion.

The Blue Job was around for several years and can be seen in the background in several photographs taken of other car proposals in the Styling Rotunda. When Grisinger moved upstairs to the Mercury production studio in 1958, the Blue Job went with him and was often used for comparison to newer designs being worked on. Although the front end of the Blue Job did influence the production 1959 Mercury, McNamara decided that future Mercurys had to use the same body panels and doors used on Fords. After the 1960 Mercury, which he designed, Grisinger recalls that it became a real challenge to design a Mercury that looked different from that year's Ford. Unfortunately, the emphasis on interchangeability didn't leave much room for design innovation, and it wasn't until the mid-1960s that things began to slowly change.

No one specifically remembers what happened to the Blue Job, but Grisinger and the other designers in his studio believe it was broken up out on the loading dock and hauled off with the trash in the early 1960s. ■

Jerry Morrison

Jerry D. Morrison

Jerry Morrison was born in Coos Bay, Oregon, and grew up in Bellingham, Washington, and Yreka, California. As a child, he developed a love for cars and drawing. Morrison attended San Jose State College for 2 1/2 years, where he studied commercial art. He then applied at the GM Institute and was accepted but had to wait for a new semester to begin. In the meantime, he was accepted at the Art Center College. Because of a lack of funds, he alternated semesters working and going to school.

In the summer of 1953, he and three other Art Center students were hired as summer interns clay modeling at Continental Division, where he worked on the Continental Mark II. At the end of the summer, he returned to the Art Center, completed his studies and was hired as a Ford designer in February 1955. With credit for three months work at the Design Department, he received his Art Center diploma in May 1955. His first assignment was in the Mercury pre-production studio, where the XM-Turnpike Cruiser was being designed and clay modeled. In May 1956, Morrison was drafted.

After 21 months in the Army, he returned to the Styling Center and was assigned to Buzz Grisinger's Mercury pre-production studio, where he helped finish the design of the Blue Job. He was later transferred to the Mercury production studio, where he worked on the 1959 through late 1960s Mercurys.

Beginning in 1967, Morrison was assigned to the Industrial Design Studio, where he designed Philco products. From 1967-69, he also helped design and became an expediter of several Ford concept cars including the Mach I, which he named. Before retiring in early 1987, Morrison was one of the designers of the Probe IV and V. ■

Edward R. Iser

Ed Iser

Ed Iser grew up in Kansas City, Missouri. In high school, he entered the Fisher Body Guild contest with model cars he built. Because he wanted to be a car designer, he enrolled in the Art Center College in Los Angeles, California. He received a GM scholarship while in school, so he started at GM as a designer in one of their advanced studios after graduation in 1954. Iser didn't feel comfortable at GM and left after several months. Dick Cox, a classmate at the Art Center, who was then working at the Ford Design Department, recruited Iser, who started there in late 1954.

Iser's first assignment was in Buzz Grisinger's Mercury pre-production studio, where he worked on the Blue Job. In 1959, Iser was assigned to the corporate advanced studio, where he worked on the Palomar. He was later transferred to the Mercury production studio and then to the Ford production studio.

Because of the long hours, Iser left Ford in 1963 to work with Alex Tremulis as an independent designer. In 1964, Iser returned to Kansas City, where he worked with several former Ford designers and eventually got into housing design and construction.

He retired in 1994. Iser is also an accomplished wood sculptor. ■

Buzz Grisinger

George Walker and the L'Avion in the Styling Center courtyard. The L'Avion was designed as a small Mercury, but it was not produced.

L'Avion

The L'Avion was proposed by Lewis Crusoe as a Mercury small car. It was certainly not the first, nor would it be the last small-car project at Ford Motor Co. Plans to produce a small Ford began with the X-92 project in 1938 and went by the wayside as World War II approached only to be revived towards the end of that war. By 1946, several prototypes had been built, a Light Car Division had been formed, and tooling was ordered. The program was ultimately canceled in 1948, and the Light Ford was sent to France, where it became the Vedette.

After World War II, Ford began a rebuilding process. That rebuilding process resulted in increased sales of existing products, and it included plans for several new cars. Those new cars included the Thunderbird in 1954, the Continental Mark II in 1955 and the Edsel in 1957. It was also planned to include the Mercury Monterey in 1955, and it almost included a new small Mercury in 1958. The small car was delayed for several years, but a new small Ford called the Falcon was introduced a year later as a 1960 model. In its first year, over 400,000 Falcons were sold. Robert McNamara is credited with the Falcon's success, and that success led to the Comet, the Fairlane, the Mustang and other smaller Ford products. There was a market and Lewis Crusoe saw it.

By 1955, Lewis Crusoe had been promoted to vice president in charge of all car and truck production at Ford Motor Co. He regularly visited every studio in the Design Department and was well liked by designers and clay modelers. But for an untimely heart attack he suffered in late 1956, Lewis Crusoe probably would have received credit for the small-car project he initiated earlier that year, and the car that became the Falcon may have been first produced as a Mercury. The L'Avion was Crusoe's small Mercury, and it was designed by Buzz Grisinger.

In 1955, Grisinger was named head of the Mercury pre-production studio. Crusoe first noticed Grisinger when he came up with the design of a continental kit for the 1956 Thunderbird. After Grisinger had been placed in charge of the Mercury pre-production studio, Crusoe proposed that Grisinger design a small Mercury Crusoe was considering for production. Grisinger quickly agreed and, when Bob Maguire was appointed head of a new Advanced Studio and asked to spearhead the Stylerama program, Grisinger was assigned to that studio. Grisinger believes the assignment was primarily to give him the space, time and a support crew needed to design the small Mercury he named the L'Avion, which in French means "the airplane." The wedge or wing design on the sides of the car was Grisinger's inspiration for the name he selected. Grisinger prepared the initial design of the car that became the L'Avion after he came to Ford, but before he was named head of the Mercury pre-production studio.

At about the time the car was named the L'Avion, Grisinger made a formal presentation to Crusoe and Walker, which included a written proposal of alternate designs for the car, photographs of various production small cars, and several other small car models Grisinger designed. Walker and Crusoe approved a proposal that included a wedge design on the sides of the car. Grisinger was authorized to design and build a full-sized fiberglass model of the L'Avion but without a seating buck or an interior. Crusoe had the Engineering Department work up package measurements for a small Mercury and, soon after that, Grisinger and the designers and modelers working with him began making a full-sized clay model. No 3/8-sized clay model of the L'Avion was built. As was Grisinger's standard practice, after the initial sketches were approved, most of the design work on the car was done as he worked on the clay model right along with the modelers.

Richard Kessler, Bob Marcks, Bob Chieda, Herb Tod and Ed Westcott were all designers assigned to work with Grisinger on the L'Avion. Jim Morrison was the layout man, and Doug McCombs was the lead clay modeler. No design engineer was assigned to the L'Avion, because no mechanicals were to be included on the car at that point.

The L'Avion was designed and built over approximately a six-month period in 1956-57. After the full-sized clay was built and approved, a plaster plug was made from which a full-sized fiberglass car was cast. Although the greenhouse was see-through, the doors, trunk and hood were scored where the openings were supposed to be, but they did not open. The L'Avion used a metal framework underneath with wheels and tires attached, but there was no running gear. As built, the L'Avion was a pushmobile. When it was finished, Grisinger had it painted a light blue-green two-tone.

Grisinger is fairly certain that if the L'Avion had been produced, it would have been renamed something else. By the time it was completed, Jack Reith's 1957 Mercury was in trouble and, by the end of August 1957, Reith had been replaced. Even before that, management had also begun to doubt the wisdom of the other part of Reith's plan — the Edsel. Crusoe supported and backed Reith, and Ernie Breech, Ford's executive vice president, blamed Crusoe for Reith's failures. Crusoe, who soon began to believe that Breech was out to get him, suffered a massive heart attack in October 1956. With Breech's backing, Robert McNamara took over Crusoe's duties.

By the time the L'Avion was finished, fortunes at Ford had changed so fast there was no chance it would be built. Ford Motor Co. was reeling from a recession, anticipating the failure of the Edsel and was retrenching. McNamara liked small cars, but his idea of a small car was different than the L'Avion. Probably the best explanation why Ford's first production small car didn't resemble the L'Avion was simply because McNamara's taste was different than Crusoe's. In late 1959, McNamara's small car was introduced as the Ford Falcon. It was designed by Don DeLaRossa and Gale Halderman. The next year the Mercury Comet was introduced, but neither it nor the Falcon took any of their design cues from Grisinger's L'Avion.

Several years later, after McNamara left Ford to become secretary of defense, the second-generation Falcon incorporated a wedge design on its sides. Grisinger feels it may have been influenced by the L'Avion. It was designed by Bob Koto who, at the time, was in charge of Grisinger's Mercury pre-production studio. ■

Bob Marcks

Robert C. Marcks

Bob Marcks grew up in Harvard, Illinois. As a child, he drew streamlined cars and, by an early age, he knew he wanted to be an automobile designer. After graduating from the Art Center College in January 1953, he interviewed at both Studebaker and Ford. He was offered a job by Frank Hershey and started as a designer at Ford in February 1953. After several months in Gil Spear's Advanced Studio, Marcks was assigned to Don DeLaRossa's Mercury studio, where he helped design the 1955 Mercury.

In November 1953, Raymond Loewy and Associates offered him more money to become a Studebaker designer. After two years at Studebaker, Marcks received an offer to return to Ford. He accepted, and in mid-1956 he was assigned to Bob Maguire's Advanced Studio, where he helped design the DePaolo, the L'Avion and the interchangeability studies. He was later assigned to Rulo Conrad's Lincoln pre-production studio, where he worked on the Envoy.

Marcks left Ford in November 1961 to form his own industrial design firm specializing in jewelry, automotive and recreational vehicle design and advertising. His firm eventually became Marcks, Hazelquist and Powers.

In 1973, Marcks joined the Chrysler design department, where he designed several of their reduced budget showcars including the 1977 Chrysler Turbine, the California Cordova, the 1982 mini-limousine and the Chrysler convertible. Before retiring from Chrysler in 1993, Marcks also worked in their marketing department and as a product planner. ■

John Orfe

The L'Avion was designed and clay modeled in the old Continental Division design studio. From left to right are Guy D'Angelo, Bill Fox, Buzz Grisinger, Joe Golden (in the background), Eugene Poshadlo, and Jim Hayes.

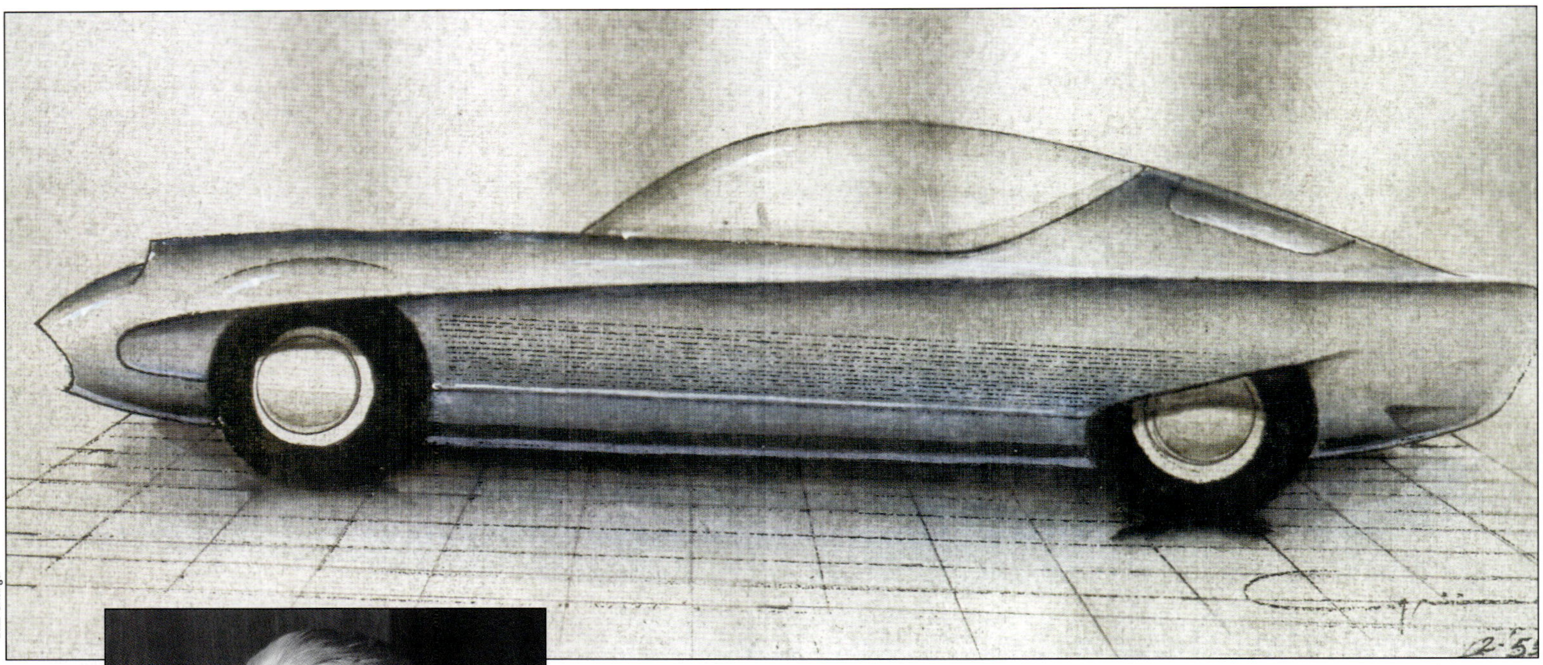

Buzz Grisinger

One of Buzz Grisinger's December 1955 renderings of the L'Avion.

Buzz Grisinger

Arnott "Buzz" Grisinger

Buzz Grisinger was born in Lincoln, Nebraska, and raised in Casper, Wyoming, where his father was stationed as superintendent of the predecessor of the Burlington Northern Railroad. From an early age, Grisinger liked to draw. After high school, he studied architecture at the University of Nebraska, but left to complete his fourth year of college at the Boeing School of Aeronautics. He graduated in 1931. In 1932, after teaching at Boeing for a year, he took a job in Chrysler's design department. While at Chrysler, Grisinger also taught night school design courses.

During World War II, Grisinger worked on the Manhattan Project developing the atomic bomb. After the war, he accepted a position as a designer for Kaiser-Frazer. In 1952, Grisinger was offered a job at Ford by Frank Hershey but chose to open his own design firm with Rhys Miller, an associate at Chrysler and Kaiser-Frazer. As independent designers, Grisinger and Miller prepared one of the six competitive designs for the Continental Mark II.

Although the Grisinger-Miller design for the Mark II did not win, their design skills encouraged Hershey to again ask them to come to work at Ford. They were both hired as Ford designers in 1955. While waiting for George Walker, the newly appointed Ford vice president and general manager of the Styling Center, to make his appointments, Grisinger designed the continental kit for the 1956 Thunderbird, while Miller did the face-lift on the 1957 Thunderbird.

Grisinger's first assignment at Ford was as head of the Mercury pre-production studio, where he designed the Blue Job. He was then transferred to the Advanced Studio, where he designed the L'Avion and the DePaolo. When those cars were completed, he was asked by Robert McNamara to prepare a study, which included several full-sized clay models demonstrating interchangeability proposals.

In 1959, Grisinger was appointed to head the Mercury production studio. In 1961, he was appointed chief designer of the Lincoln-Mercury studio and continued in that position until 1968, when he was appointed chief designer in the truck studio. He retired in 1970. ■

An alternate proposal of the L'Avion from December 1955 The primary design feature was the wedge shape on the sides of the car.

Buzz Grisinger

Edgar J. Westcott

After growing up in Bay City, Michigan, Ed Westcott attended Delta College, where he studied engineering. After graduation, he decided he wanted to be an automobile designer. Although he had no formal art education, he loved to draw cars. In 1955, he prepared a portfolio and, after being offered jobs at Ford, GM and Chrysler, he accepted an offer from Ford's Design Department, because they agreed to assign him to an advanced studio. In 1957, his design career at Ford was interrupted when he was drafted. He spent two years in the Army and then returned to the Styling Center.

While he was in Buzz Grisinger's Mercury pre-production studio, Westcott worked on the Blue Job and the L'Avion. He was then assigned to the Mercury production studio. For Ford's proposed Stylerama, Westcott designed the Missile II. Just before he left Ford, Westcott did the concept design for the Montego Sportshauler.

Because he wanted to gain broader industrial design experience, in 1961, Westcott accepted an offer from former Ford designer John Cuccio, who had established his own industrial design firm in Westport, Connecticut. As an independent industrial designer, Westcott worked on many automobile-related products and other product designs.

In 1964, he accepted an offer from Chrysler and was hired as a supervisor in the Imperial studio.

In 1967, he returned to Westport and established his own independent industrial design consulting firm. He also continued to work closely with Cuccio designing Metroliner transit cars for Budd Manufacturing. ■

Ford Design Center

(above) The L'Avion in the Styling Center courtyard.

Ed Westcott

Herbert M. Tod

Herb Tod grew up in Toronto, Canada. He was 12 when his family moved to Detroit. After he finished high school, Tod went to work at Nash-Kelvinator checking refrigerators as they came off the assembly line. Tod studied art and design at Wayne State University. In 1939, he decided to follow his life-long dream and become an automobile designer. He persuaded Hudson to hire him as an apprentice designer. To supplement his small salary as an apprentice designer, Tod worked as a milkman. He was a designer at Hudson from May 1939 until June 1941, when he left to take a job a Hi Johnson Displays.

During World War II, Tod was an Army supply sergeant and saw action in the Battle of the Bulge. From March 1946 until March 1948, he worked as a designer at Chrysler, where he formed a life-long friendship with Buzz Grisinger. From March 1948 until June 1949, Tod was a designer at Kaiser-Frazer.

In 1949, Tod was hired as a designer in the Ford studio. He was later transferred to the Lincoln studio, where he, Bill Schmidt and Rulo Conrad designed the 1956 Lincoln. Schmidt left Ford in 1955, and Tod and Conrad, as the remaining designers of the '56 Lincoln, shared the Industrial Design Institute award for the design of that car. After Schmidt's departure, Tod transferred to Grisinger's Mercury pre-production study.

Tod worked with Grisinger on the design of the Blue Job and the L'Avion. After Grisinger was appointed head of the Lincoln-Mercury studio, Tod became his design manager and then his executive designer. For health reasons, Tod took early retirement in 1970. ■

Gale Halderman

(below) From the front, the L'Avion previewed the squared-off look of the 1960s GM cars.

Buzz Grisinger

La Galaxie

In 1956, someone convinced Ernie Breech, Ford's executive vice president, that there was something wrong with the colors Company designers had been selecting for use on production Fords, Lincolns and Mercurys. Breech responded by calling on the field of high fashion for assistance. To satisfy Breech, Bob Maguire, head of the Advanced Studio, named designer Dave Ash as "Executive Stylist-Fashion Trends and Special Assignments." When Ash got his new assignment, he asked Maguire what the fashion trend part of his job was all about. Maguire told him that he was supposed to "seek out the trends of fabrics and colors in high fashion and home decorating as they might apply to and influence automotive colors and fabrics." In other words, Ash was named the designer in charge of dealing with the fashion industry. Ash pleaded with Maguire that he was just a car designer used to making cars in clay and getting them approved. He didn't know anything about high fashion and didn't want his new job or the title that went with it. Maguire told Ash he didn't have a choice, and suggested that he "play it by ear," because he didn't have the slightest idea what Ash was supposed to do either. The only direction Maguire was able to give Ash was that he was going to get some help from J. Walter Thompson, Ford's advertising agency. Ash's co-workers thought his new assignment was hilarious and, when they stopped laughing, they were glad it was Ash who was stuck with it and not them!

The help Ash got turned out to be from Virginia Van Brunt, a fashion consultant who worked for the Thompson Agency. Her job was to introduce Ash to the world of high fashion. Ash went to New York to meet Van Brunt, who introduced him to fashion designer Christian Dior. Dior suggested that, for a substantial fee, he would create a car for Ford. According to Ash, his proposal was something in brass, gold and plush red velvet. Ash passed Dior's suggestion on, but nothing came of it. Presumably, management saw the price and the suggestion itself and decided to keep looking.

With his new title, Ash was also given his own studio, but it was several miles away in the old Continental Division building then being used by the Edsel Division. Although nobody bothered him there, Ash felt isolated, and he found he had to coax his fellow designers to even come take a look at what he was doing.

After the proposal from Christian Dior, Ash was instructed to hire a fabric designer and a weaver, so he hired Helen Vincent who did both. Vincent set up a hand loom on the balcony at the Edsel plant. Because Vincent, knew nothing about automotive fabrics, Ash

The LaGalaxie at the Ford Rotunda. Its days on the show circuit were cut short when it was dropped, breaking its fiberglass body. The only part of the car that made it into production was the insignia used on the production Galaxie, Ford's top of the line model beginning late in the 1959 model year.

Ford Photomedia

got her some help from the Styling Center trim and color studio and, ultimately, one of the fabrics she made was used on a production Lincoln. While Vincent was up on the balcony making fabrics on her loom, Ash was downstairs working on the special assignments part of his job title building the most unusual concept car he had ever been involved with — the LaGalaxie.

The LaGalaxie was designed by Dave Ash with help from Bud Kaufman. The interior was designed by Jim Powers on a separate interior buck. The design engineer was Dick Noe. At the time the LaGalaxie was designed, Ash and John Najjar were studying astronomy in their off-time. Ash selected the name La Galaxie from what he was looking at through his telescope.

The car was first referred to as the Slide-A-Door, after the extra-wide parallel-action doors that were to be the primary feature on the car. Another method for entering the car was also considered. It was called the Strato-Door system. That system eliminated the A-pillar and, like the Bimini recently built at the Special Vehicles Department, the doors, roof and windshield rose up in one piece, so people could enter both the front and back seats at the same time. Although considered, the Strato-Door car was never built. The door system engineered for the LaGalaxie used the parallel-action system, which was incorporated in a working buck engineered by Ray Smith.

After several renderings, alternate full-sized elevation drawings of the LaGalaxie were placed on the studio wall, and two full-sized clay models of the car were started. When Maguire approved the project, his instructions were to include space for a small nuclear power plant in the back end of the car. Maguire would drop in from time to time and, on one occasion, he asked Noe where the nuclear power plant was supposed to go. Noe promptly took Maguire over to a full-sized elevation drawing on the wall and showed him a red-colored rectangle, where he said the nuclear power plant was supposed to go. Maguire expressed surprise, because he saw no tubes, dials or things that glowed. When Maguire asked where all those things he expected to see were, Noe told him they had been lucky enough to acquire measurements for the newest and smallest nuclear power plant available. Making it up as he went along, Noe told Maguire that the red rectangle represented the lead shielding that hid everything inside the engine, including the tubes, dials and things that glowed. Maguire left much impressed, and it was some time before he discovered he had been the butt of a practical joke.

Ford Design Center

(above) Helen Vincent and her hand loom on the balcony at the Edsel plant.

Ford Design Center

(left) At least two different LaGalaxie proposals were designed and clay modeled, before one of them was selected for the final car. This one proposed the use of parallel-action doors.

(below) The alternate LaGalaxie design from the driver's side.

Ford Design Center

It took about three months to complete the full-sized clay models. Just before they were finished, however, Ash was dispossessed of his studio and had to return to the Styling Center, where the selected design was completed. When the unfinished clay model arrived at the Design Center, it created quite a stir among those who saw it, even though only one side was done. Because a styling bridge was not available, Ash, Powers and a couple of other designers helped finish the unfinished side by eye.

When the full-sized clay was completed, Maguire approved the design, but didn't really know whether he liked it or not. He thought about it for a few days and finally asked general manager George Walker to drop by the basement studio, where the

Ford Design Center

Ford Design Center

(above) The alternate LaGalaxie design from the passenger side.

(below) The Strato-Door and the Slide-A-Door systems were both considered for the LaGalaxie, but neither was made functional, because development money in the Advanced Studio's budget ran out.

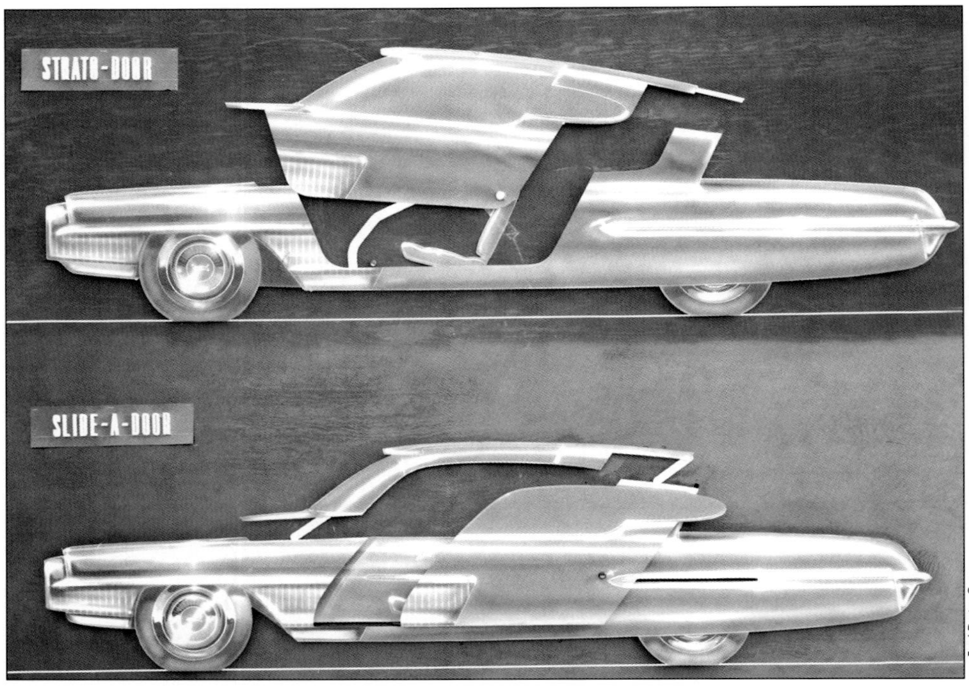

Ford Design Center

(below) The LaGalaxie on display at the Ford Rotunda. The car's two-piece trunk lid was designed to open from curbside.

Ford Photomedia

L. David Ash

Dave Ash was born in Detroit and raised in Chicago. He returned to Detroit to attend high school. He decided he wanted to be an automobile designer while still in high school and later attended Meinzinger's Art Academy in Detroit.

During World War II, Ash worked at GM's Fisher Body Division in the template shops. After the war, he was hired as a designer at GM, where he and Gene Bordinat became friends. At GM, Ash designed mostly interior hardware. Bordinat left GM to take a job at the Ford Design Department and, in October 1947, he persuaded Ash to join him at Ford.

Ash's first assignment at Ford was in the Advanced Studio, but he was soon transferred to the Ford studio, where he designed exterior ornamentation. In 1949, he designed the Ford crest. He was appointed a manager and placed in charge of all Ford exterior ornamentation. He continued in that position until the early1950s, when he became a manager in the Ford exterior studio. His first full-sized clay was of the 1955 Ford Crown Victoria. Ash is also credited with the design of the 1954-56 production glasstops.

During his career, Ash worked in the Ford, Lincoln, Mercury, truck and Edsel studios, and was also a designer at Ford of Germany. In 1956-57, under Bob Maguire, whom he considered his mentor, Ash worked in the Advanced Studio as executive stylist in charge of fashion trends. He was appointed head of the Lincoln interior studio in 1959, where he supervised the design of the interior on the 1961 Lincoln. In the mid-1960s, Ash was appointed chief designer of a special development studio. Later, he was appointed head of the Interior Design Office, where he was responsible for the interior design of all Ford Motor Co. products.

During his career at Ford, Ash worked on many concept and showcars, including the X-100, the Thunderbird, the FX-Atmos, the Diplomat, the LaGalaxie and the California showcars of the 1960s, including the Super Marauder and the Constellation. Ash also worked on the design of the 1961 Presidential Lincoln, the 1965 Mustang, the GT-40, the 1969 Continental Mark III and the Designer Series Lincolns of the 1970s.

He retired in 1981. ■

Ford Design Center

(above) As the full-sized clay model of the LaGalaxie neared completion, different C-pillar and side treatments were still being considered.

Ford Design Center

(above) This front-fender proposal was later changed. The LaGalaxie was designed and clay modeled in the former Continental Division design studio.

Ford Design Center

Just after this photograph was taken, the full-sized clay model of the LaGalaxie was moved to the Styling Center. Only after the move, did Walker decide to complete the LaGalaxie as a showcar.

Stu Denker

Keith H. Teter

Keith Teter grew up in West Virginia, where his father owned a Pontiac dealership. As a teenager, Teter learned to paint cars and became an accomplished mechanic. After high school, he enlisted in the Army. While he was stationed in Germany, he entered a Motor Trend design contest and won a scholarship to the Art Center. During his second year at the Art Center, he was awarded a Ford scholarship. When he graduated in 1958, his father wanted him to become a GM designer but, because he had received a Ford scholarship while in school, he accepted an offer from Elwood Engel to become a Ford designer.

Teter started his design career at Ford in the Advanced Studio, where he did some design work on the interchangeability studies and the LaGalaxie. He was later transferred to the Mercury production studio and, in the mid-1960s, he became a manager in the Lincoln studio.

In the late 1960s, Teter was assigned to the Industrial Design Studio, where he spent several years designing Philco products and showcars, including special projects at Ford's Concept Center in California.

In 1969, Teter was diagnosed with Multiple Sclerosis. He left Ford to move to California's warmer climate, where he accepted a position as head of the Industrial Design Department at the Art Center College. ■

Ford Design Center

After the decision was made to build the LaGalaxie as a showcar, the greenhouse was redesigned with see-through windows and a full interior.

full-sized LaGalaxie was and take a look. Walker hadn't seen it yet, but when he did, he fell in love with it! The full-sized clay model was then moved to the Styling Center shops, so a full-sized fiberglass model could be built. (Art Querfeld remembers helping complete the rear end of the LaGalaxie, after it was returned to the Styling Center.) When the fiberglass LaGalaxie was finished, the only features that really worked on the car were the interior and exterior lights. Ash said that when the lights were on, the LaGalaxie look like a spaceship.

The LaGalaxie created an even bigger stir when it was learned that Walker had authorized turning it into a showcar. After that decision was made, the greenhouse was removed so a functional interior could be included on the car. When completed, the LaGalaxie was first displayed in the main hallway of the Styling Center, where employees and designers just stood and looked at it -- some with their mouths open. Many thought Ash had spent too much time in the world of high fashion.

The La Galaxie was on the show circuit for a year. It was shown at the Detroit, Chicago and New York Auto Shows, at fairs around the country and at the Ford Rotunda. It received mixed reviews. Neither door on the LaGalaxie was operable, and the public was told that the way the doors worked was a secret. Entry into the finished showcar was gained by removing the backlight and crawling inside or by entering through a trapdoor underneath the car. Although it looked like a finished car, it was a pushmobile without running gear. Ash acknowledged that the LaGalaxie was spooky, and as far out as anything he had ever done! It was painted a pearlescent-gold. The windshield continued onto the roof over part of the front-passenger area and part way over the rear-passenger area. In the front-center of the roof was an air vent that provided fresh air to the passenger compartment. The reverse-slant backlight was hooded much like the 1958 Lincoln Continental Mark III.

In 1956, Jim Powers did a rendering of an Indianapolis racecar with a unique "X" over "O" design theme. Towards the end of the design process and to give the LaGalaxie an identity, Ash incorporated Power's unique "X" over "O" symbol into the design of the car. The "X" over "O" design theme appeared at the rear of the LaGalaxie on the back-up lights and on the outside front edges of the hood as ornaments. The hood included the top half of the front fenders.

Various interior designs were proposed by Jim Powers for the interior of the LaGalaxie.

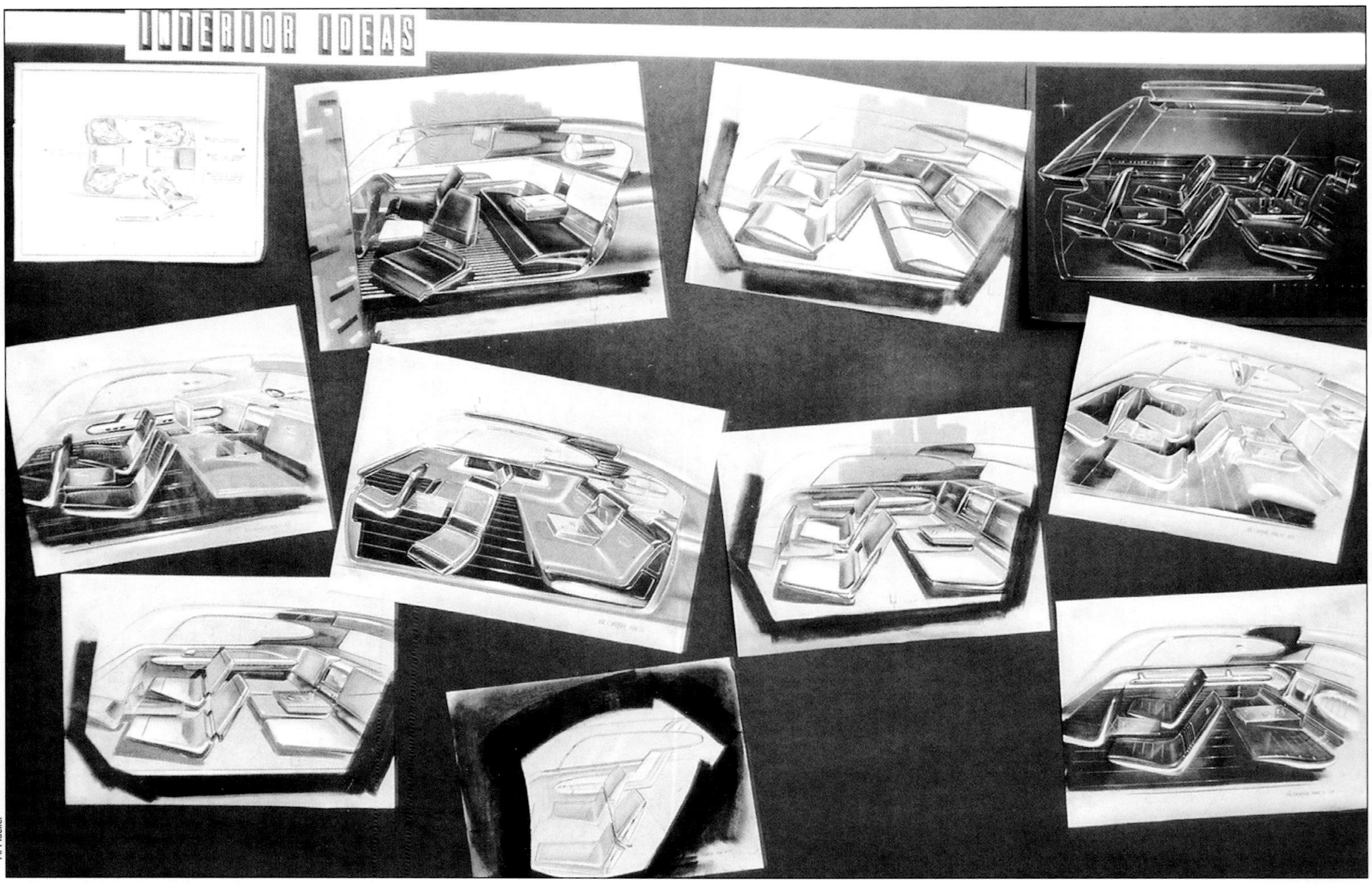

Al Mueller

The LaGalaxie had indented areas designed into the bottom of the front fenders, which were meant as air intakes for cooling the nuclear-propulsion system. The back bumpers were integrated into the body design and formed the base of the taillights. The tires on the LaGalaxie had chrome sidewalls, a feature not seen on any other Ford concept or production car before or since. The front of the car had deeply-recessed "super beam" headlights on either side and an auxiliary fluorescent lighting system across the grille area. The headlights were in recessed chromed cylinders that protruded from the front of the car and actually stuck out beyond the bumper, which was a bar below the grille. Growing out of the headlight housings were chrome spears that swept back at mid-line on the front fenders and into the door areas forming caps over the air-intake indentations on the lower front fenders and actually forming the base of the hood. The windows were flush with the outside sheetmetal of the car.

Jim Powers' rendering of a future Indianapolis race car, with its "X" over "O" rear-end design, was one of the inspirations for the LaGalaxie concept car.

The door openings extended from the front A-pillar to the back of the C-pillar. Although the doors did not open, they were designed to open wide enough to allow access to both the front and back-seat areas at the same time. Assuming the doors opened, part of the transparent roof was also designed to open, gull-wing style, to help with ease of entry. (The first operable car to use parallel-action doors was the Ford Techna concept car.) At the back of each side window, in the quarter-window area, were looping curved-glass areas that one of the designers believes may have been influenced by the engine cowling/air intake of the P-40 fighter plane.

Although they were not functional, the dual rear-deck lids were designed to be electronically operated and were accessible from the sides of the car. There were actually two trunks designed into the LaGalaxie. When opened, the deck lids exposed an area halfway down each side of the rear fenders for easy access. The split trunks were designed to be on either side of the car's nuclear engine, but at the time it was decided to turn the LaGalaxie into a showcar, that area was identified as the location of the gas tank.

The rear end of the LaGalaxie was dominated by two huge oval-shaped, chrome-bezeled taillight pods that housed massive taillights. Although the taillights were originally designed as exhausts for the nuclear power plant, they were identified in press releases simply as taillights. When they were on, the big oval-shaped taillights started out as fiery red, progressed to white and became violet the farther in they went. The design of the taillights and headlights was later copied on the Chrysler turbine car, which was designed after Elwood Engel left Ford in late 1961 to become head of design at Chrysler. The rear wheels were entirely enclosed by the quarter panels and were designed to be accessed by removing fender skirts.

The LaGalaxie on display at the Ford Rotunda. Not many people who saw the car liked it.

For the show circuit, the interior of the LaGalaxie was completely finished, even though nothing but the lights worked. According to Powers, the interior buck was built with assistance from Alden "Gib" Giberson and was intentionally designed to look as though the exterior of the LaGalaxie rolled over into the interior of the car. The LaGalaxie had three individually adjustable front seats with adjustable headrests. The outboard-front seats also had their own armrests next to the doors. The backseat accommodated three passengers, and the seat wrapped around the side panels much like the 1964 Thunderbird.

The LaGalaxie's steering wheel, and all of the instruments except the clock, were directly in front of the driver. Transmission controls and turn signals on the LaGalaxie were controlled from a rectangular-shaped steering wheel. The main instrument pod grew out of the instrument panel and was positioned close to the driver. The speedometer was contained at the end of an oval-shaped pod that started in the middle of the hood and extended back through the base of the windshield terminating almost suspended in air above the instrument pod. A matching oval-shaped pod ended in front of the passenger and contained the clock. Some of the instruments clustered near the driver were metal gauges suspended inside bubble-like clear blisters on either side of the speedometer. The LaGalaxie did not have a conventional glovebox in favor of a more spacious area in front of the front-passenger seat. The lack of a glovebox was intended to give a greater feeling of interior spaciousness with some interior storage provided for in pockets in the door panels. (Powers proposed the same glovebox and door panel treatment for the 1966 Lincoln, but it was rejected as too costly.)

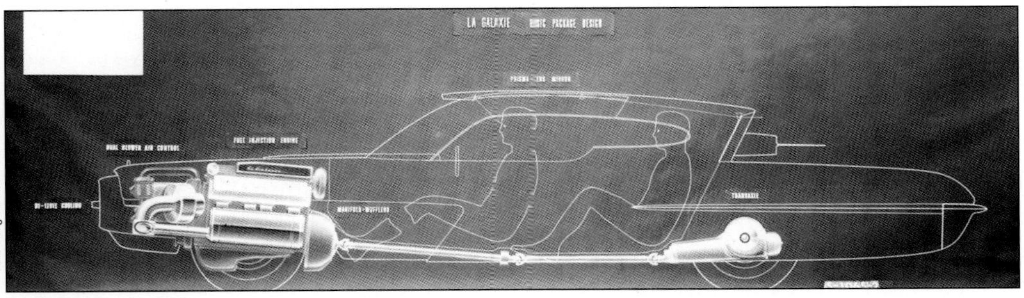

Ford Design Center

Although the LaGalaxie became a pushmobile, after it was decided not to include a make-believe nuclear power plant in it, at one time, plans called for other special features, including a prisma-lens mirror, a fuel-injected engine, bi-level cooling, manifold mufflers, dual-blower air controls and a rear transaxle.

The most unusual gadget on the LaGalaxie was a proximity warning device that press releases claimed was meant to tell the driver when he or she was too close to a car or other object ahead. A screen in front of the driver supposedly indicated the distance between the LaGalaxie and the object directly in front of it. The screen had curved-green lines at the top and bottom that moved into an amber area in the middle of the screen as the object got closer. The amber area also had an amber warning light that flashed. When the distance reached a pre-set minimum, the lines on the screen merged into the red area, and a red warning light flashed at the same time the brakes were automatically activated to slow or stop the car. The press releases said that, although the proximity warning devise wasn't available on production cars yet, it would be in the foreseeable future. Proximity devises were also being tested at Packard, until a big bird flew in front of the car almost throwing the driver through the windshield.

George Haviland, then public relations manager for the Styling Center, recalls that by the time the LaGalaxie was finished, the Styling Center had used up all of its show budget for the year, so Ford Division agreed to pay for the cost of transporting and showing the car at the various auto shows. According to Haviland, the LaGalaxie was on the show circuit for approximately one year and generated a lot of comment, not all of it favorable. After it had toured the auto shows in the United States, the LaGalaxie was shipped to Budapest, Hungary. While it was being unloaded, it was dropped, cracking the fiberglass body. It was decided not to repair the car, and it was returned to Dearborn to be destroyed.

Jim Powers was present in late 1959 or early 1960, when the La Galaxie was smashed by dropping it from a forklift to the ground. Powers was given one of the floating-metal instruments originally inside the bubble-like clear blisters on the instrument pod. He currently displays it on his desk.

Fifteen years later, the world of high fashion made a return appearance to the Design Center, this time with more success. It was Dave Ash who was again asked to work with then-current high-fashion designers to produce Lincoln's very profitable Designers' Series. ■

The LaGalaxie on display in the Styling Center Rotunda just after it was completed.

Ford Photomedia

Henry Ford Museum and Greenfield Village

(above) If a next-generation Edsel had been built, the public was led to believe it might look something like this 3/8-sized model. Shown with the Edsel 196X is its designer, Bud Kaufman (left), and George Walker (right).

After the failure of the Edsel, there were demands that Roy Brown be fired. To get him out of harms way, George Walker transferred him to Ford of England, where he designed English Fords for the next four years. Brown's life-long friend, Bud Kaufman, was appointed to replace him as head of the downgraded Edsel studio.

After he designed the full-sized Diplomat, Kaufman was transferred from the Lincoln to the Ford studio. One of his first assignments was to design what he thought the next Edsel should look like. According to Romeyn Hammond, who was in the Edsel interior studio at the time, management was afraid that any design that came out of the Edsel studio would be subject to the influence of Brown's design philosophy. Kaufman's next- generation Edsel was completed over several months and built as a 3/8-sized fiberglass model. It was named the Edsel 196X and, after it was completed, Kaufman was named executive designer for Edsels in the MEL studio.

In June 1958, photographs were released of George Walker, Bud Kaufman and a 3/8-sized model of the Edsel 196X. Although it was already obvious that the Edsel was dying, the implication of the press release accompanying the photographs was that a new-model Edsel was in the works, that the Edsel would continue into the 1960s, and that it might look something like the 196X. By November 1959, the Edsel was dead. ■

Edsel 196X

The Edsel 196X had different designs on each side. By the time the public saw the 196X, it was unlikely there would be a next-generation Edsel.

Henry Ford Museum and Greenfield Village

Quicksilver

In late 1947, when Ernie Breech asked George Walker to return to Ford as a design consultant, there were very advantageous terms built into the contract he was offered. Walker got a long-term contract for himself and for Joe Oros and Elwood Engel. At the time they returned to Ford, Oros was told by Walker that Breech insisted that he and Engel be included in the contract. Walker was also guaranteed that all publicity about design at Ford Motor Co. would be released in his name. Most importantly, Walker's contract included a provision guarantying that Oros and Engel would have the right to suggest changes to any designs in progress at the Design Department and, if there was a disagreement, Oros and Engel also had the right to produce alternate competing designs. Those contract provisions, once they became common knowledge, did not make Walker, Oros or Engel very popular, but it sure encouraged competition.

The 1952 production Ford was designed starting in 1949. Oros disagreed with the design being proposed by the Ford studio and, as was his right under the contract, he began designing a separate proposal. Oros felt strongly that the '52 Ford should lunch off the design of the X-100, then just being started. When the dust settled, more of Oros' design was included in the '52 Ford than the design prepared in the Ford studio. In 1952, when Ford studio head Frank Bianchi left for a new job at Chrysler, Bob Maguire was appointed as his replacement. Maguire was a good designer and a good administrator, but his previous design experience was primarily in interiors. He was replaced that same year by Frank Hershey, who was experienced in all aspects of design.

Ford Design Center

This full-sized clay model of the Quicksilver shown in the Styling Rotunda was very close to its final form.

Oros and Hershey worked fairly well together, but when Hershey gave two interviews that resulted in magazine articles, Walker complained long and loud. Then came the dispute over the Fairlane trim Walker talked Crusoe into putting on the sides of the Thunderbird. Hershey got it removed, but he saw the resulting confrontation with Walker as the last straw and decided to leave rather than work for Walker — assuming that Walker was going to keep Hershey as head of the Ford studio. When Walker was made a Ford vice president and general manager of the Design Department in May 1955, Hershey quit, and Walker reappointed Maguire to head the Ford studio.

When George Walker took over, he renamed the Design Department the Styling Center. He also made some significant changes in the studio structure. Although the Advanced Studio was kept, individual pre-production studios were also set up in the Ford, Mercury, Lincoln and Edsel studios. Several years later, a separate corporate advanced studio was set up downstairs with Elwood Engel as head of it. The initial purpose of the corporate advanced studio was to build competing designs with the other studios.

At first, Walker decided to keep Oros and Engel on his personal staff and gave them offices next to his, where they shared a secretary. They continued to contribute to the designs produced by the different studios and, in cases of disagreement, they produced competing designs. In March 1956, Walker was told by Breech that he should consider reducing his personal staff to just one assistant. In other words, either Oros or Engel had to blend into the Styling Center staff. Meanwhile, Ford Division was beginning to question

Maguire's qualifications as chief designer. Oros' competing design for the 1957 Ford also beat out Maguire's. Oros does not know whether it was Walker or Lewis Crusoe, head of Ford Division, who first suggested that he be appointed head of the Ford studio, but he believes it was really his qualifications that ultimately led to that appointment.

To put it bluntly, Oros was a workaholic. Designers report that he was at the Styling Center when they got there in the morning, and stayed long after they went home. Betty Oros remembers encouraging her husband to buy a boat, in the hope it would give the family time together away from the phones, and someplace her husband wouldn't be tempted to stop by the office. It only worked for a short time. Oros was an accomplished designer and practiced his art daily. During the war, he worked as an engineer and layout man, so he was thoroughly familiar with the engineering that went into each car he designed. From Walker's standpoint, Oros had also mastered the art of costing the cars he designed. One of the reasons Ford Division wanted Oros as head designer was because he always took the cost of a design into account, and consciously made an effort to design cars that looked more expensive than they really were to produce. In March 1956, Oros was appointed chief designer of the Ford studio, and Maguire became head of a new Advanced Studio formed to design the Stylerama cars.

By the time Oros took over as head of the Ford studio, the design of the 1957 Ford was practically completed. The first production car designed on Oros' watch was the 1958 Ford. Oros did not agree with Ford Division's plan to have the 1958 Ford lunch off the 1958 Thunderbird, but once the decision had been made, he supported it as best he could. The '58 production Ford lost ground. The 1959 Ford, however, was Oros' design, and Ford Division deferred to him, as he had deferred to them in designing the '58 Ford. By the fall of 1956, the design of the 1959 Ford was all but locked up.

Laura Darden

James L. Darden

Jim Darden was born and raised in California. During World War II, he misrepresented his age to enter the Army Air Corps. He became a tailgunner in a B-17 and, at age 19, was shot down. As the plane's only survivor, he was hidden by the French Resistance until D-Day and then was injured being taken through enemy lines to rejoin American forces. After the war, he spent several months in a VA hospital recovering from injuries. Between June 1946 and June 1950, Darden completed a two-year accelerated course in aeronautical engineering at USC and a four-year accelerated course in industrial design at the Art Center College. While at USC, he was also editor of the school newspaper and earned his private pilot's license.

Darden's first job as a designer was at GM's experimental styling section. He left GM in 1954 to become a senior stylist at Packard. In 1955, he joined Ford as a supervisor in the Ford Advanced Studio. After 7 months, he was transferred to the station wagon and then to the truck studio. In March 1956, he became head of the Ford pre-production studio. Darden was later appointed senior stylist in the station wagon, the Ford car, the Thunderbird, the truck interior, the Lincoln and Comet production and the Mercury Comet and Cougar studios.

Darden designed the Saturn for the Stylerama program. While he was in charge of the Ford pre-production studio, the Quicksilver was designed there. He was also one of the designers of the 1957 Ranchero, the Mustang I, the 1962 Econoline, the 1962 Fairlane and was responsible for the design of the Cougar logo.

During his free time, Darden designed boats, houses and furniture. A speedboat he designed to look like the 1960 Ford reached production.

Darden died unexpectedly in 1971 while a designer at the Styling Center. ■

Ford Design Center

Laura Darden

(above left) This three-eighths sized model ultimately became the Quicksilver. It is in Jim Darden's Ford pre-production studio. At the other end of the room was the truck pre-production studio.

(left) Jim Darden (left) and clay modeler John Amrozi are working on a three-eighths sized model of a car that eventually became the Quicksilver. After the car became a full-sized clay model, it was renamed the Quicksilver.

One of Jim Darden's early renderings of the Quicksilver.

Laura Darden

In fall 1956, Chrysler Corporation introduced its 1957 new-car lineup. Advertising referred to Chrysler's cars as the forward look, and as the cars other companies would be producing in 1960. They included, as a primary theme, big fins on the rear fenders. Originally, GM cars for 1959 were meant to be a continuation of the fatter, chrome-laden look initiated in 1958 but, when GM designers saw the new Chrysler products, they rebelled. The cars they had designed for 1958 were dumpy, when compared to the 1957 Chrysler products, so GM's Styling Section radically changed direction for 1959. Ford designers were also concerned when they got their first look at Chrysler's 1957 new-car lineup, but the Ford studio did not react in the same way GM did. Joe Oros, as head of Ford studio, was also known as strong willed, and he initially decided that the Ford studio was going to take a wait and see attitude.

As head of the Ford studio, Oros continued to work long hours but, in addition to his administrative duties, he also tried to remain an active, hands-on designer. He kept a design board in his office, which he tried to use daily. Because there was seldom enough time for him to do any of his own design work during the day, he customarily stayed after his work day was finished or came in early in the morning, so he could draw out his ideas for the designs being prepared, or that he thought should be pursued in the Ford studio. Oros says that he was one of those designers who always found it easier to think out designs with a pencil in his hand, and then use his sketches to explain his ideas to others.

After Oros found out the direction Chrysler had taken for 1957, he sketched a quick design proposal for a very low Ford, except that where Chrysler products had vertical fins, he sketched two horizontal fins that flowed from a high belt line of the car and terminated on top of the trunk. Oros showed the sketch to the designer in charge of his Ford pre-production studio, Jim Darden, and asked that he and his designers start designing a car along the lines shown in his sketch using most of the standard Ford package measurements. The Concept, as it was then named by the designers in the pre-production studio,

A later full-sized clay of the Quicksilver. The only marking on the car that identified it as the Quicksilver was the script on the front fenders.

Ford Design Center

was designed using an assumed new chassis that allowed the car to be almost three inches lower than production Fords planned for the next several years. Stu Denker recalls designing the Concept script used on the car.

In the back of Oros' mind, he believed it was only prudent to start designing a car along the lines of the new Chryslers, just in case they proved popular. In other words, Oros decided to take a wait and see approach in light of Chrysler's forward look but also decided to build the Concept as an insurance policy. The car Oros asked Darden to begin designing had a high belt line that terminated in flattened-out fins on the back end of the car and a distinctive front end with a hood that was radically sloped-off compared to current production Fords. Below the flat fins, the Concept's design incorporated Oros' Ford Division signature oval taillights that were built into a high contoured-rear bumper. Although there is some dispute, the taillights on the completed Quicksilver are more often described as oval rather than round.

Ford Design Center

(above) A full-sized clay model of the Quicksilver in the Styling Rotunda. Across the deck lid, this model was identified as a Fairlane.

Laura Darden

(left) One of Jim Darden's early renderings of the Quicksilver.

Oros also temporarily recruited George Barbaz, a designer who had worked very effectively in helping design past concept cars, to work in Darden's pre-production studio. Alex Tremulis claimed to have designed the banana nose grille used on the final design of the Concept and, at one point in his career, he did split his time between the Thunderbird and Ford pre-production studios, after his Advanced studio was abolished during cutbacks that occurred in mid-1956. Barbaz, who was a good friend of Tremulis, does not remember him working on the Concept's design. He does acknowledge, however, that Tremulis may have contributed to the design of the car. Barbaz recalls that he, Darden and other studio designers collaborated on the final design of the Concept but, as he remembers it, most of the design work was done by Darden. Stu Denker, Chuck Mashigan, Chuck Harbison and Jack Telnack were the other designers assigned to Darden's Ford pre-production studio. Telnack had just graduated from the Art Center College in Los Angeles, California, and was starting on his first assignment as a Ford designer. Ford's future vice president and general manager of the Design Center was assigned to Darden's studio, just as it was decided to turn the Quicksilver into the 1960 Ford. He recalls doing very little actual design work on the car. Denker had been at Ford for 20 months and spent a lot of his time working on the Concept even before it became the Quicksilver, mostly on the rear end. Mashigan, who soon left Ford, came up with many innovative ideas that contributed to the Quicksilver's final design.

(below) This full-sized clay model was built shortly after the decision was made to turn the Quicksilver into the 1960 Ford.

Ford Design Center

Stu Denker

Stuart R. Denker

Stu Denker grew up in Nebraska and studied art at the University of Omaha. After graduation, he joined the Air Force during the Korean Conflict. The Air Force transferred him from the military police to a presentation preparation office for the Strategic Air Command, where he prepared illustrations, charts and graphs. He was discharged in 1953 and returned to Omaha.

He was accepted at the Art Center College and was awarded a Ford scholarship after his first year. After he graduated in 1957 (with distinction), he was hired as a Ford designer. He started in Jim Darden's Ford pre-production studio, where he helped design the Quicksilver.

After 20 months at Ford, all of which were spent in Darden's studio, Denker left Ford and returned to California to get away from Michigan's cold winters. In California, he worked at a newly-formed industrial design office at North American Aviation. After the B-70 Bomber was canceled, Denker left North American Aviation and worked for several independent Southern California industrial design firms. Later, he became a partner in an architectural firm and then opened his own industrial design firm in Pasadena, California, which he still operates. ■

Darden's initial design of the Concept had flat fins, oblong-shaped taillights, a lot of relief on the sides, large front-fender vents that looked like gills and a flared-nostril front end with lights both in and outside the grille. Denker remembers a constant preoccupation with unifying the front and rear designs of the car by making the flat fins that ended on the trunk flow through to the front of the car. Attempts were also made to run the depression on the deck lid between the two fins forward into a similarly depressed area on the hood. Denker recalls that several grille designs were considered on the Concept, including the snarling-nostril look, a grille with a large snout-like opening much like the later Mustang and a pointed front end much like the one used on the 1961 Thunderbird. All designers in Darden's studio wanted to get away from what they saw as the design cliches on the production 1959 Ford, including the big round lights, the rear fender bombs and its squareness.

Joe Oros visited Darden's studio often to make suggestions and, each time Darden returned from a meeting with Oros, he brought Oros' detailed suggestions for the Concept. It was clear to everyone in the studio that Oros was very much in charge and very much involved in day-to-day design of the Concept. Denker remembers several occasions when Oros formally reviewed the Concept. His biggest complaint was always that the designers had to find a way to get more length in their design.

The story of how the Concept, reborn as the Quicksilver, became the 1960 Ford, while on its way to becoming a showcar, has some strange twists and turns.

In the mid-1950s, a research group was set up in the Ford Marketing Department to gather intelligence, including pictures, drawings, blueprints and any other material they could on competitors' future cars. Designers Bill Boyer and Ken Nelson of the Ford studio were assigned to make renderings and clay models from intelligence the marketing people were able to gather. Because these activities were clandestine and unofficial, Boyer and Nelson were paid separately by the Marketing Department and did their work at night in the Ford exterior studio, after they had completed their daily assigned duties as Ford studio designers.

After Darden's pre-production studio designers got under way on the Concept, Boyer and Nelson began making a clay model of the 1959 Chevrolet four-door sedan from measurements the marketing people had obtained from a cooperative Detroit tool and diemaker. GM had delivered the measurements to the tool and diemaker, so that tooling necessary to manufacture the car could be started. Nelson doesn't recall exactly how long it took to build the clay model of the '59 Chevrolet but, as with the other cars they built in clay for marketing, it was built in a hurry and kept under cover during the day.

The sides of Ford's clay model of the 1959 Chevy weren't entirely smoothed out, it had no trim, and the fine details weren't completed, but it did give designers and Ford Division management a good idea of what the competition was planning. For over a decade this arrangement with the same anonymous tool and diemaker was one of the best-kept secrets at the Styling Center. In the early 1990s, a hobbyist magazine ran a photograph of a full-sized Thunderbird clay model that was virtually identical to the 1966 Oldsmobile Tornado. It too was built from measurements obtained from the same source.

When Ford Division designers and management personnel got their first look at Boyer and Nelson's clay model of the 1959 Chevrolet, there was almost universal disbelief. Most who saw it assumed that GM had discovered Ford's spy game and had sent them on a wild goose chase. No one believed the competition would *ever* consider building a '59 Chevy that "looked like that thing." The batwing-flat fins on the back of the clay

One of Stuart Denker's early Quicksilver renderings.

Stu Denker

model of the 1959 Chevrolet that Ford built from the advance plans they acquired also amazed many Ford designers because of the deep panel relief in the design. Denker remembers being very doubtful that GM could build a car with as much depression in the metal as shown in the clay model of the '59 Chevrolet the designers saw, but it did encourage them to include in their designs as much sculpture as they could. As time passed, Ford studio designers began to worry that the clay model Boyer and Nelson had built might really represent what the '59 Chevrolet was really going to look like. Several months later, those doubts were confirmed by spy photos. It's an understatement to say that when Ford's Marketing Department became convinced that the finned '59 Chevrolet was for real, all hell broke loose.

By the time that Boyer and Nelson finished the clay model of the 1959 Chevrolet, it was already too late to change the 1959 Ford, but it did cause a hurried reexamination of the Ford planned for 1960. By that time, Lewis Crusoe had retired and, Robert McNamara, who replaced him as head of Ford Division, had been promoted to vice president in charge of all Ford cars and trucks. Jim Wright replaced McNamara as the new head of Ford Division, and he immediately expressed concerns about the conservative design of Oros' '59 Ford. He was also critical of the size of the round taillights but, except for Wright, management, at least initially, was very pleased with the design Oros proposed for the upcoming '59 Ford. Wright constantly referred to the '59 Chevrolet as a "swallow-tailed ground rocket."

During the 1957 model year, Chrysler products made substantial sales gains, and some in Ford Division's management began to feel that next to the finned Chrysler products and with what GM was planning for 1959, their proposed 1959 Ford looked dumpy and might be perceived by the car-buying public as out of step with the times. If there was a set of events that crystallized opposition to the design of the '59 Ford, it was Sputnik and the start of the space race. Ford management then became convinced future Ford products had to look like what Chrysler and GM were building or planning to build, or Ford would be left behind in their dust.

Meanwhile, the full-sized clay model of the Concept had been finished with alternate designs on each side. On one side it was a two-door, and on the other side it was a four-door sedan. After the car was finished and di-noced, Oros showed it to George Walker. Walker fell in love with it and suggested that it be completed as a Ford Division showcar. He also proposed that it be named the Snowflake. Oros didn't really like Walker's suggestion and proposed the name Quicksilver. Oros had no particular reason for selecting the name Quicksilver, except that he liked it. At the time he suggested it, he did not recall that the word quicksilver was an archaic name for Mercury, nor did he intend it to denote anything to do with liquid metal. His belief was that the name went well with the Concept's strong, streamlined design. Walker eventually agreed that Oros' name fit the car better than his proposal and, from then on, it was known as the Quicksilver. If the Quicksilver had been turned into a true showcar, it probably would have had a unified design theme. Oros also assumes that a Quicksilver showcar would have been built as a two-door hardtop. At about the time the decision was made to turn the Concept into the Quicksilver showcar, the proposed 1960 Ford design was being completed and had been tentatively approved by the Product Planning Committee. It was evolutionary of the 1959 Ford design.

Wright became very concerned when it was confirmed that the 1959 Chevrolet was going to have batwing fins. The production 1959 Ford had already been planned as an entirely new car. It was a very square design and, by that time, it could not be changed, but Wright proposed, and his boss Robert McNamara recommended, costly last-minute

Chuck Mashigan

Charles A. Mashigan

In 1953, Chuck Mashigan was an owner-operator of a dump truck. When his employer requested that he buy a larger truck he couldn't afford, he decided to change jobs. His wife suggested he become a car designer, since he liked to draw cars. In grade school, he had made several model cars out of soap. Mashigan, who did not attend an art school and never had an art lesson, bought some paper and brushes and prepared a portfolio on his kitchen table. He arranged for an interview at the Ford Design Department, put his drawings in a folded-over cloth bag and met with Gene Bordinat and several other Ford designers. Based on his portfolio, Mashigan was hired as a designer for a 90 day probationary period.

Mashigan was assigned to the Advanced Studio, and told that if no production studio asked for him within 90 days, he would be out of a job. Forty days later, the Ford and Lincoln-Mercury studios both asked for him.

He was then promoted to the Thunderbird studio as a manager working under Bill Boyer, where he worked on the designs of the 1956-'58 Thunderbirds. He was next transferred to the Ford pre-production studio and spent the next 20 months there. While in the Ford pre-production studio, he helped design the Quicksilver.

In 1958, Mashigan was offered and accepted a job at the Chrysler design department. After Elwood Engel was appointed vice president of design at Chrysler, Mashigan was asked to turn one of the 3/8-sized clay models he had designed at Ford into the full-sized Chrysler Turbine showcar, which he did.

In 1963, Dick Teague, who had previously worked with Mashigan at Chrysler before becoming American Motor's head designer, recruited Mashigan to head American Motors' new advanced studio. Mashigan headed the advanced studio at American Motors until 1973, when he became chief designer in the Jeep studio.

Because his wife was ill, Mashigan took early retirement from American Motors in 1982. ■

Ford Design Center

One of Chuck Mashigan's early rendering of the Quicksilver.

Ford Design Center

The Quicksilver was designed as a two-door on one side and a four-door on the other side.

John J. Telnack

Jack Telnack was born and raised in Dearborn, Michigan. As a child, he wanted to be a Ford designer. As a teenager, he watched new Fords being tested on the Ford test track. After seeking advice from Alex Tremulis and Homer LeGassey, Telnack applied and was accepted at the Art Center College in Los Angeles, California. During school, he worked summers at the Ford glass plant, as a designer for Haw Taw Engineering, and as an auto salesperson. During his last two years of school, he was awarded a Ford scholarship. He graduated in May 1958 and was hired as a designer at Ford. His first assignment was in Jim Darden's Ford pre-production studio, where he helped finish up the design of the Quicksilver. He later worked on the 1965 Mustang, in the station wagon studio and, in 1965, he was appointed manager in the Lincoln-Mercury studio.

Ford Design Center

From 1966-69, Telnack was chief designer, Ford of Australia. In 1974, he was appointed vice president of design, Ford of Europe and, in 1976, he returned to Dearborn as director of North American Light Car and Truck Design. In 1978, he was appointed director of Ford's Mid-sized Car and Interior Design Studio. From 1980-87, Telnack was chief designer for Ford North American Automotive Operations, during which time he was the driving force behind the aero design of the 1983 Thunderbird and the 1986 Taurus and Sable.

In 1987, Telnack was elected a Ford vice president and made the head of Ford World Wide Design. In the mid-1990s, Telnack was responsible for the New Edge design theme proposed for Ford products.

Telnack was named Automotive Industries *Man of the Year in 1989 and has received honorary degrees from Pratt Institute, Parsons School of Design, and the Art Center. He retired in December 1997.* ■

changes in the design to the 1960 Ford. The Product Planning Committee concurred, and a crash program was set in motion to come up with a more streamlined design for the '60 Ford and, more importantly, one that had fins. For the second year in a row, an entirely new car had to be engineered. Designers in the Ford studio feverishly began preparing renderings and then clay models, but none of the designs they came up with proved acceptable and, as Nelson recalls, the redline date was rapidly approaching.

Even though the decision was made by Walker to turn the Quicksilver into a show-car, it never got beyond the full-sized clay model that was di-noced and painted silver. Foil was used to cover the bumpers and bright work. It was a sleek, crisp and aerodynamic design but, at the time he was designing it, Darden told Ken Nelson that it was also a design Darden was positive had no chance of becoming a production model. Darden recognized that the Quicksilver went in a totally different design direction than what was originally planned for 1960. He also thought the Quicksilver was inconsistent with the type of design he expected Joe Oros to eventually select for a production proposal.

In the meantime, Barbaz recalls that, when the full-sized clay model of the Quicksilver was shown to the Styling Center staff and Ford Division management in the courtyard of the Styling Center, several secretaries complained that the car's corners could not be seen during parking. To Barbaz' surprise, Darden was asked to revamp the car so the corners were more visible to the driver.

The original proposal for the 1960 Ford used the same basic structure as the 1959 Ford, but it avoided many of the design cliches that the young designers objected to on the '59 Ford. Although the model received tentative approval from the Product Planning Committee, its final form lacked the bombs on the back fenders, and it had a pronounced open snout above the grille, much in the manner of the later Mustang. As Ford studio designers became more and more frustrated with their inability to come up with an acceptable design for the '60 Ford that Ford Division management would accept, Darden noticed that more people were stopping by his studio to look at the Quicksilver. He assumed it was because there was some specific feature on the Quicksilver that appealed to them. Wright then began stopping by Darden's studio to study the Quicksilver, and he eventually suggested that Henry Ford II take a look at it. According to Oros, it was Wright who first suggested that Henry Ford II consider turning the Quicksilver into the '60 Ford. When Henry Ford II saw the Quicksilver, he immediately decided he wanted it as the next full-sized Ford. Darden was told that the Quicksilver was being moved upstairs, and that it was going to become the 1960 Ford. As Darden told Nelson and other associates for years afterwards, it was the last thing he ever expected to happen -- and he was dumfounded by the decision.

After the Quicksilver was moved into the Ford studio, which occurred about January 1, 1957, Bill Boyer was assigned to make it fit the package set for the 1960 Ford. Several Ford exterior studio designers, including Jack Mills, John Middlestead, John Foster and Ken Nelson were also assigned to work on the car. Converting the Quicksilver into the production '60 Ford was not an easy task, and it frustrated many Ford studio designers. Oros recalls that, during that time, his studio designers were working lots of overtime trying to transform the Quicksilver into the production '60 Ford. Wright was also anxiously suggesting that more designers, including those who originally designed the Quicksilver, be put to work on it. About 2 1/2 inches in height was added to the car to accommodate the revised 1959 chassis and, to some degree, Boyer rounded out the general design. There is a difference of opinion among those who worked on the car as to just how much it was changed from the Quicksilver, but they all agree the '60 Ford and the Quicksilver looked very similar. Because of cost constraints and objections from Wright, who thought the round taillights were too big, the bumpers were simplified, which resulted in the loss of the traditional Ford round taillights except for the simulated outline in the rear bumper just below the half-oval taillights. On March 11, 1958, the job of turning the Quicksilver into the production 1960 Ford was completed, and it was given final approval by the Product Planning Committee.

Ford Design Center

(above) A full-sized clay model of the Quicksilver with a rear end proposal that was close to that of the production 1960 Ford.

By the time the redesign of the 1960 Ford was completed, a crash program was already under way to engineer the body and revise the 1959 Ford chassis to accept it. To the surprise of everyone, except Joe Oros, the more boxy '59 Ford outsold the batwinged 1959 Chevrolet, and it even won an international design award. As the time for the 1960 Ford to be introduced approached, Ford Division management began having second thoughts about their flat-finned new model. It was rapidly becoming apparent that the '60 Ford had been developed so fast that the body engineering people did not have sufficient time to eliminate excess weight from the car or to properly insure quality control.

Nelson gives as an example of how fast the car was put into production problems that arose with the laws in one midwestern state over the width of the 1960 Ford. It was 1/4 to 1/2 of an inch over the width their laws allowed and was almost required to have marker lights on the sides before it could be sold in that state. Something was ultimately worked out to avoid the side-marker lights, but that was only one example of the problems faced by the '60 Ford. As the time for the introduction of the '60 Ford approached, Ford Division management had the ad copy prepared for it stiffened up and squared-off, even though those ads were initially rejected by the Styling Center.

The 1960 Ford did not sell well. Sales of the big Ford were off by more than 500,000 units in 1960, even though combined sales of the big Ford and the Falcon almost equaled sales of the 1959 Ford. After the '60 Ford was declared a market failure, Tremulis thought he was made a scapegoat for the car's design. He also believed that the changes Boyer made to the Quicksilver made it into a different car than the one he helped design. In fact, Tremulis believed that Boyer ruined it. Darden never spoke publicly about the Quicksilver, but grumbled to his co-workers when Tremulis took credit for his design work.

Most Ford studio designers, including Oros, continue to believe that the design of the 1960 Ford was one of their better ones. They also believe that if they and the engineers had had more time to cure the weight and quality-control problems, the car would have sold better. In the haste to turn the Quicksilver into the 1960 Ford, no one now remembers what happened to the Quicksilver. It never became a Ford showcar but, by 1961, the big round taillights on it that Jim Wright thought were too big, were back. ■

John B. Foster

John Foster was born in Beliot, Wisconsin, the same hometown as his lifetime friend, Bob Koto. As a child, Foster moved to England with his family. While there, he started drawing cars. Foster's family returned to the United States, where he studied at Beliot College for a year. His family moved a lot but, after the Depression, they settled in California. His father showed Foster's artwork to a Cadillac dealer he knew, and it then came to the attention of Strother McMinn, who encouraged Foster to apply for a position at GM as a designer. To polish his talents, Foster enrolled in art classes at the Art Center and, in 1936, was hired as a designer at GM. His first assignment was in the Oldsmobile studio working under George Snyder. He was let go by GM in the recession of 1938 along with many other designers. He was a designer at Hudson for a year and at Chrysler for another year. He returned to GM in 1941, where he was reassigned to the Oldsmobile studio. That same year he was drafted. During World War II, Foster headed the Visual Arts Department at the Command and General Staff School at Fort Leavenworth, Kansas and, later, served as a counter intelligence agent in China and Southeast Asia.

Ford Design Center

After the war, Foster returned to GM. In 1951, he left to work for independent industrial designers Sundberg and Ferrar, where he worked on the Packard account. In 1954, when Packard merged with Studebaker, Foster was offered a job at Ford by Frank Hershey, who he had known at GM. At Ford, Foster was assigned to the Ford exterior studio, where he worked on the design of the front and rear ends of cars, including conversion of the Quicksilver into the production 1960 Ford. From 1964-66, Foster was manager of the Ford pre-production studio and, from 1966-68, he was manager in the Lincoln-Mercury pre-production Studio.

In 1968-69, Foster helped John Najjar form Ford's Industrial Design Studio.

In 1981, he took early retirement. ■

Ford Photomedia

Bill Ford and the Levacar Mach I at the Ford Rotunda. The Gluideair is on the track above the clock.

Levacar

On April Fool's day 1958, Andrew A. Kucher, Ford's vice president of engineering and the head of Ford's Scientific Laboratory, held a press conference at the Engineering and Research Center. Although the date on which the press conference was held was happenstance, the "car" introduced that day caused some to believe a better day could have been selected for the introduction of the Gluideair, which was designed to move on a thin film of air without wheels. Kucher and David J. Jay, senior development engineer, surprised the automotive press by demonstrating a three-foot long working model of the Gluideair. An air hose connected to the Gluideair raised it up off the ground slightly, as it skimmed along the floor. Members of the press were invited to ride a full-sized Levascooter around the room. Judging from the articles in newspapers and magazines that appeared all across the country over the next several weeks, Kucher's invention caused a sensation, with many of the newspaper and magazine articles forecasting cars of the future that would ride on a thin film of air thereby making wheels and tires obsolete.

Kucher's publicity blitz was not limited to magazine and newspaper reporters. On April 1, John Daly, television host of the game show *What's My Line,* introduced the Gluideair on WABC and, on April 6, over 15 million people read about the Levascooter and the Gluideair in a Sunday comic strip called *Closer Than We Think*. Several days later, Kucher's public relations head, Colver R. "Cog" Briggs, took the Levascooter to New York City for an appearance on Dave Garroway's *Today* show and demonstrated it to millions of television viewers, as he told them that "Henry Ford put America on wheels and now Henry Ford II was planning to take those wheels away on their future vehicles."

By mid-April 1958, over 200 television stations around the country had received a film clip from Ford public relations on the Gluideair, and reports filtering back to Ford indicated that most of the stations were airing the film clip on their news shows. Ford Motor Co. was caught up in the favorable publicity about their wheelless car and, in late April 1958, the Company purchased a two-page advertisement in *Automotive Industries* magazine forecasting that Kucher's Gluideair was nothing less than the "future successor to the wheel." As fantastic as it seems in hindsight, engineers involved with the project at the time really believed it was just a matter of time before Ford began selling vehicles that rode on air instead of wheels.

Cog Briggs, who was also an engineer, was the one who named the the wheelless vehicle the Gluideair but, just as the publicity build-up started, Boeing Aircraft notified Ford that it held a prior copyright on the name Gluideair for one of its cargo-loading systems. Briggs then came up with the name Levacar, which was a derivative of the word levitation.

The idea of the Levacar actually came from Kucher, a former General Motors and Bendix engineer who, in 1951, was hired as the first director of Ford's Scientific Laboratory (Engineering Research and Advanced Product Study Office). As a child, Kucher first noticed that cups he was drying for his mother seemed to float along on the drainboard after she washed them and placed them there to drain. Kucher wondered what caused the seemingly frictionless movement of the cups.

One of the Levacar Mach I's designers, Gale Halderman, is airbrushing a full-sized rendering of it.

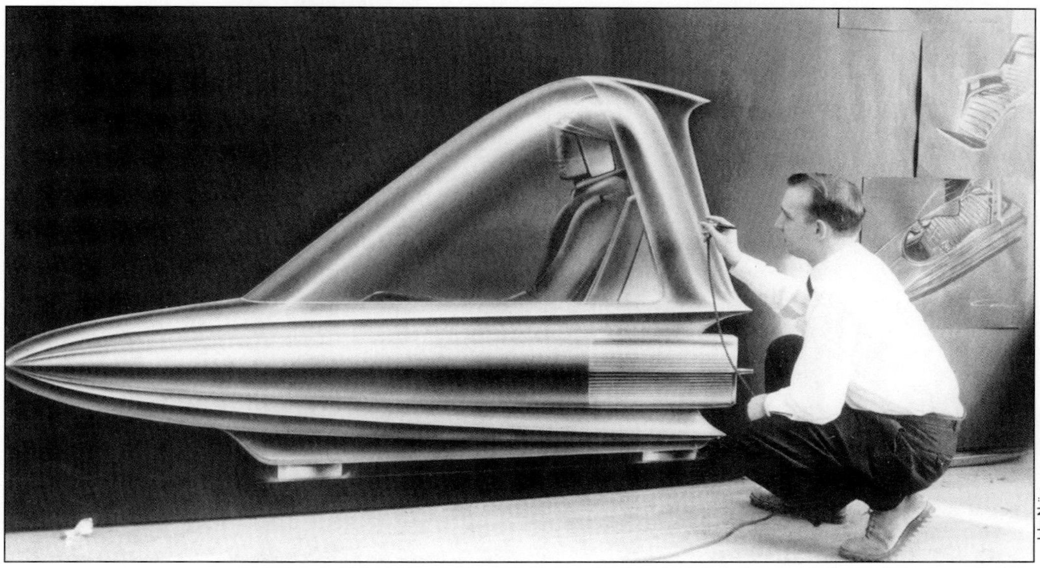

John Najjar

After 30 years of research, Kucher turned his childhood curiosity into an invention he called the levapad, which was used to create a type of frictionless movement using air as a lubricant over a smooth surface. Kucher theorized that he could generate lift with air pads and, eventually, levapads would totally do away with the limitations of wheeled vehicles. He believed that traveling on air would permit travel at a much higher speed maybe up to 500 mph.

No matter what Ford's public relations people were saying and the press was reporting, by the time the Gluideair was introduced, Kucher had already concluded that levapads could best be utilized for mass-transit trips of more than 100 miles in length. Although it was not emphasized as a part of the press buildup, by that time, Kucher thought his invention could best be utilized on wheelless trains that rode on one rail. Kucher proposed stabilizing his levatrain with a gyroscope, and his dream was to build a monorail train system operating on levapads and running from the Detroit Metro airport through Dearborn to downtown Detroit.

The 3/8-sized Gluideair was designed by Doug Miller and Norm Nebozenko, staff designers at the Scientific Laboratory. By the time the Levacar was designed, the Special Vehicles Department was being disbanded because of a lack of budgeting and, what was left of it, including designers Doug Miller and Norm Nebozenko and clay modeler Walt Jennfeldt, was absorbed by the Scientific Laboratory and generally kept busy on in-house projects. (Nebozenko soon left Ford.)

(below) The full-sized clay model of the Levacar Mach I with clay modeler Joe Seibold.

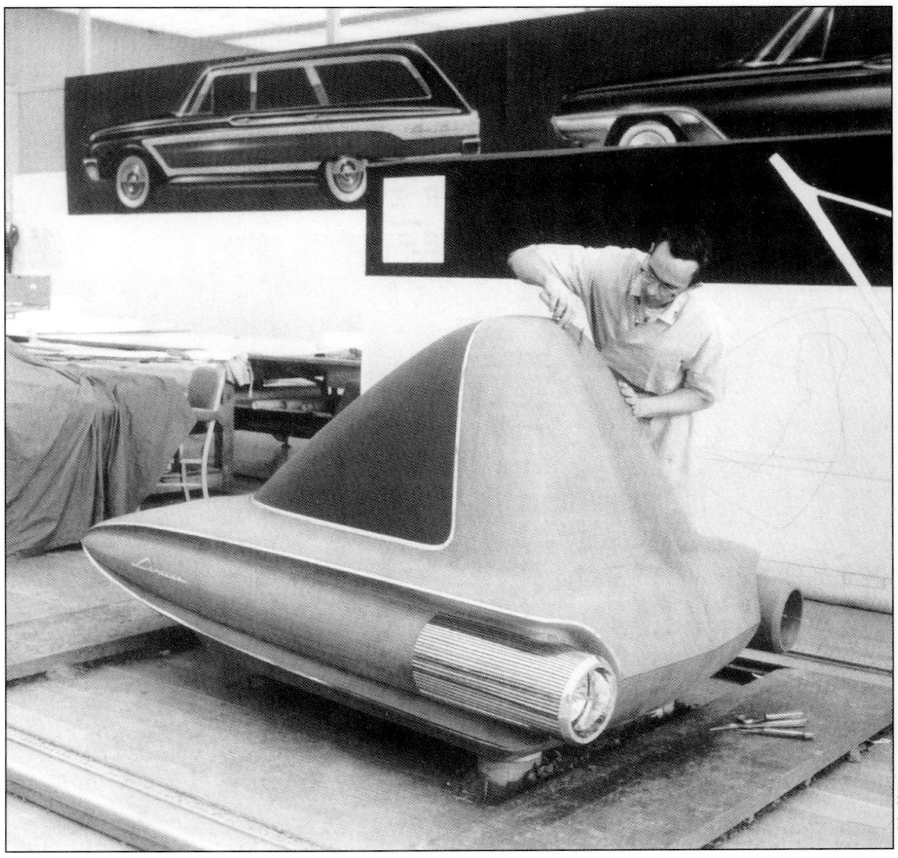

John Najjar

By 1958, Kucher's researchers solved the tendency of the levapad to vibrate at speed, and Kucher asked George Walker, head of Ford's Styling Center, to help him come up with a way of demonstrating his new invention to the public. Walker was happy for the opportunity to again do design work for the Engineering Department, and he was very anxious to help Kucher publicize his idea. When Kucher approached Walker, Walker wholeheartedly embraced the Levacar.

The project was assigned to Elwood Engel's new corporate advanced studio, and John Najjar, as Engel's executive stylist, was assigned to work closely with Kucher on the design of the Levacar. Although both Engel and Najjar oversaw the project, it was assigned to Gale Halderman, who was the project designer. Dean Beck also worked on the design of the Levacar. Bob Thomas and later Ray Smith were the design analysts on the project, and Harry Strickler, Cecil Perkins and Dick Sprague were the primary clay modelers. On the day the photographs were taken, Joe Seibold happened to be the only modeler in the studio and, although he had done little work on the Levacar, he was the one photographed working on it. Because Kucher was tall, Thomas had to make sure the one-man Levacar was tall enough for him to fit.

John Najjar

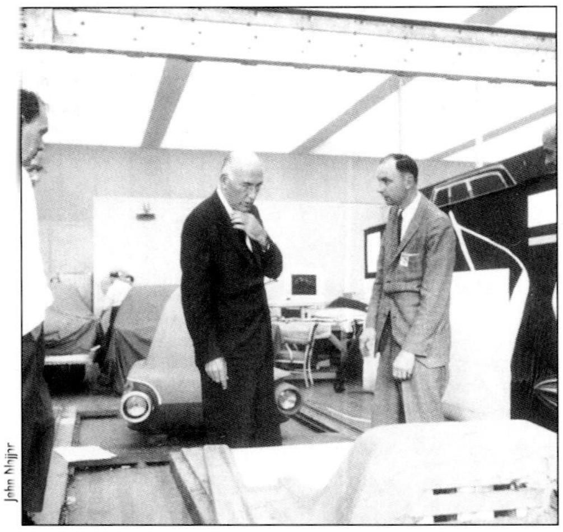

John Najjar

John Najjar

John Najjar

John Najjar

John Najjar

(top) George Walker (left) and Andy Kucher (right) inspecting the clay model of the Levacar Mach I.

(above) From left to right, JIm O'Neal (planning), Elwood Engel, George Walker and Andy Kucher watch as clay modelers make changes to one of the taillights on the Levacar Mach I.

(top) From left to right, Gale Halderman, Andy Kucher and Elwood Engel critiquing the seating buck.

(above) From left to right, Jack Malolly (Ford Rotunda), Victor Raviolo (Engineering) and Andy Kucher discuss proposed changes to the seating buck.

(top) Andy Kucher thought the fin on the rear of the Levacar Mach I needed to be bigger.

(above) Jack Malolly (left) and Andy Kucher discuss the clay model. Kucher liked the design but wanted to fine tune it.

(below) The Levacar Mach I at the Ford Rotunda. The tubes attached to the car controlled it and also furnished the air pressure needed to power it. In the background, from left to right, are the X-2000, the Nucleon and the X-1000.

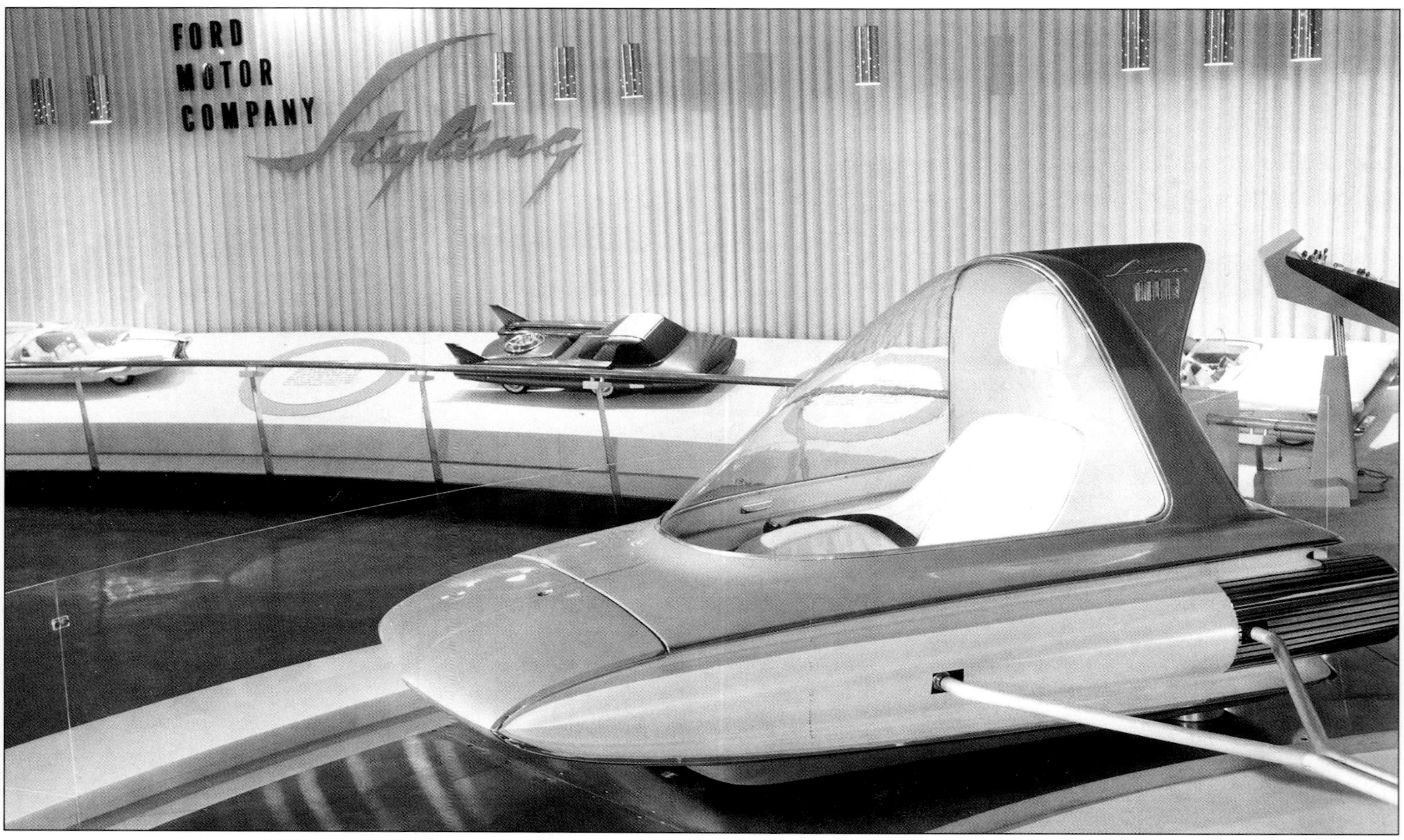

Ford Photomedia

The first thing Halderman and his designers concluded, when they started to design a full-sized one-man Levacar, was that a single levapad was not enough to provide the needed lift. Kucher's levapads were made of machined steel, were about the size of a dinner plate and about 1 1/2 inches thick. Each levapad had fine pinholes on the bottom, and a tube on top where air was pumped in. Kucher assigned engineer David J. Jay to work with the designers. Jay and Halderman soon concluded that three levapads, two in the rear and one in the front, would provide sufficient lift.

The one-man Levacar was designed to use an on-board air compressor to supply the 15-100 pounds per square inch of air needed for levitation. The levapads were designed to mount on the bottom of the Levacar on spherical bearings, which permitted rolling adjustments to the surface, which also had to be very smooth. For forward movement, the designers found that almost anything would move the Levacar, once levitation had been established. They settled on an air-powered squirrel-cage car heater fan they planned to mount behind the seat to provide the thrust needed for forward motion.

Given the monetary and weight restrictions, the single-passenger Levacar did not have an on-board power plant. The Mach 1, as the single-passenger Levacar was called, had no need for a steering wheel, because the direction of the car was controlled from the outside. The only controls on the Mach 1 were the catch that opened the rear-hinged passenger compartment, a knob to turn on the supply of air to the levapads, and another control to turn on whatever controlled forward motion.

The Mach 1 was 46.72 inches high, 93.56 inches long and 53.50 inches wide. Weight, which had to be as little as possible, was 252 pounds for the body and 160 pounds for the chassis. As the full-sized clay model neared completion, Kucher proposed a larger fin on the back of the Mach 1 for proportional balance and the designers agreed. A separate full-sized seating buck of the Mach 1 was also built.

Because the surface on which the Mach 1 ran had to be completely smooth, no suspension or springs were needed. Kucher and Walker approved the final clay model the full-sized Mach 1 was to be built from, and the Mach 1 was then built over a period of 25 days, mostly of aluminum, by Wettlaufer Engineering Corp. of Detroit. The Mach 1 used epoxy-resin plastics made by Ren Plastics of Lansing, Michigan, for the interior molded areas. When completed, it was painted two-toned red and cream and upholstered in white leather. By that time, plans were to control forward motion and to supply compressed air through a rigid-tubular tether that only allowed the Mach 1 to travel in a circle on the stainless steel track.

John Najjar

John Najjar (left) talks with George Walker about a Levatrain proposal. In the background, designer Walt Becker is at this drafting table.

On May 20, 1958, a new display featuring the Mach 1 opened at the Ford Rotunda. A press demonstration was held that night and, the next day, it was demonstrated at Ford's annual shareholders' meeting. The display was called the *New Adventures on the American Road, The Levacar* and, after Ford shareholders saw it, it was opened to the public. When visitors entered the Ford Rotunda, they saw a series of 3/8-sized models designed in the Advanced Studio, including the X-1000, the Nucleon, the X-2000, the Mexico and the Volante, arranged on a banked raised platform that ran along the side of the room.

The real show, however, occupied the entire center of the room, where a raised platform about 35-feet in diameter had been constructed with a three-foot wide flat-steel track running around its perimeter. Over the top of the round platform were arches and on top of the arches, about ten-feet higher than the track, was another round track, this one smaller and made of glass. On the raised-glass track, visitors saw the the original Gluideair Kucher had first shown to the press a month earlier. On the lower track was the spaceship-like Levacar Mach I. As visitors watched the exhibit, a Ford Motor Co. employee picked up a microphone and, as the sound of screaming jets could be heard through overhead speakers, the show got under way...

Ford Design Center

(above) After Andy Kucher decided a levatrain was more practical than the Levacar, designers and clay modelers in the corporate advanced studio produced several three-eighths sized models of proposed levatrains. This one was called the Aereon I.

Ladies and gentlemen, Ford Motor Company is proud to present its new concept of time and travel -- The Levacar Mach I. You heard the roar of the jets overhead. As they zoom by at speeds faster than sound, their whining motors blast out the refrain of the jet age. Now the jet age has come to ground travel too. This Levacar demonstrates a principle of ground travel that is as far ahead of vehicles that run on wheels as jets are ahead of propellers. Let me show you what I mean.

Take a look at the Levacar over there. (Points ...) What's the first thing you notice? Sure, no wheels. The wheel has always been considered the greatest mechanical invention man ever made. From the days of the cave man, it has figured in all our concepts of travel on the ground. But the Levacar does away with the wheel. (Picks up a levapad ...) How does it do it? The answer is right in my hand. This disk is called a levapad. It's smooth, but there are several tiny holes in the bottom. (Holds up to show bottom of pad ...) When I turn this valve, (Points to valve on top...) air compressed at high pressure blasts out of these holes. Before I turn the air on, I will move this pad on the surface of the counter. (Moves it back and forth a couple of times ...) I can move it, all right, but you can see there is a lot of friction. That's why we have always used wheels. But Ford Motor Company's scientists and engineers sometime ago began looking for something better than wheels. It is their job to search continually for better materials, better tools, better ways of doing things. A short time ago they developed a vehicle that embodies the principle of running on a layer of air. Here's the way it works. I will now turn on the air. (Does so ...) Feel it coming through the tiny holes? (Holds it towards someone in the front ...) Now see how easily I can move this levapad. (Run it back and forth on the counter ...) The reason is that the compressed air coming out of the holes is actually holding this pad a few-thousands of an inch above the ground. It forms a layer of air between the pad and this surface. The engineers call this levitation. The little car up there on the glass highway (Points ...) and the Levacar Mach I (Points ...) each have three of these levapads on the bottom. In a moment you will actually see those cars riding on air.

Ford Design Center

Included with the several different three-eighths sized levatrain proposals was this personal-sized version.

I am now going to move up to the panel that controls the small car. (Ascends steps to panel ...) You can see it there on the glass highway. The arm you see coming out from the center does not propel the car. It merely controls the direction for purposes of this demonstration. When I press this button, I will send the car, not zooming into space, but speeding along its glass highway. Here we go! (Presses button, waits until the car stops ...)

That, ladies and gentlemen, is riding on air. Some day you may be doing it.(Starts moving over to Levacar Mach I ...)

Before I go for a ride in the Levacar Mach I, I will try to answer a few of the questions you may have in mind. How would you power the Levacar? It would have its own turbojet engines which would both propel it and levitate it. Our engineers tell me it takes about fifty horsepower to levitate each ton the vehicle weighs, but merely another forty horsepower would zip it along at 100 miles an hour. And with friction practically gone, you would have tremendous fuel economy. How fast could a Levacar go? From 300 to 500 miles an hour. We call this experimental model Mach I because Mach I represents the speed of sound, which is about 750 miles an hour. The wheel is useful only up to about 150 miles an hour. After that, it looses traction, vibrates violently, and steering control is lost. But levapads are not restricted by any of these limitations. Of course no human being has fast enough reactions to control it at such speed. Radar and other automatic devices would control speed, breaking, and steering. Are there any practical uses for the Levacar? One would be a monorail. A specially designed levapad would fasten around the rail, (Show how with hands ...) and air jets would hold it away from the rail. Our engineers figure that a Levacar on a monorail could easily make a safe trip from Detroit to Chicago in less than an hour.

Ford Design Center

(above) Doug Miller's proposal of a commercial-sized levatrain. Kucher realized that the bigger the levatrain, the better chance his dream had of reaching production.

I will try to answer your questions afterwards. But now I am going to take a jet-age ride in Levacar Mach I. This is the first public demonstration of an experimental model big enough to accommodate a human being. When our engineers had finished their work, they asked our stylists to design a body. You see, the futuristic models in the previous exhibit (Point ...) do lead to some practical applications. (Pats the Levacar ...) How would you like to have a baby like this parked in your driveway?

For this demonstration we build a steel highway (Point ...) to give the perfectly smooth riding surface the levapads need. We installed a radial arm (Point ...) to control direction. And we put in a catapult (Point to it ...) to give the car its initial motion. When it gets started, I'll take over.

Now I'll open the canopy. (Does so ...) When I press this control, (Touches control ...) I levitate the car. With this knob (Touches knob ...) I will control the speed. Hang on to your hats, ladies and gentlemen. We are going to blast off into the jet age. (Gets in, takes ride ... After the ride, opens canopy ...) There you are, ladies and gentlemen. Remember, the research, engineering and styling that produced Levacar Mach I goes into every one of the present Ford Family of Fine Cars. This gives you some notion of what it is preparing for you in the future.

Thank you.

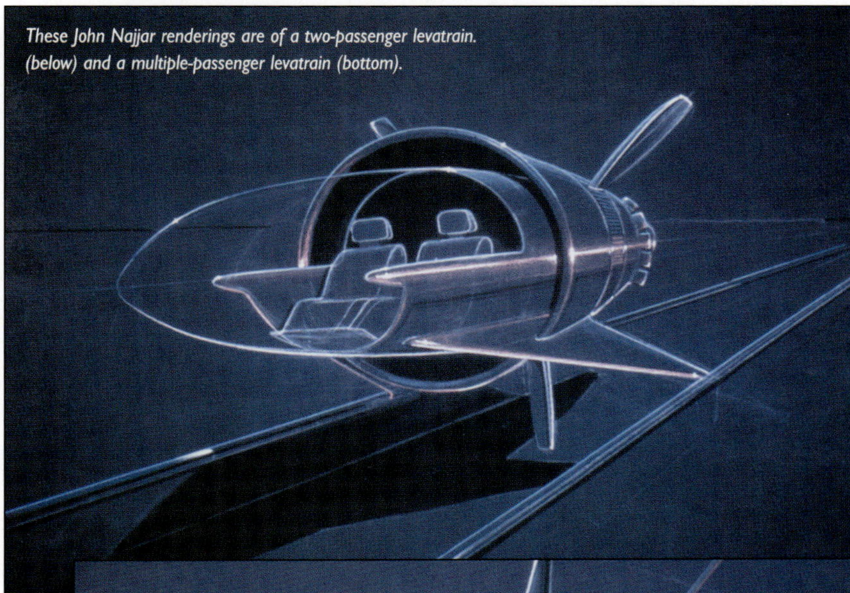

These John Najjar renderings are of a two-passenger levatrain. (below) and a multiple-passenger levatrain (bottom).

John Najjar

John Najjar

Those attending the Levacar display were given a keepsake to remind them of the levitation demonstration they had just seen. It was a little plastic model of the Levacar Mach 1. The toy Mach 1 had a mouthpiece and a small tube that attached to the model. When you blew into the mouthpiece, air pressure lifted the small model off the ground like the real thing.

Kucher originally wanted to build two other full-sized Levacars; the Fomotion, a large sixteen-passenger bus, powered by two turbo-jets, stabilized by a gyroscope and designed to run on a single rail and; a full-sized Gluideair, which was to be a smaller multi-passenger vehicle. Although designs were prepared for several full-sized Levacar-type vehicles, only the single-passenger Mach 1 was built. Several years later, a small model of a two-railed levatrain named the Aeolus was built and displayed at several events around the country.

Ford wasn't the only manufacturer developing new forms of air-powered craft in the late 1950s. In 1958, Saunders-Roe Ltd. of Great Britain, put hovercraft passenger boats into operation between England and the Continent. (A hovercraft rides on a cushion of air from six inches to several feet above the surface, while the Levacar was designed to slide along a rail or across a smooth surface on a thin film of air.) In August 1959, newspa-

Andy Kucher is about to ride in the Levacar Mach I, while George Walker looks on from the outside.

Ford Photomedia

pers in the United States and Canada reported that as many as 30 companies, including Ford and Chrysler, were working on cars, boats or trains that rode on a film of air. In September, Curtis Wright Corp of South Bend, Indiana, announced it was putting a hovercraft-type air car into production that fall, that was for use in off-road areas. In the future, Curtis Wright said it planned to produce personal-sized air cars using the same technology. Curtis Wright's air car was designed by former and future Ford designer Duncan McRae. Later that same month, the U.S. Army and the Navy held a joint demonstration of their own air car proposals and, as a result, the Defense Department awarded $224,000 in research contracts.

In October 1959, seven vehicles, or models of vehicles that rode on air, including Ford's Levacar, were exhibited (two were demonstrated) at the Forestal Research Center at Princeton University. After that, almost every newspaper in the country again carried articles forecasting that wheelless vehicles that rode on air were in our immediate future. The newspapers said that it was just a matter of time before Levacars or hovercraft-type vehicles replaced wheeled vehicles. In mid-1959, magazines all across the country, including *Boys Life,* carried two-page ads on Ford's Levacar. By December 1960, General Motors announced that it had also begun serious research on air-cushioned cars and trains.

Meanwhile, the demonstration of the Mach 1 at the Ford Rotunda showed there were problems with the Levacar. For several months, the public was permitted to ride in the Mach 1, but it was soon determined that the surface it ran on had to be constantly readjusted, so the demonstration rides were discontinued. In January 1959, the Gluideair and the Levascooter were shown at the Chicago Auto Show as part of the Edsel display. The next month, the Gluideair and the Levascooter were demonstrated in Cleveland, where the Gluideair blew a fuse at the local TV station knocking it off the air for several minutes. In March 1959, the Gluideair and Levascooter were demonstrated by Kucher to local service organizations in Miami, Florida.

By mid-year 1959, Kucher decided that his best chance of getting an actual levapad transportation system built was as a train. The Mach 1 was then modified to run on levapads attached to rails. At the same time, the model of the Gluideair was also converted to run on rails. The redesigned Mach

Andrew A. Kucher

Before he completed grade school, Andy Kucher was building model airplanes. He continued to build them for the rest of his life. By age 18, Kucher had designed and flown his first full-sized plane. After high school, Kucher worked as a designer for several east coast aircraft manufacturers, where he designed planes and developed a patented design for an aircraft that obtained its lift from the shape of the fuselage. Beginning in 1918 and, even though he had never attended college, Kucher taught aerodynamic theory and practice at the City College of New York.

Starting in 1921, Kucher was employed by Westinghouse Electric as a consulting engineer and soon became an expert on refrigeration. While a Westinghouse Electric employee, Kucher also started his own business manufacturing airplanes of his own design. He later started his own airplane design business.

During the Depression, Kucher returned to the refrigeration industry. In 1942, he took a position as director of research at GM's Bendix Aviation Division. In 1944, he was appointed vice president of research engineering at GM's Frigidaire Division..

In 1951, when Ford organized its Scientific Laboratory, Ernest Breech, who knew Kucher from his days at Bendix, recruited him as its first director. In 1957, Kucher was appointed vice president of engineering and research at Ford, a position he held until his retirement in 1962.

Kucher held over100 patents for his inventions and industrial processes, which included the sealed-unit refrigerator, new methods of aircraft construction and the residential air-conditioning system in use today. After retirement, Kucher continued working on a full-sized flying car. ■

(below) Each person who visited the Levacar display at the Ford Rotunda received this toy model of the car. It came with an air tube that was attached to the toy. When air was blown into the tube, the toy rose slightly off the ground and slid along the surface just like the real Levacar Mach I. The model shown is from Cog Briggs' collection

1 was shown running on rails at Ford's *Design For Suburban Living* display at Redondo Beach, California, in February 1960 and, in June 1960, the Mach 1 was displayed at a Denver, Colorado, shopping center. In the early 1960s, Kucher submitted a proposal to build a multi-passenger levatrain transportation system from an airport New York City wanted to build near Atlantic City, New Jersey, into New York City, but the airport project never got off the ground. At one point, Kucher also got Ford approval to purchase a square mile of property near Flint, Michigan, where he planned to build a circular test track, set up a prototype full-sized levatrain, and test it at speeds of up to 500 mph. A DC-3 fuselage was proposed as the basis for the test levatrain.

Ford Design Center

Gale Halderman in the Levacar Mach I at the Ford Rotunda.

After awhile, Kucher came to believe that Ford Motor Co. was not as interested in his levitation projects as he thought it should be. When George Walker was head of the Styling Center, he encouraged unusual and different concept cars. When Walker turned 65 in 1961, he was required to retire. His successor, Gene Bordinat, no longer encouraged the unusual type concept cars designed during Walker's regime. Henry Ford II also decided that Ford Motor Co. should not be in the business of building railway vehicles. That ended all work on the levatrain/Levacar including the full-sized test track.

Ford Motor Co. spent over a million dollars on the Levacar and, at its peek, Kucher had 12 full-time employees working on the project. Before Walker retired, John Najjar made several renderings of the multi-passenger Levacar concept vehicles Kucher wanted to build, and several levatrains were designed, but each of those projects was rejected by Henry Ford II as too expensive. In 1962, a small cigar-shaped model of a turbine-powered train called the Aeolus and named by Cog Briggs for the Greek god of wind, was designed, and 20 feet of track assembly were built at the Design Center for display at the 1962 Seattle Worlds Fair as a part of a railroad display. The Aeolus was designed by Doug Miller, just before he was transferred back to the Styling Center from the Scientific Laboratory.

Kucher's forecast for the future of levitation emphasized that, with no springs, suspension or transmission and, with much smaller engines, Levacars would really be economy cars. Kucher also forecast that roads would be cheaper to build, because there would be no need for strengthened roadbeds or bridges. According to Kucher, trains and cars that operated on the levitation system could be commonplace by the year 2000. Kucher's retirement, on March 31, 1962, meant that the Levacar concept had no champion and, eventually, interest in levitation as a means of land transportation went by the wayside.

The Levacar Mach 1 and the Gluideair were so different, they had no known influence on future Ford products. Sadly, fire destroyed the Ford Rotunda in 1962. The Mach 1 is believed to have been in storage in one of the back rooms at the Rotunda and was presumed lost in the fire. The Gluideair, the Aeolus and the tracks used for their display were stored above the wind-tunnel in the Engineering Building. Some years later, long after he had retired, Cog Briggs received a phone call from an engineering employee asking what he wanted done with the Gluideair and the Aeolus. Briggs wasn't able to take them and presumes they were destroyed. ■

The three-eighths sized Gluideair was designed by Doug Miller.

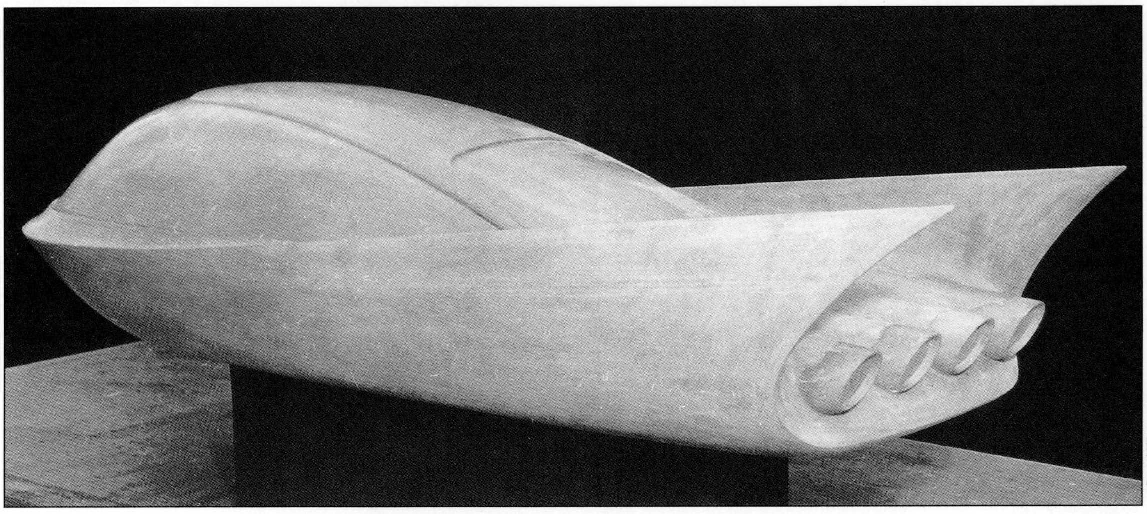
Doug Miller

Gyron

The completed Gyron in the corporate advanced studio. Giving it the once over are Gale Halderman (right) and Ray Smith (left). In the background, Colin Neale, Phil Payne and Bill Dayton are seated at their drafting tables.

One of the more unusual concept cars to come out of the corporate advanced studio was the Gyron. Although it was introduced in 1961, it was designed over the course of five years beginning in 1956. When McKinley Thompson was hired as a designer in 1956, he was assigned to Alex Tremulis' Advanced Studio. One of the tasks he was given was to prepare sketches for a proposed vehicle Tremulis wanted to build that incorporated a gyroscope to keep it upright.

Thompson's first Gyron design had four wheels and was built as a small model. His next design had two wheels with a bridge between them. Tremulis put a toy gyroscope on the two-wheeled Gyron model and wound it up with air pressure from the air hoses used to run the studio airbrushes. Tremulis delighted in airing up the gyroscope and then watching the small Gyron, stabilized on two wheels, dart around the room until the gyroscope ran down. When the full-sized Gyron showcar was built, it looked very little like the proposals Thompson had designed or the small model that Tremulis played with.

In 1952, Alex Tremulis was hired as a senior designer at Ford. A year later, he was transferred to Gil Spear's Advanced Studio as assistant manager. Tremulis soon began a school in a room upstairs in the new Design Department building, where he and Regg Bennett, who had taught GM's new designers for many years, taught Ford's new designers the Ford way of car design. When Spear became head of the new Special Products Studio in December 1953, Tremulis took over Spear's position in the Advanced Studio. Tremulis' primary interest, as head of the Advanced Studio, became aerodynamics — how a car moved through the air. Tremulis was prone to argue the merits of aerodynamics with everyone, including Gene Bordinat, who did not believe that aerodynamics played any real part in ordinary passenger car design. He also began to rub Bob Maguire the wrong way. After he became general manager of the Styling Center, Walker appointed Tremulis head of the Advanced Studio. Friction between Tremulis, Bordinat and Maguire probably encouraged the appointment. Maguire did not appreciate Tremulis' sense of humor, and it wasn't long before both Bordinat and Maguire began to question whether Tremulis belonged at Ford -- let alone as head of the Advanced Studio.

After Tremulis became head of the Advanced Studio, Maguire asked Alden "Gib" Giberson, then Tremulis' assistant manager, to keep an eye on him and try to control him.

Ford Design Center

Giberson told Maguire that he could hardly be expected to keep his boss under control, but eventually he did agree to at least try to keep Tremulis directed. Giberson's efforts came to an end when Tremulis told Giberson he would replace him if he didn't follow his orders about the projects they were going to work on. With Giberson's unsuccessful efforts to help keep Tremulis focused, Maguire became even more frustrated with Tremulis. When Giberson moved on to his next assignment, Regg Bennett was appointed as Tremulis' assistant, but Bennett was not inclined to ride herd on anyone.

Walker eventually came to believe he made a mistake in appointing Tremulis a studio head and, in March 1956, the Advanced Studio was dissolved, and Tremulis became a manager in Maguire's new Advanced Studio, which was set up to spearhead the Stylerama program and to continue making showcars as needed. Soon after the Stylerama program went by the wayside, Maguire's Advanced Studio was itself dissolved, and Tremulis was reassigned half-time to the Thunderbird studio and half-time in the Ford preproduction studio, but he was no longer a manager.

Although Tremulis and his students had worked on the concept for several years, the Gyron showcar was conceived by Tremulis as an answer to the Firebird III, a GM concept car introduced in 1959, and it was designed and built in Elwood Engel's corporate advanced studio in 1959-60, where Tremulis was next assigned as a senior designer. Although the concept was Tremulis', the final design was the joint effort of Alex Tremulis, Elwood Engel, Syd Mead and Bill Dayton. John Najjar, as executive stylist, and Gale Halderman, as manager of the studio, also share some of the credit as does Phil Payne, a new designer hired into the studio in 1960, just as the Gyron was being completed.

When the Firebird III was introduced, GM announced it was their laboratory on wheels, and that it incorporated features expected on future GM production cars, including an electronic guidance system and a unicontrol system, so that all of the car's control functions could be operated by a single control from either front seat. The Firebird III had both a turbine and a small reciprocating engine on board. Tremulis took offense to references in the press to the Firebird III as the world's most exotic car and felt he could design a car so dramatic GM would never catch up.

Tremulis had some advanced knowledge of the Firebird III, and he became serious about producing the Gyron as a response, when he heard the Firebird III described as having up to six tails, two canopies and the looks of a turbo catamaran. Tremulis was intent — almost obsessed — with outdoing GM and, early on, he began comparing his proposal for the delta-shaped Gyron to an earth reentry vehicle. He also told anyone who would listen that he thought his idea represented the ultimate in aerodynamic design. In short, Tremulis expected the Gyron to be a genuine breakthrough that would influence all future car design.

Phil Payne

Philip E. Payne

Phil Payne grew up near South Bend, Indiana, where his mother was an art teacher. In 1956, Phil followed his brother Howard to the Cleveland Institute of Arts. In 1959, he was a summer intern at the Ford Styling Center. When he graduated in 1960, he joined his brother as a designer at Ford.

Payne's first assignment was in the Thunderbird preproduction studio. In January 1961, he was transferred to the corporate advanced studio, where he helped complete the design of the Gyron. Later, he helped design the Continental Mark IX. In May 1961, he left to serve six months active duty in the National Guard. His tour of duty in the Guard was extended because of the Berlin Crisis delaying his return to the corporate advanced studio until August 1962. He was next assigned to the Ford interior studio, where he helped design the 1965 Mustang interior. He then spent the next five years in the Mercury exterior studio.

Payne was assigned to Ford's Industrial Design Studio, during which time he also became a Certified Interior Architect. He was later a designer in the Escort studio. In 1981, he accepted an offer from American Motors to head their trim and color studio.

When American Motors was purchased by Chrysler, Payne became a designer in the Chrysler truck studio. He holds the patent on the current Dodge Ram pickup, which was designed in his studio. Currently, Payne is a designer in Chrysler's international studio. ■

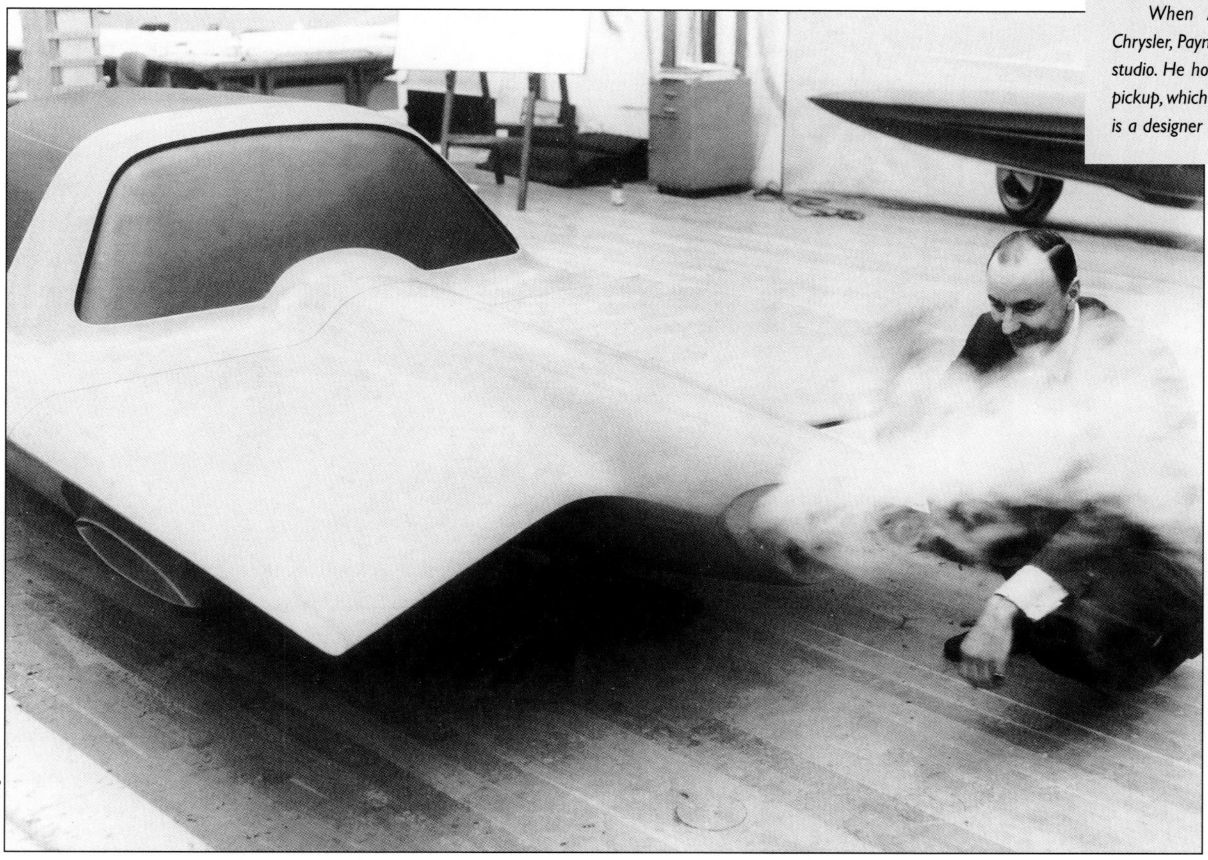

Ford Design Center

Elwood Engel wanted the rear end of the Gyron to look like the afterburner on a jet plane. He couldn't resist setting a fire in the full-sized clay model's exhaust tube just to see the smoke flowing from it.

Ford Design Center

The Gyron being readied for display at the Ford Rotunda. Bill Dayton is on the left under the car's wing.

Ford Motor Co., on the other hand, thought of the Gyron as a two-passenger showcar only, and never as a predictor of what was to come on production Fords, Mercurys or Lincolns. When Bill Ford came into the Advanced Studio one day and saw Tremulis' small two-wheeled Gyron under construction, he asked him to explain what the Gyron was and how it worked. However Tremulis explained the Gyron, it didn't convince Bill Ford the car had anything to do with what Ford Motor Co. was producing or would be producing in the near future, and Mr. Ford left shaking his head.

When Tremulis was transferred to the corporate advanced studio, Gene Bordinat asked John Najjar, Engel's executive stylist, to keep an eye on Tremulis and give Bordinat his evaluation of Tremulis' work. Najjar was told that if Tremulis could not be controlled, Bordinat wanted to get rid of him. Several designers report that, by that time, Tremulis had run the LaTosca into George Walker's office once too often, and that Walker had tired of hearing complaints from Maguire, Bordinat and other managers, who were frustrated in their efforts to redirect Tremulis.

After Bordinat told him to keep an eye on Tremulis, Najjar sat him down and told him that, if he was ever going to get anywhere at Ford, he had to start getting serious. Tremulis wasn't too happy about it, but he listened to Najjar. Najjar assigned Syd Mead to work with Tremulis on a design presentation for the Gyron. Together, Tremulis and Mead created sixteen 8"x16" drawings for a presentation book on the Gyron project. The presentation book not only described the design features of the Gyron, it also identified proposed advantages of such a vehicle. For the presentation, Najjar himself built a 1/10-sized model of the Gyron that was weighted from underneath the aluminum platform it was mounted on. When tilted, the weights kept the model vertical and gave it the same effect, on a small stationary scale, as if a gyroscope really controlled it.

When everything was ready, Najjar encouraged Tremulis to make the presentation to the studio heads and their engineering counterparts himself. The drawings from the presentation book were arranged in sequence on a blackboard. When Tremulis started up the small gyroscope-powered model he made for the presentation, the gyroscope part of the model came loose from its moorings and hit him in the chest giving him a superficial cut, which Tremulis wasn't aware of until Najjar asked him if he was okay. The rest of the presentation went well. Based on the presentation, money to build a full-sized clay model of the Gyron was authorized. At the time, the plan was to build a full-sized working Gyron. Bordinat was impressed not only with the presentation, but with what looked like Tremulis' new-found seriousness.

Sydney J. Mead

Syd Mead was a minister's son, who grew up drawing cars and reading science fiction. After high school, he enlisted in the Army. Because of his admiration for the Continental Mark II, he sent some of his drawings to John Reinhart at Continental Division and asked him to review them. Reinhart's response was encouraging. He also suggested that Mead attend the Art Center College. Before he was discharged from the Army, Mead prepared a portfolio and, upon discharge, applied at the Art Center College, was accepted, and started there in 1956. After his first year, he was awarded a Ford scholarship.

Syd Mead

He graduated in 1959 and was hired as a designer at Ford. He was assigned to the corporate advanced studio, where he helped design the Gyron. After 20 months at Ford, Mead left to take a job with Hansen and Co. as an illustrator. John Reinhart, who by that time had left Ford and was working for U. S. Steel, asked Mead to illustrate the book Concepts. *In 1964, Mead was hired to design the Ford Pavilion at the New York World's Fair. That same year, he also designed the operable Touring Pickup showcar for Ford Motor Co.*

Mead started his own illustration business in 1970 and moved it to Southern California in 1975. His illustrations of futuristic cars have since appeared in many books. He is also well known for his work in the movies, including his design of the futuristic city in Blade Runner.

Mead has also designed cars for Volvo, American Motors and Raymond Loewy and Associates. Most recently, he has prepared architectural renderings and design illustrations of futuristic boats and planes. ■

Syd Mead and Tremulis designed alternate gyro-stabilized sedans, using the research Tremulis had earlier developed. Dayton assisted in the clay modeling of Mead's design. When the basic design of Mead's car was far enough along in clay to tell what it looked like, Engel told Mead it looked too much like contemporary GM products and then broke off one of the skegs on his clay model with his foot. As Mead remembers it, it took him the rest of the day to calm down after Engel broke off the skeg.

According to Mead, the next day Engel brought in a notebook of designs he had done while in school. One of the designs in the notebook was of a delta-wing-shaped car with a large spine running down the center. Tremulis, Mead and Dayton were asked to start on a new showcar heavily based on the sketch from Engel's school notebook. The two models of the Gyron that Tremulis, Dayton and Mead were working on were scraped clean, and they started over again on a single proposal. When the design was finalized, Joe Seibold recalls that a plaster plug was made of the upper part. After that, the Gyron was turned over, and a plaster plug of the bottom part was made and, from the two plaster plugs, the fiberglass Gyron showcar was made.

McKinley Thompson

The Gyron would have likely become a fully-operable showcar except for the cost of the gyroscope. Najjar asked George Haviland, who was then in charge of public relations at the Styling Center, to find a gyroscope for the full-sized model they hoped to build. Haviland contacted Minneapolis-Honeywell Corporation, supplier of gyroscopes used on the B-52 bomber. Minneapolis-Honeywell had a gyroscope about two-feet in diameter that Najjar and Tremulis thought would work, but the cost was $60,000! After doing some additional figuring, Najjar estimated that it would cost another $50,000 to mount the gyroscope correctly, and another $25,000 to build the car itself. There just wasn't that kind of money in the styling budget.

The Gyron was 209 inches long, 45 inches high and 86 inches wide. It looked like a airplane with a delta-wing-shaped body supported by two wheels. The Gyron was meant to be stabilized both when it was moving, even at slow speeds, and at rest by a gyroscope. Tremulis claimed that the on-board gyroscope and the two-wheeled configuration permitted the Gyron to be severely banked into corners presumably permitting travel at greater speeds.

The interior of the Gyron was designed by Bill Dayton, and it was undoubtedly the most ultra-modern interior design ever included in a Ford showcar. Steering on the Gyron was by a dial that also automatically controlled speed. A ten-button panel on the instrument panel was supposed to be controlled by an on-board computer. At the time the Gyron was designed, many Ford designers and engineers thought that cars would soon be capable of safely traveling at speeds of up to 200 mph on controlled-freeway systems.

McKinley W. Thompson Jr.

From the time he saw a new DeSoto Airflow on the streets of New York City, Mac Thompson knew he wanted to be a car designer. After he graduated from high school, he served in the Army Corps of Engineers. After he was discharged, Thompson studied mechanical engineering at the Neptune Veteran's School, and then started his career as an engineering layout design coordinator for the U.S. Army Signal Corps.

Ten years later, he was working for the Signal Corps in New Jersey. He was married, had children and was buying a house, when he was awarded one of four Motor Trend magazine scholarships to study automotive design at the Art Center College in Los Angeles, California. He and his family pulled up stakes and moved to California, where Thompson started at the Art Center College in 1953. He graduated in 1956 and was hired by Alex Tremulis as a Ford designer.

Thompson's first assignment was in Tremulis' Advanced Studio, where he prepared sketches for the Gyron concept car. Thompson's real interest was in designing trucks and, after six months in the Advanced Studio, he was assigned to the truck pre-production studio, where he helped design the Light Cab Forward.

Later in his career, Thompson worked in the Falcon and the Thunderbird studios. He also did early design work on the GT-40. Because of his engineering background, Thompson established and was subsequently appointed manager of the Appearance Development Studio, where design changes proposed by Engineering were modeled for Design Center review.

He retired in 1984. ■

Mead's Gyron proposal was originally designed as a four-wheeled car. The front wheels were set very close to each other because of the triangular shape of the car. The Gyron later became the two-wheel concept Alex Tremulis had been working on since 1956.

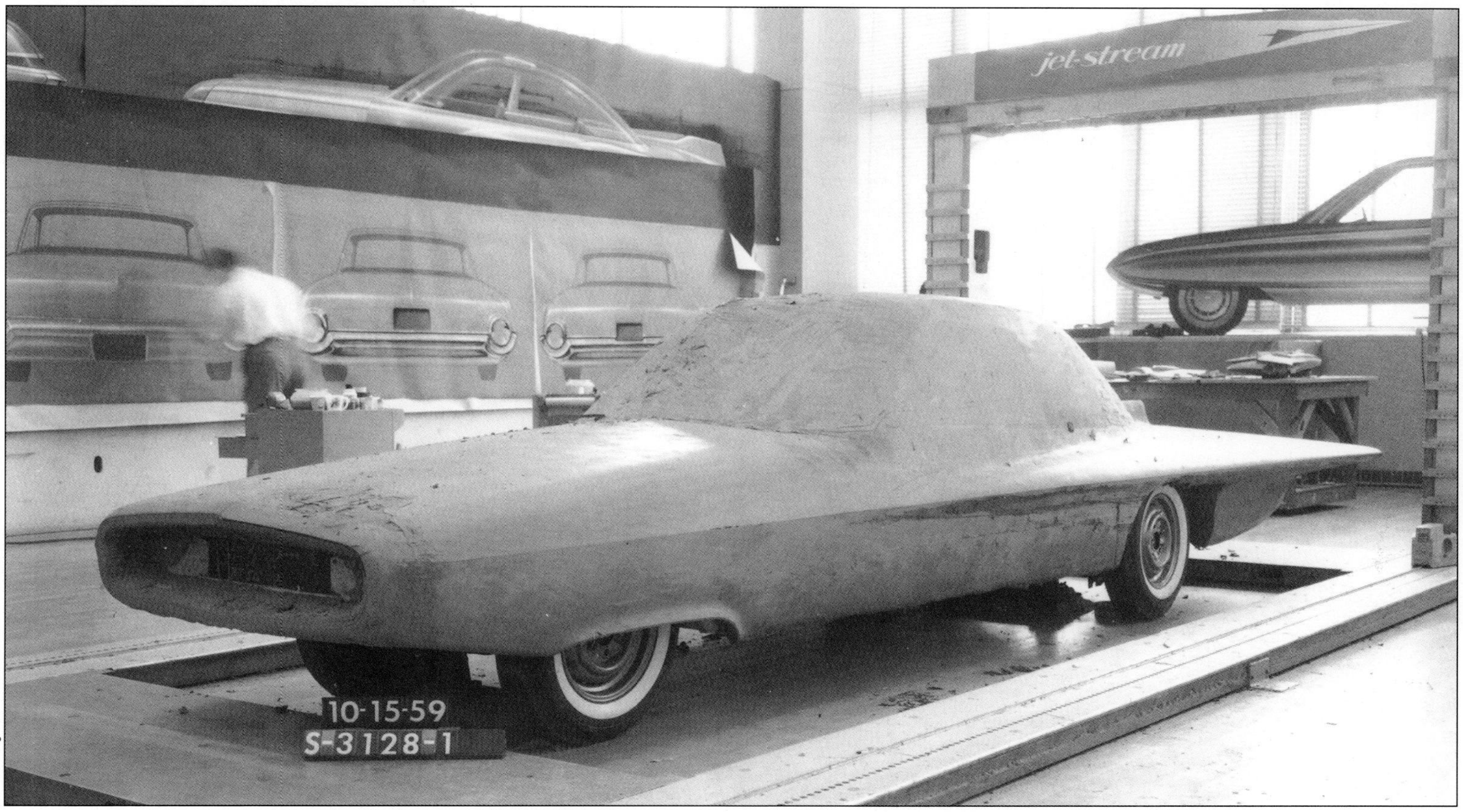

Ford Design Center

The Gyron had other unique features. The tires were designed to inflate or deflate depending on the speed at which it was going. The tire tread was meant to be wider at slower speeds and narrower at higher speeds. To compensate for the lack of a working gyroscope, two outrigger wheels were designed to automatically extend at about 30 mph and retract at speeds over that. When the gyroscope wasn't active, the Gyron was prevented from falling over by the two small outrigger wheels. Footbars with gas and brake pedals on them were located on both sides of the passenger compartment, so the Gyron could be driven by either passenger. It is unknown what kind of engine or transmission would have been used to power the Gyron. Because they never got that far, the drivetrain may not have been determined.

Engel was enthusiastic about the Gyron and, when it was finished, he insisted that all the lighting on it be battery-powered, and the car have electric steering operated by the mechanism on the center console of the car. The car could move under its own power thanks to an electric motor installed by Ray Smith. Although the top speed was only about 5 mph, with the outrigger wheels down, it could nonetheless be driven. When the car was first shown, Engel decided he was going to drive it down the hallway and into the Styling Rotunda. When he drove it into the room, the hallway behind him was filled with smoke from lighted gel Engel placed in the exhaust tube.

The first the public knew of the Gyron was from a press release dated March 31, 1961. It reported that the Gyron had a dazzling array of on-board computers that, from the sound of it, predicted what would be in cars built 30 years later. At the time it was introduced to the public, George Walker was quoted as saying that the purpose of building the Gyron was "to explore ways of designing cars that were other than rectangular in shape and didn't have four wheels." Walker explained that the Gyron was "an outstanding example of the visioneering that is at the heart of progressive automotive styling."

The Gyron was the last full-sized Ford concept car introduced by Walker, who retired several months later. At the time he said it, it is doubtful that even Walker really believed that the Gyron predicted the future direction of automotive design.

(below) From left to right, Bob Thomas and Alex Tremulis are discussing one of the design features of the Gyron. Elwood Engel is standing at the back of the car.

(bottom) The Gyron in the corporate advanced studio just after the fiberglass upper and lower sections were joined together. From right to left are John Orfe, Elwood Engel, Dick Avery and Shirley Ann Johnson, a studio secretary.

Ford Design Center

Ford Design Center

Mead also designed a proposed stage display for use with the Gyron at auto shows. Mead's display included a single rail over a shallow pool of water the two-wheeled Gyron was to be driven out over. On each wing of the Gyron, there was supposed to be a model sitting with her feet dangling over the pool. Because it was uncertain, at that time, whether the car would actually have a gyroscope in it, Halderman suggested two small outrigger wheels to keep the car upright, supposedly when it was shut down. The lack of a gyroscope in the car and the cost of the display designed by Mead meant it could not be used. The stage display used with the Gyron on the show circuit and at the Ford Rotunda was designed by Bill Dayton.

The Gyron was first publicly shown April 1, 1961, at the New York International Auto Show. In May 1961, it was displayed again at the 25th anniversary show of the reopening of the Ford Rotunda. The Ford Rotunda anniversary show featured the Gyron in a *City Of The Future* display with two-level streets that had radar directed traffic, a monorail train and a helicopter based on top of a skyscraper. According to Mrs. Tremulis, the Gyron was lost during the fire that destroyed the Rotunda in 1962.

In one of its press releases about the Gyron, Ford gave credit to James Scripps Booth, a Detroit artist-engineer who built the 1908 Bi-Autogo, a two-wheel vehicle that was balanced by a gyroscope located under the driver's seat. The gyroscope apparently kept the Bi-Autogo upright but, because of steering difficulties caused by the gyroscope when active, the Bi-Autogo was abandoned as unworkable. Ford's press release pointed out that modern hydraulics and power steering solved the problems Scripps Booth couldn't.

Bill Dayton

(above) Bill Dayton designed the seats in the Gyron to support the passenger's entire body including the legs. Ray Smith installed an electric motor and a working steering system, so Elwood Engel could drive the car into the Styling Rotunda for its first showing.

Tremulis' problems at the Design Department continued even after he was twice demoted. He lived in Ann Arbor, a college town and kept abreast of the latest scientific research. Stylists who worked with him refer to him as a free thinker and a passionate disciple of aerodynamics, who disliked production work. Tremulis wasn't reluctant to let his superiors, or anyone else, know when he disagreed with them. He advocated total commitment to aerodynamics, when other designers, including his superiors, thought aerodynamics mostly an extravagance. Even after the Gyron presentation was made, and approval to build a full-sized Gyron was obtained, Tremulis continued to play with the 1/10-sized Gyron, airing it up and running it across the floor constantly. One day, Henry Ford II was at the Styling Center, when Tremulis released the aired-up Gyron, and it almost ran into Mr. Ford as it went skittering across the floor just as he walked in. Needless to say, Walker and the studio heads were upset and embarrassed.

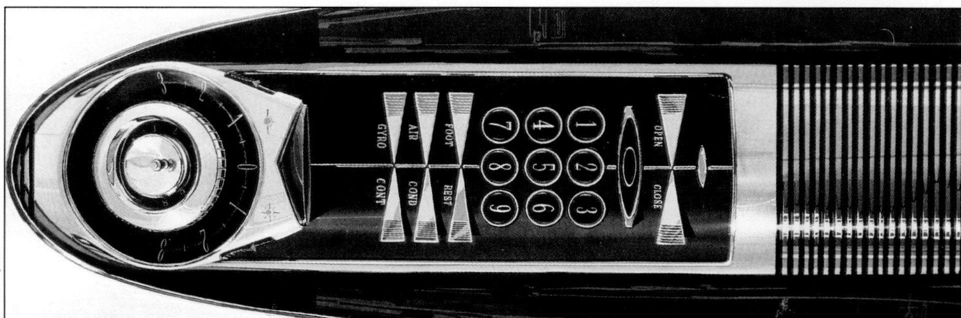

Bill Dayton

(left) All of the Gyron's controls were located between the two seats. The steering dial is on the left. It was adapted from a fighter plane design manual and actually steered the car. The butterfly-control switch marked "gyro cont" was inoperable. Switch two operated a blower to cool the inside of the car. Switch three operated the footbar. The buttons were inoperable. The right-most butterfly switch was used to open and close the top.

When Gene Bordinat replaced Walker in 1961, as head of the Design Center, his emphasis became production cars. Bordinat measured the success of a design by how many cars were sold. That was inconsistent with the way Tremulis did things and, as a result, his last years at Ford were not very happy.

So ended the saga of the Ford two-wheeled car that Tremulis considered the ultimate form of automobile design. When Elwood Engel left Ford to become design

Bill Dayton

(below) The controls on the Gyron could be used by either occupant. The brakes were operated from either side by the footbar. The radar scope at the front of the passenger compartment was supposed to show the road ahead.

Ford Design Center

(left) From left to right, Al Nesti, Amirigo Pieri, Bruce Barringer and Rolando Suffredini are preparing the full-sized clay model of the Gyron for plaster.

Ford Design Center

(right) From left to right are Syd Mead (back to camera), Gale Halderman and Alex Tremulis, with clay modelers Wolfgang Konietzko, Dick Sprague and Steve Melek, who are scribing the door openings on the clay model of the Gyron.

Ford Design Center

(left) From left to right, Syd Mead and Alex Tremulis inspect a small Gyron model (held by Tremulis) and Mead's renderings of the Gyron. In the background, Bob Thomas is seated at his drafting table.

Ford Design Center

(right) The fiberglass Gyron was built in two parts. When the top was completed, a plaster plug was made of the bottom, which was then cast in fiberglass. After that, both fiberglass parts were joined together.

director at Chrysler in 1961, Bob Maguire took over the corporate advanced studio and became Tremulis' boss. Although Tremulis saw the Gyron as the high point of his life's work, Maguire saw it as a waste of time. Tremulis was also working on other designs, like the six-wheeled Seattle-ite, which Maguire also disapproved of. Maguire assigned Tremulis to design alternate cars in competition with proposals being designed in the production studios. Tremulis disliked doing production work and kept working on new ideas that came to him, even though Maguire told him not to. When Tremulis couldn't move on to other more practical design work, as he was instructed to do, it became the cause of another demotion and, finally, his departure from Ford in 1962, when he was asked to resign. Tremulis later said he believed his work on the Gyron made him unpopular and blocked any chance for advancement. According to the designers who worked with him, that was probably an oversimplification

The Gyron influenced many of the designs Tremulis created after he left Ford. In 1964, he worked on a proposal for a gyro-stabilized train he called the LBJ-200 and the LBJ Gyronaut. Tremulis felt the LBJ-200 was capable of speeds of 200 miles an hour. His proposal was for a 200-passenger train that moved on a rail on two wheels and banked into corners. Even after the model was built, which is now in the Smithsonian, it was considered a little too advanced and unproven to build an operable train. The Japanese government wanted to use the design for the high-speed train system they were developing, but Tremulis turned down a proposal to help build a gyro-stabilized train in Japan, because he wanted his concept built first in the United States.

Tremulis spent the rest of his career trying to convince the automobile industry of the superiority of his two-wheeled streamlined vehicles. He helped build the 90-horsepower Gyronaut X-1 two-wheel racecar that Bob Leppan drove, in 1965, to a world record of 245.644 mph. Tremulis was also the aerodynamic consultant for a rocket car, the Goodyear Wingfoot Express, that reached 520 mph. Then he became a consultant for the three-wheeled rocket powered Green Eyed Monster that lost a wheel at 600 mph while on a speed run at the Bonneville Salt Flats.

In 1966, Tremulis moved to California. With Thomas O. Summers, an expert on gyroscopic stabilization and a company called Gyro Transport Systems Inc., of Northridge, California, Tremulis designed and helped build another two-wheeled gyro-stabilized car. That vehicle was called the Gyro X, and what was represented as a full-sized working prototype of it was displayed at the 1967 New York Auto Show. Tremulis also worked with NASA on a gyro-stabilized moon vehicle. Tremulis called his proposal *Operation Copernicus*. It apparently had the support of some astronauts, but it was not selected, because it cost too much to build. Tremulis worked as a consultant for Subaru until the mid-1980s and designed several three-wheeled vehicles for them, which were capable of extremely high gas mileage.

Today, Tremulis' insistence on aerodynamics in automobile design is universally accepted and is now incorporated in almost all new car designs including Ford products. Two-wheeled cars, buses and trains stabilized by gyroscopes never caught on. ■

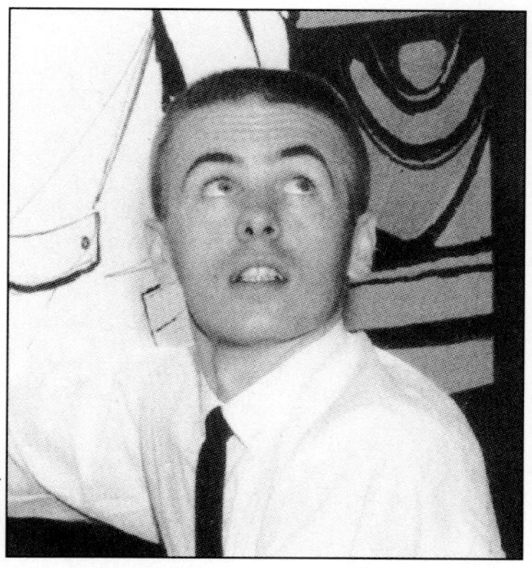

Bill Dayton

William A. Dayton

As a high school student growing up in Sweethome, Oregon, Bill Dayton saw an article in the Saturday Evening Post about the Art Center College. With a portfolio his art teacher helped him prepare, he applied at the Art Center and was accepted. He was awarded a Ford scholarship during his last two years in school. When he graduated in January 1958, Dayton was recruited by Elwood Engel as a designer in Ford's corporate advanced studio. Soon after he started at Ford, he left to serve six months active duty in the National Guard. In April 1959, he returned to the corporate advanced studio, where he designed the interior of the Gyron and the display used with it at the 1961 New York Auto Show and in the Ford Rotunda.

After the Gyron was completed, Dayton was recalled to active duty during the Berlin Crisis. He returned to the Styling Center just after Engel had been hired as Chrysler's vice president of design. He received a call from Engel asking him to come to Chrysler at double his salary and as a senior designer. Dayton moved to Chrysler in December 1961.

Dayton is currently in charge of one of Chrysler's three major design studios. He was the leader of the team that designed the first Chrysler mini-van, the LeBaron convertible, the Cirrus and Stratus and the LH series Chrysler. Dayton was also involved in Chrysler's international design effort, which included tours of duty in Australia and Europe. During a two-year assignment in Italy, he redesigned the Lamborgini. ■

Ford Design Center

Designer Dick Avery is helping Styling Center secretary Shirley Ann Johnson into the partially completed Gyron. The seats were later redesigned to include built-in headrests and more leg support.

The Envoy was supposed to be the 1961 Lincoln. It was a difficult car to design, and was radically changed at least three times during its design process. When it was completed, there were two different proposals of the Envoy, each with a different roof line. Each proposal also had different side treatments. The difficulty in coming up with an acceptable design and the different roof lines included on the Envoy represented differences of opinion among the designers and at all levels of Lincoln-Mercury Division's management about what the 1961 production Lincoln should look like.

The two Lincoln concept cars that preceded the Envoy, the Diplomat and the Attache, were originally designed as a part of the 1956 Stylerama program. Even before that program was canceled, the Diplomat was authorized as a full-sized concept car. Soon after it was completed, it was decided the Diplomat would not become the direction for the next-generation Lincoln and, when Don DeLaRossa replaced John Najjar as head of the Lincoln studio in October 1957, the process of finding a new design for the 1961 Lincoln started all over again.

Larry Conrad

This version of the Envoy had a winged-deck lid to match the roof. The winged roof was designed as Ford's answer to the cantilever roof on GM's 1959 four-door hardtops.

Howard Payne

So both roofs could be used on the same model, there was a joint at the base of the C-pillar on this full-sized model of the Envoy.

Envoy

While he was head of the Lincoln studio, Najjar assigned executive stylist John Reinhart, as head of the Lincoln pre-production studio, to begin designing the car that became the Envoy. At that time, an informal contest was held between the designers and modelers who worked on the car to name it. Whose idea it was to name the car the Envoy has been lost to posterity, but it was chosen from one of the suggestions proposed.

Reinhart immediately put studio designer John Orfe to work on the front end/grille design, while he worked on the car's rear and side design. Reinhart designed several sets of large sweeping fins on a full-sized clay model of the Envoy, but Walker rejected them. Not long before Najjar was replaced as head of the Lincoln studio, Reinhart left to take a job at U. S. Steel in their new design department. His new job was to develop design ideas that encouraged automobile manufacturers to use more of their steel. By the time Reinhart left Ford, the front end of the Envoy was in close to final form, but not much else was.

Even before Continental Division was dissolved in mid-1956, Ben Mills, who had been its assistant manager, was transferred to Lincoln Division. Former Packard executive James Nance was hired by his friend, Ernie Breech, to head Lincoln Division (later consolidated with Mercury and Edsel Division as MEL Division), but Mills was assigned responsibility for the day-to-day activity of the Division. Mills was one of the original "Whiz Kids" hired in 1946 by Henry Ford II to help him turn the Company around. His last assignment had been running Ford's aircraft operations in Chicago. Although Mills quickly developed a feel for cars, his specialty was accounting. To those born with gasoline in their blood, Mills was a bean counter, but he was decisive and effective at solving problems, where Nance had trouble making decisions.

One of the first things Mills did, when he came to Lincoln Division, was to institute a study to determine what Lincoln had been doing wrong and, more importantly, what had to be done to make Lincoln profitable. Mills set up a small internal committee, which included Don DeLaRossa as head of the Lincoln studio. DeLaRossa enlisted the help of Gene Bordinat, who was then in charge of the Mercury studio, to determine why Cadillac had been so successful and, more importantly, why Lincoln had not achieved the same level of sales or respect in the marketplace as Ford had, when compared to Chevrolet. It was obvious that Ford Division had closed the competitive gap. Lincoln, on the other hand, had actually gone backwards -- and the sales gap between Lincoln and Cadillac was still growing.

(above) The Envoy was started by John Reinhart in the Lincoln pre-production studio in 1956. The front-end design was John Orfe's. It was incorporated on the final design of the Envoy, but Reinhart's large fins were rejected by George Walker.

For their part of the study, Bordinat and DeLaRossa concluded that the primary reason Lincoln had not been as successful in the marketplace as Cadillac was because Lincoln did not have design consistency during the period 1949-57. Cadillac, they concluded, appealed to the same market year after year with cars that at least looked like they were related. Lincoln did not. In other words, Lincoln's goalposts were constantly being moved. There was some logic to Bordinat and DeLaRossa's conclusions, but Mills saw the problem as more complex. Mills was never happy with the design of the 1958 Lincoln and felt it was pushed through the Product Planning Committee by a too-influential George Walker. Mills and Ford Division head, Robert McNamara, shared a general belief that Lincoln was too big, and that it relied on too many design cliches that were likely to go out of style too quickly. Neither McNamara nor Mills felt Lincoln was appealing to its natural more conservative market. Mills very quickly concluded that the fastest way for Lincoln to become profitable was to stretch out the time between model changes, a concept that went against conventional wisdom in the industry and at Ford's Stying Center. He suggested stretching the model cycle from three to as much as nine years.

At about the time their study was completed, Bordinat and DeLaRossa played musical chairs. As a result of divisional consolidation, Bordinat was placed in charge of a consolidated Lincoln-Mercury studio, and DeLaRossa became his executive stylist. Rulo Conrad, who had been a manager in the Lincoln-Mercury studio and one of the primary designers of the award-winning 1956 Lincoln, was placed in charge of the Lincoln pre-production studio. His job was to complete the design of the Envoy, which had been started by John Reinhart. Conrad's Envoy was to become the next-generation Lincoln.

When Reinhart left, Rulo Conrad replaced him as head of the Lincoln pre-production studio. This full-sized clay model was one of Conrad's early proposals. It shows the influence of the Quicksilver. The rendering on the wall shows one of Conrad's favorite design themes.

The other designers in Conrad's pre-production studio who worked on the Envoy included John Orfe, Merle Adams, Bob Chieda, Howard Payne and Joe West. The clay modelers included Doug McCombs and Alex Aloe. Very early during the design process, Adams was drafted by the Navy in their first peacetime draft and left.

As might be expected, Conrad changed directions on the design of the Envoy except for the front grille area, which Orfe completed basically as he had first proposed it.

Larry Conrad

(above) Designers in Conrad's Lincoln pre-production studio tried to come up with an attractive winged roof, but management never liked it.

Larry Conrad

(below) The front of the Envoy was meant to show continuity with the 1958-60 Lincolns. The winged roof actually drained towards the center. The raised eyebrows on which the wipers rested were a carry over from the Diplomat.

Rulo N. Conrad

Rulo Conrad was raised on a farm near Bremen, Indiana. In high school, he was constantly drawing tractors and cars. After he graduated from high school, he attended the John Herrin Art Institute in Indianapolis on a scholarship. During his second year, his draft board required that he return home to work on the farm. After World War II was over, Conrad was drafted and served in the Philippines as a cartographer. In 1947, after his tour of duty had been completed, he returned home, prepared a portfolio and interviewed at the Chrysler and Ford design departments. Chrysler was not interested in his current car design, and told him they would have preferred to see his use of imagination on futuristic car designs. He stayed up all night redoing his portfolio and, with the revised portfolio, interviewed the next day at Ford and was hired.

Gale Halderman

Conrad's first assignment was in the Ford studio. In September 1947, after two years at Ford, Conrad left to accept a position as a stylist at Chrysler. He was rehired at the Ford Design Department in May 1952. After the original Thunderbird was moved into the Ford studio, he worked on its final front-end design. He was next assigned to the Lincoln production studio, where he, Bill Schmidt and Herb Tod designed the 1956 Lincoln. Schmidt left Ford in 1955, and Tod and Conrad, as the remaining designers of the '56 Lincoln, shared the Industrial Design Institute award for the design of the car.

In October 1957, he was placed in charge of the Lincoln pre-production studio, where he supervised the design of the Envoy. He was later assigned to the Ford and Thunderbird studios. In the mid-1960s, Conrad designed the Techna concept car.

In the 1970s, Conrad was assigned to the truck studio, where he designed Ford pickups. Because of his farm experience, he was asked to redesign Ford's tractor cabs and was later sent to France twice to assist in the design of European heavy tractors.

Conrad retired in 1981, after several years as head of the tractor studio. ■

(below) The absence of C-pillars and a roof support in the middle of the backlight gave rear-seat passengers a panoramic view. The roof overhang was designed to provide shade for them.

Larry Conrad

Conrad's design for the Envoy was meant to avoid the excesses of the Diplomat, but it was still intended to show continuity with the current production Lincolns.

With the exception of the roof line on one version of the Envoy, the car's design was very conventional. Bordinat was interested in a winged roof as a means of countering GM's new four-door hardtop cantilever-roof design. Orfe was asked to prepare renderings of an Envoy with a winged roof, and those renderings became the basis for the alternate-winged roof used on the Envoy. The winged roof was not well liked by Orfe or most other designers who saw it. Doug McCombs, one of the clay modelers assigned to work on the Envoy, remembers working with Rulo Conrad night after night trying to come up with an acceptable winged roof, but they never succeeded. Although neither Conrad nor any of his designers were pleased with the winged-roof concept, Bordinat asked that it be included as an alternate-roof design on the Envoy along with a more conventional roof. The winged roof allowed for more glass area and better visibility unlike the more formal wide C-pillar of the more conventional roof, but the winged roof always looked a bit odd.

Ford Design Center

(above) The Envoy was the first Ford product designed with sequential-turn signals, but many designers disliked the billboard taillight.

Ford Design Center

(above) From the front, the final version of the Envoy showed too much continuity with the 1958 Lincoln to be acceptable to management.

(below) This proposal of the Envoy came close to production approval.

Ford Design Center

Robert A. Chieda

Elizabeth McPherson

As a child growing up in Akron, Ohio, Bob Chieda received encouragement from his teachers to develop his artistic talent. As a teenager, Chieda enrolled in the Akron Art Institute. In 1951, he applied and was accepted at the Cleveland Institute of Arts, where he majored in industrial design. While in school, Chieda worked in community theater helping design and build sets. Although he had no prior interest in automobiles, he was recruited as a designer by Ford, when he graduated from the Cleveland Institute of Arts. He and Dick Kessler both started at Ford in June 1955.

Chieda's first assignment was in Buzz Grisinger's Mercury pre-production studio, where he worked on the Blue Job. He was then transferred to the Advanced Studio, where he worked on the L'Avion. He was later transferred to the Lincoln interior studio, where he helped design the interior of the 1958 Lincoln Continental Mark III. He was next assigned to the Lincoln pre-production studio, where he worked on the Envoy. After that, he was assigned to the corporate advanced studio.

In early 1960, Chieda decided he wanted to broaden his experience as an industrial designer and accepted an offer to join John Cuccio's industrial design firm in Westport, Connecticut. Later, Chieda established his own industrial design firm in Westport. In 1991, he moved to Chicago to work in former Ford designer Bob O'Neal's industrial design firm.

He later worked as an architectural consultant and rehabilitated several architecturally significant homes in Akron, Ohio.

Chieda retired in 1995. ■

Merle Adams

Merle J. Adams

Merle Adams grew up in Erie, Pennsylvania, where his father owned and operated a bicycle shop.As a child,Adams liked to draw cars. His uncle, Wilbur Henry Adams (designer of the Divco truck), saw one of his drawings, and suggested to his father that Adams attend the Cleveland Institute of Arts. Adams graduated from the Cleveland Institute of Arts in 1955 and was hired as a designer at Ford Motor Co. His first assignment was in the Lincoln production studio, where he worked on the finish details of the 1957 Lincoln and the design of the 1958 Lincoln. Adams was later assigned to Rulo Conrad's Lincoln pre-production studio, where he helped design the Envoy.

In 1957,Adams was drafted by the Navy and spent the next two years painting signs. In 1959, he returned to the Styling Center and was assigned to the Canadian and international studio to work under Jim Sipple. Adams designed the grilles for the 1959 and '60 Canadian Meteors.

In 1960,Adam's father decided to retire. Adams left Ford, bought his father's bicycle shop and still sells Schwinn bicycles in Erie, Pennsylvania. ■

The Envoy was ultimately built so that it could be shown with either roof on it. There were joints in the body so that one roof could be removed, and the other one placed on the car. One version of the full-sized clay model of the Envoy also had an overlay on the deck lid shaped to match the wing-like roof. The Envoy had no running gear and was a push-mobile. By the time a full-sized fiberglass Envoy was built, the winged overlay on the deck had been eliminated. Conrad and all studio designers and modelers were relieved when the winged roof was finally abandoned as unworkable.

Even though the Envoy was intended to avoid what were seen as excesses on the Diplomat, many of the features designed into it were taken directly from the Diplomat, including the general design of the headlights, front end and the eye brows at the bottom of either side of the windshield against which the windshield wipers rested. Although the two sides of the Envoy were designed differently, both were much cleaner designs than the Diplomat. Also eliminated on the Envoy were the Diplomat's sculptured bullets that ended behind the front wheelwells, the dorsal fins, the overly-busy rear end and the unusual C-pillars.

The final rear-bumper design on the Envoy encircled the taillight panel with a Lincoln star in the middle. John Orfe came up with the idea of sequential taillights shortly before the Envoy was finished, but because the mechanicals necessary to operate sequential taillights had not yet been engineered, Orfe used a car battery hidden under the clay model and clock motors to time the lights. Sequential taillights were first used on the production 1965 Thunderbird.

When the Envoy was finished, it was painted pearlescent-white, and had Plexiglas windows, but the interior was not finished below the beltline. The lights, including the sequential taillights, worked, but the doors, trunk and hood were only scribed lines.

The Envoy was shown to Henry Ford II and other Product Planning Committee members in the Styling Rotunda and in the courtyard. Bordinat and DeLaRossa hyped the Envoy as the general direction of the next generation Lincoln. The response was lukewarm, but several designers believe the Envoy was given tentative approval by the Product Planning Committee for production as the 1961 Lincoln. Ben Mills was on the Product Planning Committee at the time and does not believe the Envoy, in any of its configurations, ever received production approval as the '61 Lincoln. He recalls that many on the Product Planning Committee, himself included, firmly believed the Envoy did not represent the direction they thought Lincoln should go and were frustrated with the lack of alternatives.

Several of Envoy's designers, including Howard Payne and John Orfe, shared the belief that the Envoy did not represent the direction the next generation Lincoln should take. That led to the Marlin and, several months later, Robert McNamara suggested that a rejected proposal for the 1961 Thunderbird be reworked as a four-door 1961 Lincoln. That decision, although generally greeted with relief by the designers in the Lincoln studio, meant the end of the Envoy, which was soon broken up and forgotten. ■

Ford Design Center

The preferred version of the Envoy was shown in the Styling Center courtyard.The driver's side was slightly different than the passenger side and showed more continuity with the previous Lincoln model.

Small fins and skegs at the rear of the Marlin were concessions by the car's designers to the belief that the 1961 Lincoln must show at least minimal continuity with the previous model Lincoln.

Lincoln Marlin

The 1961 Lincoln Continental was such a radical departure from the design of previous Ford Motor Co. products, there has been much speculation about where it originated. The '61 Continental was a product of the corporate advanced studio. As head of that studio, Elwood Engel was the primary motivating force behind the design, but designers in that studio insist that their studio had not been in existence long enough for the car to be the result of a natural design evolution that occurs in every studio. The Continental Mark II and Mark III Berline were part of the evolutionary process that led to the '61 Continental -- but at least one other car also influenced it.

As an assistant to George Walker, Engel had free roam of all studios in the Styling Center. Day and night, he roamed from studio to studio looking at everyone's designs, absorbing and borrowing the best from what he saw. That didn't make him very popular, and more than a few designers still believe he stole their designs in his midnight visits to their studios without giving them credit and, sometimes, taking the credit himself. Those stories include tales of designers coming to work the next day, only to find the designs they had been working on torn from their drafting tables, leaving only the corners of the paper where it had been taped to their boards. Because the separate studios were required to remain locked, many designers suspected, but never knew for sure, that Engel had taken their drawings after they'd gone home the previous night. Every once in awhile, a clay modeler would tell a designer that he had seen his design on a car in another studio. Less often, a door would be left unlocked, and designers were able to sneak a guarded look at clay models that incorporated their designs.

In the early 1950s, before Walker was placed in charge of the Styling Center, there were stories of designers in the Lincoln studio trying to get Engel barred from their studio, so its probably understandable that most of his roaming occurred at night, after everyone else had gone home. After Walker became general manager of the Styling Center, in May 1955, Engel became a Ford employee and Walker's assistant manager. More than anything else, that legitimized his activities including the borrowing. In fairness, it must also be acknowledged that, by combining the best designs produced at the Styling Center, Walker and Engel insured that more of them made it into production.

Part of Engel's job, as Walker's assistant, required frequent trips to Europe to oversee Ford's design studios in Cologne, Germany, and Dagenham, England. One of those trips to Cologne took place in June 1958. Wes Dahlberg had recently been assigned as chief designer at Ford of Germany and, when Engel arrived, Dahlberg was designing the

Valli Slate

Raymond N. Slate

Ray Slate was born in Ann Arbor, Michigan and grew up in Nebraska. From an early age, he developed a proficiency at drawing. After he graduated from high school, he attended the University of Nebraska, where he played on their football team. When his education was interrupted by World War II, he enlisted in the Army Air Corps and became a navigator in the Flying Tigers.

After the war, Slate attended Pratt Institute on the G.I. Bill. He graduated in 1949 and was hired as a designer at Ford. His first assignment was in Gil Spear's Advanced Studio. He was then transferred to the Ford studio.

In 1952, Slate left Ford to work for Moffatts in Toronto, Canada, designing appliances. Two years later, he was hired as a designer in the Chrysler design department. In 1955, after several months at Chrysler, he returned to the Ford Styling Center, where he was assigned to the Edsel studio. He later worked in the Ford and Thunderbird studios before being assigned to Elwood Engel's corporate advanced studio, where his designs were used on the Lincoln Marlin. He also did design work on the first Mustang.

In 1962, Slate left Ford and moved to San Fransisco, California, where he designed furniture and became an accident reconstruction artist and expert witness. He retired in 1986. ■

John Orfe's later grille design on the Marlin was used on the production 1964 Lincoln. The power dome on the hood appeared on the 1966 Lincoln.

Ford Design Center

Taunus 17M. Dahlberg's head clay modeler was Fred Hoadley, who had accompanied him to Germany. Both Dahlberg and Hoadley recall that during his visit to their Cologne studio, Engel expressed unusual interest in the just completed 3/8-sized theme model of the Taunus 17M. Dahlberg remembers that Engel returned to the studio several times, specifically to get another look at the Taunus 17M, and that he asked numerous questions about its design.

Dahlberg doubts the Taunus 17M had a direct influence on the design of the 1961 Continental. Hoadley, however, believes there is a connection, although he is unaware of any direct statement by Engel that confirms his belief. Hoadley returned home for a vacation about a year after Engel's visit and, while at the Styling Center, saw the '61 Continental for the first time. Hoadley thought then, and continues to believe, that the Taunus 17M had a direct influence on the '61 Continental.

Because of what Engel told other designers in the studio, there's little doubt the the Continental Mark II influenced the '61 Continental, and engineer Harold Johnson confirms the influence of the proposed Continental Mark III Berline. There are other designers who believe that the design of the '61 Continental was also heavily influenced by a car designed by Lincoln studio designers John Orfe and Howard Payne, two designers who were frustrated with what their own studio was proposing for the next-generation Lincoln. Even before the production 1958 Lincoln was introduced, many designers in the Lincoln

The Marlin was an attempt by frustrated Lincoln studio designers to come up with an entirely new Lincoln for 1961 featuring slabsides and cleaner lines.

Howard Payne

studio — especially the younger ones — were convinced that Lincoln badly needed a new design direction. Although they did not realize it at the time, Lincoln Division management had come to the same conclusion.

In mid-1957, Orfe and Payne told Lincoln studio chief designer John Najjar and his executive stylist John Reinhart, they did not agree with the design direction proposed for the next-generation Lincoln. To their surprise, Najjar and Reinhart told them they could build a full-sized clay model of their idea for the next-generation Lincoln. Payne and Orfe were given a small makeshift studio under a staircase in a room that had previously been a coffee shop. Fitting it in between other assignments, they designed and then, with clay modeler Joe Seibold, built the car themselves over a period of about six months. Both Orfe and Payne remember staying late one night and gluing on ball-bearings to finish the grille for the car. Neither Payne nor Orfe recall naming the car they designed and built. Seibold, however, recalls that, at some point, the car was named the Marlin. It is referred to here as the Marlin because Payne and Orfe acknowledge that, at one time or another, the car could have been known as the Marlin.

After the full-sized clay was completed, it was favorably received by management. In fact, Walker and Engel liked it enough to order two full-sized plaster and fiberglass models built, both with four doors on one side and two on the other. After the clay model was completed, but before it was cast in plaster prior to building the fiberglass model, Ray Slate suggested that stubby fins be added to the top of the front fenders and dropped over the grille area of the car dividing the dual headlights on either side. Slate's idea was incorporated on the clay models, although the stubby fins were later removed during one of the design reviews of the car.

One model of the Marlin was painted white, and the other was painted black. John Najjar designed the stylized "L" used on the deck lid. At the time the stubby fins on the front fenders and grille were removed, rocker and reveal moldings were added on the black car along with a chrome cap on top of the small fins on the rear fenders, which wrapped around the backlight to join the fins on both fenders.

Orfe and Payne both recall that many designers who saw the car mentioned how nice it would be if Lincoln Division produced something like the car they had designed. Between them, Orfe and Payne had a total of three years design experience at Ford. Neither had a sponsor who could champion their design. Almost immediately after Payne and Orfe finished the Marlin, Engel began work on the alternate 1961 Thunderbird that eventually became the 1961 Lincoln.

The squared-off look of Payne and Orfe's Marlin, the blade sides capped by the Quicksilver's stainless molding, the Mark II's greenhouse and taillights and the Mark III Berline's suicide doors became the primary design features of the 1961 Continental. As many designers point out, design is an evolutionary process.

A revised grille proposal Orfe designed for the Marlin eventually made it into production on the 1964 Lincoln. After Engel moved to Chrysler, the rear end of the Marlin also influenced his Chrysler Turbine showcar. ■

This full-sized version of the Marlin was built in the Styling Center shops.

Ford Design Center

Howard Payne

Howard E. Payne

Howard Payne grew up near South Bend, Indiana. His mother, an art teacher, encouraged her sons' artistic talent. Payne received a scholarship to attend the Fort Wayne Art School. While there, he read a magazine article about George Walker, an alumnus of the Cleveland Institute of Arts. He was also encouraged by relatives to attend the Cleveland Institute of Arts. Payne applied and was accepted.

He spent the summer after his third year in school as an intern in Alex Tremulis' Advanced Studio, and was offered a job at Ford when he graduated.

In 1956, he started his design career in John Najjar's Lincoln studio. His first assignment was designing stacked headlights for the Continental Mark II in the hope it would be continued for one more year. (It was not.) Payne was then assigned to the Advanced Studio and worked on the LaGalaxie. After several months, he was reassigned to the Lincoln exterior studio, then transferred to the Lincoln interior studio and then to the Lincoln pre-production studio, where he worked with Rulo Conrad designing the Envoy.

Payne and John Orfe were permitted to design their own proposed 1961 Lincoln, the Marlin, because they were unhappy with the design direction of the Envoy.

After the Marlin was completed, Payne returned to the Lincoln exterior studio, where he worked with John Orfe changing the 1961 Thunderbird proposal designed in the corporate advanced studio into the production four-door 1961 Lincoln Continental. Payne designed the Continental script used on 1961 and subsequent year Lincolns. He was later transferred to the Mercury studio, where he worked on the design of the Mercury Marauder.

Most of Payne's career was spent in the Lincoln-Mercury studios, but he also worked in the international studio, the advanced concepts studio and in Ford's German design studio.

Payne was one of the designers of the 1988 Lincoln Machette showcar and the 1993 Lincoln Mark VIII.

In 1987-88, Payne was a manager in the luxury car studio and, while assigned to Ford's English design studio, he supervised the design of the 1990 Lincoln Town Car.

At the time he retired in January 1997, Payne was manager in the trim and color studio. ■

1961 Continental

The 1961 Lincoln Continental first came into existence as an alternate proposal for the 1961 Thunderbird that failed to make the cut. It was rejected in favor of the more sporty Thunderbird proposal designed in Joe Oros' Ford studio. That rejected Thunderbird proposal, however, was unexpectedly reborn as the '61 Lincoln Continental, because of a compromise worked out between Robert McNamara, by that time Ford's vice president in charge of all cars and trucks, and Ben Mills, assistant manager of Mercury-Edsel-Lincoln (MEL) Division. The '61 Lincoln Continental, produced as a result of that compromise, went on to win the prestigious Industrial Design Institute award and became *Motor Trend* magazine's Car of the Year. It also restored the fortunes of Lincoln, and gave it a design direction that is still followed to this day.

To understand how an alternate 1961 Thunderbird proposal became the 1961 Lincoln first requires an explanation of the difficulties Joe Oros' Ford studio designers had in coming up with an acceptable design for the '61 Thunderbird, and how close Lincoln came to extinction because of the market failure of the 1958-60 Lincolns.

By 1958, the new Ford Division head, Jim Wright, decided that the proposed 1961 Thunderbird had to be even more sporty than the previous model. George Walker didn't agree but, since Wright had the last word, Joe Oros directed his designers to begin work on a sporty new four-passenger Thunderbird. The design of the '61 Thunderbird did not come easy, but after many renderings and 3/8-sized clay models, the design direction of the '61 Thunderbird was reached during an all-night session in which designers Bill Boyer and Jim Powers developed a full-sized clay model that established its basic lines. Boyer left around midnight and, before Powers knew it, he and clay modeler Walt Amrozi were meeting people coming to work the next morning. Jack Telnack recalls that Powers' side of that clay became the basis for the '61 Thunderbird. The model Boyer and Powers developed had huge shark fins on each rear fender. When the design was approved, no objection was made to the fins, but they were later removed for cost reasons.

From the time George Walker became general manager and Ford's vice president of styling in May 1955, until early 1958, Elwood Engel acted as Walker's trouble shooter and assistant manager. He had free roam of the Styling Center and, if he saw a project that needed his attention, the project got just that. Engel usually favored car designs that were futuristic, and many designers who worked with him considered his designs a little too far out. Every once in awhile, however, Engel surprised his co-workers with a very conservative design.

At the end of 1957 and during the first half of 1958, the Ford studio lost several months settling on a design direction for the 1961 Thunderbird. That led Engel to begin

One of Colin Neale's original renderings of Engel's proposed 1961 Thunderbird. This rendering influenced the rear-end design used on the 1961 Continental.

Colin Neale

thinking about designing his own competing proposal. For whatever reason, in early 1958, Engel decided he wanted his own studio, so he could design an alternate, more formal 1961 Thunderbird proposal that took its design cues from the recently-discontinued Continental Mark II.

Engel's alternate 1961 Thunderbird proposal.

Ford Design Center

Ever since the last one came off the assembly line in 1957, the Continental Mark II was being increasingly referred to as a classic and timeless design. In late 1955, the Mark III Berline was almost chosen as the 1958 Continental but, for cost reasons, it was rejected. The design direction of the 1958 Lincoln was set by Ben Mills, and although he wasn't completely happy with the final design, by early 1958, his Division was well on its way to a $60 million loss for the model years 1958-60. In a strange way, the public's rejection of the 1958-60 Lincolns made the design of the Continental Mark II and the proposed Mark III Berline even more appealing.

Although George Walker has been given credit for initiating the more-formal alternative 1961 Thunderbird designed in Elwood Engel's studio, it was probably Engel himself who first proposed designing an alternate Thunderbird proposal using the Continental Mark II as a benchmark. From the beginning, even the renderings prepared of the car refer to it as a Continental-style Thunderbird. According to a narrative statement he gave the Henry Ford Museum in June 1984, after Engel returned from a trip to Europe to visit Ford's design studios there, Engel asked Walker if he "could have a studio to develop an idea for a new Thunderbird." He was given his own studio in a small room in the basement of the Styling Center. According to Engel, the room left barely four feet on either side of a full-sized clay model. Engel's studio was nicknamed the stiletto studio by designer Colin Neale and, even though it was later claimed that the name referred to the long, narrow layout of the room, Neale now acknowledges that the name was coined as an irreverent jab at the production studios they were competing with. When the studio was formed, it was named the Advanced, International and Tractor Studio, but it was commonly referred to as the corporate advanced studio.

The designers assigned to Engel's studio were all hand-picked by him. Those who worked on the 1961 Thunderbird/Lincoln proposal, included John Najjar, Engel's executive stylist, manager Gale Halderman, designers Colin Neale and John Orfe and design analyst Bob Thomas. Designers Phil Payne and Dick Avery came into the studio as the car was nearing completion, but they did very little work on it. Orfe was the designer assigned to supervise the clay modeling. Harry Strickler was the lead modeler on the alternate 1961 Thunderbird proposal, and on all the changes the car went through in Engel's studio on its way to becoming the 1961 Lincoln. Studio clay modelers Joe Seibold, Cecil Perkins, Ross

Engel's alternate 1961 Thunderbird proposal had a stylized continental tire designed into the rear end. To the left, in the background, is the Ford studio Thunderbird proposal. To the right is the Lincoln Marlin.

Ford Design Center

Ed Albright

Edward E. Albright

Ed Albright grew up in California and, as a child, loved to draw cars. In 1953, he started his design education at the Art Center in Los Angeles, California. After Albright's fourth semester and, as a part of their schooling, he and other Art Center students spent the summer at the Ford Design Department as interns in Alex Tremulis' Advanced Studio. He returned to school at the end of the summer. In 1955, he and nine other Art Center students were recruited by Continental Division to help design the Mark III Berline. After seven months at Continental Division, Albright was drafted and spent two years in the Navy.

After he was discharged from the Navy, Albright returned to Ford, where he worked in the Lincoln-Mercury-Edsel Interior studio. He helped design the 1960 Edsel interior, and then worked on the interior of the 1961 Lincoln. Albright left Ford in 1959, to finish his schooling and graduated from the Art Center College in 1961.

After graduation, Albright was hired as an appliance and product designer by GM's Frigidaire Division. In 1965, he was recruited to work as a product designer in Raymond Loewy's Paris office, where he remained until 1971.

In 1971, Albright opened his own industrial design office in Paris but left in 1975 to return to California to care for an aging parent. Since 1975, he has worked as an independent industrial designer in Southern California. ■

John Orfe

John C. Orfe

John Orfe grew up in Philadelphia. In high school, he prepared art work for Indy car racetracks on the east coast and midwest. During World War II, while he was an Infantry soldier in the South Pacific, Orfe sent several of his drawings to automobile manufacturers asking for information about becoming a car designer. He was invited to contact them after the war. After he was discharged in 1946, Orfe returned home but was not able to find a job as a designer. He spent several years as an artist and lumberjack in Massachusetts, before he was hired as an artist and illustrator at Piasecki Helicopter Company in Morton, Pennsylvania. Over the next several years, Orfe wrote a comic strip about auto racing for the Illustrated Speedway News, *had one of his paintings of the Novi racecar used for the cover of* Speed Age *magazine and drew a series of profiles of racing personalities for* Speed Age.

In 1955, Orfe saw a Ford ad recruiting new designers. At the time, Orfe was head of Piasecki's art department, but felt he had gone as far as he could so he applied at Ford. After seeing his portfolio, Ford sent him an airline ticket for an interview. Orfe interviewed with several of Ford's studio heads, but decided he could not afford to change jobs. Before returning home, he had lunch at the Dearborn Inn, where he met Alex Tremulis, who convinced him to change his mind. Tremulis also got him a higher offer.

Orfe, as an experienced designer, was first placed in the Lincoln pre-production studio working with John Reinhart and then for Bud Kaufman in the Lincoln interior studio. After Kaufman was appointed head of the Edsel studio, Orfe worked for Rulo Conrad in the Lincoln pre-production studio, where he helped complete the Envoy. In a small, makeshift studio, Orfe and Howard Payne designed the Lincoln Marlin, which was their idea of what the next-generation Lincoln should look like. In 1958, Orfe was assigned to the corporate advanced studio and was one of designers of the proposed 1961 Thunderbird/Continental and the Mark IX.

Orfe also worked on the Gyron, the Allegro, the Median, the Stiletto and several other concept cars, before he left Ford in 1963 to return to work for Frank Piasecki, his former boss, on a flying car project. In 1969, Orfe was recruited by Colin Neale to become a designer at Chrysler but left Chrysler in 1974. Between 1974 and 1987, when he retired, Orfe worked for Autodynamics, Pontiac Engineering, and Volkswagen of America. ■

Ford Design Center

(above) The Thunderbird insignia on Engel's alternate 1961 Thunderbird proposal was designed into and a part of the car's door handle.

(below) Engel's more conservative 1961 Thunderbird proposal being revised in the stiletto studio. Alternate rear-end designs were considered for the 1961 Lincoln, but the design on the passenger side of this clay model was tentatively selected.

Ford Design Center

(below) This rear-end design on Engel's more conservative 1961 Thunderbird proposal was not used, but it is similar to the rear-end design of the proposed Mark IX, which was designed later.

Ford Design Center

(above) This was one of 12 renderings John Orfe prepared for the Product Planning Committee presentation books. Robert McNamara missed the meeting where the Ford studio proposal was chosen as the 1961 Thunderbird. By the time Engel's proposed '61 Thunderbird was shown to the Product Planning Committee, it only needed four-doors, a more formal roofline and a different rear-end design to become the 1961 Lincoln Continental.

John Orfe

McLean, Art Rockall, Wolfgang Konietzko, Ray Trombley, Rolland McDonald and Toby Melone worked on the car during its various stages. Design analyst Ray Smith, and several other designers, were assigned to the studio but did not work on Engel's '61 Thunderbird proposal. Although Engel retained all of his other duties as Walker's eyes and ears, the designers in the corporate advanced studio confirm that he spent much of his time as a hands-on designer in that studio.

When Najjar was replaced as head of the Lincoln studio, he was reassigned to the truck studio, and thought the transfer meant the end of his career at Ford. He later came to see it as a blessing in disguise and, with the benefit of hindsight, now feels that the design work he did after that transfer, including the 1961 Thunderbird/Lincoln, was the most productive of his career. When Engel got his own studio, he requested that Najjar become his executive stylist. Recalling the 1958 Lincoln they had worked on together, Engel and Najjar agreed that the alternate Thunderbird proposal they were designing had to be clean with no garbage on it. Engel told the designers in his studio that he wanted their Thunderbird proposal to be similar in design to the Continental Mark II -- but taken to the extreme.

Ford Design Center

Engel's final 1961 Thunderbird proposal on display in the Styling Rotunda. It had the round taillights customarily used on Fords.

Engel began by asking Thomas to prepare a package layout for the car, which included blade sides and a Mark II greenhouse (including curved glass) set on top of the area between the blades. Thomas worked on the original Continental Mark II and was delighted to be working on any car that gave homage to it. He immediately went to the Ford studio to get the package dimensions, and the project got under way. According to Thomas, the design analyst in the Ford studio told him that the only crucial measurement on the 1961 Thunderbird was the cowl area.

Based on what Engel said he wanted, Thomas designed the body structure of the car on a large blackboard. The wheelbase was set at 113 inches, with a cowl width of 75 inches and an overall length of 205 inches. Following Engel's directions, Thomas laid out the greenhouse, the roof, the curved-glass side windows and tumblehome, the body structure, and the shape of the front and rear-wheel cutouts. Thomas recalls that the shape of the high front-wheel cutouts was a matter of form following function. Engineering required that the front-wheel cutouts accommodate the tires at full turn and at full joust. Thomas

Ford Design Center

(above) After Engel's 1961 Thunderbird proposal was rejected, the full-sized clay model was returned to the stiletto studio. The roof was removed and several new rear-end proposals were considered including the alternate designs shown here.

says he copied the original design of the Mark II windshield to avoid the dogleg he thought made it harder to enter and exit cars with severe-wrap-around windshields.

Thomas suggested to Engel that the car be designed in a wedge-shape, and he recommended that the width of the car at the C-pillar be as wide as possible to emphasize the Continental-style greenhouse he designed to sit on top of the blades. Engel agreed, and when the package drawings were finished, the plan-view (from above) drawing on the blackboard showed a Mark II greenhouse straddling an area bordered with two outer blades that extended front to back like a catamaran. Unfortunately, the width of the car at the C-pillars turned out to be too wide for the package measurements set by Engineering. What Thomas didn't realize in laying out the package of the car was that the measurements he set down were five inches wider at the C-pillars than the package called for. Thomas recalls that several people, including George Walker, asked him if the design was within package and, although he always said that it was, it must have caused him to double check the measurements, because he discovered the error the night before the car was to be shown to the studio heads. By working almost all night, Thomas, Engel and the clay modelers were able to reconstruct the full-sized clay model to a maximum width of 75.4 inches. Engel wasn't happy about the mistake, and Thomas recalls being cussed out by Engel several times over the course of what was a long night.

According to clay modeler Joe Seibold, Engel's way of designing a new car was to order clay added to an armature until he said "stop." Then he would direct the clay modelers the way he thought the design should go. According to Najjar, Engel was inclined to cheat on the package measurements by hammering down the wooden pegs installed in the clay model to establish minimum heights and clearances. Najjar was constantly on guard and, when Engel's alternate Thunderbird design was being clay modeled, if the designers and modelers couldn't come up with a better idea for making the car look lower, Engel was inclined to solve the problem his way. Once the full-sized clay model was shown, it

(below) The taillights on the ends of the blade sides on the 1961 Lincoln proposal shown in the photograph above came from this rendering Colin Neale made for Engel's proposed 1961 Thunderbird.

Colin Neale

would come back to the studio, the measurements would be checked, and the modelers would then reset the pins or redo whatever part of the car was involved and, supposedly, no one would be the wiser.

The designers in Engel's studio actually completed two finished full-sized clay model Thunderbird proposals. They were ready by late June 1958, and both were very different from Oros' proposal. The grille on the first of Engel's proposals consisted of fine horizontal lines that wrapped around the front of the car's blade-like sides, a roofline somewhat different than the Mark II, and a low-slung, squared-off continental spare-tire hump built into the rear-deck sheetmetal and bumper. The tube sides ran from front to back, but with the Continental Mark II hopup on the ridge line just behind the door openings. Once the proposal was completed, it was di-noced and painted dark green.

This is the proposal Robert McNamara saw when he visited Engel's studio. The rear-grille nacelle with taillights on the ends of the blade sides had already replaced the round Ford-style taillights.

On the more conservative proposal, Colin Neale designed the rear end with large Ford taillights and very simple rounded ends on each blade that he capped with brightmetal on the ends of the rear fenders. Najjar suggested a grille he asked Orfe to design for the car. It was similar to a design theme that had previously been considered on several cars, and was commonly referred to as a Schick razor front end. Najjar's suggestion of a horizontal bar in the middle of the Schick razor grille made it distinctive. Najjar suggested that grille blocks be used inside the grille to lessen the electric shaver look. Although the final design of the grille was completed by Orfe, according to Neale, the clay modeling of it was completed on a day Orfe was home sick. On the more conservative proposal, there was no hopup on the belt line. At Engel's suggestion, the belt line on both proposals were capped by a narrow brightmetal cap. Engel told designer John Orfe he got the idea of the brightmetal cap at the top of the belt line from the Quicksilver, which had been completed several months earlier in the Ford pre-production studio.

Orfe was actually responsible for the similar front end/grille design on both the 1961 Thunderbird and the 1961 Lincoln. He recommended against the blunt front-end design Engel wanted on his Thunderbird proposal. To Orfe, the tube sides looked great, but he felt they should come to a point at the front of the car rather than ending bluntly, which he thought made the car look chopped-off and unfinished. Because the blunt front-fender ends bothered him, he drew multiple renderings of proposed front ends with pointed fenders. While Orfe was drawing new front ends for Engel's proposed Thunderbird, Walker came into the corporate advanced studio, one day, looking for ideas for the Ford studio Thunderbird proposal. He liked one of Orfe's renderings with a distinctive grille and pointed fenders and took it to the Ford studio, where it became their Thunderbird front end and grille proposal.

Before the cars were presented to the Product Planning Committee, Walker asked for Oros' support for Engel's proposal over the one designed in his studio. Oros knew Walker wanted Engel to replace him when he retired in three years and was doing everything possible to encourage his promotion. Oros supported Walker's efforts, but he told him he couldn't support Engel's Thunderbird proposal, because he honestly believed the Ford studio proposal was the better of the two. Under pressure from Walker, Oros told him the best he could do was remain silent.

The more conservative of Engel's Thunderbird proposals was shown in competition with the Ford studio proposal at a Product Planning Committee meeting held in late July 1958. One of the first people to comment on Engel's design was Bill Ford, who said the car was "too nice for a Thunderbird," and it "should be the new Continental."

During the presentation, Walker argued persuasively for Engel's proposal. Jim Wright, head of Ford Division, was also very persuasive in urging that the Ford studio proposal should be the next Thunderbird, and Henry Ford II seemed to be in a real quandary over which one to select. Henry Ford II then turned to Oros and told him he had not heard from him yet. Oros hesitated and then tried to avoid a direct answer, but Mr. Ford sensed what was happening and asked Oros for a direct answer. When Oros was pinned down, he acknowledged that the formal looks of Engel's proposal would look fine as a Lincoln, but he favored the Ford studio proposal because it continued Thunderbird's sporty look. Henry Ford II, who had probably already come to the same conclusion, agreed, and Engel's rejected Thunderbird proposal went back on the freight elevator.

Harry M. Strickler

Harry Strickler is a Detroit native, who started as a clay modeler at Chrysler in 1954. Before he became a clay modeler, he was a taxidermist and had served two years in the Army as an engineer. He had no formal training as a clay modeler and learned modeling as a result of on-the-job training. In 1955, he was hired as a clay modeler at Packard, where he worked on the Packard Predictor and the proposed 1957 production Packard. When Studebaker and Packard merged, Strickler moved to South Bend, Indiana, where he worked for Duncan McRae before leaving in 1957.

Strickler interviewed at the Styling Center with John Najjar. In November 1957 he was hired as a clay modeler in Elwood Engel's corporate advanced studio. He worked on the Levacar and was lead modeler on the 1961 Thunderbird that became the 1961 Continental. He later worked as the clay modeling supervisor on the Firefly/Astrion, the Mark IX and did some of the clay modeling on the Gyron. Later, he clay modeled many of Bob Thomas' concept cars, including the Allegro, Avventura and the Aurora. Before retiring in 1991, Strickler worked in the Ford studio for 12 years and later in the truck studio as a unit supervisor. ■

Harold Johnson

Harold W. Johnson

Harold Johnson grew up in the Detroit area. He attended Wayne State University graduating in 1940 with a degree in mechanical engineering. He started his automotive career at Briggs Manufacturing. During World War II, Johnson built aircraft assemblies at Briggs.

In 1946, he left Briggs to work for Graham-Paige, where he assisted Bill Stout in building a fiberglass Scarab car proposal. The body and frame of the Scarab were built entirely from fiberglass. The Scarab was one of the first fiberglass cars. When Graham-Paige merged into Kaiser-Frazer, Johnson became a Kaiser-Frazer chassis engineer and eventually the project engineer on the Darrin-designed 1951 Kaiser.

In 1953, Jesse Richards, a former Kaiser-Frazer body engineer, who had taken a job as a body engineer at Ford's Special Projects Division, (later Continental Division), recruited Johnson as the Mark II's chassis engineer. In 1955, when Harley Copp left to take a job as chief engineer at Lincoln Division, Johnson took his place as Continental Division's chief engineer. Johnson was responsible for engineering the proposed Mark III Berline.

After Continental Division was dissolved in July 1956, Johnson went to Lincoln Division as executive engineer-advanced. Johnson did much of the engineering work on the four-door 1961 Lincoln Continental.

In 1965, Johnson transferred to Ford's Research and Development Department, where he was in charge of developing Ford's catalytic convertor. He retired in January 1975. ■

Ben Mills was also a member of the Product Planning Committee. When he first saw Engel's Thunderbird proposal, he immediately thought the car was a natural for the next Lincoln. He did not like the shovel-nosed proposal from the Ford studio, but recalls secretly hoping it would be selected as the next Thunderbird. If it was, Mills decided, on the spot, he was going to requisition Engel's alternate design as the new Lincoln. Although Mills voted for Engel's design as the next Thunderbird, when Oros' proposal won, he immediately told Engel he wanted his proposal as the next Lincoln

Although Engel's bid to design the 1961 Thunderbird failed, when he got back downstairs to his studio, the modelers in the studio were surprised to see that he was elated. Engel told them that the design from Oros' studio had been chosen as the next Thunderbird, but that both Bill Ford and Ben Mills thought the car they had designed "was too beautiful to be a Thunderbird, and it should become the next Lincoln." At that point, Engel had good reason to believe there was something else in store for his proposed Thunderbird. He instructed the designers and modelers in his studio to replace the big round Ford taillights on the clay model with large chrome caps on the ends of the rear-fender blades, to design a squashed-horizontal tube in the area between the blades, and to include a separate grille at the back of the deck lid. After that, the car just sat awaiting either destruction or rediscovery.

Robert McNamara missed the meeting at which the Ford studio Thunderbird proposal was selected by the remaining seven members of the Product Planning Committee. Mills does not recall what if anything he told McNamara about Engel's proposal but, within the next week, McNamara went to the Styling Center, and Walker took him around to the studios to review the selections the other members of the Committee had made. Most of the designers thought McNamara, although a bean counter, had a good sense of design. When he got to Engel's studio, the rear end of the alternate Thunderbird proposal had already been restyled to eliminate the Ford-style taillights and incorporate capped blades and a grille nacelle. McNamara looked at the proposal for a few minutes and then started asking Walker and Engel questions about it. One of the questions he asked was whether it could be made into a four-door Lincoln. Engel had already decided it could. Several designers recall that McNamara was also told it would take about two weeks to turn the model into a four-door Lincoln. McNamara then left.

McNamara was new to his job of vice president in charge of all cars and trucks having recently replaced Lewis Crusoe, who retired in 1957 because of ill health. Having been the head of Ford Division before his promotion, McNamara was familiar with its plans, but he wanted a full explanation as to what was planned for Lincoln, so he instituted a series of bi-weekly meetings with Lincoln's management. Those meetings were held in a big conference room at the Styling Center and were attended by Ben Mills, assistant to James J. Nance, head of the newly combined Mercury-Edsel-Lincoln (MEL) Division, Harold McDonald, Lincoln's chief engineer, Emmett Judge, Lincoln's chief product planner and support staff. Apparently no one from the Styling Center was included in the meetings. Because McDonald expected that McNamara would ask questions about advances planned for future Lincolns, he asked Harold Johnson, his executive engineer-advanced, to attend the meetings. The meetings followed a set format with different subject matters selected for discussion by McNamara at each meeting. The subjects for discussion at the first meeting included sales, finance and projections. The second meeting dealt with new products and planned technical features. Johnson remembers being very involved in the presentations made at the second meeting, which included proposed suspension advances, how to install curved glass in future production Lincolns, etc. By the time the third and fourth meetings were held, they had become presentations with little discussion, except when McNamara asked questions.

During the fourth meeting, that all changed. McNamara stopped whoever was making the presentation, and said it was time to "clear the deck and put all the cards on the table." Everyone except Mills,

The front of this 1961 Thunderbird proposal had the grille shape, but not the grille texture used on the 1961 Lincoln.

Ford Design Center

(above) John Orfe's suggestion for the front end of Engel's 1961 Thunderbird featured a more-pointed grille and pointed front-fender ends. This grille proposal was picked by George Walker for use on the production 1961 Thunderbird.

Ford Design Center

McDonald and Johnson were asked to leave the room, and then McNamara announced that, based on Lincoln's dismal financial projections, he had decided to recommend that the car line be terminated. He didn't like the direction Lincoln was heading, didn't like the fact Lincoln hadn't *ever* made a profit, and he hadn't heard anything from Lincoln's management that changed his mind. For a few seconds, Johnson says you could have heard a pin drop in the room. Everyone in the room took McNamara, who was not known for his sense of humor, as dead serious. Then, Ben Mills, who had come to Ford in 1946 with McNamara as one of the "Whiz Kids," asked, "Bob, you can't really do that, can you?" McNamara answered, "You bet I can do it." Then a very heated discussion took place, with everyone in the room arguing against McNamara's proposed recommendation, but none of the arguments made a dent on McNamara, who gave no sign of changing his mind.

After what seemed like an eternity to Johnson, it was Mills who was finally able to persuade McNamara to back off, something that was not seen very often. Mills argued that the Ford family was not likely to agree to such a proposal, especially since the Continental Division had been dissolved only a year and a half earlier, and the Edsel was likely to be next. Besides, Mills argued, it was unfair to employees, dealers and customers to stop making Lincolns, cold turkey, without some sort of advance warning. The point of Mills argument was that, if McNamara was going to recommend that Ford Motor Co. stop building Lincolns, it was only fair to go through at least one more design cycle, give some warning, and then make the decision only after a lot more thought and study -- and the involvement of the Ford family.

McNamara told Mills that, with great reluctance, he would agree to one more Lincoln production cycle, but only on certain conditions. One of the conditions was that the next Lincoln had to be a lot smaller. Even today, McNamara emphasizes that his biggest concern was that Lincoln finally make a profit. McNamara thought the 131-inch wheelbase of the 1958 Lincoln was too big. He told Mills, McDonald and Johnson that he had recently seen a full-sized clay model of a formal, Continental-style Thunderbird downstairs in Elwood Engel's studio that had been rejected in favor of a more sporty Thunderbird proposal. McNamara told them that if the clay model in Engel's studio could be made into a four-door and profitably built as the new Lincoln, "without significantly increasing its size," he would "go with the program." According to Johnson, at that time, no one else in the room, except possibly Mills, knew anything about the particular clay model McNamara was referring to but, as the meeting broke up, Mills turned to McDonald and Johnson and said, "You'd better get down there ASAP, your jobs are cut out for you!"

It was late when the meeting broke up so, early the next morning, Johnson and Bill Davis, an MEL Division department head who was one of Johnson's assistants, returned to the Styling Center and immediately proceeded downstairs to Engel's studio. They looked at the car, decided they didn't really know what McNamara meant about not making it significantly bigger, picked up the package drawings for the car and returned to Lincoln advanced engineering, where they laid the drawings out on a large desk and, after studying them, decided not to take a chance, but to only lengthen the car the bare minimum needed to make it a four-door with the bare minimum seating capacity needed for a Lincoln. Johnson and Davis prepared rough package drawings for a proposed four-door Lincoln and returned to Engel's studio, where they asked him to lengthen the car by ten inches, add two more doors and to also build a separate seating buck using their package drawings. That meant going from a 113-inch wheelbase to 123-inch wheelbase and from 205 inches long to 213 inches long. Johnson was afraid to add any more length to the car, because McNamara's original instructions were to make it a four-door car, but not to appreciably lengthen it from the two-door size it started out as. As Johnson recalls it, McNamara never mentioned the additional length he added to the car.

One week later, Johnson got a call from someone in Engel's studio, probably John Najjar, telling him the seating buck had been completed and the clay model had been enlarged to a four-door. Johnson and Davis headed over to the Styling Center that after-

Phyllis Wong

Paul Wong

Paul Wong grew up in northern California. As a child, he loved to draw cars. After he graduated from City College in San Fransisco, his art teacher suggested he attend the Art Center College. After sending his portfolio, Wong was accepted at the Art Center and received a Ford scholarship after his first year. In 1955, Wong was one of ten Art Center students recruited by John Reinhart of Continental Division to help design and clay model the Continental Mark III Berline. Before the Continental Mark III Berline was completed, Wong returned to the Art Center and graduated in 1956 with honors. He then returned to Continental Division as a designer.

When the Continental Division was dissolved in 1956, Reinhart recommended Wong for a job at the Styling Center, where he was assigned to the Edsel exterior studio. Wong was later transferred to the Lincoln interior studio, and was one of the designers of the 1961 Lincoln interior. He also did some design work on the Gyron and later worked in Ford's Industrial Design Studio.

Wong died during surgery in 1989. At the time, he was a supervisor in Ford's luxury car studio. ■

When Engel's 1961 Thunderbird proposal was shown to the Product Planning Committee, it had no interior. Orfe's presentation book, however, included this rendering of a proposed interior.

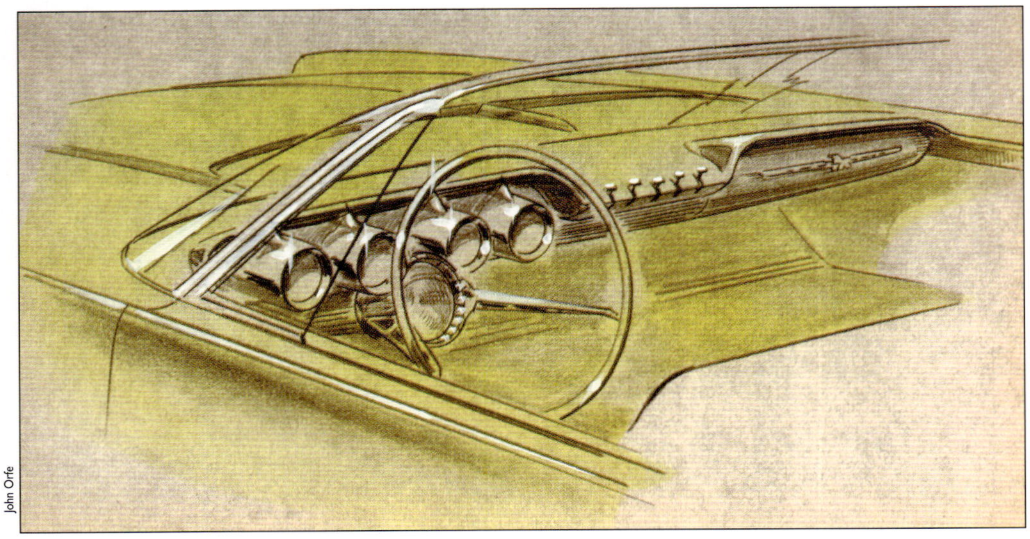

John Orfe

noon. When the clay model was rebuilt as a four-door, the hopup was put back in the beltline just in front of the termination of the rear doors. At the same time, Engel decided to eliminate the chrome caps, so Neale cut them in half, incorporated the lower part in the bumper and designed taillights, similar to those on the Mark II, into the top part of the caps at the ends of the blade sides.

The seating buck was the first thing Johnson and Davis looked at. It had four-doors, but minimum entry room was what was being determined. The seats were built to Lincoln seating specifications, but were more rounded at the corners in an effort to gain every advantage possible for entry and exit. The front seat was fine but, when Johnson got into the back seat, he couldn't get his foot back out without kicking the door. At that point, both Johnson and Davis realized a Lincoln with a wheelbase short enough to please McNamara would never work with conventionally-configured doors.

Johnson had formerly been chief engineer at Continental Division and had engineered a Mark III Berline proposal with back doors that were hinged at the rear. He suggested that the back doors of the 1961 Lincoln proposal be hinged at the rear, so the thickness of the doors would no longer present an egress problem. When he proposed suicide-rear doors, Johnson also proposed to do away with the B-pillars entirely and to lock the doors to the floor and to each other, as he had done on the Mark III Berline proposal. He ultimately decided against eliminating the B-pillars on the 1961 Lincoln, because he was concerned that McNamara might still suggest cancellation of Lincoln if this model became too complicated or too costly.

After the full-sized four-door clay model with the "barn doors," as Johnson called them, and a new four-door seating buck had been completed, they were shown to Lincoln's chief engineer Harold McDonald, who told Johnson and Davis they would never sell the idea of barn doors to McNamara let alone the Product Planning Committee — even if it was the only way to save Lincoln. With McDonald's permission, Johnson and Davis worked up an alternate plan, just in case the barn doors proved unacceptable. Although the full-sized clay model of the Mark III Berline and the diemodel had been long destroyed, Johnson saved the plans and was prepared to propose that the Mark III Berline be considered as the 1961 Lincoln if, for any reason, Engel's proposal didn't work out. They never had to make that proposal, and McNamara never knew there was an alternative plan, because the 1961 Lincoln, with its barn doors, was enthusiastically supported by Bill Ford and approved by McNamara. The car was again shown to the Product Planning Committee, which approved it for production as the '61 Lincoln. Once the decision was made to produce Engel's alternate Thunderbird proposal as the '61 Lincoln, the clay model was moved upstairs to Gene Bordinat's MEL studio to prepare it for production.

Engel and Bordinat were like oil and water. Although they had a correct professional relationship, they were fierce competitors. Bordinat was suave, smooth and politically astute. Engel, on the other hand, is most often described as a little crude and more at home throwing clay balls or pitching pennies with the clay modelers. To discourage Bordinat from making changes to the car, something that was thought likely, Bill Ford visited his studio to emphasize that he wanted the car built just as it was. There were several changes that had to be made to the exterior design of the car after it was moved upstairs. The most significant change was to the taillights. Although the Mark II-style taillights were designed before the car was moved upstairs, they had to be enlarged to meet legal standards.

(below) The Barcelona chair featured a chrome framework, which at one point was considered for the 1961 Lincoln seats.

(bottom) The seats planned for the 1961 Lincoln were influenced by the Barcelona chair designed by Mies van der Rohe.

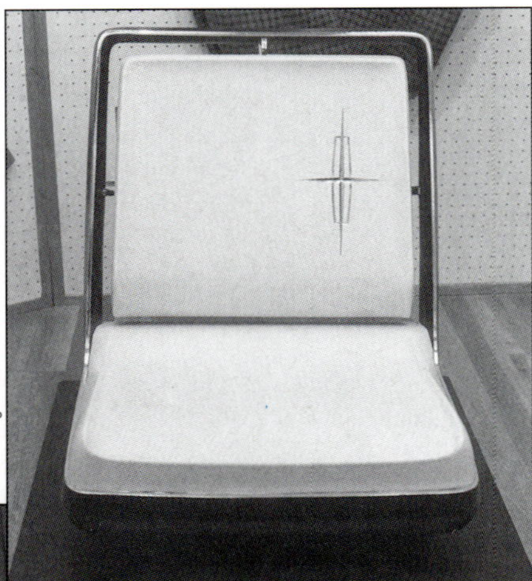

Ford Design Center

Ford Design Center

Harry Strickler, head modeler in the corporate advanced studio, recalls that even before the car went upstairs to the Lincoln studio, an instrument panel buck was started in Engel's studio in competition with the proposal being prepared in Dave Ash's Lincoln interior studio. At one point, Ash brought Engel a proposal for the instrument panel, but Engel told him he had his own ideas. Although it was unusual and would not have been tolerated if anyone else had proposed it, Engel insisted on following through on the entire design of the 1961 Lincoln Continental including the interior.

Engel's idea for the instrument panel was to group the air-conditioning and radio controls between two boxes housing the instruments and the glovebox. Neale and Thomas both recall Engel placing boxes and tubes on a table to demonstrate how he wanted the instrument panel designed. Although the basic design of the instrument panel is attributed to Engel, Art Miller and Bob Zokas, of the Lincoln interior studio, are credited with refining Engel's box and tube design into the 1961 Lincoln's instrument panel. Again, it was Bill Ford who got behind Engel's instrument panel design, and his support was crucial in its being selected over the Ash proposal.

The designers in the Lincoln interior studio included Paul Wong and Ed Albright, both of whom had started at Ford as Art Center students recruited to help design the Continental Division Mark III Berline proposal. They convinced Ash to let them try a new seating design for the 1961 Lincoln. Many of the younger designers in the Lincoln interior studio, especially those trained at the Art Center College, were intrigued by the Barcelona chair designed by Mies van der Rohe for the German Pavilion at the 1926 Barcelona Worlds Fair. The chair had shallow curves on both the bottom and the back and was generally upholstered in small leather squares. A Barcelona chair was even purchased, the legs cut off, and it was placed in the seating buck to see how it looked. Several renderings were done of an interior with Barcelona chair-style seating, and the finished seats of the '61 Lincoln have more than a passing resemblance to the Barcelona chair.

After Engel's proposed two-door Thunderbird was modified and accepted as the four-door 1961 Lincoln, other decisions McNamara made also influenced the production car. McNamara's bi-weekly meetings with the engineers continued, but the topic of discussion was almost exclusively the new Lincoln and how to solve past problems, especially those that had resulted in the large losses. The 1958 Lincoln was originally planned with a 126-inch wheelbase, but it had to be increased to 131 inches, so the engine could be moved five inches forward to make the transmission hump lower. Because the '61 Lincoln was to be shorter, the engine and transmission had to be moved back as close to the firewall as possible. That meant that the transmission hump in the front-passenger compartment became huge, threatening to make the car a four-passenger instead of a six-passenger sedan, something unacceptable for a Lincoln. According to Johnson, the tunnel was "so big, you could put a saddle on it." Again, technology developed at Continental Division for the Continental Mark III Berline was dusted off to solve the problem. Johnson became chief engineer of Continental Division after the Mark II entered production. One of the problems he inherited was a tendency for the Mark II driveline to vibrate, so he asked Dana Corporation to come up with an inexpensive constant-velocity double-cardin universal joint. It wasn't developed in time for use on the Mark II, because the car was canceled after only two years, but it was incorporated in the proposed Mark III Berline. Johnson contacted Dana, and suggested their universal joint proposal be used to solve the tunnel-height problem on the '61 Lincoln. Use of the double-cardin universal joint on the '61 Lincoln allowed the transmission and the propeller shaft to be angled, so that the end of the propeller shaft was right at ground clearance.

At one of McNamara's bi-weekly meetings, he brought up the subject of Lincoln's quality — or what he saw as the lack thereof.

Robert F. Zokas

Bob Zokas grew up in Boston, Massachusetts. Almost from infancy, he loved to draw cars. During World War II, he was a Navy combat crewman flying off carriers in the backseat of a Hellcat dive bomber. After the war, Zokas attended Boston's Vesper George School of Art, where he studied industrial design and then switched to advertising. After he graduated in 1949, he went to work as a commercial artist at Gale Hall Engineering and as a type designer for Photon Advertising Agency.

Bob Zokas

Zokas always wanted to be a car designer. With his boss' encouragement and, after several years designing type, Zokas prepared and submitted a portfolio to GM and Ford. In July 1955, he was interviewed and hired by Ford as a designer in the Edsel preproduction studio. He was soon transferred to the Edsel interior studio. In 1958, when the Edsel studio was cut back, Zokas was assigned to the Lincoln interior studio, where he was one of the designers of the instrument panel on the 1961 Lincoln. He also designed the 1962 Comet instrument panel.

From 1965-68, Zokas was Ford's corporate design manager with responsibility for coordinating and redesigning all signage, logos and advertisements used by Ford on its properties and by Ford dealers. In 1969-70, Zokas was executive designer in the Fairlane and Maverick studio and, from 1978-79, he was in charge of Ghia and their showcar program.

In the early 1980s, after heading the luxury car studio, Zokas became one of the primary designers of and the manager of the studio where the Taurus and Sable were designed. For three years before his retirement in 1987, Zokas was director of Ford's interior and trim and color studio. ■

This early interior buck still shows the chrome framework on the seats and an instrument panel layout similar to the instrument panel used on the D-524.

Ford Design Center

With McDonald and Johnson's enthusiastic support, vendor cooperation and new testing methods, a 24,000-mile warranty program and even universal oil standards were pushed through and became standard features on the 1961 Lincoln. Although the '61 Lincoln sold only slightly more cars than the 1960 Lincoln, market share rose by 60%. More importantly, the profit it made, the praise it received, the design awards it won, and increasing sales over the next decade assured that Lincoln would survive. Besides, even though McNamara became president of Ford Motor Co. in 1960, in 1961, he left Ford to become John F. Kennedy's secretary of defense.

The Bordinat-DeLaRossa study concluded that Lincoln had not been as successful in the marketplace as Cadillac because of a lack of design continuity. McNamara believed that lack of continuity was only a small part of the problem, and that quality and sensible design were far more important. Time has proven McNamara right. Bill Ford's Continental Division may not have survived, but the Mark II and the proposed Mark III Berline his division designed became the primary influence for all Lincolns, starting in 1961. ■

Even after the rear-grille nacelle was finalized for the proposed 1961 Lincoln, Engel still wanted other rear-end proposals to consider, including this one.

Ford Design Center

(right) By January 1959, what started out as a two-door 1961 Thunderbird proposal, had four-doors, a new grille texture and was badged as a Lincoln Continental. Judging from the mannequin shown in the driver's seat, this proposal probably had a seating buck installed.

Ford Design Center

Ford Design Center

(left) This 1961 Continental proposal is in close-to-final form. According to Lincoln studio designers, the chrome hash marks on the front fender, behind the wheelwell, were proposed by Elwood Engel. Lincoln studio head Gene Bordinat was ultimately successful in having them removed.

Almost immediately after Elwood Engel and the designers in his corporate advanced studio finished converting their 1961 Thunderbird proposal into the 1961 Lincoln Continental, it was moved upstairs to the Lincoln production studio. At the same time, Engel was given additional studio space on the first floor of the Styling Center in room 138. At least for the time being and because room 138 was only a single bay, some of the designers remained downstairs in the smaller stiletto studio.

Engel's studio had responsibility for the design of the advanced, international and Canadian cars, tractors and some competing production proposals. Before the move upstairs, advanced designers were in the small studio downstairs, while designers assigned to tractor studio were in another room, and designers in the international studio were in a third room at the other end of the building. After the move, Engel's studio was spread out in at least four different rooms, but the biggest space Engel had was in the new upstairs bay sandwiched in between existing exterior production studios.

(below) This rendering by John Orfe was included in the presentation book he prepared at the time the proposed Continental Mark IX was shown to the Product Planning Committee. A Sports Phaeton was only one Mark IX model proposed for 1964.

(bottom) This rendering by John Orfe shows a proposed Mark IX Club Sedan.

Ford Design Center

Ford Design Center

1964 Mark IX

One of the first projects undertaken in the corporate advanced studio, after the 1961 Continental was delivered to the Lincoln production studio, was a next-generation slabside Lincoln Continental. It was designated the Mark IX, and it was designed in room 138. If a successive Continental model was produced each year, the Mark IX would have been produced in 1964, the assumed target date for the next-generation Lincoln.

Designers in the corporate advanced studio, at the time, recall that after the 1961 Lincoln was approved for production, George Walker suggested to Engel that he also design the next-generation Lincoln. It was somewhat unusual that the Mark IX project was started in 1959, almost five years ahead of the planned production date, but Walker was nearing retirement age, and he wanted Engel to succeed him. Engel's primary competition was Gene Bordinat, who was then head of the Lincoln-Mercury studio. Engel had bested Bordinat in head-to-head competition for the design of the 1961 Lincoln, and Walker felt by doing the same thing with the next-generation Lincoln, it would go a long way towards cementing Engel's position as the front runner.

After the Mark IX project was authorized, Engel asked Gale Halderman, one of his design managers, to take charge of it. The primary designers Halderman assigned to the Mark IX project were Colin Neale and John Orfe, who at the time were probably two of the most imaginative and experienced designers in the studio. As with the 1961 Thunderbird/Lincoln, Orfe designed the grille and front end, while Neale designed the rear end of the Mark IX.

Neale, in particular, enjoyed designing large cars. Until he emigrated to the United States, in late 1958, he was head of the design department at Ford of England. In England, he never had the luxury of designing what he considered to be a large car. Orfe was Engel's designer of choice when it came to designing grilles and front ends that were distinctive, but could be built within budget. Orfe and Neale worked well together and were

Gale Halderman

Gale L. Halderman

Gale Halderman grew up in Ohio and graduated from the Dayton Art Institute in 1954. He was hired by Gene Bordinat as a designer in the Lincoln-Mercury studio. On his first day on the job, Bordinat sent him upstairs to help out and get acquainted in Alex Tremulis' Advanced Studio. He was in the Advanced Studio for two weeks and was then sent to the Ford studio because of a crisis there. After several weeks in the Ford studio, Halderman was assigned to the Ford pre-production studio working with Damon Woods and Bill Boyer on the Mystere.

Halderman's next assignment was in the truck studio and, after that, he was reassigned to the Ford studio.

In 1958, he was assigned to Elwood Engel's corporate advanced studio as a manager. While in the corporate advanced studio, he helped design the Levacar, the Mark IX, the X sixty five, the Astrion and the Gyron. Halderman also supervised the design of production car proposals in competition with the other studios.

Halderman transferred back to the Ford Studio in the early 1960s and was one of the designers of the 1965 Ford and the original Mustang. In 1968, he replaced John Najjar as head of the truck studio. Five months later, Halderman was asked by new Ford president Bunkie Knudsen to design cars, although he remained in the truck studio. When Joe Oros became chief designer, Ford of Europe, Halderman replaced him as head of the Ford studio.

In 1968, Halderman replaced Buzz Grisinger as head of the Lincoln-Mercury studio. At the time of his retirement in 1994, he was director of the small car studio, the large and luxury car studio, and the trim and color studio for all cars and trucks.

Halderman and John Aiken were the primary designers of the Lincoln Continental Mark VI. Halderman also led the team that designed the Lincoln Mark VII and VIII. ■

Ford Design Center

(above) After the 1961 Continental was moved to the Lincoln studio, designers in Engel's corporate advanced studio designed the next-generation Continental Mark IX. It was well-liked and initially planned for production in 1964.

(below) The Mark IX featured landau bars, the owner's initials on front-fender badges, and fully-integrated front and rear bumpers.

Ford Design Center

(below) The rear-end design of the Mark IX came from one of Colin Neale's 1961 Thunderbird renderings. The Mark IX had a padded-vinyl roof and deck lid and unique Continental-insignia taillights.

Ford Design Center

able to create a unified look on the cars their joint efforts produced. Halderman actively participated in the design of the Mark IX and is credited with updating and translating the slabside look of the 1961 Lincoln into the Mark IX.

Contrary to what happened with the 1961 Lincoln, which started as a Thunderbird, the Mark IX was originally designed as a Lincoln. As Halderman recalls, the Mark IX looked bigger than it really was but, in fact, the Mark IX was actually smaller than the 1961 Continental. The dimensions for the Mark IX were: 211.70-inch overall length, 76.34-inch width, 53-inch height and a 120-inch wheelbase.

The Mark IX was even more slabsided than the 1961 Lincoln. It featured pushbutton doors with fingerholds hidden in the creased ridge line. Until Buzz Grisinger suggested moving the parking lights forward on the front fenders, they were closer to the windshield.

Ford Design Center

The Mark IX incorporated several design features Neale had earlier proposed for the 1961 Thunderbird/Lincoln, which were not used, including chrome caps on the ends of thin rear fender blades, and wide catwalk-like panels between the rear fenders and the deck lid. The twin grille, bumper and dual-headlight design on the front of the Mark IX were designed by John Orfe, including the domino-shaped grille texture. Orfe also designed the rear taillights, incorporating the Continental symbol in them. The general overall design of the Mark IX's box made use of several features developed by Tony Yuki, one of the studio's better idea men. Bob Thomas and Ray Smith were the design engineers on the Mark IX.

Ford Design Center

(left) The sides of the Mark IX were extremely flat.

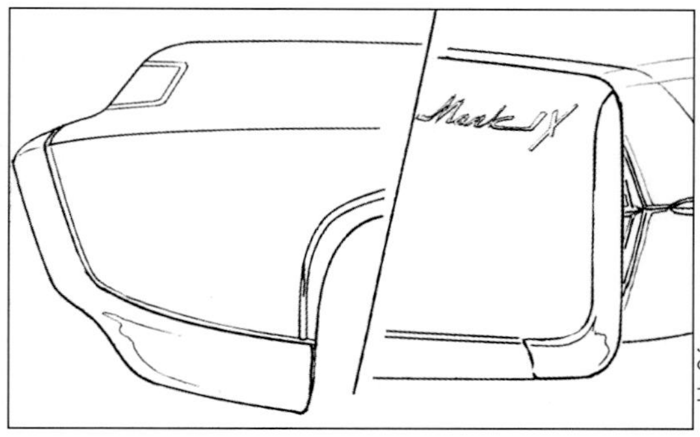

John Orfe

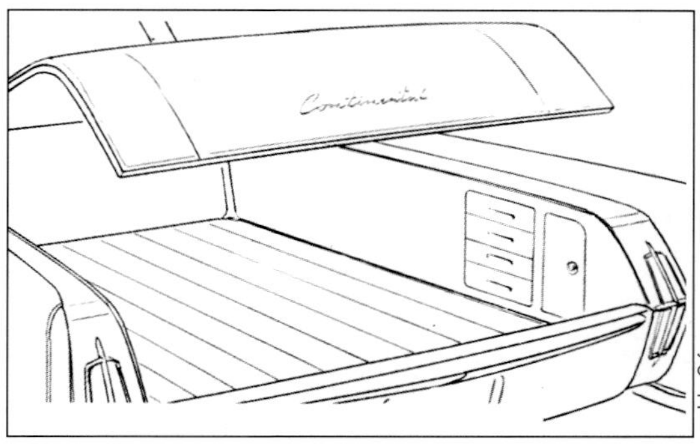

John Orfe

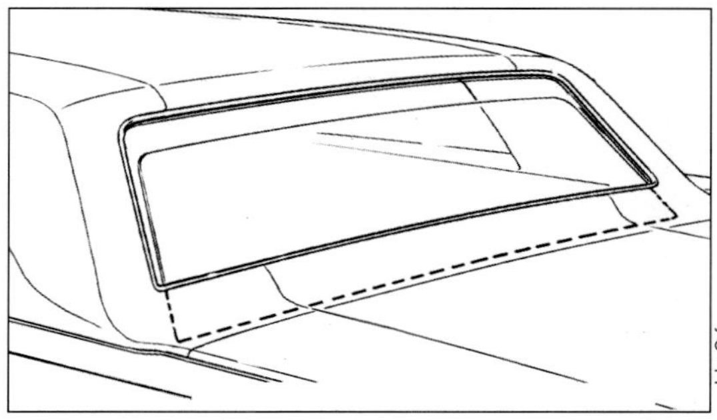

John Orfe

By the time the Mark IX was designed, it was no longer customary to build a 3/8-sized model first and, according to Halderman, no 3/8-sized model of the Mark IX was ever built. Halderman, Neale and Orfe all worked, along with the clay modelers, to develop the full-sized clay model. Bill Dayton, who was on active duty with the National Guard when the 1961 Thunderbird/Lincoln was designed, had since returned to the corporate advanced studio and, as one of his first projects, designed the color-coded milk can-type wheelcovers on the Mark IX. Phil Payne, who started as a new designer in the corporate advanced studio in January 1961, remembers the Mark IX as having slabsides that were slightly coke-bottled, an innovation he had not seen before.

As the full-sized clay model of the Mark IX was nearing completion, Buzz Grisinger, who then headed the Mercury production studio, visited the corporate advanced studio, and was asked by Engel to critique the Mark IX's design. Grisinger suggested that the stubby fins on the belt line of the car that stopped at the front doors be extended out to the ends of the front fenders. Orfe was asked by Engel to follow up on Grisinger's recommendation, and he designed the Mark IX's aircraft-type leading-edge turn signal/parking lights that were extended almost to the tips of the front fenders

Engel came up with a list of accessories he wanted on the car including a full-vinyl top that also covered the rear deck lid and landau bars. One of the other features incorporated on the Mark IX was a small door in the hood used to check engine fluids. That fea-

Ford Design Center

Although the Mark IX looked longer than the 1961 Continental, it was actually three inches shorter.

Ford Design Center

Anthony Yuki

Tony Yuki was born in Japan, where his father was a revered master architect. He came to Michigan to attend school and wanted to eventually become an automobile designer. When World War II broke out, Yuki, who did not believe in the old ways, chose to stay in the United States. He attempted to enlist in the U.S. Army but was turned down. Because he was unable to work during the war, he had to drop out of school.

In the years immediately after the war, he was unable to find a design job. In 1948, Yuki was hired at Ford's Rouge plant to work on the assembly line. In 1950, he enrolled in Bill Leverenz' clay modeling class. After he completed the class, he was hired by Charlie Stobar as a clay modeler in the Ford studio.

In 1953, Charlie Waterhouse transferred Yuki to Gil Spear's Advanced Studio to become a designer. After that, he worked in the Ford interior, the Mercury exterior and the Ford exterior studios. He was one of the primary designers of the 1959 Ford.

In 1958, Yuki was assigned to Elwood Engel's corporate advanced studio, where he helped design the Mark IX. Later, Yuki was transferred to the Industrial Design Studio, where he developed design sketches that became the front end and side designs for the Mustang.

During the time he was a designer at Ford, Yuki was known as an excellent artist and an idea man. He was one of the first designers to use narrow black tape to make blackboard car designs, a practice that soon became widespread throughout the industry. He retired in 1978. ■

ture was suggested by John Najjar and later used on the 1965 Techna concept car. In the presentation book Orfe prepared for the Product Planning Committee, he listed several pages of the Mark IX's special features, including;

- leading-edge turn signals;
- flush-type bumpers;
- stainless rocker panels;
- a service access panel on the hood that permitted checking the oil, water, battery, power steering, etc without opening the hood;
- a split-radiator grille front end;
- concealed windshield wipers with air-deflector flaps. The wipers were hidden by little doors in the hood that opened when the wipers were activated;
- pushbutton doors with hidden fingerholds that were recessed into the beltline of the car;
- sculptured landau bars that had a built-in safety night lights;
- a backlight that retracted up to five inches for ventilation;
- a padded vinyl-clad roof, deck lid and rear-bumper area;
- a spacious trunk with built-in compartments;
- personalized monogram plates on the front fenders. On the prototype, Engel used his own initials;
- floating-type wheelcovers and;
- unique Continental-type taillights.

Walker liked the car, as did all of the design executives who reviewed it. It also received excellent reviews from Lincoln Division management and the Product Planning Committee. As a result, a full interior was designed, and the fiberglass model was then rebuilt with a full interior and doors that opened. The designers working on the car took that as an indication that it was being seriously considered for production. Ray Smith, who developed the battery-powered electric motors for the Gyron and the Firefly, was asked to install a similar electric motor and steering on the Mark IX, so the car could be seen in motion. Like the Gyron and Firefly, the battery-powered Mark IX only went about 5 mph.

In 1960, when Neale left Engel's corporate advanced studio to start an advanced interior studio, the Mark IX was completed and had been built in fiberglass, but it had no interior below the beltline. According to Neale, the car was painted a deep maroon, but the black vinyl earlier installed on the top and deck lid had been removed.

According to Halderman, if the decision to produce a next-generation Lincoln had been made in 1961, there's not much question it would have been the Mark IX. As 1964 drew closer, several factors worked to defeat production of the Mark IX. In November

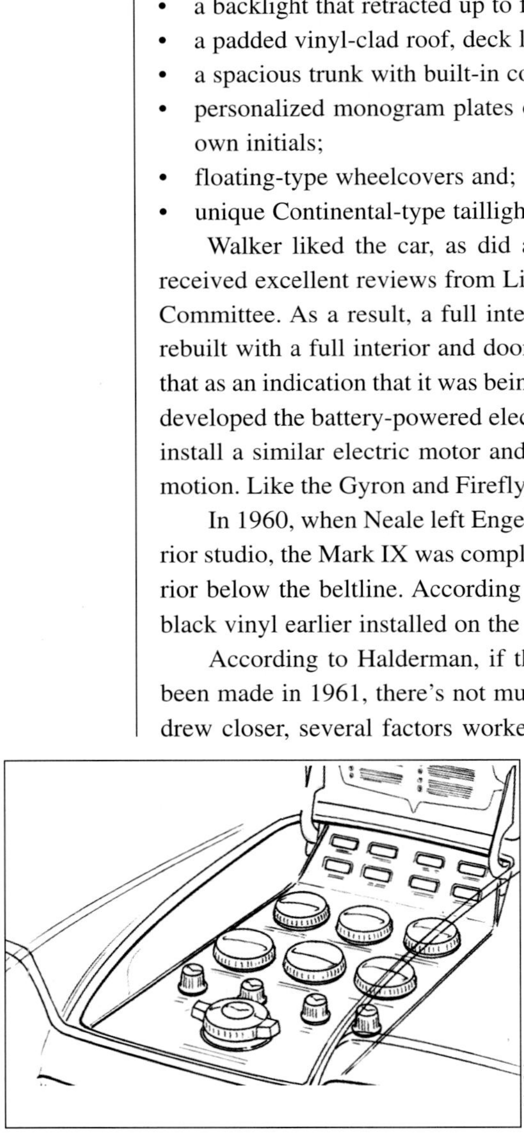

John Orfe

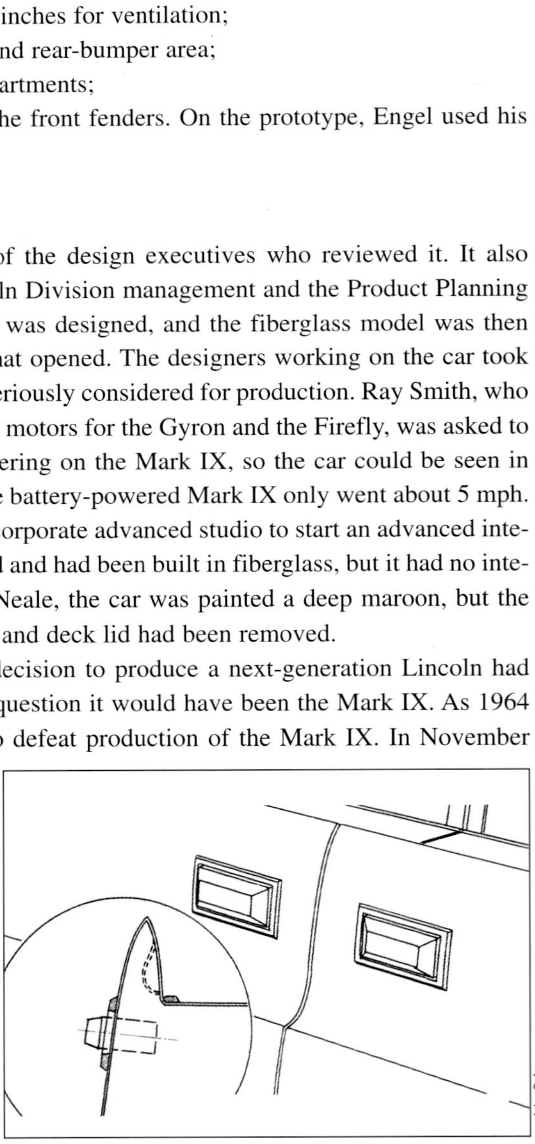

John Orfe

1961, several months after Walker retired, Engel left Ford to become Chrysler's vice president of styling. He asked Neale to join him at Chrysler as head of Chrysler's interior studio. Neale accepted and offered Dayton a job at Chrysler, which Dayton accepted.

When Engel left Ford, he took many of the designs produced in the corporate advanced studio with him. Designers in that studio recall that Engel had copies made of hundreds of designs over the several week period before he left. Nightly, they remember seeing boxes of photographs and copies of drawings taken from the building and loaded into Engel's car. More than a few eyebrows were raised at Ford, in late 1963, when the 1964 Imperial was introduced, because the design similarities between it and the Mark IX were obvious. By that time, other Chrysler products, including the Chrysler Turbine showcar, had been introduced. The Turbine car closely resembled designs developed by Engel and the designers in his studio, while Engel was still at Ford. As a result, Ford Motor Co. decided not to produce copy cat cars that Chrysler's new chief designer could claim he had designed or designs that mimicked Chrysler production cars. That, as much as anything, killed the Mark IX.

When Ben Mills came to Lincoln Division, his job was to make it profitable. For too many years Lincoln had been loosing money, and Mills was given the job of not only stopping the losses, but making Lincoln a profit center. McNamara's recent threat to cancel Lincoln gave Mill's task a new urgency. Mills decided that there were two ways Lincoln could make a profit. The first was to drastically increase sales, although such efforts had not worked in the past. The second was to stretch Lincoln's model run, which would allow a longer time to amortize tooling costs. Needless to say, lengthening the model run to a longer period was exactly what Mills did, making the lack of change a feature of Lincoln's advertising -- and kind of a reverse snob appeal. Although an increased model run left room for the type of changes that occurred with Lincoln in 1966, it eliminated an all new model — in this case the Mark IX.

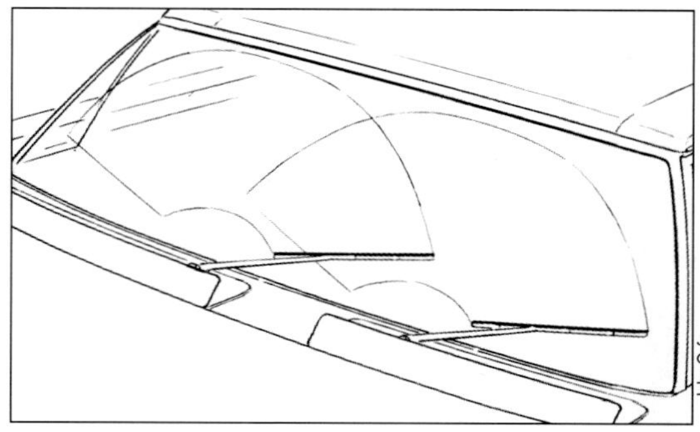

John Orfe

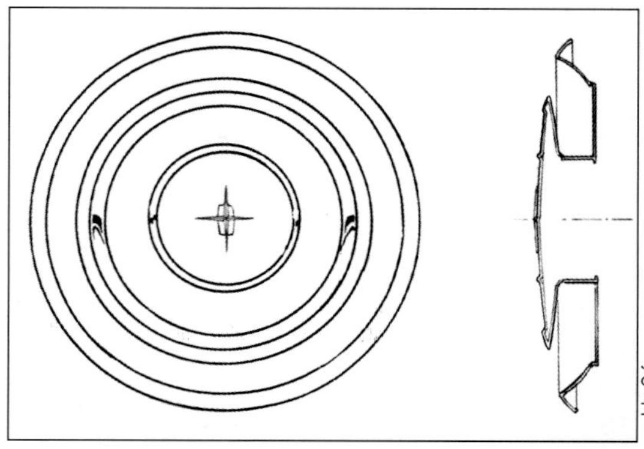

John Orfe

Ford Design Center

(left) In January 1961, an alternate full-sized clay model of the Mark IX was built with conventional doors. If the Mark IX had been produced, its designers believe the suicide doors would have been retained.

(below) Full engineering studies were done for the proposed Mark IX. Lincoln Division engineers were concerned that a split grille would interfere with cooling, so the designers included a hole between them and camouflaged it by including the Continental script in the middle. By the time this Mark IX proposal was built, the grilles and other parts of the body had also been squared-off.

Ford Design Center

At the time a longer model run was adopted by Lincoln, no one really knew if it would work or not. If Lincoln sales increased dramatically, it also meant that an all-new body style was not needed. The very success of the 1961 Lincoln Continental eliminated the need for the Mark IX.

According to Halderman, cost may have also been a factor in the decision not to produce the Mark IX. By the mid-1960s, the Product Planning Committee decided to devote more of the available production money to mid-sized cars. Lincoln underwent a successful facelift in 1966. The next generation Lincoln however, was the 1970 Lincoln, which had little similarity to the Mark IX. It was designed in Buzz Grisinger's Lincoln-Mercury studio primarily under the direction of design manager John Aiken, who started his design career in Elwood Engel's corporate advanced studio. ■

His co-workers generally liked working for Elwood Engel. Engel often spent his lunch hours with designers in his studio, or pitching pennies with clay modelers in the back rooms or hallways. At times, he could be crude in his speech, and he had a bawdy sense of humor, but clay modelers also remember times when he worked all night right along with them, finishing up a clay model needed for a show the next day. During design reviews in his studio, designers and modelers describe Engel as often walking around the car, redesigning as he went by slicing at the clay with his modeling knives, all the while throwing clay balls at the nearest waste basket or, if he was in a mischievous mood, at designers or modelers. Sometimes they were thrown back.

X sixty five

During the years Gale Halderman worked with Engel, he developed an admiration for his design ability, and the way he found design inspiration in almost everything. Although some designers disagree, Halderman remembers Engel as a good artist but with no set design philosophy. Halderman probably comes closest when he identifies Engel's primary motivation as a constant quest for marketable designs. If it sold because the public liked it, Engel felt it was good design. Engel also thought that good design ought to have flair.

Even Halderman acknowledges that it's hard to fathom how Engel could be responsible for both the 1958 Lincoln and the 1961 Lincoln, two cars that are as different as night and day. According to Halderman, at the time each car was designed, Engel felt they were both exactly what the public wanted. Engel designed the '61 Lincoln specifically because he knew he had misjudged the marketability of the '58 Lincoln.

Ford Design Center

Elwood Engel and Gale Halderman created the X sixty five by reworking the full-sized clay model of the Continental Mark IX, after the plaster mold was pulled from it.

To many of his contemporaries, Engel was also a design Jekyll and Hyde. Every time he came up with a clean and beautiful design, designers in his studio joked among themselves that his next design proposal would regress to something unusual if not bizarre. Halderman believes that Engel was often misunderstood. Engel was a showman, and he felt that his designs got noticed if he got noticed. In addition, Engel felt that almost any design was worth trying at least once. Like everyone else, Engel had his down days, and the pressure got to him but, on the whole, he thoroughly enjoyed his job and position.

Engel encouraged his designers to produce nice, clean cars but, after a design was completed, they always expected him to sock something on it that was meant to make everyone say "wow!" Engel preferred designs with things on them. His rationale was that hood ornaments, scoops and miscellaneous trim pieces increased profit, whether they had design merit or not. Over the course of time, designers in his studio were inclined to complete their designs the way they wanted them — and then put an ornament or scoop where they thought Engel would put it. According to Halderman, Engel also believed it easier to back-off on a design than to have to tune it up.

Engel constantly pushed his designers to come up with "something noo," as he pronounced it. If creativity lagged, Engel would come to work in the morning with boxed toy models of spaceships for his designers to build, or he would send them to J.L. Hudson's department store or the local art museum to find inspiration for new designs. At times, Engel took designers with him to the kitchen section of local hardware stores to search out skillets, pots, pans, or anything else with unique shapes that could inspire fresh designs. On more than one occasion, Engel came to work with pie and cake pans he thought would make great shapes on some part of a clay model then being worked on. Designers who traveled with Engel recall that, even when driving down the street, he drew inspiration from everyday things he saw like garbage trucks, wheelbarrows, lawn tools, houses, ladders, etc.

The X sixty five's greenhouse was inspired by the LaGalaxie. The front-fender bombs, although much modified, were later included on the 1964 full-sized production Ford.

By the time the Mark IX was cast as a fiberglass model, Engel felt the design was clean, attractive and marketable. Management confirmed that judgment with high praise for the car, and it was seriously considered for production. In typical Engel fashion, when the design of the Mark IX was finished, he told Halderman that "now that we've done a pure and clean car, let's do one that really shocks them." Engel instructed Halderman to take the full-sized clay model of the Mark IX from which the plaster plug had been pulled, and redesign it into something that even he thought was too aggressive. That aggressively redesigned clay model of the Mark IX became the X sixty five.

Engel suggested that Halderman begin by putting a greenhouse on the car like the one on the LaGalaxie showcar, including the vertical backlight he liked so well. As the design progressed, Engel also suggested that big bombs be added to the front fenders with headlights at the front of the bombs, cooling vents on the sides and ending in points on the doors. Not much was changed on the rear end or the quarter panels, but the front end was redesigned without any grille, something designers were trying to convince management could be attractive, while, at the same time, pulling in needed air to cool the radiator through vents below the bumper.

The name X sixty five was probably meant to represent the direction Engel thought car design was likely to take by 1965. According to Halderman, Engel was famous for trying to place names and lettering on scoops or places he thought looked like it called for lettering. On the X sixty five, the name was on the back of the bombs on the doors, where Engel told Halderman to put it.

The X sixty five was never considered for production, but it did what it was supposed to — it experimented with new shapes and forms and combined them with existing features. Besides, Engel thought it in his best interest to gradually acquaint management with his ideas for future automobile designs -- even if he was wrong.

One of features from the X sixty five was included on the production 1964 Ford. Shortly after he completed the X sixty five, Halderman was transferred to the Ford studio. The front-fender bombs on the '64 Ford, although greatly diminished in size and effect, were designed by Halderman, and they came directly from the X sixty five.

When Engel left Ford in November 1961 to become Chrysler's head of design, he asked Halderman to become his assistant design director. Halderman accepted, but later changed his mind. Going to Chrysler meant long hours, and a child's illness altered his priorities. In 1986, when Engel was on his deathbed, Halderman visited him, and they spent an afternoon reminiscing about their days at Ford. By that time, Halderman had been head of the Ford studio and, at the time, was head of the Lincoln studio. As they were looking through the binders of photographs Engel had of cars they had done at Ford, he stopped and told Halderman he had made the right decision when he decided to stay at Ford. Halderman left feeling that Engel saw his time at Ford as the highlight of his design career. ■

Very few modifications were made to the rear- quarter panel and deck area on the X sixty five. The name was located in script at the end of the fender bombs on each door.

Dom Orlando

The Mercury Palomar at the Chicago Auto Show. It was Ford's first station wagon showcar, and it featured a flying-bridge third seat.

Mercury Palomar

By 1957, station wagons accounted for 15% of all new-car sales industry wide -- and the market was increasing every year. A separate station wagon design studio was created as a part of Joe Oros' Ford studio, and Damon Woods even wrote a book on station wagon design. Station wagons had become such an important part of the market that, by the early 1960s, two showcars were built as station wagons, something that hadn't happened before. The first of those vehicles was the Mercury Palomar, which was designed in the fall of 1960. The second was the Aurora, designed several years later by Bob Thomas, Ed Jaquet, Ray Beamer and Ray Smith. The Palomar was designed and clay modeled, full-sized, in the Canadian and international portion of Elwood Engel's studio by Dom Orlando with assistance from his studio manager, Jim Sipple, designers Dick Avery and Ed Iser and design analyst Ray Smith.

When Engel was given his own studio in early 1958, it was a catch-all studio. Engel's studio not only designed advanced cars and proposed production cars in competition with the Ford and Lincoln-Mercury studios, it also had responsibility for the design of tractors and cars intended for the Canadian and international markets. During the first several months Engel's studio was in existence, advanced cars and cars designed to compete with proposals from other studios were all designed in the small basement room referred to as the stiletto studio. After the move to larger quarters upstairs, the Canadian and international part of Engel's studio was relocated to one end of the new studio. Although Engel and his executive stylist, John Najjar, were in charge of the entire studio, the corporate advanced and the Canadian and international parts each had separate managers. Even though they worked together on occasion, Gale Halderman was manager in the corporate advanced studio, and Jim Sipple was manager in the Canadian and international studio.

Engel's upstairs studio had room for ten full-sized styling bridges. They were, more or less, lined up in a row with the Canadian and international part of the studio at one end of the room and the advanced part at the other end. All the styling bridges were constantly busy, and it wasn't uncommon for advanced and Canadian and international designs to be mixed all throughout that studio -- and sometimes tractors also.

Even the designers assigned to the advanced part of the studio spent most of their time designing alternate proposals of production cars but, even though concept and showcars were only a small part of what the corporate advanced studio produced, they were seen by all designers in all parts of Engel's studio as the fun part of their jobs. Although it was not really business as usual, the Palomar was designed and clay modeled in the Canadian and international part of Engel's studio. As Dom Orlando explains it, from time-to-time, Engel tried to let all designers in his studio spend time designing concept cars, and all designers were constantly encouraged to submit ideas for concept and showcars.

According to Orlando, the idea of a new Mercury showcar originated with George Walker, who sold it to Lincoln-Mercury Division. Designers and clay modelers generally credit design analyst Ray Smith with the idea of making the new Mercury concept car a station wagon. Smith was constantly involved in developing new and innovative seating and entry concepts. He thought vans and station wagons would only increase in popularity in coming years, and felt that Ford should be ready to produce more varied models. Smith is also credited with developing the Palomar's integrated front and rear bumpers.

It was Engel, however, who decided that the Palomar should be built in the Canadian and international studio and, because Orlando came up with the idea, he was named its lead designer. Orlando began the design of the Palomar by making several alternate renderings of the car and, from these renderings, Engel chose the design features he wanted. Canadian and international designer Dick Avery was asked to assist Orlando with the design work of the Palomar. At the time, both Orlando and Avery were avid boating enthusiasts, and they both eventually left Ford to become boat designers.

Ford Design Center

Even before the ill-fated Tucker '48, designers have been fascinated with Cyclops-type headlight proposals. Both GM and Ford designers tried to incorporate four-headlight clusters in their designs. This headlight configuration on an early full-sized clay model of the Palomar probably amused corporate advanced studio designer Alex Tremulis, one of the designers of the Tucker.

The Palomar had a wheelbase of 119 inches and a height of 55.88 inches, but its width and length were flexible. It was designed as a proposed 1966 Mercury, and as a new type of open car that was being considered as an alternative to the traditional convertible model. From its conception, the flying bridge was the Palomar's primary design feature. According to Orlando, he borrowed the idea from the boating industry. He envisioned the flying bridge as a third seat that could be raised so the seat's occupants sat with their heads and shoulders above the open rear-roof section. Orlando's first design of the flying bridge included a section of the rear roof that could be retracted and stored under the floor using the retractable mechanism originally developed at Continental Division. Orlando submitted the idea of a retractable flying bridge on a station wagon, and it was approved by Walker and Engel. Orlando does not recall whether the retractable top went by the wayside because it was too expensive or too complicated to incorporate on a station wagon. Ray Smith developed the roll-top desk-type sliding-roof section that was actually used on the Palomar. The roll-top retracting-roof mechanism was similar to an idea earlier developed by Gil Spear, but it is not known whether Smith was aware of Spear's prior design. To protect passengers seated in the raised flying bridge from wind and bugs, a Plexiglas windscreen designed by Alex Tremulis was installed on the roof in front of the third-seat passengers. The Palomar had a complete interior, which was designed by Jim Sipple. Sipple was previously the head of the Edsel interior studio, so interiors were a natural for him.

The car was named the Palomar after the California observatory located on Mt. Palomar. Neither Orlando nor Avery recall being aware that a 3/8-sized model had been designed by Jim Powers several years earlier, which was also named the Palomar.

The sculpted sides, bumpers that were integrated into the body and the retracting-roof proposal were included in the Palomar's design almost from the beginning.

Ford Design Center

According to Orlando, Engel specified a front-end design that featured four headlights grouped together in the middle of the grille, and the full-sized clay model of the Palomar originally did feature four headlights bunched together just below the hood in the center of an oval-shaped integrated bumper/grille. According to John Orfe, the saddle beneath the four-headlight grille was designed for the Palomar by Ray Slate.

Shortly before the Palomar was designed, Orfe, as one of the corporate advanced studio's better artists, was recruited by John Najjar to meet with a man Orfe describes as the spitting image of Sherlock Holmes. Orfe remembers the man's British accent and that he smoked a pipe. Orfe was told the man had frequent access to GM's design studios. His meetings with the man took place in an out-of-the-way unused office in the old EEE Building. Every two weeks,

Dom Orlando

Domenico E. Orlando

Dom Orlando was born in Windsor, Ontario, Canada. His family moved to Detroit when he was a child. At the end of World War II, he enlisted in the Navy and was discharged in 1948. In 1951, his father, a 46-year Ford employee, helped him get a job as a sweeper in a Ford glass plant. Because he was always drawing cars, boats and planes, Orlando soon applied for and was accepted in an illustration class offered by the Ford Engineering Department. In 1953, when Orlando heard that Charlie Waterhouse, general manager of the Design Department, was looking for designers, he prepared a portfolio and submitted it. He was accepted and assigned to Gil Spear's Advanced Studio.

After six months in the Advanced Studio, Orlando was assigned to the Ford pre-production studio, where he helped design Thunderbirds. He later worked in various Ford and Mercury pre-production and production studios, and was later temporarily assigned to the corporate advanced studio, where he was one of the designers of the Palomar, the Seattle-ite and the Grande showcars.

In 1968, Dick Avery, a former Ford designer, hired Orlando as a boat designer at Chris Craft. After three years with Chris Craft, Orlando left to become a general contractor. After several years as Toyota's U.S. design director, he retired in 1977. ■

Ford Design Center

(above) The Palomar's designers tried several different designs on the sides of the car. This quickly discarded proposal featured boat-type vertical strips along the Palomar's belt line.

Ford Design Center

Ford Design Center

(above) A plaster plug was made from this full-sized clay model. Using the plug, the Styling Center shops built the showcar's fiberglass body. The di-noced wood paneling and the surrounds on the front fenders were later changed.

(left) The license plate used on the full-sized clay model identified the Palomar as a product of Canadian Styling.

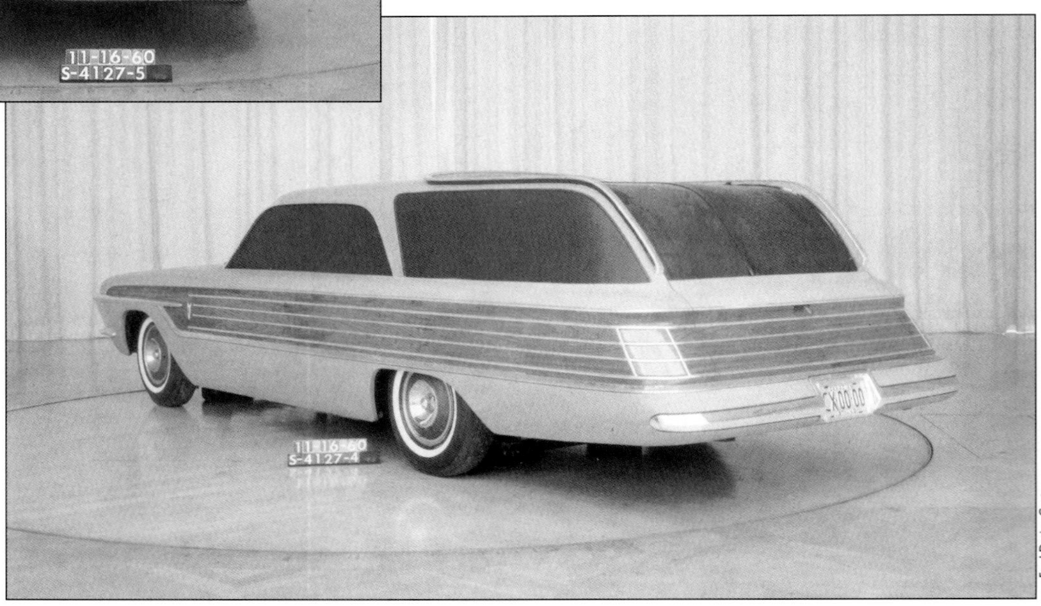

Ford Design Center

(right) The wood paneling on the sides of the Palomar emphasized a modified wedge shape.

Orfe would secretly meet with the man and listen to him describe the full-sized clay models GM was building. From those descriptions, Orfe would sketch the cars until the man approved the drawings. When Orfe finished his sketches, he delivered them to Najjar, who would share them with Engel. Although the four headlights in the center of the grille design never made it into production at either GM or Ford, it was likely included in the designs sketched by Orfe and passed on to Engel, who tried it on the Palomar. The only GM car Orfe specifically remembers sketching was the 1963 Buick Riviera, which he did some time in 1960.

Ford Design Center

By October 1960, the four-headlight grille had to be abandoned, not because of design problems, but because of objections from the Engineering Department and questions raised about whether the engine could be properly cooled with four headlights blocking the radiator. Orlando came up with the final design used for the Palomar's grille, which he credits to the electric shaver-style grille popular at the time and also used on the 1961 Continental.

A modified wedge shape was used on the sides of the Palomar, accentuated by di-noc wood paneling and brightmetal surrounds running from the front fenders back along the sides of the car and wrapping around the tailgate area.

The Palomar was built as a fiberglass model in November 1961. In finished form, it was a four-door hardtop on the driver's side with suicide doors operated by pushbuttons, and a two-door hardtop on the passenger side. All the doors on the Palomar worked. Ray Smith developed the driver's side suicide doors, which eliminated the B-pillars by locking the doors to the floor with the same mechanism he helped develop for the Mark III Berline. The completed Palomar was painted silver-blue metallic and was upholstered in matching colors.

Sipple recalls that the carpet he first ordered for the car was the wrong shade. When reordering, he made a deal with the supplier to purchase the incorrect color carpet himself rather than return it. For many years, he used that carpet in his home, and only got rid of it after he retired and moved from the Dearborn area.

James H. Sipple

Jim Sipple is a Detroit native, who spent World War II in the Navy as a plane captain. After he was discharged, he attended Pratt Institute on the GI Bill and a GM scholarship. He graduated in 1949 and started his design career at GM, where he eventually became assistant studio head in Ned Nickel's Buick studio and then in Paul Gillan's Pontiac studio. In 1954, Sipple was placed in charge of recruiting new designers, when Jules Andrade retired. Budget cuts caused by a recession resulted in his transfer to the Cadillac interior studio.

In June 1955, Sipple left GM to accept a position at the Ford Design Department, where he was assigned to the Special Projects Studio as executive stylist designing Edsel interiors. In 1958, he was transferred to the Ford interior studio, and then to Elwood Engel's studio as a manager in charge of the design of Canadian and international vehicles. While in the Corporate Advanced, Truck, Tractor and International Studio, he designed the interior of the Palomar.

Later in his career, Sipple was assigned to the Lincoln-Mercury studio and, in 1964, he was placed in charge of the interior studio at Ford's facilities in Cologne, Germany. In 1966, he returned to Dearborn to replace Ted Hobbs, when he retired as head of the Ford trim and color studio. In 1968, he returned to the German design studio as its chief designer. In 1970, he returned to Dearborn as chief designer in the truck and tractor studio, where he remained until retirement in 1980. ■

(below) The Palomar on display at the Ford Rotunda. The car had a complete interior and the doors opened. The third seat, the sliding roof and the car's lights worked but, otherwise, it was a pushmobile.

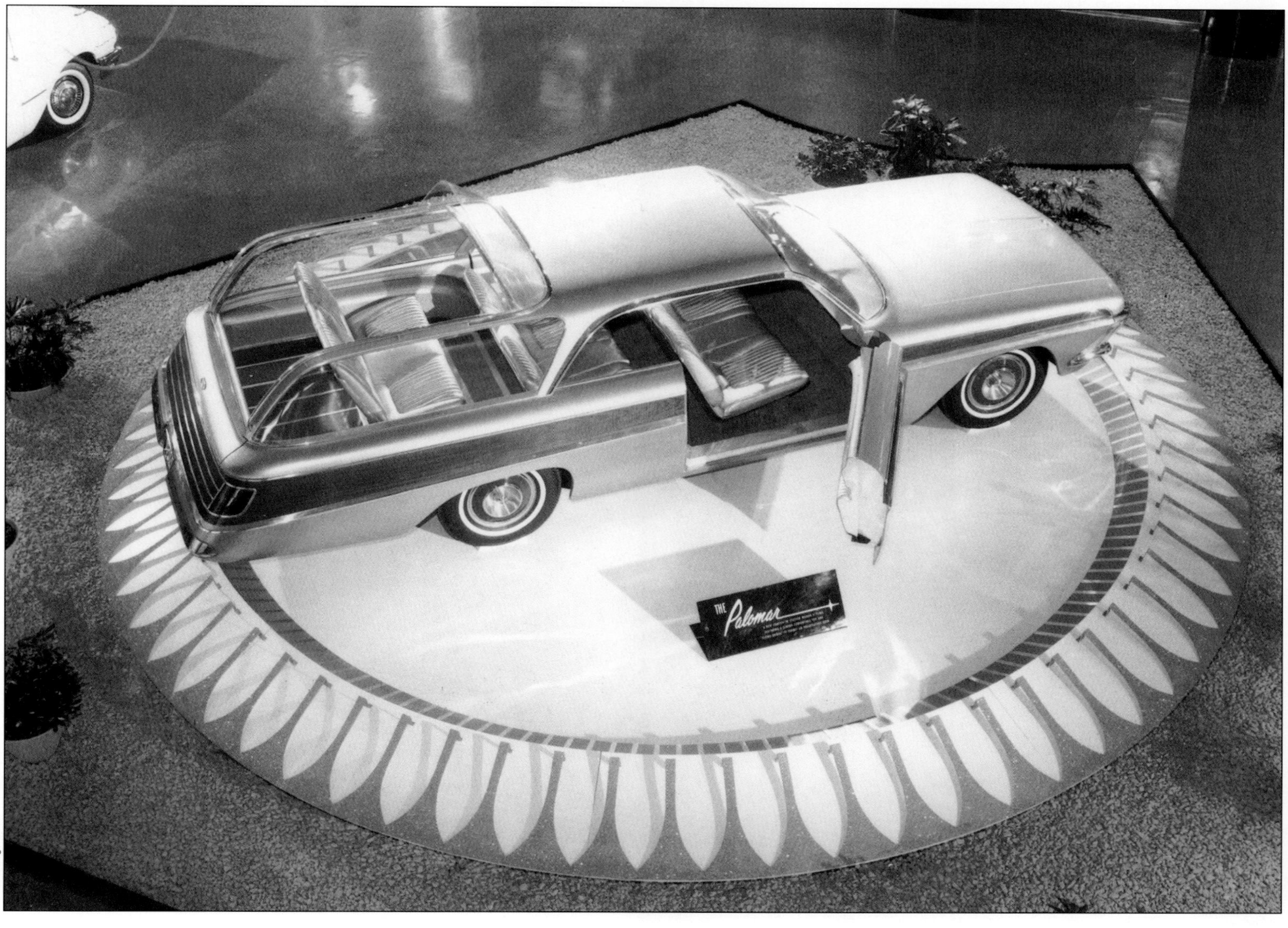

Ford Design Center

Dick Avery

Richard D. Avery

Dick Avery was born in Lakewood, Ohio. As a child, he liked to draw cars and boats. As a teenager, he built several custom cars. While in college at Miami University, one of his father's friends showed Dick his son's rendering. (The son was a designer at Chrysler, and the renderings were of the proposed 1957 Chrysler.) After he saw the renderings, Avery decided he wanted to be an automobile designer, so he applied and was accepted at the Cleveland Institute of Arts. During his last year, he won a design award from Ford Motor Co.

Joe Oros hired Avery as a Ford designer in 1958. His first assignment was in the Ford trim and color studio. Because of his interior design experience, he was transferred to Elwood Engel's corporate advanced studio in 1959 to work on the design of the instrument panel and seats for the 1961 Lincoln. He also worked on the design of the Gyron and the Palomar and was later transferred to the Lincoln-Mercury studio.

In his garage during his off-hours, Avery built small boats, including the original wave runner. When Chris Craft Industries advertised for a boat designer in a Detroit newspaper, Avery applied and was hired. He left Ford in February 1962 and, for the next 25 years, he designed boats for Chris Craft, Glastron, Hatteras, Murray, Wellcraft, Triumph and Stingray. He also designed interiors on van conversions for Mark III Industries, before retiring in 1992. ■

Ford Design Center

The Palomar had no running gear, and only the lights, the third seat that could be raised and lowered and the sliding top worked. Before the Palomar was shown publicly, the tires were changed from wide whitewalls to narrow whitewalls.

According to its designers, the Palomar was built only as a showcar, and was never considered for production. It was on the show circuit for only a short while, although publicity photographs were distributed and the car was featured in several publications. The Palomar was shown at the Chicago Auto Show and may have been displayed at other auto shows. It was on display at the Ford Rotunda with the Gyron and was probably destroyed in the fire that engulfed the Ford Rotunda in October 1962. ■

Ford Design Center

Ford Photomedia

(top) The third seat on the Palomar could be raised about 30 inches. Passengers seated in the flying bridge felt like they were in an open convertible.

(above) On the driver's side, the Palomar had suicide doors that were opened by pushbuttons.

(left) Studio secretary Jean Brown is seated in the Palomar's flying bridge. The roof is in the open position. When the third seat was collapsed, it became a part of the floor and gave the Palomar as much room as any other station wagon.

Unitron

The Unitron was certainly not business as usual at Ford's Styling Center. Although cab-forward pickups had previously been designed there, the Unitron was Ford's first cab-forward concept recreational vehicle. It was designed in fall 1960 in the corporate advanced studio by Dean Beck, a recent graduate of the Art Center College, who was hired by Elwood Engel while he was visiting the school on a recruiting trip in early 1959. Design work on the Unitron was supervised by Beck's studio manager, Gale Halderman. As Beck designed it, the Unitron turned out to be a sleek-looking, rear-engined, cab-forward minivan. By late 1961, the Unitron was a full-sized fiberglass model with a full interior. After it was completed, it was motorized by one of Ray Smith's 5 mphelectric motors attached to one wheel.

When Engel asked Beck to design a cab-forward streamlined van-type vehicle, he was told it was intended for use primarily as a recreational vehicle. Beck was also told that management was trying to justify the Unitron as a proposed executive office on wheels. Unfortunately, the Unitron turned out to be a design in search of a justification -- which it never found.

The idea for the Unitron did not come from Engel or any of the designers in his studio. According to Halderman, George Walker received a request from someone in Ford Division management, who thought the company should get in on the ground floor of the coming recreational vehicle market. That request was met with skepticism but, to be as versatile as possible, the Unitron was designed and built as a passenger vehicle on one side, with a full set of side windows, and a panel van on the other side, with closed-in panels behind the door window. At the time the Unitron was designed, there was no real market for recreational vehicles or portable offices, and most of Ford's engineers and designers thought that any need for more passenger room could be adequately served by station wagons -- and what limited market there was for vans was about to be filled by the 1962 Econoline.

According to Beck, the name Unitron came from Engel. The spaceage had begun in earnest several years earlier with the launch of the Russian Sputnik, and everyone at the Styling Center was caught up in spaceage styling. Beck recalls that Engel liked electronic and spaceage names and was inclined to propose that type of name anyway for the cars

In its final form, the Unitron had doors that opened from the center of the windshield.

Ford Design Center

Ford Design Center

(above) Dean Beck is shown with a full-sized clay model of the Unitron. Unfortunately, Beck moved his head at the time this photograph was taken.

created in the corporate advanced studio. Beck does not know where Engel got the name, but believes it came directly from his fertile imagination.

According to Beck, he started out designing a box and, to break it up, he included an outward-curving horizontal crease all the way around the beltline of the vehicle except at the rear end, which was relatively flat and curved in towards the bottom. The front end was veed at the beltline and had an integrated bumper and grille. On the first version of the Unitron, there were air vents on the rear corners on either side of the backlight presumably for ventilation. Beck's first version of the Unitron also had a series of vertical nerf bars that wrapped under the rear end. The nerf bars were later removed, and a narrow back bumper was placed across the beltline at the back of the car. In its final form, the bumper was moved from the beltline and placed between the taillights, and a small band of brightmetal trim was added across the back and along the sides at the beltline. Otherwise, except for the trim around the windshield, backlight and windows, and the wide front grille and integrated bumper, the Unitron was devoid of brightmetal trim.

The rear end of the Unitron featured extra-large Ford-style taillights. According to Halderman, Engel believed that all Fords should have big round Ford taillights. While the Unitron was still in the clay modeling stage, Engel came to work one morning carrying a box with different sized pie and cake pans. He put two of the pie pans on the lower-rear end of the clay model, and told the modelers he wanted the same sized round taillights right where he had placed the pie pans. The pie pan-sized taillights became a part of the car, even though several designers in the studio were critical of their shape on what they considered as an otherwise classically-beautiful design.

Although running gear was never installed in the Unitron, it was designed for a rear-mounted engine. The dimensions of the Unitron indicate it was probably designed for use on a modified car chassis. It was 200 inches long, 56 inches high, and had a wheelbase of 95 inches.

When the full-sized clay model of the Unitron was completed, it was shown to the Styling Center's executive stylists and then to the Product Planning Committee in both the Styling Rotunda and in the courtyard. The first time the Unitron was moved to the Styling Rotunda, shop personnel used a lift truck to help move it. When the blades from the lift truck were placed underneath the back end of the clay model, they slipped and ripped off many of the vertical nerf bars. After the Unitron was delivered to the Styling Rotunda, Beck and several other designers and clay modelers had to scramble to reattach the bars and straighten out the scars in the clay.

Dean W. Beck

Dean Beck was born and raised in Peoria, Illinois. From childhood, he wanting to be a car designer. After high school, he enlisted in the Navy. After a Far East tour and, while his ship was stationed in San Diego, Beck contacted the Art Center College, submitted a portfolio and was accepted. In 1956, after his enlistment was completed, he started school. When he graduated in early 1959, he was recruited by Elwood Engel as a designer in the corporate advanced studio, where he worked on the Levacar and the Unitron.

After a year in the corporate advanced studio, Beck was reassigned to the Mercury pre-production studio, where he worked on the Cougar. Later, he also worked in the Lincoln production studio, the Corporate Projects Studio and the interior studio.

Beck was one of the designers of a special Lincoln limousine Henry Ford II had built for Jackie Gleason in the late 1960s. He also helped design the 1968 and 1969 Mercurys, the 1972 Mark IV, the 1986 Taurus and the 1994 Mustang. Beck retired in 1992. ■

Dean Beck

(right) One of Beck's renderings of the Unitron's exterior.

Dean Beck

Before it was built in fiberglass, the Unitron went through several modifications, especially to the front doors. In December 1960, the A-pillars on the full-sized clay model were located on the windshield side at the front corners of the car, so the door openings included a part of the front end and front bumpers. By January 1961, the doors and the A-pillars had been relocated to a more conventional location on the sides of the van, but the doors did not include any part of the front of the car or the front bumpers.

By the time the Unitron was completed as a full-sized fiberglass van in August 1961, the A-pillars had been eliminated, and the doors opened from the center of the windshield using an articulated double-pivot hinge mechanism developed by Ray Smith. The door-opening mechanism on the Unitron was slightly different than the one used on the Firefly. On the Unitron, some of the car's metal structure below the windshield was included with each half of the windshield for strength. Ford Division management never got used to the old-fashioned division bar in the center of the windshield.

Before the A-pillars were eliminated by doors that opened from the center of the windshield, conventionally opening doors were considered on the Unitron.

Ford Design Center

As Beck remembers it, the Unitron always got good reviews, but production was never authorized because of concerns about the lack of a market for it.

Beck designed several alternative interiors for the Unitron and a full interior was eventually installed. Either just before or just after the Unitron was built in fiberglass, the decision was made to turn the Unitron into a show vehicle. That decision was short-lived, and the next thing designers in the corporate advanced studio remember is that plans were abruptly changed, and another interior was designed and installed in the Unitron. The Unitron was finished not as a showcar, but apparently for use as someone's executive office. There are conflicting reports as to whether the Unitron was actually made into an executive office, or whether the office configuration was only a proposed alternative use.

Ford wasn't sure there was a market for the Unitron. Otherwise it might have been produced in addition to the Econoline. (In front of the Unitron is the recently repainted Mystere.)

Ford Design Center

When designer Phil Payne was discharged from the military in late 1962, he returned to the Styling Center to resume his career. He remembers seeing the completed Unitron with a full interior including a long couch along one side. Shortly after he got back, he was introduced to his wife to be, who was a secretary in the corporate advanced studio. Early in 1963, she and Payne were offered a ride in the electric-powered Unitron. They were slowly being driven around the ponds in front of the Styling Center, when the right front wheel fell off. The Unitron was towed to the Design Center shops, where it stayed for several weeks, until it could be repaired. Management asked a lot of questions about how the wheelwell and underside of the Unitron got chewed up, but it was eventually fixed and the damage forgotten. As Payne remembers it, the Unitron was then stored for awhile in an old building, somewhere in Detroit, Ford rented from the Detroit Public Works department. That's the last memory anyone has of the Unitron, and it was probably destroyed sometime in the mid-1960s.

From Beck's standpoint, he wishes he had been asked to design the Unitron 20 years later. By that time, a market had developed for recreational vehicles and minivans, and he believes the Unitron would have had an excellent chance for production. ■

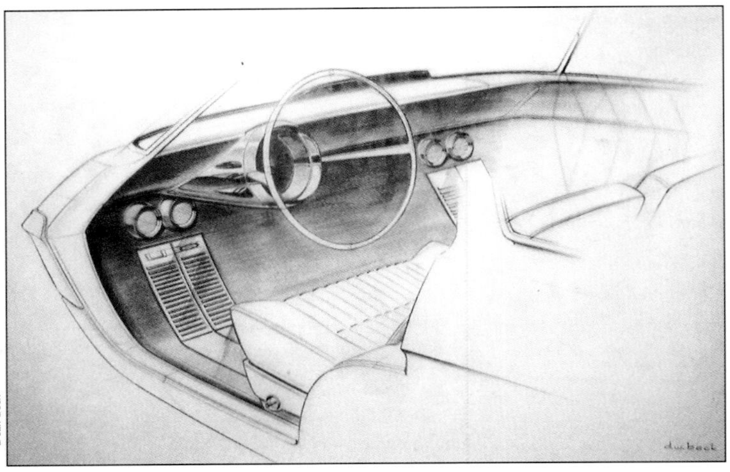

Dean Beck

(left) One of Dean Beck's rendering for the Unitron's interior. As designed, the interior featured a speedometer in the center of the steering wheel hub, as on the Futura, and a flat floor with no transmission hump.

Ford Design Center

(above) Although they liked the entry room the center-opening doors provided, management never got used to the pillar in the middle of the Unitron's windshield.

Ford Design Center

(above) Elwood Engel made the taillights on the Unitron the size of pie pans, but the nerf-bar bumpers were later redesigned.

Ford Design Center

(above) At various times, the Unitron had doors that opened from the center of the windshield, from the front of the van and from behind the A-pillars.

Ford Design Center

(above) A full interior was installed in the Unitron, and a small electric motor was attached to one of the rear wheels, so management could see the vehicle in motion.

Ford Design Center

(right) In its final form, the Unitron was built as a closed panel van on the driver's side and with full windows on the passenger side.

In March 1954, Colin Neale, who, at the time, was head of the design department at Ford of England, came to Dearborn to better acquaint himself with the way the "Yanks" designed cars, as he puts it. On that visit, he spent about a month at the Design Department and became friends with many of the designers there, including Gil Spear, Wes Dahlberg and Elwood Engel. From then on, Neale's visits to Dearborn became a annual event.

As they got to know each other better, Engel began suggesting to Neale that he move to the United States and become a designer at the Styling Center. At first, Neale had no interest in leaving England but, as time passed, he became more frustrated with what he saw as resistance to any effort to modernize the design process at Ford of England. Neale felt that Ford of England and British designers were generally ahead of the United States in aerodynamics, but he liked many of the design practices he observed in Dearborn during his visits. At Ford of England, the design department was still under Body Engineering, the same way it had been at Dearborn prior to 1955, when George Walker became general manager of an independent Styling Center. Every time Neale butted heads with the body engineers at Ford of England, he lost.

The Astrion was later renamed the Hummingbird after he Astrion and the Firefly were built in fiberglass.

Ford Design Center

Engel kept up the pressure on Neale to move to Dearborn by sending him ads for cars, appliances, and even groceries listed in the Detroit and Dearborn newspapers. Prices were higher in England and, soon, the attraction of cheap and readily-available consumer goods became very enticing, especially in light of Neale's inability to make any headway in his efforts to modernize the design process at Ford of England.

In October 1958, Neale gave up and moved to the United States. He was unable to arrange a transfer and had to start all over again as a beginning designer at the Styling Center, but at least he had Engel's promise of a job and rapid advancement. He was assigned to Engel's corporate advanced studio, where he soon became a senior designer working on the alternate 1961 Thunderbird proposal that became the 1961 Continental.

Firefly/Astrion

After the 1961 Continental and the Mark IX were completed, one of the next projects Neale was asked to work on was an electric car proposal, one half of which became the Firefly, and the other half of which became the Astrion. By that time, the corporate advanced studio had been given additional space upstairs in addition to the small stiletto studio downstairs. Neale was one of the designers who moved upstairs, and the Firefly/Astrion was designed there.

Several Ford engineers came from England to Dearborn at about the same time as Neale and, at least one of them, Roy Lunn, was assigned to advanced engineering. In mid-1960, Lunn was asked to develop a chassis for a proposed electric car. (Lunn later left Ford to become vice president of engineering at American Motors.) As the chassis was being developed, Lunn's boss suggested he approach the Styling Center about designing

Gil Treweek

Colin G. Neale

Colin Neale grew up in wartime England. He quit school at age 17 to join the Royal Air Force. When the war was over, he found work as an engineer. In 1950, he entered a design competition. His airbrushed rendering won third prize and a job at Briggs Ltd. as a design apprentice. In 1953, Briggs Ltd. was purchased by Ford, its major customer.

In 1954, Neale was appointed head of the Ford of England design department. Because new-car designs prepared by Ford of England required approval from Dearborn, it was customary for Neale to make yearly trips to Dearborn. Beginning in 1954, Neale worked in Dearborn about eight weeks each year.

In 1958, when Elwood Engel was visiting Ford of England, he renewed an invitation for Neale to become a designer at the Styling Center in Dearborn. In October 1958, Neale emigrated to the United States and was assigned to Engel's corporate advanced studio, where he helped design the 1961 Thunderbird/Continental, the Mark IX and the Firefly.

In 1960, Neale was asked to start an advanced interior studio at the Styling Center and, in 1961, he was assigned to the Thunderbird and Fairlane studio.

In late1961, Engel left Ford to take a job as head of the Chrysler design department. In early 1962, he invited Neale to become head of the Chrysler interior design studios. Neale was later assigned responsibility for both exterior and interior design for Chrysler's international operations and retired from Chrysler in 1977.

Since his retirement, Neale has continued to work as a independent design consultant mostly designing advanced interiors. ■

a show-car for use with the chassis. Lunn contacted his friend Colin Neale, and Neale introduced him to Engel. Engel correctly assumed that Engineering was trying to build its project partially on his studio's budget, but he was nonetheless intrigued with the idea of designing an electric car. Ultimately, Walker agreed to let Engel's corporate advanced studio design a body for the electric car proposal, because it was anticipated it would at least become a showcar. At the same time the electric car project was assigned to Engel's studio, he had to make an oversight trip to one of Ford's European design departments. Before he left, he called the designers in his studio together and briefed them on projects that were to be worked on during his absence, including the new electric showcar. As he was telling the designers about the project, he noticed a small plaster model Neale had on his desk, and suggested that they "do something like that." Engel asked Alex Tremulis and Neale to design the electric car, and studio manager Gale Halderman was to supervise the project. After that, Engel left, telling everyone he'd be back in nine days.

When things were slow in the corporate advanced studio, Neale would make what he now refers to as "aerodynamic hand sculptures." The hand sculptures were really shape exercises about 15 inches long and made from clay. Once completed, Neale had his shape exercises cast in plaster and painted exotic pearlescent colors. The most recent one Neale had completed was designed without A or C-pillars but, instead, had support pillars in the center of the windshield and backlight. There were also small B-pillars on the hand sculpture located just behind the door openings. When he finished it, Neale thought it was an interesting idea, but it wasn't really meant as a serious proposal, or one he had given much thought to. As Neale remembers it, the small plaster model had just recently been painted and had just been placed on his desk.

Ford Design Center

(above right) The driver's side of the clay model became the Firefly. The Firefly did not have A or C-pillars in the traditional sense. Its designer, Colin Neale, claimed it was safer because it had a built-in roll bar.

Colin Neale

(right) This is one of Neale's renderings of the Firefly. As built, the Firefly had doors that opened from the center of the windshield, but they were not hinged at the B-pillars as shown in this rendering.

(right) The Firefly/Astrion was originally designed as alternate versions of a small electric car. It was completed during Elwood Engel's nine-day trip to Europe.

Ford Design Center

Ford Design Center

(left) As originally designed, both sides included doors that opened from the center of the windshield, eliminating the A-pillars.

(right) When George Walker reviewed the full-sized clay model, he suggested that Neale de-emphasis the sweep over the rear fender on the Firefly side, but both cars were generally accepted as designed.

Ford Design Center

Ford Design Center

(left) Alex Tremulis and Gale Halderman designed the passenger side (Astrion), and Colin Neale designed the driver's side (Firefly), on the same full-sized clay model.

Ford Design Center

(right) The Astrion side of the car reflected Halderman's more conservative influence and probably came closer to production. The Astrion also featured hidden headlights.

Ford Design Center

The front end of the Firefly was more streamlined than the Astrion.

According to Neale, Engel was an impulsive designer, who was often inclined to redirect a completed or partially completed design the way he thought it ought to look on the day he saw it, even if on the next day he changed his mind and asked that the original design be restored. When Engel visited the various studios as assistant manager of the Styling Center, he would often pick up a knife and slice up a clay model to indicate the way he thought the design should be. For that reason, designers and clay modelers — even those in Engel's own studio — tried to finish what they were working on, before Engel had a chance to experiment on them, if that was possible.

When Engel said he was going to Europe, Neale and Tremulis recognized that they had a golden opportunity, if they could finish the full-sized clay model of the electric car before he returned. Neale and Tremulis each decided to design one side of the same full-sized clay model, and they worked as much overtime as possible during the next nine days to make sure they were finished before the boss returned.

Neale's side was identical to his hand sculpture, and Tremulis' side was of an alternate aerodynamic design. Neale thought Tremulis' design was much more conservative than his was. Because the car was planned for electric power, the package measurements did not require the normal hood height required to accommodate the radiator and carburetor, so both sides were built with severely-sloped hoods. Neale and Tremulis named their sides of the car the Firefly and the Astrion respectively. When completed, the car was 50 inches high, 50 inches wide, 155 inches long and had a wheelbase of 82 inches.

Although it took the full nine days to design and build the full-sized clay model, it was completed before Engel returned. Halderman pitched in to help Tremulis finish his design on time, which probably resulted in the more conservative Astrion. When Engel saw the designs, he was delighted with both of them. The next day, Engel asked George Walker to review them. Walker suggested lengthening the character line over the rear-wheel opening on Neale's side of the car, so that it started farther forward, but Neale was able to persuade Engel and Walker not to change the design as much as Walker suggested, and both sides were approved almost as they were designed. Walker also suggested that two separate clay models be built, and that both designs then be cast in plaster and built as full-sized fiberglass cars as a prelude of choosing one as the showcar.

Ford Design Center

(right) This three-quarter rear view of the Firefly shows the wide field of vision available because of the absence of traditional A and C-pillars.

(below) The doors on the Firefly were designed by Ray Smith to open from the center of the windshield.

Ford Design Center

When the two-sided clay model was completed, the doors on each side were designed to open from the center of the windshield using a double-pivot hinge developed by Ray Smith.

The doors actually eliminated the A-pillars and opened from the center of the car's windshield. According to Neale, management wasn't comfortable with the pillar in the center of the windshield, so Smith's center-opening doors were never seriously considered. Neale thought it also gave the car a retro-look at a time before it was popular.

The Styling Center shops built both cars from the same clay model. When the fiberglass cars were built, the one named the Astrion was renamed the Hummingbird, and it was built with conventional doors. Later, a third model with a wider, more formal C-pillar was built in fiberglass as an alternate Hummingbird model. (In the mid-1960s, the name Hummingbird was reused by Bob Thomas for a proposed world car.)

Ford Design Center

George Walker initially wanted the character line over the rear wheelwell lengthened on the Firefly.

While the fiberglass models of the Firefly and the Hummingbird were still in the Styling Center shops being finished, Lee Iacocca, who was then vice president in charge of all cars and trucks and on his way to becoming president of Ford, walked through one day on his way to somewhere else with product planner Will Scott. Iacocca saw the Firefly and Hummingbird and asked what they were. After he looked them over, he suggested to Scott that maybe they should reconsider the sporty-car plans they had just placed on the back burner. As a result of Iacocca's revived interest, Scott soon visited Engel's studio and asked him to expand the measurements of the car to a 2+2 package. As Neale recalls it, Ray Smith drew the new and the old packages next to each other on large blackboards. When the new package measurements were approved, several studios were asked to make full-sized clay models using the revised package. That car eventually became the Mustang.

Ford Design Center

(left) The package measurements permitted a slope to the front end of the Firefly/Astrion that was greater than usually allowed by the Engineering Department, because it was designed as an electric car.

(below) After the full-sized fiberglass Astrion was built, it was renamed the Hummingbird. This version of the Hummingbird had pod headlights on the hood. The full-sized fiberglass Firefly, which is just ahead of the Hummingbird, was one of 32 design proposals shown in the Styling Center courtyard that day.

Ford Design Center

There are two postscripts to the story of the Firefly and Astrion. After the cars were built in fiberglass, both had full interiors installed. The Firefly was painted red, and the renamed Hummingbird was painted yellow. Ray Smith then made at least one of them into a running model by installing a car battery and an electric motor that powered a rear wheel. Smith was also able to rig-up steering on the car. On an afternoon when Smith and Bob Thomas had the running car in the Styling Center courtyard, Bill Ford saw it through a window, and came outside to find out what it was. Thomas and Smith explained what they had done and, the next thing they knew, Mr. Ford accepted an offer to take a ride in it with Thomas. Soon, the two of them were bouncing across the cobblestone courtyard at a top speed of

Ford Design Center

about 5 miles per hour in a little two-passenger car with makeshift seats and no suspension. As Thomas remembers it, Mr. Ford had a great time.

Although the Mustang project was started before George Walker's retirement, the leadership transition at the Styling Center was not an easy one. At that time, Ford policy required retirement at age 65, and Walker turned 65 in 1961. Walker's request for a one-year extension as general manager of the Styling Center was turned down, although he was asked to remain on the Ford payroll during the following year and to tour Ford's European plants and design departments and then to report on his trip to Henry Ford II.

Ford Design Center

(top) An alternate fiberglass version of the Hummingbird was built with more formal C-pillars.

(above) The formal C-pillars used on this alternate version of the Hummingbird made it into production on the 1966 Thunderbird.

Walker had been grooming Engel to become his successor. When his request to stay on for an extra year was turned down, he stepped up his campaign to have Engel appointed to replace him. Bill Ford, on the other hand, decided it was time to replace Walker and make a fresh start with Gene Bordinat -- and not Elwood Engel or Joe Oros.

At the time Walker's request for an additional year was turned down, Henry Ford II had recently been diagnosed with mononucleosis and was instructed by his doctors to get lots of bed rest, so the job of selecting Walker's successor fell primarily to Bill Ford. Engel was disqualified, ostensibly because of a billing discrepancy on his travel expense account. Joe Oros was then told he had been selected as Walker's replacement, only to be told the next day that Gene Bordinat had been appointed. Walker was furious, and felt he had been betrayed!

Walker started on his retirement tour, but it was cut short when Henry Ford II recalled him, because of remarks critical of Henry Ford II that Walker made during the tour. Walker returned to Bordinat's newly-renamed Design Center one last time to prepare a report on the portion of the trip he had completed. He was assisted in preparing his report by John Najjar and then left the Design Center never to return again.

Walker was bitter and remained so for the rest of his life. In late 1961, he arranged for Engel to become head of Chrysler's design department. He also tried to arrange a job for Oros as head of American Motors' design department. Oros met with American Motors' management, but decided there was still room for him to grow and advance at Ford. Until, and even after, he was named vice president of design, Ford of Europe in 1968, Oros continued to work well with Gene Bordinat. ■

(below) This full-sized fiberglass model of the Hummingbird was probably equipped with an electric motor and driven around the Styling Center courtyard by Bob Thomas and Bill Ford.

Ford Design Center

Glossary

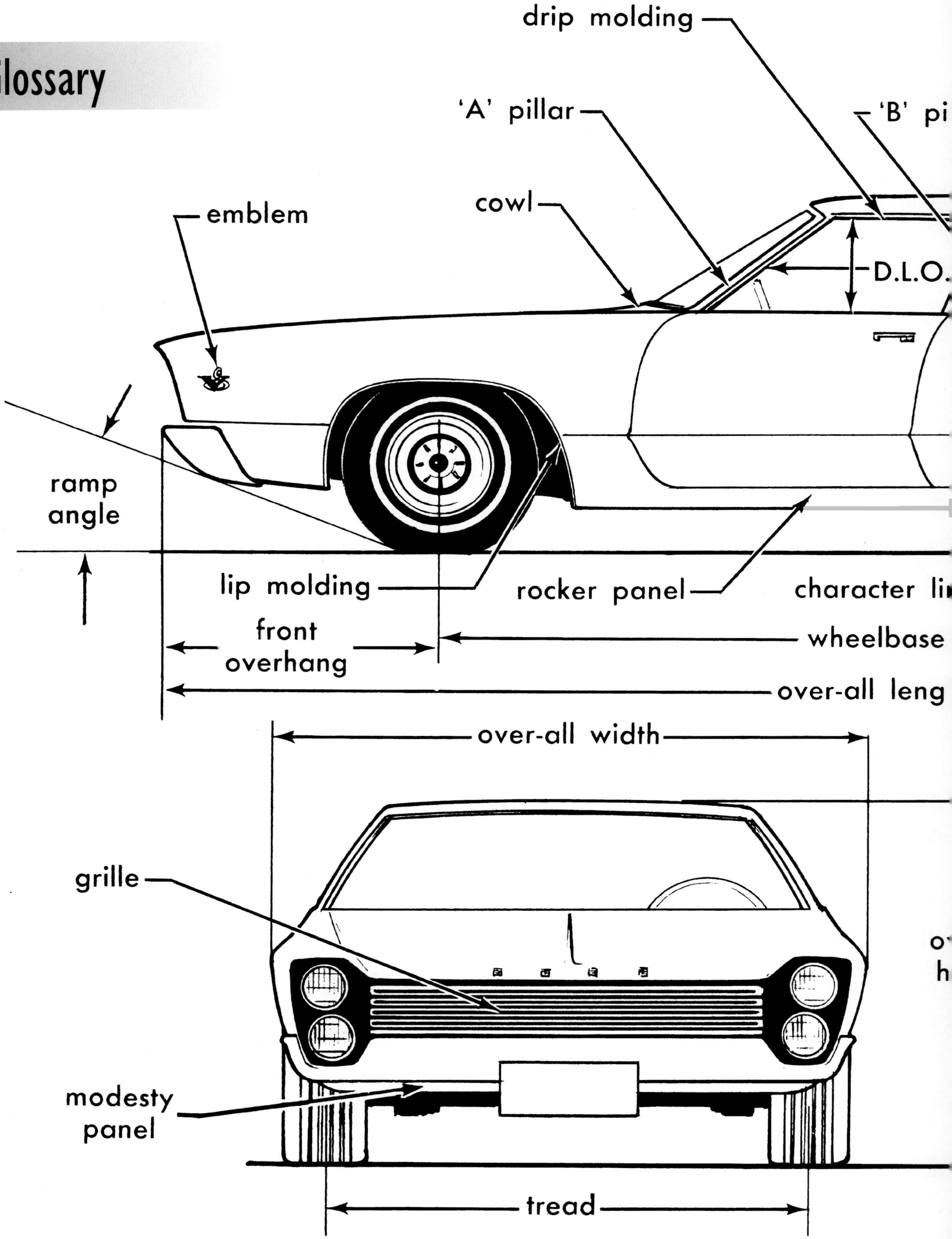
drip molding
'A' pillar
'B' pi
cowl
emblem
D.L.O.
ramp
angle
lip molding
rocker panel
character li
front
overhang
wheelbase
over-all leng
over-all width
grille
o
h
modesty
panel
tread

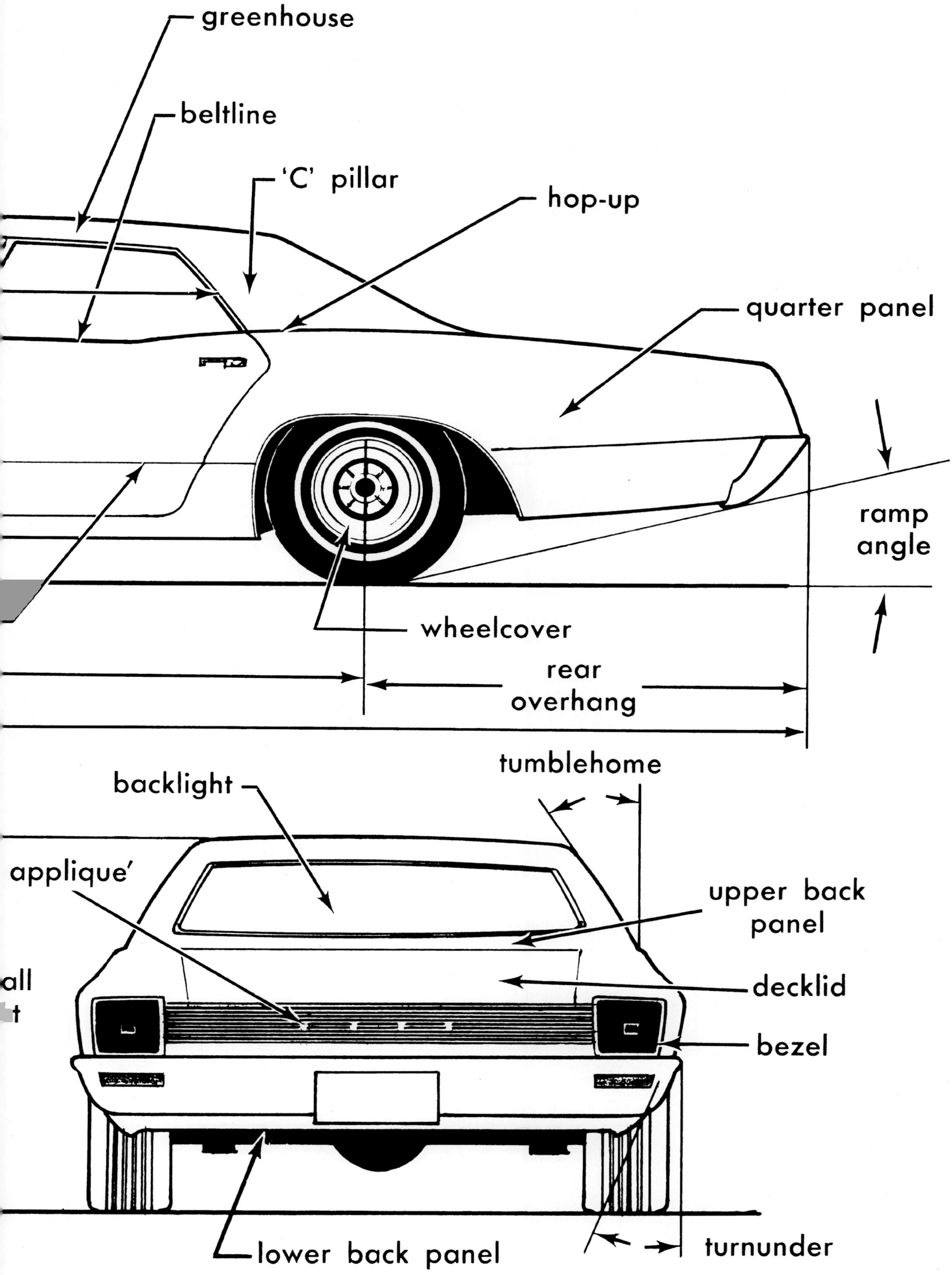
greenhouse
beltline
'C' pillar
hop-up
quarter panel
ramp angle
wheelcover
rear overhang
all
t
tumblehome
backlight
applique'
upper back panel
decklid
bezel
lower back panel
turnunder

Bibliography

1) *Primary Sources:*

A) *Interviews And Correspondence:*

Clare (Mrs. C.E. "Bud") Adams, Merle Adams, Robert Adomeite, John Aiken, Ed Albright, Walt Amrosi, Bob Anderson, Ed Asadoorian, Rose Marie (Mrs. L. David) Ash, Laura (Darden) Atkinson, Dick Avery, Bennie Bailey, John Bailey, Bill Balla, George Barbaz, Bennie Barbera, George Barris, Fred Beamish, Dean Beck, Raymond Behmer, Richard Beneicki Jr., Carol (Conrad) Berndt, Dave Beyreis, Frank Bianchi, Tony Boland, Bill Bonner, Joe Bortz, Bill Boyer, Mrs. Bill (Phyllis) Braathan, Cog Briggs, Dan Brooks, Roy Brown, Mrs. Bill Brownlie, Kathryn (Mrs. Gordon) Buehrig, Roy Butler, Carl Cameron, Jerry Capizzi, Bob Casey, Andrew Cheida, Roger Clement, Dave Cole, Ruth Conrad, Larry Conrad, Ray Cosh, Ross Cousins, Dave Crippen, John Cuccio, Wes Dahlberg, Rusty Davis, Bill Dayton, Stuart Denker, Sally (Mrs. Elwood) Engel, Joe Farkas, Edsel Ford II, William Clay Ford, John Foster, Werner Framke, Jim Fulton, Charles H. Gale, Gib Giberson, Hans Goeppenger, Bill Grayell, E. T. Gregorie, David Grainger, Buzz Grisinger, John Gump, Bob Gurr, George Hackett, Jesse Haines, Gale Halderman, Romeyn Hammond, Chuck Hannah, R. Hass, George Haviland, John Hay, John Hazelquist, Frank Hershey, Fred Hoadley, John Hollowell, Mark Hollowell, Starky Hoot, Pat Housch, Fred Hunter, M. Rey Isip, Ed Iser, Ed Jaquet, John Jaeger, Walt Jennfeldt, Harold Johnson, Bob Jones, Lloyd Kamm, Mark Kelly, Karen (Mrs. Charles) Keresztes, Esther (Mrs. Ed) Kessler, Steve Koppen, Allan Kornmiller, Tak Kosakura, Mildred (Mrs. Holden) Koto, David Koto, George Krispinski, Freida (Mrs. Ben) Kroll, Robert Kroll, Tom Land, Paul Lorenz, Homer LeGassey, Sherry (Wilson) Luka, Gary MacNorius, Robert McNamara, Tucker Madawick, Jerry Magayne, Clint Mahlke, Bob Marcks, Marty Martino, Chuck Mashigan, Doug McClure, Doug McCombs, Glenn McGrove, Elizabeth (Cheida) McPherson, Syd Mead, John Middlestead, Doug Miller, Rhys Miller, Ben Mills, Mrs. Jack Mills, Chuck Mitchell, Joe Mooradian, Agnes (Mrs. James) Morrison, Jerry Morrison, Al Mueller, John Murphy, Alex Musichuck, John Najjar, Colin Neale, Norm Nebozenko, Ken Nelson, Dick Nesbitt, Andy Olinick, Bob O'Neal, Emmett O'Rear, John Orfe, Dom Orlando, Joe Oros, Ralph Ortiz, Hiram Pacific, Dolph Painter, Jim Pappas, Mark Patrick, Howard Payne, Phil Payne, Terry Peterson, Ken Pheasant, Al Phillips, Ron Perry, Charlie Phaneuf, Al Phillips, Linda Phillips Esq., Bob Pitchford, Jim Powers, Bob Puffer, A. V. Pulleyblank, Art Querfeld, Jim Quinlan, Richard Quinn, A. J. Reganhart, Pete Renz, Jesse Richards, Frank Ruff, John Samsen, Joe Santoni, Filipo Sapino, Dick Schierloh, John Schmidt, Arby Schuman, Bill Shannon, Larry Shinoda, Mike Schultz, Joe Seibold, Jim Sipple, Valli (Mrs. Ray) Slate, Ben Smith, Barbara (Mrs. Ray) Smith, Ed Spagnolo, Tom Spear, Gil Spear, Ken Spencer, Dick Steiger, Harry Strickler, Peter Sullivan, Jr., Christine Stobar Swift, Leonard Stobar Jr., Don Stumpf, Art Tarabusi, Jack Telnack, Mrs. Keith Teter, Bob Thomas, McKinley Thompson Jr., Stan Thorwaldsen, Florence (Mrs. Herb) Tod, Richard Tod, Bob Todd, Chrissie (Mrs. Alex) Tremulis, Steve Tremulis, Gil Treweek, Sharon (Slate) Tuttle, Helen Conrad Vance, John Van Tilburg, Paul Wagner, Larry Waterhouse, Don Weiland, Bob Weiland, Suzanne (Mrs. Robert) Weiland, Betty (Mrs. John) Weisbrod, Ed Westcott, Bill Wheeler, Warren Williams, Jerry Windle, Phyllis (Mrs. Paul) Wong, Gary Wright, Dick Yagami, Tony Yuki, Bob Zokas.

B) *Henry Ford Museum Oral Histories:*

George E. Adams, David Ash, William Boyer, Eugene Bordinat, Herman C. Brunn, Gordon Buehrig, A.E. Conn, Ross Cousins, Don DeLaRossa, Elwood Engel (statement of), William Clay Ford, John B. Foster, Don Frey, Alden Giberson, Chalmers Goyert, E.T. Gregorie, Gale Halderman, Franklin Q. Hershey, Fred Hoadley, Holden H. Koto, Irving Kaufman,

Tucker Madawick, Fred Mayhew, John Middlestead, John Najjar, Emmett O'Rear, Charles Phaneuf, William Schmidt, Adrian G. Spear, Ken Spencer, Jack Telnack, Alex Tremulis, George Walker.

C) Organizations, Archives, Libraries and Museums:

Art Center College, Automobile Manufacturers Association Library, Blackhawk Automobile Museum Library, Cleveland Institute of Arts, Collier Automotive Museum, Detroit Public Library National Automotive History Collection, Guild of Automotive Restorers, Henry Ford Museum and Greenfield Village, Ford Design Center, Ford Highland Park Services, Ford Industrial Archives, Ford Photomedia.

2) Secondary Sources:

A) Books, Periodicals, Newspapers, and Statements:

A Century of Automotive Style, Lamm and Holls; Arnold, Marvin, interview with E.T. Gregorie, Art and the American Automobile, Nick Georgano; Auto Age, various issues; Autobiography, Charles Sorensen; Automobile Quarterly, various issues; Automotive News, various issues; Boss, The Bill Stroppe Story, Tom Madigan; Bob Bourke designs for Studebaker, John Bridges; Car Classics, various issues; Car Life, various issues; Cars Detroit Never Built, Edward Janicki; Cars That Never Were, Editors Consumer Guide; Cisitalia, Catalogue Raisonne, Automobilia; Classic American, 5/96; Collectible Automobile, various issues; Complete Book of Corvette, Langworth; Confessions of an Automobile Stylist, Bob Thomas; Crown Victoria Club, 1990 speech by David Ash given at; Design Department/Styling Center/Design Center, miscellaneous publications; Detroit News, various issues; Ferrari, Lehbrink and Schlegelmilch; Victor Favot, statement of; Ford at Fifty, Ford Motor Co; Ford Convertibles, Loren Sorensen; Ford, Decline and Rebirth, Nevins and Hill; Ford, The Dust and the Glory, Leo Levine; 55 Years of Mercury, Gunnell; Ford,The Men and the Machine, Robert Lacey; Ford Times, various issues; Ford World, various issues; Harley Earl and the Dream machine, Bayley; Highlights (Ford Design Center), various issues; Ghia, Automobilia; Henry Ford's Airport, T.J. O'Callaghan; Henry's Lieutenants, Ford Bryant; Lincoln, 1945-1995, Gregory Von Dare; Lincoln Continental 1958-69, James Howell; Lincoln and Continental, Marvin Arnold; Edward Martin, statement of 9/3/51; Lincoln-Mercury Messenger, various issues; Lincoln-Mercury News, various issues; Lincoln-Mercury Times, various issues; Management (Ford) magazine, various issues; Motor, various issues; Motor Life, 5/54; Motor Trend, various issues; Motor World, 6/5/53; Mustang Monthly, 8/82; National Speed Sports News, various issues; New York Times, 4/4/66; Pace Cars of the Indy 500, Riggs; Road and Track, various issues; Radco Automotive Review, 11/33; Research and Engineering Center News (Ford), various issues; Rolling Sculpture, Gordon Buehrig; Setting The Pace, Oldsmobile's First 100 Years, Earley and Walkinshaw; Seventy-One Society newsletter, summer 1973; Skyliner, Jerry Magayne; Special Interest Autos, various issues; The American Automobile, A Centenary, Nick Georgano; The Art of American Car Design, C. Edison Ami; The Fords, Horowitz and Collier; The Ford Factory, Loren Sorensen; The GM Motorama, Bruce Berghoff; The Insolent Chariots, John Keats; The Lincoln Continental, Ocee Rich; The Lincoln Motorcar, Tom Bonsall; The Motor, various issues; The Thunderbird Story, Personal Luxury, Richard Langworth; Thunderbird 1955-66, Tast; The X Cars, Henry B. Lent; The Way of the Zephyr, various issues; Thoroughbred and Classic Cars, 3/76; V-8 Times, various issues.

B) Persons who reviewed all or a part of this book prior to publication:

Dave Cole, Buzz Grisinger, Gale Halderman, Fred Hoadley, Mike Lamm, Tucker Madawick, John Najjar, Howard Payne, Jim Powers, Joe Sherlock, Gil Spear, Jack Telnack, Gil Treweek, and Jerry Windle.

Index